# World Sustainability Series

**Series Editor**

Walter Leal Filho, European School of Sustainability Science and Research, Research and Transfer Centre "Sustainable Development and Climate Change Management", Hamburg University of Applied Sciences, Hamburg, Germany

Due to its scope and nature, sustainable development is a matter which is very interdisciplinary, and draws from knowledge and inputs from the social sciences and environmental sciences on the one hand, but also from physical sciences and arts on the other. As such, there is a perceived need to foster integrative approaches, whereby the combination of inputs from various fields may contribute to a better understanding of what sustainability is, and means to people. But despite the need for and the relevance of integrative approaches towards sustainable development, there is a paucity of literature which address matters related to sustainability in an integrated way.

**Notes on the quality assurance and peer review of this publication**

Prior to publication, the works published in this book are initially assessed and reviewed by an in-house editor. If suitable for publication, manuscripts are sent for further review, which includes a combined effort by the editorial board and appointed subject experts, who provide independent peer-review. The feedback obtained in this way was communicated to authors, and with manuscripts checked upon return before finally accepted. The peer-reviewed nature of the books in the "World Sustainability Series" means that contributions to them have, over many years, been officially accepted for tenure and promotion purposes.

Walter Leal Filho · Sandeep Kautish · Tony Wall ·
Sonja Rewhorn · Sanjoy Kumar Paul
Editors

# Digital Technologies to Implement the UN Sustainable Development Goals

Springer

*Editors*
Walter Leal Filho
Faculty of Life Sciences
Hamburg University of Applied Sciences
Hamburg, Germany

Tony Wall
Liverpool Business School
Liverpool John Moores University
Liverpool, UK

Sanjoy Kumar Paul
UTS Business School
University of Technology Sydney
Ultimo, NSW, Australia

Sandeep Kautish
Chandigarh University
Mohali, Punjab, India

Sonja Rewhorn
Geography Department
The Open University
Chester, UK

ISSN 2199-7373 ISSN 2199-7381 (electronic)
World Sustainability Series
ISBN 978-3-031-68429-6 ISBN 978-3-031-68427-2 (eBook)
https://doi.org/10.1007/978-3-031-68427-2



This Springer imprint is published by the registered company Springer Nature Switzerland AG
The registered company address is: Gewerbestrasse 11, 6330 Cham, Switzerland

# Preface

In a world increasingly defined by rapid technological advancements, the role of digital technologies in shaping our future cannot be overstated. The convergence of innovation and global imperatives offers unprecedented opportunities to address some of the most pressing challenges humanity faces today. At the forefront of this endeavor are the United Nations Sustainable Development Goals (SDGs), a universal call to action to end poverty, protect the planet, and ensure peace and prosperity for all by 2030.

This book, "Digital Technologies to Implement the UN Sustainable Development Goals," emerges from the recognition that achieving the SDGs necessitates harnessing the power of digital technologies. From artificial intelligence and blockchain to the Internet of Things and big data, these technologies are not merely tools but transformative forces capable of accelerating progress across all 17 goals.

Our journey through the chapters of this book will explore the multifaceted ways in which digital technologies intersect with the SDGs. We delve into practical case studies, innovative projects, and cutting-edge research to illustrate how technology can drive sustainable development. Each chapter is a testament to the collaborative efforts of experts, practitioners, and policymakers who are leveraging technology to create impactful solutions that transcend traditional boundaries.

The intent of this book is twofold. First, it aims to provide a comprehensive understanding of how digital technologies can be strategically employed to address specific targets within various SDGs. By offering detailed insights and practical examples, we hope to inspire stakeholders across various sectors to adopt and adapt these technologies in their initiatives. Second, it seeks to foster a dialogue among academia, industry, government, and civil society on the ethical, social, and economic implications of digital transformation in the context of sustainable development.

As readers navigate through the pages, they will encounter a rich tapestry of ideas and innovations that highlight the critical role of technology in building a sustainable future. The challenges we face are complex and interconnected, requiring holistic and integrated approaches. This book is a call to action for a global partnership where knowledge and technology are shared for the greater good.

We are living in an era of profound change, where the decisions we make today will shape the legacy we leave for future generations. It is our hope that this book serves as both a guide and an inspiration, empowering readers to contribute to the realization of the SDGs through the thoughtful application of digital technologies.

With immense gratitude to the contributors and a deep sense of optimism, we present this book as a valuable resource for anyone committed to making a difference in the world through technology and sustainable development.

Hamburg, Germany — Walter Leal Filho
Mohali, India — Sandeep Kautish
Liverpool, UK — Tony Wall
Chester, UK — Sonja Rewhorn
Ultimo, Australia — Sanjoy Kumar Paul

# Contents

# From Efficiency to Sustainability: Examining the Paradox of Digitalization of Energy

**Aliaksandr Novikau, Amal Mersni, and Nedim Redžović**

**Abstract** In today's world, characterized by growing energy consumption, depletion of nonrenewable energy resources, and increasing risk to energy services, attaining sustainable energy is critical. The increasing use of information technologies, also known as digitalization, has led to fundamental transformations in energy systems. Digitalization enhances the stability of energy systems, enables the integration of power generation with end users, reduces peak demand, speeds up energy supply restoration after disruption, reduces operating expenses, and facilitates the integration of renewables into energy systems. Thus, the digitalization of energy can contribute to achieving many of the United Nations' Sustainable Development Goals, particularly Goal 7, "Affordable and clean energy." While the potential benefits of such digitalization are widely recognized, its challenges, especially those impacting sustainable development, are less frequently discussed in academic literature. The study aims to analyze the implementations and benefits of digital technologies across various aspects of the energy industry. It specifically discusses the impact of the digitalization of energy on achieving Sustainable Development Goals, highlighting the benefits and challenges it can bring.

**Keywords** Energy · Digitization · Digitalization · Smart grids · SCADA · Blockchain · Artificial intelligence · SDGs

A. Novikau (✉)
Department of Political Science and International Relations, International University of Sarajevo, Hrasnička Cesta 15, 71210 Ilidža, Bosnia and Herzegovina
e-mail: anovikau@ius.edu.ba

A. Mersni · N. Redžović
Department of Engineering, International University of Sarajevo, Hrasnička Cesta 15, 71210 Ilidža, Bosnia and Herzegovina
e-mail: amersni@ius.edu.ba

 
W. Leal Filho et al. (eds.), *Digital Technologies to Implement the UN Sustainable Development Goals*, World Sustainability Series,
https://doi.org/10.1007/978-3-031-68427-2_1

## 1 Introduction

Energy is crucial for every aspect of human life – from thermal comfort to industrial production, transportation, communication, and entertainment. Therefore, energy systems are a cornerstone of economic and social progress, exerting a profound influence on the achievement of objectives outlined in the United Nations Sustainable Development Goals (SDGs) (Fuso Nerini et al. 2017).

Traditional energy systems, which rely mainly on nonrenewable fossil fuels, are rarely sustainable and often result in environmental degradation that negatively affects human and planetary health (Novikau 2022). In combination with increasing energy demand, unsustainable energy is one of the major factors that have been slowing the path toward achieving the United Nations' SDGs (Leite De Almeida et al. 2021; Leal Filho et al. 2023).

The role of energy in sustainable development has been recognized since the concept was first introduced in 1987 (Gunnarsdottir et al. 2021). However, energy was explicitly acknowledged only in Agenda 2030 as SDG 7, which expressly mandates to "ensure access to affordable, reliable, sustainable and modern energy for all." SDG 7 has specific targets with a focus on universal access to affordable, reliable, and modern energy services, increasing the share of renewables in the energy mix, and improving energy efficiency. Moreover, SDG 7 explicitly emphasizes the role of infrastructure and upgraded technology in such energy services (United Nations 2023).

At the same time, recent decades have witnessed the proliferation of digital technologies, also known as digitalization, marking a time of unprecedented transformation of every aspect of human life. Digitalization has unlocked new possibilities, which hold immense potential for creating a more efficient and sustainable society. This rapid increase in digitalization has exerted profound and far-reaching impacts on sustainable development, influencing its economic, social, political, and environmental dimensions (Chien et al. 2021).

Since energy and sustainable development are inevitably linked, there must also be a connection between the digitalization of energy and sustainable development. However, an exhaustive investigation into the tangible effects of digitalization on these SDGs is still in its nascent stages. Although some studies emphasize the potential benefits of digitalized energy systems (Elavarasan et al. 2021; Brenner and Hartl 2021; Husaini and Lean 2022; Çelik et al. 2022), particularly in terms of enhanced energy efficiency and subsequent environmental advantages, a considerable gap exists in academic literature regarding the potential drawbacks of the digitalization of energy. These drawbacks, linked to energy systems' increased complexity and vulnerability to cyberattacks, have remained largely unexplored. Moreover, the role of the most recent trends in the digitalization of energy systems, such as the Internet of Things (IoT), Cloud Computing, Artificial Intelligence (AI), and Machine Learning (ML), has not been analyzed to a significant degree.

Consequently, the central question about the implications of the rapid digitalization of energy on global and local efforts to attain the SDGs and the specific opportunities and challenges it introduces remains largely unanswered. Little is known about whether and to what extent the digitalization of energy could accelerate or, on the contrary, hinder the achievement of the SDGs. It is crucial to underscore the significance of this question from both a policymaking and theoretical standpoint.

This study, therefore, aims to comprehensively analyze the impacts of the digitalization of energy on SDGs. By doing so, it seeks to shed light on the intricate interplay between the digitalization of energy systems and the broader global pursuit of sustainable development, thereby providing insights for future policy formulation.

The paper is structured as follows. The following section discusses digital technologies used in the energy sector, highlighting their main benefits. The next section discusses the effects of digitalization of energy on SDGs. The last section concludes and offers some suggestions for further research.

## 2 Digitalization of Energy

The digitization of the energy sector represents a transformative shift from traditional practices of using separate analog energy systems to more interconnected and smart networks. This section explores the integration of advanced digital technologies, from the automation capabilities of smart grids to the decentralized transactions enabled by blockchain and other digital technologies. It highlights their individual and collective impact on improving operational efficiency, energy sustainability, and consumer satisfaction in the energy business.

### *2.1 An Overview of the Digitalization of the Energy Sector*

Energy plays a crucial role in digitization, converting information into a digital format, because the process ultimately requires electrical power. This transformation from analog to digital is fundamentally based on electrical principles, where information is processed, stored, and transmitted through electrical circuits. The importance of digitization is underscored by the continuous operation of digital technologies that depend on a steady electricity supply, from data centers to personal devices. Digitalization, however, refers to a broader application of digital technologies that uses digitized data to revolutionize energy sector operations. It relies on electricity and innovates its generation, distribution, and consumption. Energy systems are increasingly digitized, with digital applications increasing efficiency and introducing renewable solutions across all energy sources and the electricity sector. This digital evolution is critical because it requires instantaneous responses in systems where electricity moves at the speed of light. In addition, energy companies are using Information

Technologies (IT) in various business processes, from communications to security (Chebotareva 2021; Światowiec-Szczepańska and Stępień 2022).

The World Economic Forum emphasizes the vast economic potential of IT solutions in the energy sector, asserting that digitalization could unlock 1.3 trillion USD in value for the electricity sector by 2025. It could be realized through heightened productivity, cost reductions, and enhanced customer experiences. The global market for digital energy services is expected to double from USD 32 billion in 2017 to USD 64 billion in 2025. This growth is expected to be propelled by a rising demand for distributed energy resources, sophisticated energy management systems, and advanced grid optimization solutions (Bloomberg 2017).

In the same way, the European Union (EU), recognizing the transformative power of this digital wave, launched an action plan in 2022 to digitalize the energy system. The initiative aims to bolster connectivity and interoperability, advocate coordinated investments in smart grid technologies, empower consumers, enhance cybersecurity measures, call for increased efficiency, and orchestrate effective governance via collective planning. The EU has two renewables and energy efficiency goals, "Fit for 55" and REPowerEU (European Council 2023).

## 2.2 *Applications of Digital Technologies in Energy Systems*

In the rapidly changing energy sector, digital technologies such as smart grids and Supervisory Control and Data Acquisition (SCADA) systems play an essential role. These technologies, functioning as both standalone innovations and applications of broader digital advancements, have significantly transformed energy management. Furthermore, the digitalization of the energy industry leverages advanced technologies like AI and ML, whose practicality and economic viability vary across different areas of the energy sector. Central to this transformation are such technologies as Big Data Analysis (BDA), Big Data Mining (BDM), IoT, Cloud Computing, blockchain, and Robotic Process Automation (RPA), as illustrated in Fig. 1.

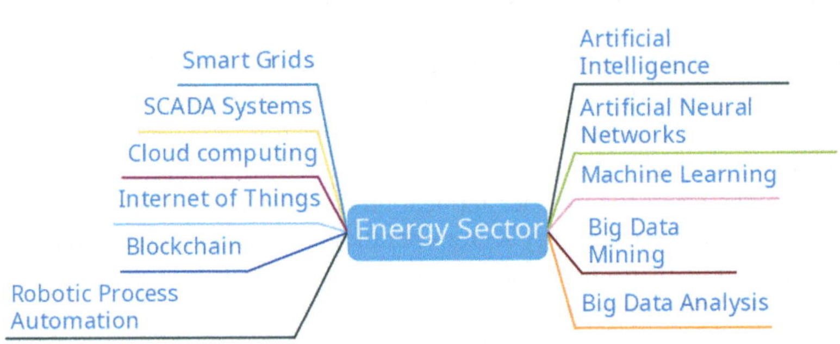


**Fig. 1** Digital technologies integration in the energy sector

Driving the sector toward greater automation and data-centric operations, these prominent technologies (Table 1) have proven invaluable across the energy supply chain, finding uses from power network design to energy sales (Legner et al. 2017; Weigel and Fischedick 2019; Światowiec-Szczepańska and Stępień 2022).

### 2.2.1 Smart Grids and SCADA Systems

Smart grids represent a crucial advancement in the digitalization of energy. Smart grids employ digital technologies for real-time, two-way communication and sensing along advanced networks. They integrate the electric power grid with multidirectional communication network systems. This integration includes the generation and delivery of electricity to consumers. Smart grids optimize power consumption at the consumer end through methods such as smart meters and microgrids while transforming traditional power grids into more efficient, sustainable, and responsive networks (Pearson 2011; Moreira et al. 2016).

The adoption of smart grid technologies enhances energy system efficiency and reduces fossil fuel dependence, significantly contributing to environmental sustainability. They are beneficial for developing renewable energy sources and distributed energy resources (DERs), offering flexibility and cost-effectiveness that improve grid performance and ensure accurate, real-time billing. Smart grids provide efficient, secure, and inexpensive energy management through distributed architecture (Ma et al. 2018).

SCADA systems are vital in smart grids because they provide robust control, monitoring, diagnostics, safety, maintenance, and adaptability for digital energy systems. Initially used in large-scale power systems, SCADA has evolved to be integral in smart grid development, shifting from mere data transmission to a central role in grid operation and management. Future designs will enhance this functionality, supporting decentralized decision-making and reflecting the strategic behavior of various stakeholders in the energy sector. SCADA's application spans traditional power grids, renewable energy facilities like wind farms and solar plants, and industrial power generation and distribution processes. Its critical role extends to infrastructure such as oil and gas pipelines, water treatment plants, and nuclear power stations, showcasing its versatility in ensuring efficiency, reliability, and safety across different energy systems (Harpe and Thomas 2009; Vale et al. 2013; Thomas and McDonald 2015; Rahman et al. 2016).

### 2.2.2 Cloud Computing

Cloud Computing is critical to the energy industry's digital transformation by centralizing the management of energy to improve visualization and operational efficiency. Its secure and cost-effective data storage and transfer capabilities address crucial data safety and economic needs within the sector. It ensures dependable and secure information exchange throughout the energy infrastructure, which is vital as automation

**Table 1** Applications and benefits of digital technologies for energy systems

| Technology/application | Description | Main benefits |
|---|---|---|
| Smart grids | • Integration of IT with traditional energy grids<br>• Real-time two-way communication<br>• Decentralized power generation<br>• Enabling Distributed Energy Resources<br>• Helping small renewable producers<br>• Supporting cloud computing integration | • Increased efficiency<br>• Reduced dependence on fossil fuels<br>• Environmental sustainability<br>• Distribution flexibility and cost efficiency<br>• Improved grid performance<br>• Accurate billing in real time |
| SCADA systems | • Used in smart grids for control, monitoring, and diagnostics<br>• Applications in traditional grids and other energy industries | • Improvement of system efficiency, reliability, and safety<br>• Supporting real-time operational decision-making |
| Cloud computing | • Energy system management centralization<br>• Big data analytics for energy operations<br>• Automation of energy data processing<br>• Predictive analytics for system forecasting<br>• Artificial intelligence for managing renewable energy<br>• On-site data processing using edge computing | • Improving energy system data security and integrity<br>• Reduced costs<br>• Renewables integration optimization promotes environmental sustainability<br>• Enhanced energy operations through predictive analytics |
| Internet of Things | • Virtual energy storage solutions<br>• Optimization of household energy<br>• Advanced monitoring with smart meters and intelligent battery storage<br>• AI for strategic energy supply and demand management<br>• Assistance with regulatory compliance for energy companies | • Improved energy efficiency and system sustainability<br>• Energy storage and usage flexibility<br>• Financial incentives for clients<br>• Minimize downtime<br>• Remote energy asset automation and control |

(continued)

**Table 1** (continued)

| Technology/application | Description | Main benefits |
|---|---|---|
| Artificial Intelligence | • Application of machine learning and deep learning<br>• Implementing advanced algorithms for utility energy control<br>• Intelligent renewable energy integration software for autonomous control | • Increasing the efficiency and reliability of renewable energy generation<br>• Improving demand forecasting and energy management accuracy<br>• Strengthening cybersecurity measures within smart grid infrastructures<br>• Effective decision-making and efficient energy trading |
| Artificial Neural Networks | • Neural models for energy data analysis<br>• Energy demand forecasting<br>• Automating energy system protection<br>• Managing energy loads<br>• Selecting low-cost energy suppliers | • Improved forecasting of energy demand and reliability<br>• Increase automation and efficiency of energy systems<br>• Optimized energy production costs<br>• Streamlined and accurate energy billing |
| Blockchain | • Enabling direct energy market transactions<br>• Facilitating secure tracking of the energy supply chain<br>• Direct connection of consumers to several energy providers | • Efficient and cost-effective energy transactions<br>• Increased security<br>• Increased energy choice for consumers<br>• Better energy management and consumer awareness |

expands in energy enterprises. Additionally, Cloud Computing plays a crucial role in utilizing Big Data for the energy sector, automating data processing, and enabling predictive analytics to anticipate system developments and prevent emergencies. These Cloud Computing functions demonstrate their essential role in enhancing the energy sector's sustainability and operational efficiency (Walnum and Andrae 2016; Konovalov et al. 2020). Cloud Computing advances the energy sector's capabilities, helping stabilize energy markets as reliance on renewable sources increases. It uses analytics and AI to optimize electricity delivery and manage the variability of renewables, ensuring demand matches the intermittent supply. Edge computing, a complement to cloud technology, processes data on-site, improving real-time grid management with low latency and preserving data privacy by minimizing long-distance data transfer. This technology also supports the development of energy markets for trading excess power. Localized data processing enables precise energy usage forecasting and efficient management of renewable energy, which is integral for both providers and consumers (Adib 2023).

### 2.2.3 Internet of Things

IoT has transformative applications in the energy sector, bolstering the efficiency and sustainability of energy management systems. Sensors are the cornerstone of IoT networks. In smart grids, IoT can improve energy management monitoring (Baidya et al. 2021). For instance, smart meters can be used for a demand response initiative, enabling intelligent load control via smart thermostats and compensating customers for contributing their flexibility. Additionally, they optimize household energy flow, promoting the use of self-generated electricity and seamlessly integrating with diverse systems like photovoltaic (PV) systems, electric vehicle charging stations, and heat pumps. Beyond these, IoT implementations have expanded to include smart meters for meticulous energy monitoring and intelligent battery storage systems, furthering the innovative management and resilience of energy resources (Paukstadt and Becker 2021). The use of IoT in the energy sector extends beyond monitoring and control, encompassing various innovative applications. It facilitates real-time grid adjustments to minimize outages and predictive maintenance to address potential equipment failures proactively. AI-enhanced IoT devices enable strategic energy supply and demand management while taking automation to new heights with remote control of energy facilities. Furthermore, IoT is the foundation for peer-to-peer energy trading platforms that utilize blockchain technology for transactions. Additionally, it is essential for assisting companies in adhering to regulatory compliance, which is crucial in adapting to the industry's ever-changing standards (Podorozhko 2023).

### 2.2.4 Artificial Intelligence and Artificial Neural Networks

AI, including ML and deep learning (DL), is transforming the energy sector. In addition to improving energy planning and control for utilities, AI optimizes renewable energy generation, such as solar PV and hydrogen power, by maximizing power point tracking. It improves demand forecasting across energy sources and supports predictive maintenance. AI also strengthens the cybersecurity of smart grid infrastructure, streamlines the integration of renewable resources, and facilitates efficient energy trading. By providing intelligent software for autonomous system control, AI supports robust and optimized decision-making, driving significant advances in the sustainability and operational efficiency of the sector (Ahmad et al. 2021).

Another fast-growing technology applied to the energy sector, Artificial Neural Networks (ANNs), computational models that emulate the human brain's functionality, introduce a novel approach to information analysis in the industry. ANNs are interconnected processes that can learn, retain, analyze, and generate solutions like human neurons. During the power network design phase, ANNs aid in predicting energy demand and assessing generator equipment reliability. They are also vital in automating protection systems, managing production and transmission overload, and optimizing equipment operation costs for energy generation. In transmission and sales, these networks enable automated selection of the most strategic or cost-effective suppliers. Additionally, they can optimize energy prices dynamically based on seasonality and customer habits. It is worth highlighting that in terms of bills, ANNs assist in automating billing processes, guaranteeing precision and efficiency in charging customers (Sozontov et al. 2019; Chebotareva 2021; Świątowiec-Szczepańska and Stępień 2022).

### 2.2.5 Blockchain

Blockchain technology can potentially transform the energy sector by simplifying its complex structure through direct transactions and ensuring secure supply (Mengelkamp et al. 2018; Strüker et al. 2019; Krajnakova et al. 2019). One of the most significant uses of this technology is in facilitating direct transactions between energy suppliers and customers. The decentralized design of blockchain removes the necessity for intermediaries, allowing the end user to choose between competing energy suppliers available in the market. This removal of the middleman simplifies transactions, potentially reducing administrative fees and duties while increasing transparency and saving consumers money. Furthermore, utilizing blockchain to track and verify each stage in the transaction process guarantees supply security by offering comprehensive data on the energy's source. It also helps the end-users who wish to know the source of their energy supply or track their household energy consumption to be better able to manage their energy consumption habits (Świątowiec-Szczepańska and Stępień 2022).

In summary, the digital transformation of the energy industry represents a significant milestone, offering unprecedented improvements in resource utilization, energy efficiency, and user participation. The strategic deployment of technologies such as smart grids, SCADA systems, cloud computing, IoT, AI, ANNs, and blockchain is pivotal in driving this sector into a new era of innovation. Collectively, these technologies contribute to a more robust, adaptable, and sustainable energy infrastructure, ensuring the industry not only meets the demands of the present but is also well-equipped for the challenges of the future. The dynamic interplay between digital advances and energy systems is captured in Fig. 2. which provides a visual representation of their positive advances and the potential challenges they pose.

**Fig. 2** Impact of digitalization on the energy sector

# 3 Effects of Digitalization of Energy on SDGs

The digitalization of energy significantly impacts progress towards achieving the SDGs. This influence is multifaceted, as energy is critical in many SDGs. The complex interactions between technological advancement and sustainable development objectives present positive and negative effects.

## *3.1 Effects of Digitalization Energy on SDGs: Benefits*

Digitalization of energy systems not only positively impacts many, if not all, SDGs, as shown in Fig. 3, but its effects are particularly evident in SDG 7. This transformation can enhance all the targets within this goal. For example, by ensuring universal access to affordable, reliable, and modern energy services, digitalization aligns with and supports Target 7.1.

First, the digitalization of energy systems improves the accessibility of energy services to consumers. While it does not increase energy reserves, digitalization facilitates the exploration of such resources, thereby improving their availability for energy consumers. Energy companies widely adopt digital technologies to explore and make energy resources more accessible (Lu et al. 2019). The digitalization of energy systems contributes to improved accessibility of energy services by fostering the electrification of areas previously devoid of such services. Thus, remote and underserved regions, especially in developing countries, would gain access to previously unavailable energy resources. For instance, the deployment of off-grid solar

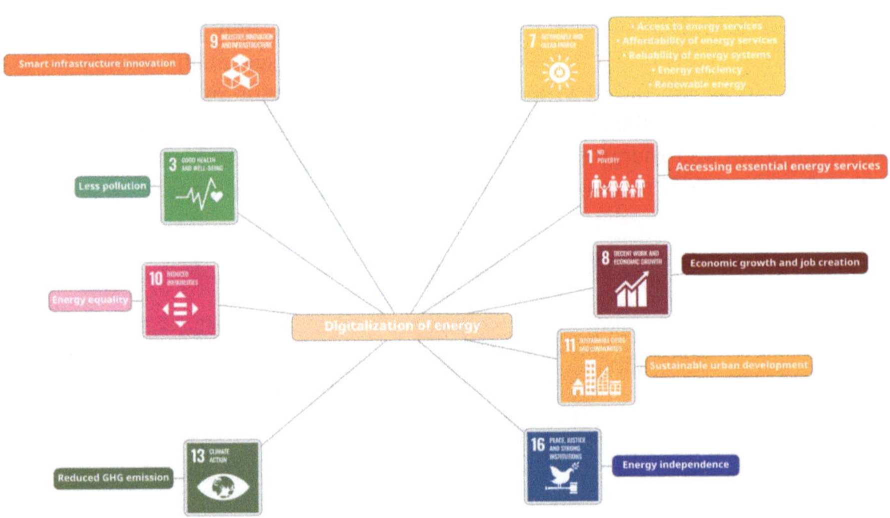


**Fig. 3** Positive effects of digitalization of energy on SDGs

microgrids equipped with smart metering systems allows for the provision of reliable electricity to remote communities. Furthermore, the integration of digital platforms enables efficient management of decentralized energy sources, optimizing energy distribution and enhancing accessibility for end-users who were previously without electricity access.

Second, digitalization makes energy services more affordable by decreasing costs for the operation, maintenance, monitoring, and diagnostics of energy systems. Although initial investments in digital infrastructure are needed, in the long run, digital systems are more cost-effective and lower energy prices for end users (Thomas and McDonald 2015).

Third, digitalization makes energy systems more reliable because digital transformation reduces the chances of human error since data is received, recorded, and analyzed automatically (Adams et al. 2021). In addition, the security of energy systems is a pressing issue because the nature of digital systems makes them vulnerable to cyberattacks. Blockchain technologies can play a crucial role in ensuring the security of energy systems against those threats (Baidya et al. 2021).

Digitalization of energy systems can benefit Target 7.2, which focuses on the increasing share of renewable energy in the global energy mix. Digitalization helps to solve many challenges caused by the introduction of renewables. Indeed, renewable technologies and energy have different technological requirements. For instance, a decentralized generation typical for renewables often poses a challenge in existing energy grids, which can be solved through digital technologies that enable new real-time dispatch activities (Loock 2020).

The digitalization of energy is crucial for achieving Target 7.3, which prioritizes enhancing energy efficiency. Objective evidence suggests that digitization improves the efficiency of energy systems, affecting both the generation and consumption stages. Digital technologies enable energy production and consumption to be efficiently coordinated and allow energy operators to monitor energy demand in real-time and consequently adjust the energy supply (Pearson 2011; Goldbach et al. 2018; Loock 2020). Digitalization also allows the adjustment of the energy-consumption behavior of end energy users. Although some consumers recognize the importance of energy efficiency, many choose personal convenience over broader environmental benefits. Demand-side management (DSM) leverages digital technologies to optimize energy consumption. DSM employs real-time data, automated systems, and consumer notifications to promote energy stability and efficiency. Digital thermostats, smart meters, and other advanced devices can help final consumers conserve energy. In addition, dynamic energy tariffs help to adjust energy consumption behavior (Martiskainen and Coburn 2011; Pearson 2011; Goulden et al. 2014; Goldbach et al. 2018). Using AI-powered energy management systems facilitates detailed data analysis, resulting in better energy usage patterns and the exposure of novel energy-saving prospects (Baidya et al. 2021).

In addition to SDG 7, the digitalization of energy systems directly and indirectly affects other SDGs.

For instance, the digitalization of energy can contribute to SDG 1 by providing access to essential energy services, especially to vulnerable populations. The shift to renewables and the overall use of energy resources because of digitalization can reduce hazardous chemicals and pollution and thus contribute to SDG 3. Digitalization of energy systems can create new jobs and foster economic growth aligned with SDG 8. Digital energy systems support the development of smart infrastructure and innovation, aligning with SDG 9. Digital energy solutions can mitigate energy poverty, promoting more equitable access to energy resources and reducing other inequalities aligning with the objectives of SDG 10. Digitalization contributes to the optimization of energy consumption in smart, sustainable urban spaces and, therefore, positively affects SDG 11. A shift to renewable energy sources backed by digitalization results in decarbonization and, thus, contributes to SDG 13 (Pérez-Martínez et al. 2023). Finally, the digitalization of energy can contribute to domestic stability and reduce risks related to the geopolitics of energy, thus contributing to SDG 16.

## 3.2 *Effects of Digitalization Energy on SDGs: Challenges*

The transition to digitalization, however, poses challenges. Some SDGs and their indicators show a negative correlation with digitalization (Pérez-Martínez et al. 2023). Thus, paradoxically, the digitalization of energy can impede progress toward SDG 7 and its targets (Fig. 4).

For example, the digitalization of energy presents challenges in enhancing the accessibility of energy services. Remote regions lacking advanced infrastructure and connectivity face disadvantages, as the digitalized energy solutions rapidly adopted in urban centers remain inaccessible to less developed areas. The widespread adoption of such technologies can unintentionally escalate overall energy consumption, posing obstacles to achieving sustainable energy goals.

From an economic perspective, integrating digital technologies like smart grids requires significant capital investment. It has implications for end consumers who may experience heightened energy costs, deviating from the affordability goal of SDG 7. The digitalization of energy can potentially make energy services less affordable due to the high costs caused by initial investments, including materials, training,

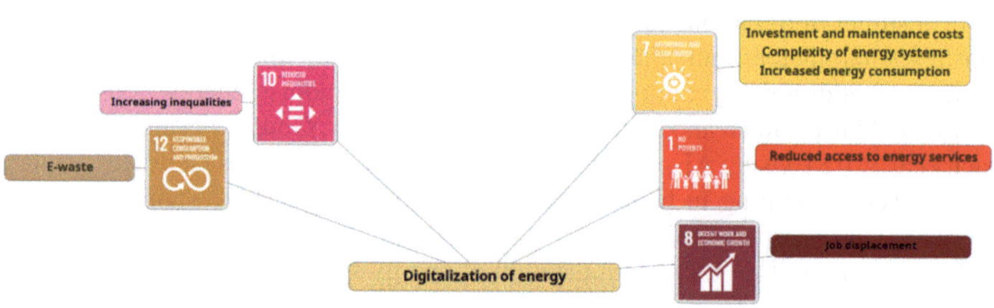


**Fig. 4** Negative effects of digitalization of energy on SDGs

and maintenance of digitalized energy systems. The complexity of digitalized energy systems can lead to errors, operational instability, and challenges in maintenance and repair. All this can slow down the progress towards Target 7.1.

Digitalization can be a potential challenge to achieving Target 7.2. Despite the progress in renewable energy, a considerable part of digital infrastructure, including server farms, depends on fossil fuels due to their availability.

There are concerns about how the digitalization of energy would affect Target 7.3. Although energy systems are usually characterized by increased energy efficiency, computer technologies require energy for their development, manufacturing, and functioning. The rapid digitalization of energy systems, akin to digitalization in other facets of human existence, may lead to heightened energy consumption. Should the energy derive from fossil fuels, this could subsequently give rise to an increase in greenhouse gas emissions, threatening environmentally focused SDGs (Belkhir and Elmeligi 2018; Lange et al. 2020).

Digitalization of energy systems can negatively affect other SDGs again because of the interconnected nature of SDGs (Elavarasan et al. 2022).

For instance, the lack of access to modern digital technologies can slow down the introduction of digitalized energy systems and, therefore, access to the essential services identified in SDG 1.

The digitalization of energy systems can help to create new jobs. At the same time, as automated energy systems require fewer workers, it can lead to unemployment and therefore jeopardize the progress towards SDG 8.

The digitalized energy sector makes overall critical infrastructure vulnerable to cyberattacks that can disrupt energy supplies and have considerable economic implications (Pearson 2011; Eder-Neuhauser et al. 2017; Xu et al. 2021) that can negatively affect the progress towards SDG 9.

Because of the existing digital divide, digitalization can result in even more unequal access to energy services and, therefore, further increase inequalities between and within countries, a primary concern of SDG 10.

Finally, the apparent negative consequence of the digitalization of energy is the production of e-waste related to energy production and use, and therefore, challenges the progress towards SDG 12. E-waste includes hundreds of toxic and nontoxic materials, the heavy metal and microplastic content of which is highly concerning. Since digital parts and tools become obsolete with distressing speed, the impact of such waste on the environment and human health must be carefully thought through (Hossain et al. 2015).

The shift towards digitalization in energy systems shows promise, but it also presents numerous challenges. Such challenges, discussed comprehensively earlier, have notable consequences for SDGs:

- Negative correlation with certain SDGs;
- Limited accessibility in less developed areas;
- Escalation in overall energy consumption;
- Economic burden and affordability concerns;
- Operational complexity and stability issues;

- Fossil fuel dependency on digital infrastructure;
- Rise in greenhouse gas emissions;
- Exacerbation of the digital divide;
- Potential increase in unemployment;
- Vulnerability to cyberattacks;
- Production of e-waste.

## 4 Conclusion

Energy plays a crucial role in achieving the SDGs. The use of digital tools, systems, and applications enhances energy systems' stability, enables the integration of power generation systems with end-users, reduces peak demand, speeds up energy supply restoration after disruption, reduces operating expenses, and facilitates the integration of renewables into energy systems. Thus, the digitalization of energy can contribute to achieving SDGs through increased efficiency, improved stability, and lower cost of energy systems. Although the digitalization of energy would primarily affect SDG 7, it can also positively affect other SDGs by indirectly eradicating poverty and hunger, facilitating healthcare, education, and water access, driving economic growth, and protecting the environment.

Yet, in some instances, the digitalization of energy systems can negatively affect the progress toward SDGs through increased initial costs, energy consumption, electronic waste, and overall complexity of energy systems. Nevertheless, the net effect of digitalization of energy systems on SDGs is likely positive and, therefore, should be promoted.

Several promising directions for further research about the relationship between digitalized energy and the achievement of the SDGs can be explored. First, future research could investigate how energy policies can be strategically designed to advance multiple SDGs simultaneously. It would require a comprehensive analysis of the trade-offs and complementarities between energy-related goals and other SDGs, focusing on optimizing policy frameworks for maximum positive impact. Second, the digitalization of energy is an umbrella term that includes numerous IT technologies and approaches. Consequently, their individual effects on energy systems and, accordingly, on SDGs can vary significantly. Therefore, a detailed investigation of specific IT technologies in energy systems and their impact on SDGs could be further investigated. Finally, research on relationships between global energy policies and the SDGs and their practical implementation on national and local community levels is critical. Such research could analyze the challenges and opportunities associated with translating high-level energy objectives into actionable initiatives that resonate with the specific goals and needs of local communities.

## References

Adams S, Kuch D, Diamond L et al (2021) Social license to automate: A critical review of emerging approaches to electricity demand management. Energy Res Soc Sci 80:102210. https://doi.org/10.1016/j.erss.2021.102210

Adib D (2023) Edge computing in smart grids: Reshaping the energy landscape. In: STL Partners. https://stlpartners.com/articles/edge-computing/smart-grids-edge-computing. Accessed 11 Nov 2023

Ahmad T, Zhang D, Huang C et al (2021) Artificial intelligence in sustainable energy industry: Status Quo, challenges and opportunities. J Clean Prod 289:125834. https://doi.org/10.1016/j.jclepro.2021.125834

Baidya S, Potdar V, Pratim Ray P, Nandi C (2021) Reviewing the opportunities, challenges, and future directions for the digitalization of energy. Energy Res Soc Sci 81:102243. https://doi.org/10.1016/j.erss.2021.102243

Belkhir L, Elmeligi A (2018) Assessing ICT global emissions footprint: Trends to 2040 & recommendations. J Clean Prod 177:448–463. https://doi.org/10.1016/j.jclepro.2017.12.239

Bera S, Misra S, Rodrigues JJPC (2015) Cloud computing applications for smart grid: A survey. IEEE Trans Parallel Distrib Syst 26:1477–1494. https://doi.org/10.1109/TPDS.2014.2321378

Bloomberg (2017) Market for digitalization in energy sector to grow to. https://about.bnef.com/blog/market-digitalization-energy-sector-grow-64bn-2025/. Accessed 5 Nov 2023

Brenner B, Hartl B (2021) The perceived relationship between digitalization and ecological, economic, and social sustainability. J Clean Prod 315:128128. https://doi.org/10.1016/j.jclepro.2021.128128

Çelik D, Meral ME, Waseem M (2022) Investigation and analysis of effective approaches, opportunities, bottlenecks and future potential capabilities for digitalization of energy systems and sustainable development goals. Electric Power Syst Res 211:108251. https://doi.org/10.1016/j.epsr.2022.108251

Chebotareva G (2021) Digital transformation of the energy sector: A case of Russia. E3S Web Conf 250:01001. https://doi.org/10.1051/e3sconf/202125001001

Chien F, Anwar A, Hsu C-C et al (2021) The role of information and communication technology in encountering environmental degradation: Proposing an SDG framework for the BRICS countries. Technol Soc 65:101587. https://doi.org/10.1016/j.techsoc.2021.101587

European Council (2023) Fit for 55. In: European Green Deal. https://www.consilium.europa.eu/en/policies/green-deal/fit-for-55-the-eu-plan-for-a-green-transition/. Accessed 5 Nov 2023

de la Harpe B, Thomas I (2009) Curriculum change in universities: Conditions that facilitate education for sustainable development. J Educ Sustain Dev 3:75–85. https://doi.org/10.1177/097340820900300115

Eder-Neuhauser P, Zseby T, Fabini J, Vormayr G (2017) Cyber attack models for smart grid environments. Sustainable Energy, Grids and Networks 12:10–29. https://doi.org/10.1016/j.segan.2017.08.002

Elavarasan RM, Pugazhendhi R, Jamal T et al (2021) Envisioning the UN Sustainable Development Goals (SDGs) through the lens of energy sustainability (SDG 7) in the post-COVID-19 world. Appl Energy 292:116665. https://doi.org/10.1016/j.apenergy.2021.116665

Elavarasan RM, Pugazhendhi R, Irfan M et al (2022) A novel Sustainable Development Goal 7 composite index as the paradigm for energy sustainability assessment: A case study from Europe. Appl Energy 307:118173. https://doi.org/10.1016/j.apenergy.2021.118173

Fuso Nerini F, Tomei J, To LS et al (2017) Mapping synergies and trade-offs between energy and the Sustainable Development Goals. Nat Energy 3:10–15. https://doi.org/10.1038/s41560-017-0036-5

Goldbach K, Rotaru AM, Reichert S et al (2018) Which digital energy services improve energy efficiency? A multi-criteria investigation with European experts. Energy Policy 115:239–248. https://doi.org/10.1016/j.enpol.2017.12.036

Goulden M, Bedwell B, Rennick-Egglestone S et al (2014) Smart grids, smart users? The role of the user in demand side management. Energy Res Soc Sci 2:21–29. https://doi.org/10.1016/j.erss.2014.04.008

Gunnarsdottir I, Davidsdottir B, Worrell E, Sigurgeirsdottir S (2021) Sustainable energy development: History of the concept and emerging themes. Renew Sustain Energy Rev 141:110770. https://doi.org/10.1016/j.rser.2021.110770

Hossain MdS, Al-Hamadani SMZF, Rahman MdT (2015) E-waste: A challenge for sustainable development. J Health Poll 5:3–11. https://doi.org/10.5696/2156-9614-5-9.3

Husaini DH, Lean HH (2022) Digitalization and energy sustainability in ASEAN. Resour Conserv Recycl 184:106377. https://doi.org/10.1016/j.resconrec.2022.106377

Konovalov Y, Vaygachev A, Levin N (2020) Cloud computing and their role in digitalization electric power. Mod Technol Sci Technol Prog 2020:203–204

Krajnakova E, Svazas M, Navickas V (2019) Biomass blockchain as a factor of energetical sustainability development. JESI 6:1456–1467. https://doi.org/10.9770/jesi.2019.6.3(28)

Lange S, Pohl J, Santarius T (2020) Digitalization and energy consumption. Does ICT reduce energy demand? Ecol Econ 176:106760. https://doi.org/10.1016/j.ecolecon.2020.106760

Leal Filho W, Viera Trevisan L, Simon Rampasso I et al (2023) When the alarm bells ring: Why the UN sustainable development goals may not be achieved by 2030. J Clean Prod 407:137108. https://doi.org/10.1016/j.jclepro.2023.137108

Legner C, Eymann T, Hess T et al (2017) Digitalization: Opportunity and challenge for the business and information systems engineering community. Bus Inf Syst Eng 59:301–308. https://doi.org/10.1007/s12599-017-0484-2

Leite De Almeida CM, Bergqvist E, Thacker S, Fuso Nerini F (2021) Actions to align energy projects with the Sustainable Development Goals. Discov Sustain 2:16. https://doi.org/10.1007/s43621-021-00020-3

Loock M (2020) Unlocking the value of digitalization for the European energy transition: A typology of innovative business models. Energy Res Soc Sci 69:101740. https://doi.org/10.1016/j.erss.2020.101740

Lu H, Guo L, Azimi M, Huang K (2019) Oil and Gas 4.0 era: A systematic review and outlook. Comput Ind 111:68–90. https://doi.org/10.1016/j.compind.2019.06.007

Ma Y, Zhao F, Zhou X, Gao Z (2018) Summary of Cloud Computing Technology in Smart Grid. In: 2018 IEEE International Conference on Mechatronics and Automation (ICMA). IEEE, Changchun, pp 253–258

Martiskainen M, Coburn J (2011) The role of information and communication technologies (ICTs) in household energy consumption—prospects for the UK. Energ Effi 4:209–221. https://doi.org/10.1007/s12053-010-9094-2

Mengelkamp E, Notheisen B, Beer C et al (2018) A blockchain-based smart grid: Towards sustainable local energy markets. Comput Sci Res Dev 33:207–214. https://doi.org/10.1007/s00450-017-0360-9

Moreira N, Molina E, Lázaro J et al (2016) Cyber-security in substation automation systems. Renew Sustain Energy Rev 54:1552–1562. https://doi.org/10.1016/j.rser.2015.10.124

Novikau A (2022) Energy, and human and planetary health: Is the mutual link inevitable? In: Leal Filho W (ed) Handbook of Human and Planetary Health. Springer International Publishing, Cham, pp 101–112

Paukstadt U, Becker J (2021) Uncovering the business value of the internet of things in the energy domain – a review of smart energy business models. Electron Markets 31:51–66. https://doi.org/10.1007/s12525-019-00381-8

Pearson ILG (2011) Smart grid cyber security for Europe. Energy Policy 39:5211–5218. https://doi.org/10.1016/j.enpol.2011.05.043

Pérez-Martínez J, Hernandez-Gil F, San Miguel G et al (2023) Analysing associations between digitalization and the accomplishment of the Sustainable Development Goals. Sci Total Environ 857:159700. https://doi.org/10.1016/j.scitotenv.2022.159700

Podorozhko Y (2023) IoT energy: 6 Ways the IoT benefits the energy sector. In: MadAppGang Blog. https://madappgang.com/blog/iot-in-energy-sector/. Accessed 11 Nov 2023

Rahman MA, Jakaria AHM, Al-Shaer E (2016) Formal analysis for dependable supervisory control and data acquisition in smart grids. In: 2016 46th Annual IEEE/IFIP International Conference on Dependable Systems and Networks (DSN). IEEE, Toulouse, France, pp 263–274

Sozontov A, Ivanova M, Gibadullin A (2019) Implementation of artificial intelligence in the electric power industry. E3S Web Conf 114:01009. https://doi.org/10.1051/e3sconf/201911401009

Strüker J, Albrecht S, Reichert S (2019) Blockchain in the energy sector. In: Treiblmaier H, Beck R (eds) Business Transformation through Blockchain. Springer International Publishing, Cham, pp 23–51

Światowiec-Szczepańska J, Stępień B (2022) Drivers of digitalization in the energy sector—The managerial perspective from the catching up economy. Energies 15:1437. https://doi.org/10.3390/en15041437

Thomas MS, McDonald JD (2015) Power system SCADA and smart grids. CRC Press, Boca Raton, FL

United Nations (2023) Ensure access to affordable, reliable, sustainable and modern energy. https://www.un.org/sustainabledevelopment/energy/. Accessed 7 Aug 2023

Vale Z, Morais H, Faria P, Ramos C (2013) Distribution system operation supported by contextual energy resource management based on intelligent SCADA. Renewable Energy 52:143–153. https://doi.org/10.1016/j.renene.2012.10.019

Walnum HJ, Andrae ASG (2016) The internet: Explaining ICT service demand in light of cloud computing technologies. In: Santarius T, Walnum HJ, Aall C (eds) Rethinking Climate and Energy Policies. Springer International Publishing, Cham, pp 227–241

Weigel P, Fischedick M (2019) Review and categorization of digital applications in the energy sector. Appl Sci 9:5350. https://doi.org/10.3390/app9245350

Xu L, Guo Q, Sheng Y et al (2021) On the resilience of modern power systems: A comprehensive review from the cyber-physical perspective. Renew Sustain Energy Rev 152:111642. https://doi.org/10.1016/j.rser.2021.111642

# AI-Based Predictive Maintenance Strategies for Improving the Reliability of Green Power Systems

**Anu Sayal, Chaithra N, Janhvi Jha, and Nithin Allagari**

**Abstract** The adoption of AI-based predictive maintenance techniques increases the dependability of green power systems. Using AI, researchers are examining a vast array of potential approaches to revamping the maintenance of renewable energy systems. The use of supervised learning algorithms enables the classification and prediction of faults using historical data from green power systems, enabling the detection and categorization of distinct failure types. In addition, sensor data collected from renewable energy sources can be utilized by unsupervised learning algorithms for anomaly detection and failure identification. Moreover, reinforcement learning algorithms optimize the scheduling and allocation of maintenance resources by utilizing the system's current state. The combination of Internet of Things (IoT) sensors and data analytics tools makes real-time monitoring of renewable power systems feasible. To better predict the likelihood of equipment failure and the remaining useful life, it is necessary to analyze sensor data for patterns, trends, and anomalies that may indicate impending failures or performance degradation. In addition, there is an increase in the use of digital twin technology, which creates digital duplicates of green energy systems to model and foresee their operation in order to facilitate condition monitoring, predictive maintenance planning, and performance enhancement. Integrating real-time sensor data with digital twin models enables continuous system monitoring and diagnostics, providing the groundwork for timely maintenance interventions. Predictive maintenance of hydraulic systems is the primary focus of this paper, which demonstrates the practical application of AI-based predictive maintenance strategies via intriguing case studies and results. The dependability and performance of renewable energy sources illustrate the effectiveness and advantages of AI-driven maintenance procedures.

A. Sayal (✉)
School of Accounting and Finance, Taylor's Business School, Taylor's University, Subang Jaya, Malaysia
e-mail: anu.sayal.07@gmail.com

C. N · J. Jha · N. Allagari
Department of CSE (AI & ML), JAIN (Deemed to Be University), Bangalore, Karnataka, India

 
W. Leal Filho et al. (eds.), *Digital Technologies to Implement the UN Sustainable Development Goals*, World Sustainability Series,
https://doi.org/10.1007/978-3-031-68427-2_2

**Keywords** Artificial Intelligence · Green Power Systems · Machine Learning · Predictive Maintenance

## 1 Introduction

Renewable energy has replaced fossil fuels since their depletion. We use more alternative energy each year. Complex and hybrid energy-generating technologies are being developed to fulfill requirements and secure energy. Infrastructure must be improved to use renewable energy. AI can improve energy infrastructure and decision-making. Predictive maintenance using AI, machine learning, and sensor data makes data-driven decisions. AI algorithms can find trends in real-time and historical data to forecast green power system failures. Preventive maintenance reduces unforeseen problems and optimizes resource utilization. AI-based solutions can also illuminate issue sources, enabling more targeted interventions and better system performance (Afridi, Ahmad, & Hassan, 2022; Bangroo, 2023).

Machine learning algorithms applied to sensor data analytics have optimized energy distribution networks and systems in recent years. These algorithms have shown promise in AI-based preventative maintenance plans for green power installations (Jha et al., 2023; Sayal et al., 2023). Machine learning is used to renewable energy sector concerns including fault classification, anomaly detection, predictive modeling, and the recently proposed Digital Twin technology. Machine learning algorithms applied to sensor data analytics have optimized energy distribution networks and systems in recent years. These algorithms are promising for AI-based green power facility preventative maintenance programs. Machine learning is used to renewable energy sector concerns including fault classification, anomaly detection, predictive modeling, and the recently proposed Digital Twin technology. IoT sensors and data analytics platforms provide real-time green power system monitoring. The Internet of Energy (IoE) can connect the energy sector to the growing ICT sector. This chapter discusses how the renewable energy business may use IoT to better monitor assets, discover abnormalities, and optimize system performance.

The chapter also examines sensor data for patterns, trends, and anomalies that may indicate imminent breakdowns or reduced performance. Machine learning and deep learning identify and classify abnormalities, reducing outages and repair time. This chapter covers sensor data prediction models, an important issue. Predictive maintenance uses real-time data analysis to evaluate machinery lifespan. Machine learning and historical data optimize maintenance plans, save costs, and prevent unplanned outages. Digital Twin technology models and predicts system performance. Stakeholders may better understand green power systems, analyze situations, and make choices by modeling real-world infrastructure in a digital environment.

The chapter's last case study applies data analysis and machine learning to hydraulic system predictive maintenance. The study provides in-depth data purification, exploratory analysis, preprocessing, model training, assessment, and performance analysis. This chapter discusses AI-based predictive maintenance's future.

These include machine learning algorithm improvements, peripheral computing and artificial intelligence integration, sensor technologies, big data analytics, digital twin applications, and cybersecurity precautions. These advancements may increase fault classification, prediction accuracy, system dependability, and long-term viability in green power systems.

This chapter primarily focuses on the utilization of artificial intelligence (AI) for forecasting and upkeep of sustainable energy systems. The Sustainable Development Goals (SDGs) that are relevant to this context are SDG 11: Sustainable Cities and Communities, SDG 13: Climate Action, SDG 7: Industry, Innovation and Infrastructure, and SDG 9: Affordable and Clean Energy. Implementing predictive maintenance techniques can improve the reliability and effectiveness of renewable energy sources like wind and solar power. Advocating for the adoption of sustainable energy sources is a means to contribute to the attainment of SDG 7. A technological infrastructure that combines sensor data analytics and machine learning can provide additional support in achieving SDG 9 by efficiently distributing and monitoring energy. To effectively achieve SDG 11, which focuses on creating environmentally sustainable and resilient urban areas, the adoption of solar-powered smart grid systems offers a practical solution. Utilizing artificial intelligence, and green power infrastructure optimization solutions can assist in achieving SDG 13, which seeks to address climate change by promoting the adoption of cost-effective clean energy systems.

## 2 Machine Learning Algorithms

Current trends suggest autonomous software will oversee energy distribution and decision-making. This program will be customized for these tasks. The program will control energy demand and supply. Optimizing energy distribution networks and systems requires advanced ML technology. Wide area monitoring systems (WAMS) use AI to analyze synchronized data from phasor measuring devices in smart grids. Machine-learning technologies use supervised, unsupervised, and reinforcement learning. These methods train and improve machine-learning models. Labeled examples help the model understand patterns and make predictions. Unsupervised learning trains the model using unlabeled data to reveal hidden patterns and structures. Reinforcement learning uses incentives and punishments to teach the model. Labeled data is used during supervised learning training. Unsupervised learning groups incoming data based on preset criteria without using training data labels. The analysis's clustering criterion determines the number of clusters. Reinforcement learning is a computational approach that includes an agent interacting with an environment, getting rewards or penalties, and utilizing this knowledge to enhance its performance through iteration. These three learning concepts have inspired many theoretical procedures and implementations (Qiu et al. 2016). Neural networks, decision trees, support vector machines, and deep learning assess power system transient stability. These techniques are used to generate mapping functions from offline data. This permits online time series and fresh instance analysis.

## *2.1 Supervised Learning Algorithms for Fault Classification and Prediction Based on Historical Data from Green Power Systems.*

Random Forest (RF) is a popular ensemble learning technique that is being used more and more in the fields of economics and health. Collaborative learning surpasses solitary learning in terms of the ability to apply knowledge to a wide range of situations. Many students worked on this. Random Forest (RF) training involves generating distinct subsets through bootstrapping sampling. Each student receives personalized instruction. A new training set is generated by constructing a regression tree and randomly selecting m features from each of the w training subgroups. The regression tree decision tree model is a powerful tool for conducting regression analysis. In his 2001 paper, Breiman demonstrates the correlation between the division of the input space and the resulting output value at each node of a regression tree partition. Finding the most suitable values for the parameters of an RF model is a difficult task. This issue can be resolved by leveraging the out-of-bag (OOB) error. The reliable and accurate measure for quantifying error in the process of generating random forests is the out-of-the-box (OOB) error. Finding the most suitable values for the parameters of an RF model is a difficult task. This issue can be resolved by leveraging the out-of-bag (OOB) error. The reliable and precise metric for assessing the error in the process of generating a random forest is the out-of-the-box (OOB) error. The online prediction of TSM was examined by Gao, Mi, and Wang. The Random Forest (RF) algorithm was utilized for conducting regression analysis. The ensemble regression trees model utilizes multiple base learners to establish a correlation between initial conditions and CCTs. To address the dimensionality catastrophe problem, one can utilize bootstrapping and randomly select variables during the training process, even without employing feature selection. The study conducted by Mi et al. (2021) evaluated power engineering solutions using ridge regression and step lasso.

Solar and wind energy availability has been predicted using several methods. Support vector machines outperform other modeling methods in precision prediction, according to the study. SVMs are notable for their efficiency, reliability, and accuracy (Zendehboudi, Baseer & Saidur 2018; Cortes & Vapnik 1995). Statistical learning theory and structural risk minimization underpin this technique. Nonlinear mapping translates the input space, which is nonlinear, into a higher-dimensional space. This method locates a hyperplane in newly generated space. SVMs are recommended for classification, pattern recognition, and regression analysis. They supplanted traditional statistical models. SVMs may approximate functions and regression. SVR is achievable (Huang et al., 2002). SVM models have several kernel functions. Polynomial, Exponential, Radial, Sigmoid, and Linear functions are offered.

Shamshirband et al. (2016) provide a novel approach for estimating daily diffuse solar radiation on a flat surface. SVM and WT are combined in the method. These two methods are intended to increase estimation precision and reliability. The clarity index is the sole measure used to evaluate dispersed radiation. This study assessed

MABE, RMSE, and R. After a thorough investigation, SVM-WT outperformed SVM-RBF, ANN, and an empirical model in predicting diffuse solar radiation.

Baser and Demirhan introduced a novel worldwide horizontal sun radiation estimation method. They use fuzzy regression and SVM. Their model improves solar radiation computation precision and reliability. Researchers hope to improve predicting performance by combining both strategies. This unique solar radiation estimating approach might improve energy systems and solar radiation estimation. FRF-SVM uses linear, polynomial, Gaussian, and sigmoidal kernel functions to boost performance and versatility. This study compared FRF-SVM models to adaptive neuro-fuzzy system (ANFIS) and robust coplot-supported genetic programming (GenProg) models. This study employed temperature, humidity, longitude, latitude, and altitude. FRF-SVM and ANFIS input these variables. This page describes local weather and geographical features. Coplotting the variables ensured a full examination. This study found that the FRF-SVM method, utilizing a Gaussian kernel function, estimated horizontal global solar radiation more accurately than previous models. This analysis used Turkey-specific data (Baser & Demirhan 2017).

Time series prediction is difficult. Our study used deep neural networks to solve this problem. We forecasted well using numerous trials and meticulous analysis. SVR and other models were compared. This paper's case studies show that support vector regression performs similarly to deep neural networks in accuracy and error. The result highlights support vector machines' generalization potential. Meteorological data, projections, and power generation have been used to estimate solar energy system performance (Sharma et al. 2011).

Jiang and Dong developed QKSVM-FR forward regression. This technique primarily addressed variable selection while predicting global horizontal radiation in the Tibet Autonomous Region. Quadratic regression was used to find the most important input factors for an accurate estimate. The suggested technique generated predictions with more precision than existing reference models (Jiang & Dong 2017).

In addition to linear regression, Urraca et al. used random forests and support vector regression (SVR). The investigation was conducted at 39.11°38′N and 26°13′W in Southeast Spain. The researchers examined 1-h data at this site. Static and dynamic models were used to assess the research topic. SVMs forecast solar irradiance better than other models (Urraca et al. 2016).

Abbas et al. optimized renewable energy generating capacity with genetic algorithms and storage systems. Chance-constrained modeling handled renewable energy system challenges. Genetic algorithms addressed the problem. A Western China scenario proved the model's validity (Abbas et al. 2018).

Chen et al. developed an ensemble learning system to estimate home power usage. The researchers used an ensemble learning model to predict household energy use in this data-driven design. This study used ridge regression, feed-forward deep networks, and an extreme gradient-boosting forest (Chen et al. 2018).

## 2.2 Unsupervised Learning Algorithms for Anomaly Detection and Early Fault Identification in Sensor Data Collected from Renewable Energy Sources

Gallagher et al. investigated how machine learning may reduce energy savings measurement and verification mistakes. Machine learning algorithms are being studied for energy savings in industrial operations. Fine-tuning the model's parameters improved its forecasts. Machine learning (ML) models outperformed conventional approaches in the research (Gallagher et al. 2018).

The researchers presented a novel approach for achieving power stabilization and generation scheduling in a hybrid system consisting of wind turbines and marine current. This study presents novel approaches for reducing wind-induced damage. Oceanic currents exhibit a higher level of predictability compared to atmospheric winds. The methods provided endorse the efficient scalability of the hybrid arrangement. The Harmonic Analysis Method employed ocean current velocities as a model to enhance the training of artificial neural networks. A variety of neural networks were trained to forecast wind speeds at various time intervals. The model ensures the scheduling of power generation dispatch, leading to cost reduction and the mitigation of power fluctuations (Anwar et al., 2016).

To project global radiation time series, one can utilize the innovative Kalman filter models developed by Voyant et al. Their approach eliminates the necessity of prior data to attain accurate predictions. Unlike these methods, these studies employed data-driven models that incorporated variable time intervals. The root mean squared error (RMSE) has emerged as the prominent metric. The findings confirmed the enhanced predictive capacity of the model (Voyant et al. 2017).

This research demonstrates that the predictive accuracy of the ARIMA and ANN models is influenced by temporal patterns. By examining the error distributions of both models, the empirical cumulative distribution function yielded valuable insights. The outcomes illustrate the differences among the models. During the first week of testing, the ANN model exhibited superior performance in comparison to the ARIMA model. The artificial neural network (ANN) model consistently produced cumulative errors that were consistently below 5%. When it comes to resuming data analysis after interruptions, Artificial Neural Networks (ANNs) perform better than traditional methods. To enhance the precision of ARIMA predictions, it is imperative to train the model using a substantial amount of historical load data. An Artificial Neural Network (ANN) can gain information about the desired output solely based on the input data models. ANN demonstrates higher accuracy than ARIMA, as stated by Tarmanini et al. (2023).

### 2.3 Reinforcement Learning Algorithms for Optimizing Maintenance Scheduling and Resource Allocation Based on Real-Time System Conditions.

Reinforcement learning (RL) is a branch of machine learning that focuses on how autonomous agents can maximize their overall reward by optimizing their interactions with the environment. This study examines the degree to which agents can gain knowledge from their surroundings by receiving positive or negative reinforcement after their actions. The agent is driven to enhance its performance and decision-making within the environment as a result of this procedure.

Through the utilization of a reinforcement learning algorithm, an intelligent agent acquires the capacity to adapt its decision-making process in accordance with changing environments. The essential elements of the algorithmic framework enhance the agent's learning process. The agent's decision-making is influenced by tactics, regulations, and rules, while its performance is evaluated through incentive systems (Arulkumaran et al., 2017). Enhancing the agent's efficiency and adaptability can be achieved by developing improved decision-making skills.

1. The intelligent agent learns from its surroundings and makes educated judgments. The agent seeks to maximize benefits from its interactions.
2. The agent operates in a physical domain, simulated realm, or any system with specified states and behaviors.
3. An agent's current status or perception of the environment informs decision-making.
4. Using its learning process and current state, the agent chooses an action to change states.
5. After an action, the agent receives numeric input from the environment called a reward, which can be positive, negative, or neutral, indicating the action's desirability and consequences. This dynamic interaction between agents, environment, states, actions, and rewards underpins a complicated and goal-oriented system of intelligent decision-making.

Reinforcement learning algorithms find an optimum policy that matches states and actions to maximize cumulative rewards over time (Sutton & Barto, 2018). The algorithm investigates the environment, assesses its effects, and modifies its policy depending on rewards.

Below are reinforcement learning methods used to optimize maintenance schedules and resource allocation depending on system conditions:

1. Q-Learning is an environment-independent reinforcement learning algorithm. It seeks an environment-optimal decision-making policy. Q-tables record the predicted cumulative rewards for each state-action combination in the algorithm. Q-Learning iteratively refines the policy by exploring and exploiting Q-values. This iterative procedure lets the program assess the current state and choose actions, improving performance over time (Sutton & Barto, 2018).

2. Deep Q-Network (DQN) advances reinforcement learning by handling complex state spaces. DQN enhances Q-Learning with deep neural networks. The Q-network, a complex neural network, approximates Q-values in DQN. Experience replay and a target network helps the algorithm stabilize and accelerate learning. DQN reduces correlation by replaying prior events. The target network, independent from the Q-network, stabilizes learning by providing a stable reference point. DQN uses this integrated technique to navigate complicated state spaces and optimize reinforcement learning (Mnih et al., 2015).
3. PPO optimizes policy functions directly. Iteratively changing the policy to maximize projected cumulative reward while putting limits prevents major policy changes. PPO uses sample data effectively, especially in continuous action spaces (Schulman et al., 2017). PPO optimizes policy in a variety of situations by balancing exploration and exploitation.
4. Trust Region Policy Optimization (TRPO) optimizes policies within a trust region restriction. This limitation helps TRPO avoid instability and sudden, harmful policy changes by keeping policy updates close to the prior one. TRPO is prominent in continuous control optimization due to its theoretical guarantees. It works well in such situations and is widely accepted (Schulman et al., 2015).
5. Model-based reinforcement learning method Proximal Value Iteration (PVI) combines value and policy iteration. It uses value iteration to estimate the state-value function using a learned environment model. PVI also improves policy using value estimations. This approach is useful when system dynamics are well understood or modeled. PVI optimizes reinforcement learning when system dynamics are understood or can be properly modeled (Richardson, Berant, & Kuhn, 2018).
6. Monte Carlo Tree Search (MCTS) is a tree-based decision-making technique. Iteratively extending alternative actions and modeling their results develop a search tree. MCTS excels in complex fields like game-playing and planning. MCTS effectively navigates the search tree to help solve complicated issues by making smart decisions (Winands & Lee, 2019).

Reinforcement learning methods can optimize maintenance schedules and resource allocation in real time. Each algorithm has pros and downsides, therefore choosing one depends on the situation and needs. Algorithm selection is crucial to effective maintenance scheduling and resource allocation.

## 3 Sensor Data Analytics

In the era of the Internet of Things (IoT) and the proliferation of networked devices, sensor data analytics has become a crucial topic. Sensors play a crucial role in the acquisition of real-time data in a variety of contexts, including environmental conditions, industrial operations, infrastructure monitoring, and personal devices.

Massive opportunities exist to optimize system performance, make better decisions, and obtain actionable insights from the abundance of sensor data.

### 3.1 Integration of IoT Sensors and Data Analytics Platforms for Real-Time Monitoring of Green Power Systems

The Internet of Energy (IoE) is an innovative concept in energy research that combines the information and communication technology (ICT) revolution with conventional energy infrastructure. The Internet of Things is commonly abbreviated as IoE. The Internet of Things (IoT) facilitates the connection of various devices and systems through the internet, in combination with other modern methods. The Internet of Things marks the onset of the third wave of informatization. The Internet of Things (IoT) improves the automation and intelligence of various systems. Real-time monitoring of intricate systems is feasible. Power grids serve as the central hubs of energy ecosystems. Components of the Internet of Everything are essential for the functioning of smart grids. Having conducted an inquiry into the Internet of Things (IoT) as it pertains to the electrical grid. Renewable energy sources (RES), commonly known as green energy, can potentially reduce greenhouse gas emissions, cleanse air pollution, and reduce reliance on fossil fuels. The specific publication being referred to is Saleem et al. (2019). Investing in solar and wind farms in the renewable energy industry is expensive. An abundant supply of quickly deployable power generation is equally important. The Internet of Things (IoE) greatly revolutionizes wind and solar energy initiatives. A considerable proportion of grid components, especially renewable sources, are not under the control of individuals unless they have access to an internet connection and a communication platform. Real-time data transmission from all components to national and regional hubs is necessary. Data supervision is essential. A wind and solar farm comprises a variety of wind turbines and solar panels. Within a system characterized by an excessive abundance of moving components, executive operators may experience a state of being overwhelmed. Time, fuel, and remuneration are necessary to maintain everything in operational condition. These elements undermine the resilience of the grid. RES issues are resolved using real-time status monitoring. Implementing end-to-end Internet of Everything (IoE) systems can be achieved by utilizing cellular modules, such as routers that are equipped with integrated Subscriber Identity Modules (SIMs). The primary objective of this is to provide a source of energy for wind and solar farms. The ubiquity of the Internet of Everything (IoE) can be addressed by companies specializing in Renewable Energy Solutions (RES). Many of these challenges can be easily and inexpensively addressed by implementing the Internet of Things (IoT) in the renewable energy sector, particularly in wind and solar power. A cohesive and user-friendly system is necessary for the grid. They possess a high level of skill in promptly recognizing problems and swiftly finding resolutions. The deployment of IoT technology has facilitated the remote surveillance of solar and wind resources (Zhang et al., 2021; Shahinzadeh et al., 2019).

Intelligent power systems can manage energy consumption and delivery. Intelligent power systems help provide and conserve energy. Real-time environmental monitoring automates energy allocation and conservation. Grid-connected power plants are often located near diesel or renewable energy sources in rural locations. Thus, bulk transmission networks carry power across long distances, often across countries. Thus, grid electrical voltage rises. Energy reduction is difficult. Internet and smart city gadgets may be energy efficient. IoT sensors and other devices collect crucial data about urban infrastructure including roads, bridges, railroads, communications, water systems, electricity grids, and skyscrapers. Data science and predictive analysis help people make better judgments. Smart cities integrate renewable energy sources like the utility grid to an autonomous and widespread IoT platform. IoT and AI determine energy distribution. Eliminating heat loss saves electricity. Allocation and demand balancing algorithms using machine learning (Hossein Motlaghet al., 2020; Sagiroglu, Terzi, Canbay, & Colak, 2016).

The "Green smart grid" study develops AI-assisted electrical infrastructure management. Smart grid systems are large, sophisticated networks that need regulation, monitoring, tracking, and management of numerous networked devices and assets. Smart grid systems can help clients, estimate future demands, and generate new ideas using big data. Smart meter faults and use statistics are further instances. Recent big data analytics breakthroughs are thought to solve these issues. "Big data" refers to data collections that are too large to gather, handle, or analyze using typical methods (Moradi et al., 2019).

Big data analytics has revolutionized the power systems business by collecting, analyzing, and interpreting massive amounts of data. AI, ML, DM, and time-series forecasting are used. When dealing with large-scale, real-world instances, power system decision-makers have faced many variables, complex procedures, huge computing loads, inconsistent findings, mistakes, and low model accuracy. Big data analytics seeks relevant information from enormous datasets. Its various uses include electricity markets, microgrid operation, active distribution network control, energy theft detection, renewable energy use, and stability evaluation. Big data analytics requires widespread IoT device adoption, easy cloud storage, and blockchain technology. Smart sensor networks will boost data volume and variety in electricity infrastructures. This data requires continual management and interpretation, but it can improve the electricity grid (Jaradat et al., 2015; De Benedetti et al., 2018).

### 3.2 Analysis of Sensor Data to Identify Patterns, Trends, and Anomalies that Can Indicate Potential Failures or Performance Degradation.

The study (Himeur et al., 2021) proposed finding anomalies by comparing the actual and predicted AC power supplied by a PV system. An ANN model trained using data from the under-observation PV facility predicted AC power production. Comparing real-time PV system data with model output and examining the residual vector for abnormalities yielded daily predictive maintenance warnings. The program collected all residuals and ran them through the Triangular Moving Average (TMA) method with an algorithmically derived window size to find out-of-threshold samples and system deterioration trends throughout the day. The investigations' positive predictive detection rate approached 90%, showing high anomaly identification. By highlighting abnormalities and sending maintenance notifications, the algorithm supported decision-making.

Sub-meters and smart sensors in homes can reveal energy usage abnormalities and save money. Unsupervised detection approaches assume anomalies are less than 20% of the data and aim to find previously unknown abnormal consumption patterns. Supervised anomaly detection, which requires labeled datasets for machine learning classifier training, is limited by the absence of labeled data. Ensemble approaches may manage power consumption data complexity and discover abnormalities by applying many models to progressively smaller groups of observations. Using proper measurements and modeling consumption patterns in new areas, feature extraction improves anomaly identification. Hybrid or semi-supervised anomaly detection uses normal footprint annotations and deep autoencoder architectures to find outliers (Zaher et al., 2009).

Plant owners can achieve operational goals with CM systems. These devices help operators make maintenance choices by delivering vital machinery data. This extreme departure from standard maintenance approaches is a major step toward condition-based maintenance, which detects and fixes early symptoms of degradation or new issues before catastrophic breakdowns. Well-maintained turbines can run without planned maintenance to decrease downtime. Many sectors use Supervisory Control and Data Acquisition (SCADA) to identify, diagnose, and prognosticate equipment failures. High-tech algorithms automatically identify turbine degradation, compressor leak band failure, fuel supply system faults, combustion liner burn-through, and sensor faults within acceptable limits in turbine engine diagnostics. Real-time SCADA data helps monitor turbine effectiveness. Performance monitoring can also isolate component defects for more accurate maintenance (Kim et al., 2011; Abu-Samah et al., 2015).

### 3.3 *Development of Predictive Models Based on Sensor Data to Forecast Equipment Failure Probabilities and Remaining Useful Life.*

Maintenance should only be done when necessary, such as when warning indications suggest system failure. Predictive or condition-based maintenance is this method. Unlike preventative maintenance, this strategy adapts maintenance periods to the system state. Predictive maintenance research has grown due to the industrial Internet of Things and sensor technology. Predictive maintenance limits maintenance staff to essential tasks to cut expenses, system downtime, and maintenance time. Predictive maintenance requires continuous machinery status data. These sensors improve wind turbine maintenance and reduce unscheduled outages (Zhang & Zhang, 2015; Yurek & Birant, 2019).

Machine learning (ML) can help predictive maintenance embrace new approaches. Predictive models identify components by their likelihood to fail within a certain time period and utilize past data to forecast RUL. Predicting with a continuous input and outcome requires regression. Thus, several RUL estimation studies have used it. One research (Yurek and Birant, 2019) predicts RUL using regression analysis. On the C-MAPSS dataset, one of two ways computes the RUL before running the regression models. First, the machine's current time is calculated by subtracting the fault time. RUL maximum value for each failure determines runtime in the second procedure. Azure Learning Machine used many feature selection and machine learning methods. The investigation includes 72 models with different feature selection and machine learning methods. After examining the data, Decision Forest Regression gave the most accurate estimates (Ouda, Maalouf, & Sleptchenko, 2021).

## 4 Digital Twin Technology

Through the utilization of digital twin technology, it is possible to create a highly precise digital copy of an object, process, or system, including all of its components, behavior, and characteristics. A digital twin is a virtual version of a physical object that closely mimics its behaviors and characteristics. They can be used to gain deeper insights, perform comprehensive analyses, and optimize the actual system over its entire lifespan. By implementing a novel approach, we can improve our understanding of the physical object being portrayed, carry out a thorough examination of it, and maximize its efficiency, all with the goal of achieving its highest potential (Grieves & Vickers, 2017).

A digital twin is made up of numerous parts, including:

1. The physical system embodies the tangible entity, whether it be a machine, structure, or infrastructure, which finds its reflection in the Digital Twin.

2. Sensors strategically integrated within the physical system perform the vital task of capturing real-time data, encompassing measurements, conditions, and operational parameters. This valuable information is then transmitted to the Digital Twin, where it undergoes meticulous analysis and forms the foundation for modeling.
3. Leveraging the wealth of collected data, the Digital Twin constructs a virtual model that faithfully mirrors the intricacies and characteristics of the physical system. The fidelity of the model can range from a simplified representation to a sophisticated simulation, tailored to meet the specific requirements and desired level of detail.
4. The process of data integration and analytics plays a pivotal role in the realm of Digital Twins. By amalgamating and analyzing the data gathered, a comprehensive understanding of the behavior, performance, and condition of the physical system can be obtained. This integration and analysis not only provide insights but also facilitates real-time monitoring, predictive maintenance, optimization, and informed decision-making.
5. Moreover, Digital Twins often offer interactive interfaces and visualization tools, empowering users to engage with the virtual model, explore various scenarios, and gain valuable insights into the system's behavior. This interactive capability fosters improved decision-making processes and enables the formulation of effective optimization strategies (Jones et al., 2020).

### *4.1 Creation of Virtual Replicas (Digital Twins) of Green Power Systems to Simulate and Predict System Behavior.*

To accurately depict green power systems in digital form, it is crucial to create accurate digital models, simulate system behavior, acquire a thorough understanding of the system, and collect real-time data (Grieves, 2019). The digital siblings have unmatched analytical, optimization, and decision-making abilities.

1. Consider the broad picture: the green power system is comprised of solar panels, wind turbines, energy storage devices, inverters, and grid connections; therefore, it is essential to understand as much as possible about each of these components and how they interact.
2. By accumulating data from sensors and monitoring equipment, measurements of solar irradiance, wind speed, energy generation, battery state, and grid characteristics can be obtained in real-time.
3. Using mathematical equations, physics-based models, or machine learning techniques, develop precise digital models that accurately represent the dynamics of the green power system.
4. To ensure the model's accuracy and dependability in simulating the behavior of the physical system, its parameters must be calibrated and validated using historical data.

5. Utilize the calibrated digital twin model to simulate the behavior of the green power system, taking into consideration variables such as weather, load profiles, and control techniques, in order to make performance predictions and identify potential issues in advance.
6. Applying cutting-edge algorithms to the digital twin model and deducing optimal configurations, control techniques, and resource allocation will maximize energy output, system efficiency, and stability.
7. Utilize the information provided by the digital counterpart to aid in system expansion, maintenance planning, investment estimation, and performance evaluation.

By utilizing digital twins to enhance the performance, efficiency, and dependability of green power systems, stakeholders can enhance the sustainability, efficiency, and effectiveness of renewable energy infrastructure (Bae et al., n.d).

## 4.2 Utilization of Digital Twins for Condition Monitoring, Predictive Maintenance Planning, and Performance Optimization.

Digital twins have numerous applications, including condition monitoring, predictive maintenance, and efficiency improvements. To monitor the physical world in real-time, digital siblings incorporate sensors. The Digital Twin performs an analysis of the data and compares it to expected thresholds and patterns. By continuously monitoring critical variables such as temperature, vibration, and pressure, the Digital Twin can detect irregularities, deviations from normal operation, and impending failures. This form of preventative condition monitoring can reduce maintenance expenses and disruptions.

Digital siblings employ sophisticated analytics to forecast when physical systems may require maintenance. By analyzing both historical and real-time data, component failure or degradation trends can be uncovered. Businesses can plan preventive maintenance in advance to make better use of available resources and decrease the likelihood of disruptions. Utilizing Digital Twins for predictive maintenance improves machine utilization, lowers maintenance costs, and increases asset dependability (Gupta et al., 2023).

With the aid of their digital counterparts, physical systems can be improved. The models and algorithms of Digital Twin can enhance a variety of environments. Variable configurations, operational strategies, and parameter modifications are attempted as part of this procedure. The Digital Twin is an exact replica of the actual system that demonstrates how various configurations affect key performance indicators such as efficiency, throughput, and quality. Digital twins enable data-driven decisions, optimized operational settings, and enhanced system performance (Kim et al., n.d).

### *4.3 Integration of Real-Time Data from Sensors with Digital Twin Models for Continuous System Monitoring and Diagnostics.*

Real-time sensor data and digital twin models provide system monitoring and diagnosis. Data from physical sensors in the system is continuously sent to their digital twins. Thus, real-time system performance analysis, monitoring, and diagnostics are available.

1. To start this process, strategically placed physical sensors in the real-world system measure variables like temperature, pressure, vibration, humidity, or other relevant characteristics dependent on the system being studied. These sensors capture data on measurements, circumstances, and system behavior and performance (Sana Kazilbash). The digital twin concept receives raw sensor data through wired or wireless networks.
2. Preprocessing sensor data ensures its quality and compliance with the digital twin model. The digital twin approach may need data cleansing, filtering, or transformation. The digital twin then uses preprocessed sensor data to update its internal state and imitate the real-world system's behavior (Drobnjakovic et al., 2023).
3. After integration, the digital twin model monitors. It rigorously compares real-time sensor data to its model's predicted behavior and performance. The digital twin can detect physical system abnormalities, deviations, and difficulties through constant examination. It diagnoses, notifies, or initiates actions based on rules or algorithms.
4. User interfaces, dashboards, and reports may exhibit digital twin evaluation findings and analysis. These visual representations help operators, engineers, and decision-makers understand the system's performance, identify areas for improvement, and make educated decisions to improve operations (Kim et al.).

## 5 Case Study: Predictive Maintenance of Hydraulic Systems

The purpose of the model is to perform data cleansing, generate and compare diverse classification models, and graphically display the results (Schneider, Helwig, & Schütze, 2017; Helwig, Pignanelli, & Schutze, 2015a; Helwig, Pignanelli, & Schütze, 2015b). In addition to Scikit-learn and Seaborn, the tpot, scipy, zipfile, numpy, pandas, matplotlib, and os libraries are also utilized. This predictive maintenance case study aims to analyze sensor data to predict the health of the hydraulic system and identify potential failure points. According to available information, the following steps comprise the procedure:

| | PS1 | PS2 | PS3 | PS4 | PS5 | PS6 | ESP1 | FS1 | FS2 | TS1 | ... | TS4 | VS1 | CE | CP | SE | Cooler_Condition | Valve_Condition | Internal_Pump_Leakage | Hydraulic_Accumulator | Stable_Flag |
|---|---|---|---|---|---|---|---|---|---|---|---|---|---|---|---|---|---|---|---|---|---|
| 0 | 160.673492 | 109.466914 | 1.991475 | 0.0 | 9.842169 | 9.728098 | 2538.929167 | 6.709815 | 10.304592 | 35.621983 | ... | 31.745250 | 0.576950 | 39.601350 | 1.862750 | 59.157183 | 3.0 | 100.0 | 0.0 | 130.0 | 1.0 |
| 1 | 160.603320 | 109.354890 | 1.976234 | 0.0 | 9.635142 | 9.529488 | 2531.498900 | 6.715315 | 10.403098 | 36.676967 | ... | 34.493867 | 0.565850 | 25.786433 | 1.255550 | 59.335617 | 3.0 | 100.0 | 0.0 | 130.0 | 1.0 |
| 2 | 160.347720 | 109.158845 | 1.972224 | 0.0 | 9.530548 | 9.427949 | 2519.928000 | 6.718522 | 10.366250 | 37.880800 | ... | 35.646150 | 0.576539 | 22.218233 | 1.113217 | 59.543150 | 3.0 | 100.0 | 0.0 | 130.0 | 1.0 |
| 3 | 160.188088 | 109.064807 | 1.946576 | 0.0 | 9.438827 | 9.337429 | 2511.541633 | 6.720565 | 10.302678 | 38.879050 | ... | 36.579467 | 0.569267 | 20.459817 | 1.062150 | 59.794900 | 3.0 | 100.0 | 0.0 | 130.0 | 1.0 |
| 4 | 160.000472 | 108.931434 | 1.922707 | 0.0 | 9.358762 | 9.260636 | 2503.449500 | 6.690308 | 10.237750 | 39.803917 | ... | 37.427900 | 0.577367 | 19.787017 | 1.070467 | 59.455267 | 3.0 | 100.0 | 0.0 | 130.0 | 1.0 |
| 5 | 159.920210 | 108.887682 | 1.913285 | 0.0 | 9.301160 | 9.206877 | 2501.007067 | 6.699023 | 10.178720 | 40.659450 | ... | 38.212067 | 0.572683 | 19.149683 | 1.072083 | 59.563333 | 3.0 | 100.0 | 0.0 | 130.0 | 1.0 |
| 6 | 159.672675 | 108.676466 | 1.888100 | 0.0 | 9.233942 | 9.143320 | 2494.416900 | 6.698573 | 10.140810 | 41.463633 | ... | 38.932100 | 0.573033 | 18.666383 | 1.081683 | 59.789900 | 3.0 | 100.0 | 0.0 | 130.0 | 1.0 |
| 7 | 159.614452 | 108.651745 | 1.874894 | 0.0 | 9.194159 | 9.105058 | 2489.421933 | 6.678027 | 10.095978 | 42.215267 | ... | 39.558967 | 0.572000 | 18.178433 | 1.078700 | 59.590617 | 3.0 | 100.0 | 0.0 | 130.0 | 1.0 |
| 8 | 159.475745 | 108.529738 | 1.858120 | 0.0 | 9.144616 | 9.057067 | 2484.419067 | 6.671652 | 10.039710 | 42.891983 | ... | 40.080533 | 0.567067 | 18.334867 | 1.115083 | 59.608883 | 3.0 | 100.0 | 0.0 | 130.0 | 1.0 |
| 9 | 159.437997 | 108.510885 | 1.841063 | 0.0 | 9.104831 | 9.019265 | 2480.434867 | 6.659990 | 9.997762 | 43.532833 | ... | 40.612550 | 0.571683 | 18.205733 | 1.129133 | 59.473733 | 3.0 | 100.0 | 0.0 | 130.0 | 1.0 |

**Fig. 1** Portion of the data frame

### 5.1 Data Preparation

To begin, numpy arrays containing data from multiple files are imported. The sensors gather data regarding pressure (PS1, PS2,...), temperature (TS1, TS2,...), vibrations (VS1), and additional parameters (EPS1, FS1, FS2, CE, CP, SE). To create a DataFrame (df) containing the average values, the mean of each array is calculated along the rows. To include target labels, the DataFrame is expanded by adding extra columns that represent various hydraulic system conditions. The columns contain the values Stable_Flag, Cooler_Condition, Valve_Condition, Internal_Pump_Leakage, and Hydraulic_Accumulator (Fig. 1).

### 5.2 Exploratory Data Analysis (EDA):

The subsequent procedure entails employing a heatmap, as depicted in Fig. 2, for the purpose of ascertaining the interdependencies among the various features present in the DataFrame df_final. By sorting the correlation matrix according to how it correlates with the target labels, the state of the system can be better understood.

### 5.3 Data Preprocessing:

The data undergoes preprocessing procedures to be adequately prepared for both model training and evaluation. A partition exists between the DataFrame and the desired features and labels. In order to obtain the desired labels, the data is partitioned into two distinct sets: the training set and the testing set. The StandardScaler module is employed for the purpose of feature scaling, wherein the data is normalized by ensuring uniformity in the size of all features. Subsequently, the anticipated labels are forecasted employing multiclass classification methodologies such as Logistic Regression, k-Nearest Neighbors, Support Vector Machine, Naive Bayes, Decision Tree, and Random Forest. In the given code, there are four target variables being considered. These targets are:

Cooler_Condition, represents the condition of the cooler.
Valve_Condition, represents the condition of the valve.
Internal_Pump_Leakage, represents the internal pump leakage.

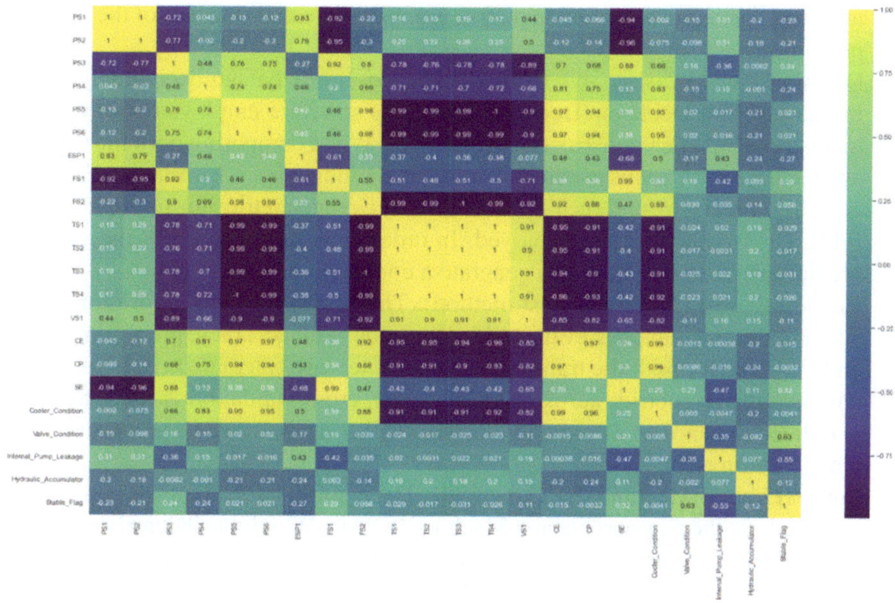


**Fig. 2** Heatmap of the data

Hydraulic_Accumulator, represents the hydraulic accumulator.

## *5.4 Model Evaluation*

The evaluation procedures employed in modeling encompass:

1. Logistic Regression (logistic_regression method): The program utilizes the Logistic Regression classification algorithm to evaluate its functionality across multiple metrics. The F1 score, accuracy, precision, and recall are computed using two distinct datasets, one allocated for training and the other for testing. By examining these indicators. System maintenance specialists can evaluate the accuracy of the model in accurately representing the current state of the system. The code computes the mean F1 score, recall, precision, and accuracy for each data fold using cross-validation. The utilization of seaborn allows for the creation of a confusion matrix, which visually represents the model's predictions and misclassifications.
2. k-Nearest Neighbors (knearestneighbors method): The code utilizes and assesses the efficacy of the k Nearest Neighbors (kNN) classification algorithm. Similar to Logistic Regression. It calculates accuracy, precision, recall, and F1 score for both the training and testing datasets. Cross-validation is employed to obtain standard evaluation metrics for various data folds. Additionally. Seaborn is

utilized to generate a confusion matrix. Enabling visual inspection of the model's predictions.

3. Decision Tree (decision_tree method): The algorithm that has been implemented makes use of a Decision Tree structure. This structure employs entropy as the main criterion when constructing a classifier. When computing both the training and testing sets. Various metrics are utilized, including but not limited to F1 score, accuracy, precision, and recall. To obtain the average metrics for multiple testing folds. Cross-validation is employed. In order to visually represent the models' predictions Seaborn is used to generate a confusion matrix.
4. Random Forest (random_forest method): The Random Forest algorithm's classification performance is evaluated by the code. The accuracy, precision, recall, and F1-score for both the training and testing sets are calculated. To obtain standard evaluation metrics across multiple folds, cross-validation is used. Seaborn is utilized to generate a confusion matrix displaying the model's predictions and classification errors.
5. Support Vector Machine (SVC method): The classification algorithm utilized in this study is the Support Vector Machine (SVM). Which has been shown to be a reliable method for pattern recognition tasks. To evaluate the performance of the SVM, a radial basis function (RBF) kernel is employed within the code. In order to determine how well the model performs. Cross-validation techniques are utilized. This involves splitting our dataset into training and test sets and calculating various evaluation metrics such as accuracy, precision, recall, and F1 score. By utilizing cross-validation. We obtain a more robust estimation of the model's performance on unseen data. Furthermore in order to gain insights into the SVM's predictive capabilities and identify any misclassifications a confusion matrix is plotted using the Seaborn library.
6. Naive Bayes (naive_bayes method): The employed code utilizes the Gaussian Naive Bayes classification algorithm and assesses its performance using precision, recall, and F1-score on both training and testing data sets. Cross-validation is utilized to calculate average metrics for evaluation across multiple folds. Seaborn generates a confusion matrix to provide a visual representation of the model's predictions and misclassifications.
7. best_model method: The best_model method combines the evaluation results of all classification models into a dataframe for comparison purposes. For each model, the dataframe contains evaluation metrics such as accuracy, precision, recall, and F1-score. By analyzing these metrics collectively, maintenance professionals can determine which model predicts the condition of the hydraulic system most accurately (Figs. 3, 4, 5, 6, 7, 8, 9, 10, 11, 12, 13).

The Cooler_Condition, Valve_Condition, Internal_Pump_Leakage, and Hydraulic_Accumulator target variables' predicted K-Nearest Neighbors (KNN) values are displayed in a dataframe (refer to Fig. 14).

In conclusion, the code's model evaluation section ranks the quality of each classifier using metrics such as accuracy, precision, recall, and F1 score. Confusion matrices can be visually examined to gain an in-depth understanding of the model's

**Fig. 3** Understanding the Confusion Matrix of the TPOTClassifier Model (target-1)

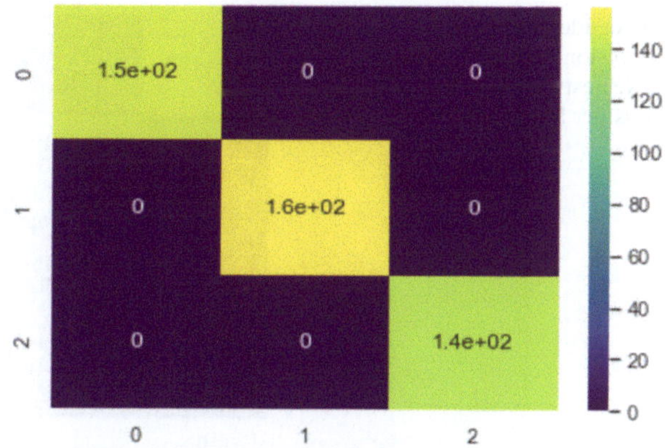


**Fig. 4** Understanding the Confusion Matrix of the TPOTClassifier Model (target-2)

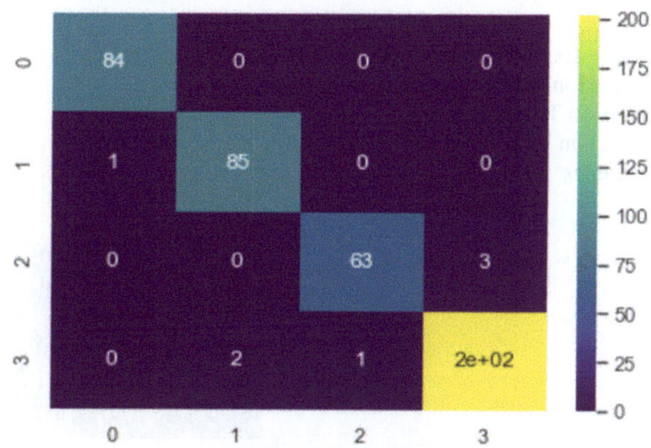


**Fig. 5** Identifying the Confusion Matrix of the Logistic Regression Model (target-3)

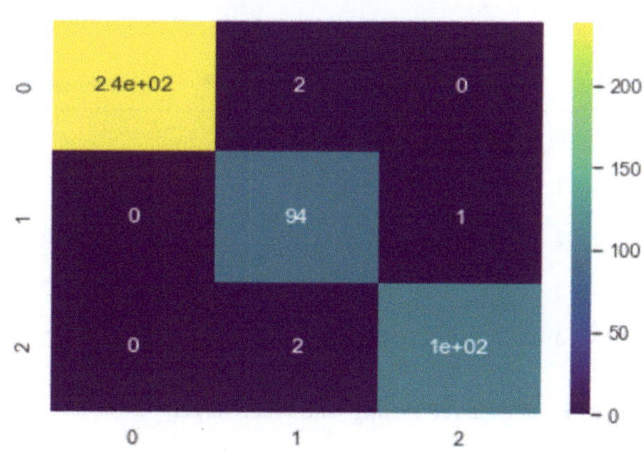

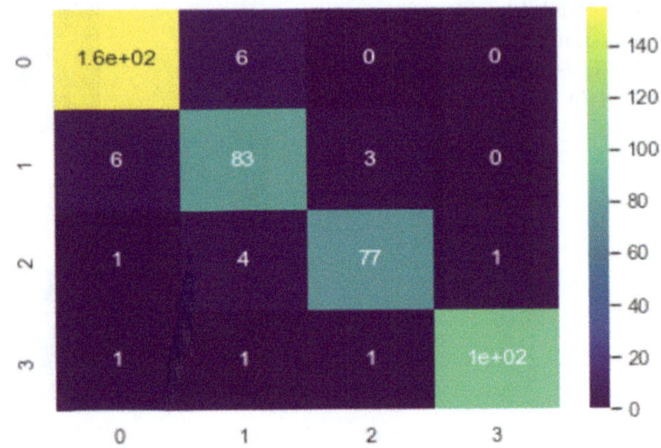


**Fig. 6** Identifying the Confusion Matrix of the K-Nearest Neighbors (KNN) Classifier Model (target-4)

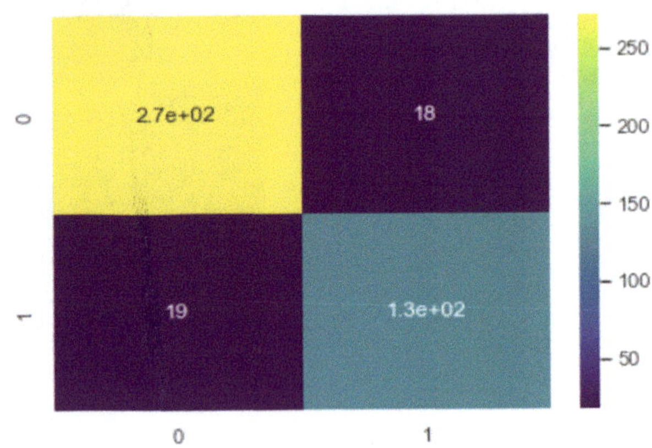


**Fig. 7** Exploring the Confusion Matrix of the Decision Tree Classifier (DTC) on Final Target Variables

| | Valve_Condition | Internal_Pump_Leakage | Hydraulic_Accumulator |
|---|---|---|---|
| **0** | 100.0 | 0.0 | 90.0 |
| **1** | 73.0 | 1.0 | 90.0 |
| **2** | 80.0 | 0.0 | 115.0 |
| **3** | 100.0 | 0.0 | 90.0 |
| **4** | 90.0 | 2.0 | 90.0 |

**Fig. 8** Final Data for Testing with Final Target

predictive abilities and misclassifications, and cross-validation ensures accurate evaluation outcomes. Using these evaluation metrics and visuals, maintenance professionals can determine which predictive maintenance model for hydraulic systems is most appropriate.

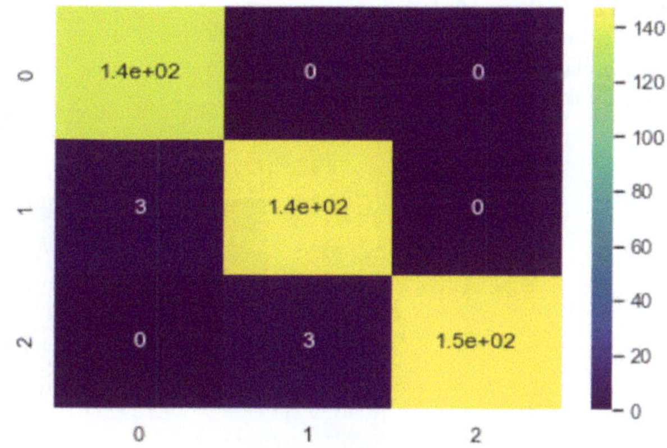


**Fig. 9** Exploring the Confusion Matrix of the K-Nearest Neighbors (KNN) on Target 1

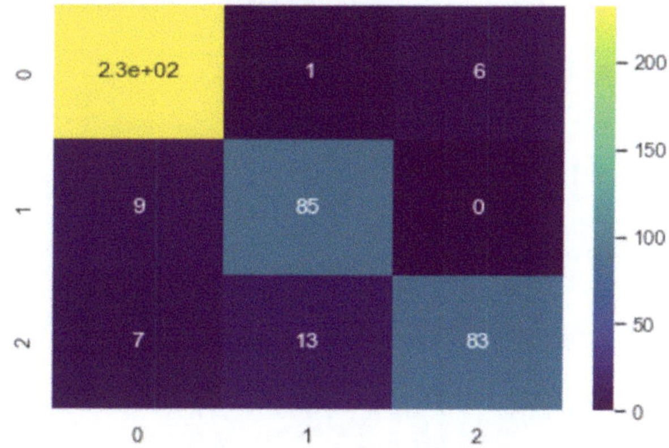


**Fig. 10** Analysis of Confusion Matrix for Multiclass Classification Model on Target 3

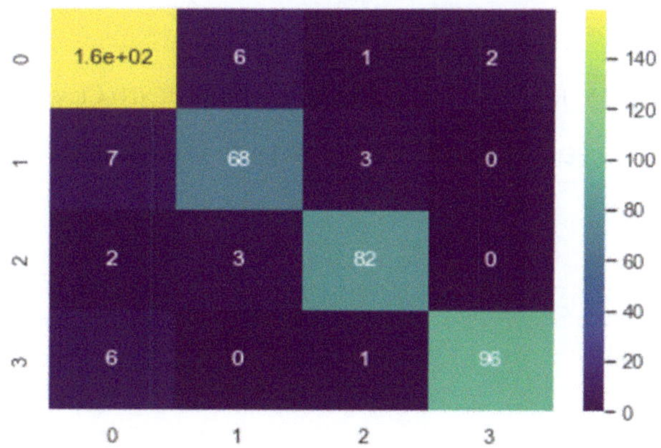


**Fig. 11** Exploring the Confusion Matrix of the Multiclass Classification Model on Target 4

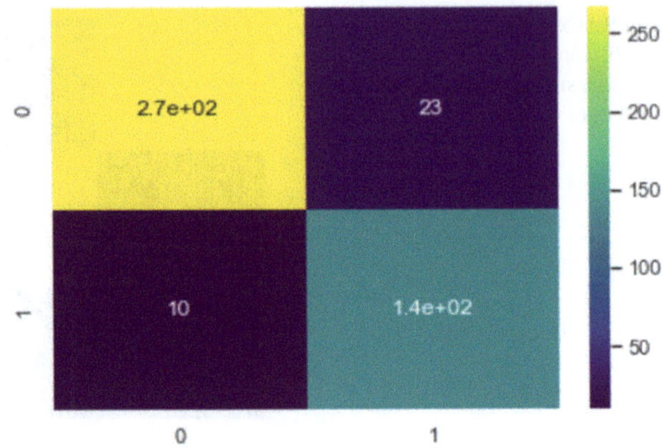


Fig. 12 Analysis of Confusion Matrix for Single-class Classification Model on Final Target

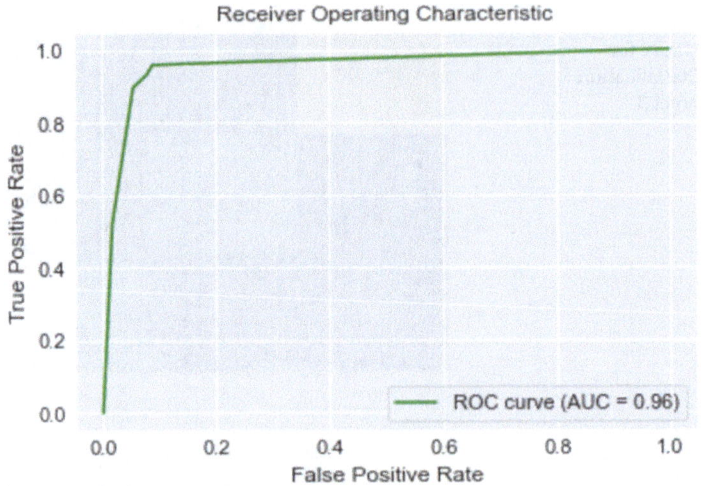


Fig. 13 ROC curve for K-Nearest Neighbors (KNN) Classifier

| | Cooler_Condition | Valve_Condition | Internal_Pump_Leakage | Hydraulic_Accumulator |
|---|---|---|---|---|
| 0 | 100.0 | 90.0 | 2.0 | 115.0 |
| 1 | 3.0 | 100.0 | 1.0 | 100.0 |
| 2 | 20.0 | 73.0 | 2.0 | 100.0 |
| 3 | 20.0 | 100.0 | 1.0 | 100.0 |
| 4 | 3.0 | 90.0 | 0.0 | 100.0 |

Fig. 14 Final predicted values for Cooler Condition, Valve Condition, Internal Pump Leakage, and Hydraulic Accumulator based on the model

The final results of the Stable Flag prediction are as follows:
Percentage of predicted values classified as "0" (Stable): 67.20%
Percentage of predicted values classified as "1" (Unstable): 32.79%

The research produced a dataset containing 2,177 rows and 5 columns. The columns indicate the status of the vehicle's cooler, valves, internal pump leakage, hydraulic accumulator, and stable flag.

This study included the training of multiple models for independent target variables. For the target variable "Cooler Condition," a k-nearest neighbors' model was developed and evaluated with an accuracy of 99.36% on the training set and 98.62% on the test set. A TPOT classifier was trained for the target variable "Valve Condition," and a precision score was calculated using test data. For additional targets, knn also produced the outcomes shown in the table below, with accuracy, precision, and recall measurements taken from both training and testing sets (Table 1).

According to the findings of this study, the k-nearest neighbors and TPOT classifier models performed well in predicting the Cooler Condition and Valve Condition variables. These models can be used to assess the condition of hydraulic system components such as coolers and valves.

This research yields several significant conclusions that can be inferred:

1. Model Performance: The findings indicate that the Cooler_Condition classification problem was effectively addressed using the K-Nearest Neighbors algorithm. This information can be utilized by technicians to ascertain the optimal models that accurately forecast the state of specific components within the hydraulic system. One can perform a comparative analysis of multiple models in parallel and opt for the most suitable one that aligns with their specific requirements.
2. Suitable Models: To successfully implement predictive maintenance strategies for hydraulic systems cooler. Professionals must have a thorough understanding of both their strengths and weaknesses. Multiple factors such as precision, comprehensibility, and computational demands need to be considered when determining the best approach.

**Table 1** Performance Metrics for Multiple Target Variables and Models

| Target | Train Accuracy | Test Accuracy | Train Precision | Test Precision | Train Recall | Test Recall |
|---|---|---|---|---|---|---|
| Cooler_ Condition | 0.9937 | 0.9862 | 0.9937 | 0.9864 | 0.9937 | 0.9862 |
| Valve_ Condition | 1.0 | 0.8303 | 1.0 | 0.8381 | 1.0 | 0.8302 |
| Internal_ Pump_Leakage | 0.9362 | 0.9174 | 0.9359 | 0.9182 | 0.9362 | 0.9174 |
| Hydraulic_ Accumulator | 0.9511 | 0.9289 | 0.9518 | 0.9296 | 0.9511 | 0.9289 |
| Final Target (Stable Flag) | 0.9299 | 0.9243 | 0.8637 | 0.8553 | 0.9380 | 0.9315 |

3. Decision Support: This research aims to improve current maintenance procedures by introducing a systematic approach that incorporates machine learning techniques. By utilizing these techniques for failure prediction and optimization of maintenance procedures. Teams can expect improved system reliability along with reduced downtime and expenses.

The study concludes with an application of machine learning models that predict potential failures as well as assess the operational condition of hydraulic systems. A structured methodology consisting of data preprocessing exploratory data analysis allows valuable insights regarding system condition extraction through model development evaluation fostering effective decision-making in choosing the ideal predictive maintenance strategy. However, it is not recommended to use global variables for storing evaluation metrics in larger projects despite their utilization in the given code. As they can limit the establishment of a predictive maintenance framework for hydraulic systems. The utilization of various data storage and management techniques facilitates enhanced stability and scalability of responses. The implemented approach in the code optimizes operational efficiency, minimizes downtime, and mitigates maintenance expenses for hydraulic systems (mayank1897, 2020, 2021).

## 6 Future Scope

The future of AI-based predictive maintenance in green power systems holds immense potential for advancements across various domains. One significant area lies in the continuous development and enhancement of machine learning algorithms. Researchers can explore sophisticated algorithms, such as deep learning architectures, ensemble methods, and hybrid models. By doing so, they can further improve fault classification, prediction accuracy, and overall system reliability (Hamid and Ganne, 2023).

Another promising avenue lies in the integration of artificial intelligence with edge computing. The emergence of edge computing offers an opportunity to directly incorporate AI algorithms into edge devices that are employed in green power systems. This integration facilitates real-time data analysis, thereby reducing latency and enhancing the responsiveness of predictive maintenance strategies. The fusion of edge computing and AI algorithms empowers the system to detect anomalies, predict failures, and optimize maintenance operations more efficiently (Rao et al., 2023).

In summary, the future scope of AI-based predictive maintenance in green power systems encompasses advancements in machine learning algorithms, edge computing and AI integration, enhanced sensor technologies, integration of big data analytics, improved digital twin applications, and strengthened cybersecurity measures. These advancements hold the promise of more accurate, efficient, and reliable maintenance strategies, ultimately leading to improved system reliability, performance, and sustainability (Salehi, 2023).

## 7 Conclusion

This chapter explored the capacity of artificial intelligence-driven predictive maintenance solutions to improve the reliability of green power systems. Reinforcement learning, supervised learning, and unsupervised learning are the fundamental components of machine learning, as stated earlier. The application of these methodologies is essential for optimizing energy distribution networks and infrastructure.

Supervised learning techniques, such as Random Forest and Support Vector Machines, have shown significant effectiveness in defect classification and prediction for green power systems. These algorithms leverage tagged data to improve the performance and reliability of the system through precise prediction and pattern recognition. Unsupervised learning techniques have been successfully applied to detect anomalies and diagnose early failures in sensor data collected from renewable energy sources. By analyzing unannotated data to uncover hidden patterns and rectifying measurement errors, these algorithms enhance the efficiency of power generation scheduling and promote power stability. The integration of the current system state into reinforcement learning algorithms has successfully optimized the maintenance schedule and resource allocation. These algorithms make mistakes as part of their learning process and then correct or strengthen them. They gather data from their surroundings and utilize it to make well-informed choices that improve the system's availability by optimizing resource allocation and carrying out preventative maintenance.

The possibility of combining data analytics platforms with internet of things (IoT) sensors to facilitate immediate monitoring of renewable power systems has also been considered. A digital twin is a visual representation of a system that helps in understanding, monitoring, and making informed decisions about its behavior, modeling, and prediction in real-time. Predictive maintenance solutions based on artificial intelligence (AI) have the capacity to greatly enhance the reliability, effectiveness, and longevity of green power systems. Organizations that utilize machine learning algorithms to analyze sensor data through the Internet of Things can potentially achieve a decrease in unscheduled outages, improved system performance, and reduced expenses. These advancements have the potential to significantly improve the reliability and environmental sustainability of the power grid, thereby greatly enhancing the renewable energy industry.

## References

Abbas F, Habib S, Feng D, Yan Z (2018) Optimizing generation capacities incorporating renewable energy with storage systems using genetic algorithms. Electronics 7(7):100

Abu-Samah A, Shahzad MK, Zamai E, Said AB (2015) Failure prediction methodology for improved proactive maintenance using Bayesian approach. IFAC-PapersOnLine 48(21):844–851

Afridi YS, Ahmad K, Hassan L (2022) Artificial intelligence based prognostic maintenance of renewable energy systems: A review of techniques, challenges, and future research directions. Int J Energy Res 46(15):21619–21642

Anwar MB, El Moursi MS, Xiao W (2016) Novel power smoothing and generation scheduling strategies for a hybrid wind and marine current turbine system. IEEE Trans Power Syst 32(2):1315–1326

Arulkumaran K, Deisenroth MP, Brundage M, Bharath AA (2017) A brief survey of deep reinforcement learning. ArXiv https://doi.org/10.1109/MSP.2017.2743240

Assouline D, Mohajeri N, Scartezzini JL (2017) Quantifying rooftop photovoltaic solar energy potential: A machine learning approach. Sol Energy 141:278–296

Bae KH, Feng B, Kim S, Lazarova-Molnar S, Zheng Z, Roeder T, Thiesing R (n.d.). A CASE STUDY OF DIGITAL TWIN FOR A MANUFACTURING PROCESS INVOLVING HUMAN INTERACTIONS.

Bangroo IS (2023). AI-based Predictive Analytic Approaches for safeguarding the Future of Electric/Hybrid Vehicles. arXiv preprint arXiv:2304.13841.

Baser F, Demirhan H (2017) A fuzzy regression with support vector machine approach to the estimation of horizontal global solar radiation. Energy 123:229–240

Breiman L (2001) Random forests. Mach Learn 45:5–32

Chaithra N, Jha J, Sayal A, Gupta V, Gupta A (2023). A Paradigm Shift towards Computer Vision. In 2023 International Conference on Device Intelligence, Computing and Communication Technologies, (DICCT) (pp. 54–58). IEEE.

Chen K, Jiang J, Zheng F, Chen K (2018) A novel data-driven approach for residential electricity consumption prediction based on ensemble learning. Energy 150:49–60

Cortes C, Vapnik V (1995) Support-vector network-. machine learning20: 273–297. Portfolio Select J Global Optim 43:2–3

De Benedetti M, Leonardi F, Messina F, Santoro C, Vasilakos A (2018) Anomaly detection and predictive maintenance for photovoltaic systems. Neurocomputing 310:59–68

Drobnjakovic M, Shao G, Nikolov A, Kulvatunyou B, Frechette S, Srinivasan V (2023). Towards Ontologizing a Digital Twin Framework for Manufacturing. In IFIP International Conference on Advances in Production Management Systems (pp. 317–329). Cham: Springer Nature Switzerland.

Gallagher CV, Bruton K, Leahy K, O'Sullivan DT (2018) The suitability of machine learning to minimise uncertainty in the measurement and verification of energy savings. Energy and Buildings 158:647–655

Grieves M, Vickers J (2017). Digital twin: Mitigating unpredictable, undesirable emergent behavior in complex systems. Transdisciplinary perspectives on complex systems: New findings and approaches, 85–113.

Grieves MW (2019). Virtually intelligent product systems: Digital and physical twins.

Gupta A, Chaithra N, Jha J, Sayal A, Gupta V, Memoria M (2023). Machine Learning Algorithms for Disease Diagnosis using Medical Records: A Comparative Analysis. In 2023 4th International Conference on Intelligent Engineering and Management (ICIEM) (pp. 1–6). IEEE.

Hamid M, Ganne A (2023). ARTIFICIAL INTELLIGENCE IN ENERGY MARKETS AND POWER SYSTEMS.

Helwig N, Pignanelli E, Schutze A (2015). Condition monitoring of a complex hydraulic system using multivariate statistics. 2015 IEEE International Instrumentation and Measurement Technology Conference (I2MTC) Proceedings. https://doi.org/10.1109/i2mtc.2015.7151267

Helwig N, Pignanelli E, Schütze A (2015). D8.1 - detecting and compensating sensor faults in a hydraulic condition monitoring system. Proceedings SENSOR 2015. https://doi.org/10.5162/sensor2015/d8.1

Himeur Y, Ghanem K, Alsalemi A, Bensaali F, Amira A (2021) Artificial intelligence based anomaly detection of energy consumption in buildings: A review, current trends and new perspectives. Appl Energy 287:116601

Hossein Motlagh N, Mohammadrezaei M, Hunt J, Zakeri B (2020) Internet of Things (IoT) and the energy sector. Energies 13(2):494

Huang C, Davis LS, Townshend JRG (2002) An assessment of support vector machines for land cover classification. Int J Remote Sens 23(4):725–749

Jaradat M, Jarrah M, Bousselham A, Jararweh Y, Al-Ayyoub M (2015) The internet of energy: smart sensor networks and big data management for smart grid. Procedia Computer Science 56:592–597

Jha J, Vishwakarma AK, Chaithra N, Nithin A, Sayal A, Gupta A, Kumar R (2023). Artificial Intelligence and Applications. In 2023 1st International Conference on Intelligent Computing and Research Trends (ICRT) (pp. 1–4). IEEE.

Jiang H, Dong Y (2017) Global horizontal radiation forecast using forward regression on a quadratic kernel support vector machine: Case study of the Tibet Autonomous Region in China. Energy 133:270–283

Jones D, Snider C, Nassehi A, Yon J, Hicks B (2020) Characterising the Digital Twin: A systematic literature review. CIRP J Manuf Sci Technol 29:36–52

Kim K, Parthasarathy G, Uluyol O, Foslien W, Sheng S, Fleming P (2011). Use of SCADA data for failure detection in wind turbines. ASME 2011 5th International Conference on Energy Sustainability, Parts A, B, and C. https://doi.org/10.1115/es2011-54243

Kim S, Feng B, Smith K, Masoud S, Zheng Z, Szabo C, Loper M BUIDING A DIGITAL TWIN FOR ROBOT WORKCELL PROGNOSTICS AND HEALTH MANAGEMENT.

mayank1897. (2021). Predictive_maintenance_of_hydraulic_system. Kaggle. https://www.kaggle.com/code/mayank1897/predictive-maintenance-of-hydraulic-system

mayank1897. (2020). Predictive maintenance of Hydraulics System. Kaggle. https://www.kaggle.com/datasets/mayank1897/condition-monitoring-of-hydraulic-systems

Mi D, Wang T, Gao M, Li C, Wang Z (2021) Online transient stability margin prediction of power systems with wind farms using ensemble regression trees. Int Trans Electr Energy Syst 31(11):e13057

Mnih V, Kavukcuoglu K, Silver D, Rusu AA, Veness J, Bellemare MG, Hassabis D (2015) Human-level control through deep reinforcement learning. Nature 518(7540):529–533

Moradi J, Shahinzadeh H, Nafisi H, Marzband M, Gharehpetian GB (2019). Attributes of big data analytics for data-driven decision making in cyber-physical power systems. In 2020 14th international conference on protection and automation of power systems (IPAPS) (pp. 83–92). IEEE.

Ouda E, Maalouf M, Sleptchenko A (2021). Machine Learning and Optimization for Predictive Maintenance based on Predicting Failure in the Next Five Days. In ICORES (pp. 192–199).

Qiu J, Wu Q, Ding G, Xu Y, Feng S (2016) A survey of machine learning for big data processing. EURASIP J Adv Signal Process 2016:1–16

Rao PS, Yaqoob SI, Ahmed MA, Abdinabievna PS, Yaseen SM, Arumugam M (2023) Integrated artificial intelligence and predictive maintenance of electric vehicle components with optical and quantum enhancements. Opt Quant Electron 55(10):1–19

Richardson K, Berant J, Kuhn J (2018). Polyglot semantic parsing in APIs. arXiv preprint arXiv: 1803.06966.

Sagiroglu S, Terzi R, Canbay Y, Colak I (2016). Big data issues in smart grid systems. In 2016 IEEE international conference on renewable energy research and applications (ICRERA) (pp. 1007–1012). IEEE.

Saleem Y, Crespi N, Rehmani MH, Copeland R (2019) Internet of things-aided smart grid: technologies, architectures, applications, prototypes, and future research directions. IEEE Access 7:62962–63003

Salehi M (2023). AI-Enhanced Renewable Energy: Revolutionizing Monitoring & Driving Sustainable Progress. Available at SSRN 4416913.

Sayal A, Jha J, Chaithra N, Gupta V, Gupta A, Memoria M (2023, April). Blockchain: Its Applications And Challenges. In 2023 International Conference on Computational Intelligence, Communication Technology and Networking (CICTN) (pp. 96–100). IEEE.

Schneider T, Helwig N, Schütze A (2017) Automatic feature extraction and selection for classification of Cyclical Time Series Data. Tm - Technisches Messen 84(3):198–206. https://doi.org/10.1515/teme-2016-0072

Schulman J, Levine S, Abbeel P, Jordan M, Moritz P (2015). Trust region policy optimization. In International conference on machine learning (pp. 1889–1897). PMLR.

Schulman J, Wolski F, Dhariwal P, Radford A, Klimov O (2017). Proximal policy optimization algorithms. arXiv preprint arXiv:1707.06347.

Shahinzadeh H, Moradi J, Gharehpetian GB, Nafisi H, Abedi M (2019, February). Internet of Energy (IoE) in smart power systems. In 2019 5th Conference on Knowledge Based Engineering and Innovation (KBEI) (pp. 627–636). IEEE.

Shamshirband S, Mohammadi K, Khorasanizadeh H, Yee L, Lee M, Petković D, Zalnezhad E (2016) Estimating the diffuse solar radiation using a coupled support vector machine–wavelet transform model. Renew Sustain Energy Rev 56:428–435

Sharma N, Sharma P, Irwin D, Shenoy P (2011, October). Predicting solar generation from weather forecasts using machine learning. In 2011 IEEE international conference on smart grid communications (SmartGridComm) (pp. 528–533). IEEE.

Sutton RS, Barto AG (2018) Reinforcement learning: An introduction. MIT press, Cambridge

Tarmanini C, Sarma N, Gezegin C, Ozgonenel O (2023) Short term load forecasting based on ARIMA and ANN approaches. Energy Rep 9:550–557

UNITED NATIONS. (2023). Retrieved from https://sdgs.un.org/goals

Urraca R, Antonanzas J, Alia-Martinez M, Martinez-de-Pison FJ, Antonanzas-Torres F (2016) Smart baseline models for solar irradiation forecasting. Energy Convers Manage 108:539–548

Voyant C, Motte F, Fouilloy A, Notton G, Paoli C, Nivet ML (2017) Forecasting method for global radiation time series without training phase: Comparison with other well-known prediction methodologies. Energy 120:199–208

Winands MH, Lee N (2019). Monte-Carlo Tree Search.

Yurek OE, Birant D (2019, October). Remaining useful life estimation for predictive maintenance using feature engineering. In 2019 Innovations in Intelligent Systems and Applications Conference (ASYU) (pp. 1–5). IEEE.

Zaher ASAE, McArthur SDJ, Infield DG, Patel Y (2009) Online wind turbine fault detection through automated SCADA data analysis. Wind Energy 12(6):574–593

Zendehboudi A, Baseer MA, Saidur R (2018) Application of support vector machine models for forecasting solar and wind energy resources: A review. J Clean Prod 199:272–285

Zhang Z, Zhang P (2015) Seeing around the corner: an analytic approach for predictive maintenance using sensor data. J Manage Anal 2(4):333–350

Zhang X, Manogaran G, Muthu B (2021) IoT enabled integrated system for green energy into smart cities. Sustain Energy Technol Assess 46:101208

# AI-Powered Solutions Advancing UN Sustainable Development Goals: A Case Study in Tackling Humanity's Challenges

**Jhansi Bharathi Madavarapu, Balaram Yadav Kasula, Pawan Whig, and Sandeep Kautish**

**Abstract** The integration of Artificial Intelligence (AI) with digital technologies has emerged as a powerful and promising avenue for achieving the United Nations Sustainable Development Goals (SDGs). This book chapter delves into the multifaceted landscape of AI's role in advancing the SDGs and offers an in-depth exploration of its potential contributions, challenges, and ethical considerations. We begin by discussing the fundamental principles of AI and its capacity to process, analyze, and interpret vast amounts of data, enabling more informed decision-making in diverse sectors. With a specific focus on SDGs, we examine how AI can be harnessed to alleviate poverty, foster environmental sustainability, enhance healthcare delivery, and optimize educational systems. Throughout these discussions, we draw on real-world examples and case studies that highlight successful AI applications and their impact on SDG progress. However, the chapter doesn't solely focus on the benefits of AI; it also critically analyzes the ethical and societal implications of AI integration within the SDGs framework. We underscore the importance of responsible AI development, emphasizing transparency, fairness, and accountability to ensure that AI-driven solutions align with the principles and values enshrined in the SDGs. This chapter provides a comprehensive understanding of how AI, as a digital technology, can be leveraged to address global challenges, accelerate progress toward the SDGs, and contribute to a more sustainable and equitable future. It underscores the need for collaborative efforts among policymakers, technologists, and society to harness the full potential of AI while navigating its associated complexities responsibly.

J. B. Madavarapu · B. Y. Kasula
Department of Information Technology, University of the Cumberlands, Williamsburg, KY, USA
e-mail: jmadavarapu0693@ucumberlands.edu

P. Whig (✉)
Dean Research, Vivekananda Institute of Professional Studies-TC, New Delhi, India
e-mail: pawanwhig@gmail.com

S. Kautish
LBEF Campus, Kathmandu, Nepal

 
W. Leal Filho et al. (eds.), *Digital Technologies to Implement the UN Sustainable Development Goals*, World Sustainability Series,
https://doi.org/10.1007/978-3-031-68427-2_3

**Keywords** AI · Sustainable Development Goals · Digital Technologies · Poverty Eradication · Environmental Sustainability · Healthcare · Education · Responsible AI · Ethical Implications · Global Challenges

## 1 Introduction

The integration of Artificial Intelligence (AI) into our rapidly evolving digital landscape has ushered in a new era of possibilities, with profound implications for addressing some of the world's most pressing challenges (Bachmann et al., 2022). Among these challenges, the United Nations Sustainable Development Goals (SDGs) stand as a universal blueprint for building a more equitable, sustainable, and prosperous future for all. The marriage of AI and the SDGs represents a dynamic and transformative synergy that holds the potential to revolutionize our approach to achieving these global objectives (Neuerburg 2022). In this chapter, we embark on a journey to explore the role of AI and digital technologies in advancing the SDGs. With their capabilities to process vast datasets, make predictions, and automate complex decision-making, AI technologies have already begun to leave an indelible mark on a wide array of sectors (Astobiza et al., 2021). From eradicating poverty to promoting environmental sustainability, enhancing healthcare access, and optimizing educational systems, the influence of AI is both far-reaching and impactful (Truby, 2020, and Mercier-Laurent, 2021). This chapter not only aims to celebrate the remarkable contributions of AI but also to critically examine the ethical and societal considerations that accompany its integration within the SDGs framework (Pedro et al., 2019). Responsible AI development, transparency, fairness, and accountability are essential principles that must guide the evolution of AI in the context of sustainable development (George 2023).

As we delve into this exploration, we will provide real-world examples and case studies, shedding light on the tangible successes and challenges encountered in harnessing AI for the SDGs. By the end of this chapter, readers will gain a comprehensive understanding of how AI, as a digital technology, can propel us toward the achievement of the SDGs while emphasizing the imperative of responsible and equitable progress (Sætra 2023). The chapter concludes by emphasizing the importance of collaboration among policymakers, technologists, and society at large to unlock the full potential of AI for a sustainable and inclusive future.

## 2 Artificial Intelligence (AI) Fundamentals

In our quest to explore how Artificial Intelligence (AI) can advance the United Nations Sustainable Development Goals (SDGs), we embark on a journey through the fundamental principles that underpin AI, its diverse technologies and capabilities, and its potential to revolutionize the SDGs. AI, with its capacity to process, analyze, and interpret vast datasets, offers unprecedented opportunities to address some of the world's most pressing challenges (Mhlanga, 2021).

## *2.1 Understanding AI*

Artificial Intelligence, in essence, is a field of computer science dedicated to creating systems that can perform tasks typically requiring human intelligence (Khakurel et al., 2018). These tasks encompass a wide range of activities, from solving complex problems and making predictions to understanding natural language and recognizing patterns. Understanding AI begins with recognizing its two primary categories:

- **Narrow AI (or Weak AI):** This form of AI is designed to perform specific tasks or solve particular problems. It excels in applications like virtual personal assistants, language translation, and recommendation systems, all of which rely on a narrow focus of expertise.
- **General AI (or Strong AI):** In contrast, General AI aspires to achieve human-level intelligence. It encompasses the ability to understand, learn, and apply knowledge across a broad spectrum of tasks, exhibiting cognitive flexibility and adaptability akin to human intelligence.

Within the SDGs context, it's primarily Narrow AI that finds practical applications (Leal Filho et al., 2023). The ability to process and analyze large volumes of data efficiently empowers AI to make predictions, automate decisions, and optimize processes—capabilities that can profoundly impact SDG progress.

## *2.2 AI Technologies and Capabilities*

AI technologies encompass a rich and varied landscape of tools and techniques. These include:

- **Machine Learning (ML):** A subset of AI, ML is a data-driven approach in which algorithms learn from data to make predictions and decisions. Supervised learning, unsupervised learning, and reinforcement learning are three common paradigms within ML.
- **Natural Language Processing (NLP):** NLP empowers machines to understand, interpret, and generate human language. It's a cornerstone of applications like language translation, sentiment analysis, and chatbots.
- **Computer Vision:** Computer vision enables AI systems to process and analyze visual information, allowing tasks such as image and video recognition, object detection, and facial recognition.

- **Deep Learning:** A subfield of ML, deep learning uses artificial neural networks to model complex patterns and representations in data. It has revolutionized applications like image and speech recognition.
- **Reinforcement Learning:** This approach is centered around training AI agents to make sequences of decisions, learning from trial and error. It's prevalent in applications such as game playing and autonomous robotics.

The combination of these technologies provides AI with a toolkit to solve a diverse range of problems, from predicting disease outbreaks based on health data to optimizing supply chain logistics for food distribution (Mhlanga, 2022). Their adaptability and scalability make them invaluable in addressing the multifaceted challenges posed by the SDGs.

## 2.3 AI's Potential in Advancing the SDGs

The United Nations Sustainable Development Goals comprise 17 interconnected objectives aimed at addressing global challenges, including poverty, inequality, climate change, and more. AI, through its data-processing prowess, can play a pivotal role in advancing these goals:

- **Poverty Eradication:** AI can be harnessed to improve financial inclusion, facilitate access to credit for underserved populations, and enable more efficient social safety net programs through predictive analytics.
- **Environmental Sustainability:** AI aids in environmental monitoring by analyzing satellite data, managing natural resources more efficiently, and enhancing climate modeling and prediction for better preparedness.
- **Healthcare:** AI-driven diagnostic tools, telemedicine, and predictive analytics improve healthcare access, enhance disease detection, and enable personalized treatment regimens, particularly in resource-constrained regions.
- **Education:** AI can facilitate personalized learning experiences, adapt curriculum based on individual needs, and increase access to quality education through online platforms.

Moreover, AI's potential is not limited to individual SDGs; its cross-cutting capabilities offer synergistic solutions across multiple objectives. For instance, by optimizing energy consumption in buildings through smart systems, AI can simultaneously contribute to both environmental sustainability (SDG 13) and affordable and clean energy (SDG 7) (Kasinathan et al., 2022).

While the potential is vast, it's essential to navigate the path forward responsibly. The ethical implications of AI, issues of fairness, transparency, and accountability, must be addressed to ensure AI-driven solutions align with the values of the SDGs. Responsible AI development will be central to harnessing the full potential of AI in the context of sustainable development (Isabelle and Westerlund, 2022).

In the subsequent sections, we will delve deeper into the application of AI in specific SDG domains, exploring real-world case studies and analyzing the ethical considerations that must guide our journey towards a more sustainable and equitable future.

## 3 AI and Poverty Eradication

In the realm of global challenges, poverty stands as a formidable adversary. The United Nations Sustainable Development Goals (SDGs) highlight the imperative of eradicating poverty in all its forms, making it a central objective of our collective efforts. Artificial Intelligence (AI), with its ability to process data, analyze patterns, and automate decision-making, presents innovative pathways toward this noble goal (Goralski and Tan, 2023; Jungwirth and Haluza, 2023). This chapter explores how AI contributes to poverty eradication, with a focus on AI-powered economic empowerment, targeted aid and assistance, and real-world case studies that illuminate the practical impact of AI in this domain.

### *3.1 AI-Powered Economic Empowerment*

Economic empowerment is a fundamental tool for poverty eradication. AI technologies empower individuals and communities in various ways:

- **Entrepreneurship and Micro-Enterprise Support:** AI-driven financial platforms provide access to microloans, empowering small businesses and entrepreneurs to thrive. These platforms utilize AI algorithms to assess creditworthiness and reduce the risks associated with lending, making financial services accessible to those who were previously excluded.
- **Job Market Enhancement:** AI can match job seekers with employment opportunities more efficiently, improving access to income-generating activities. Platforms like LinkedIn use AI to recommend jobs to users based on their skills and experience.
- **Rural Development:** AI can facilitate precision agriculture, optimizing crop yields and promoting sustainable farming practices. This, in turn, can empower rural communities to increase their income through more productive agriculture.

### *3.2 Targeted Aid and Assistance*

AI's data processing capabilities enable the targeted deployment of aid and assistance, ensuring that resources reach those who need them most:

- **Predictive Analytics:** AI algorithms can predict regions and communities most vulnerable to specific crises, such as food shortages, natural disasters, or disease outbreaks. This predictive capability allows aid organizations to preposition resources and respond more effectively.
- **Resource Allocation:** AI optimizes the allocation of resources by analyzing data on community needs. For example, it can help distribute food aid more efficiently in regions with high malnutrition rates.
- **Disaster Response:** AI-driven analysis of real-time data, including satellite imagery and social media posts, facilitates rapid response during disasters. It can identify affected areas, assess damage, and coordinate relief efforts.

## *3.3 Case Studies*

To illustrate the practical impact of AI on poverty eradication, we delve into a selection of compelling case studies:

- **Grameen Foundation's Community Agent Network (CAN):** This case study demonstrates how AI-powered mobile applications and data analytics help community agents in rural areas provide essential financial services to unbanked populations. CAN extends microloans, collects savings, and offers insurance services, fostering economic empowerment and financial inclusion (Jose et al., 2021).
- **WFP's HungerMap LIVE:** The World Food Programme (WFP) employs AI and machine learning to predict food shortages and malnutrition hotspots, enabling proactive measures to address hunger in vulnerable regions. HungerMap LIVE is a pioneering initiative that showcases the potential of AI in targeting aid and assistance (Razak et al., 2023).
- **Zipline's Medical Drone Delivery:** AI-equipped drones by Zipline deliver medical supplies, including blood and vaccines, to remote and hard-to-reach areas in countries like Rwanda and Ghana. This case study demonstrates how AI can revolutionize the delivery of critical healthcare resources, saving lives and improving healthcare access in underserved communities (Rakha, 2023).

These case studies underscore the tangible outcomes of integrating AI into poverty eradication efforts. However, it is crucial to recognize the need for responsible AI development in this context, ensuring that economic empowerment and aid allocation are conducted equitably, transparently, and in accordance with ethical principles.

As we navigate the multifaceted landscape of AI's contributions to poverty eradication, it becomes evident that the synergy between technology and social progress offers unprecedented opportunities for a more inclusive and prosperous world. Nevertheless, the path forward requires vigilant ethical considerations and responsible deployment to ensure that AI genuinely transforms the lives of the world's most vulnerable populations.

# 4 AI for Environmental Sustainability

Environmental sustainability is one of the most pressing global challenges of our time, with the preservation of ecosystems, biodiversity, and the mitigation of climate change at its core. This chapter explores how Artificial Intelligence (AI) serves as a powerful ally in the quest for environmental sustainability. With its data processing capabilities and ability to make predictions, AI contributes significantly to environmental monitoring and conservation, climate change mitigation, and sustainable resource management (Banifatemi et al., 2021).

## *4.1 Environmental Monitoring and Conservation*

Environmental monitoring and conservation are vital aspects of safeguarding the planet's delicate ecosystems. AI's role in this domain includes:

- **Wildlife Preservation:** AI-driven camera traps and acoustic sensors help monitor and protect wildlife populations. Machine learning algorithms can identify and track endangered species, contributing to their conservation.
- **Deforestation Detection:** AI-powered satellite imagery analysis is invaluable in identifying illegal logging activities and monitoring deforestation rates. This data aids conservationists and governments in taking timely action (Whig et al., 2024a, 2024b).
- **Ocean Conservation:** AI is used to process vast amounts of data from underwater sensors and remotely operated vehicles, aiding in the study of marine life and the preservation of ocean ecosystems.

## *4.2 Climate Change Mitigation*

AI plays a pivotal role in mitigating the effects of climate change through:

- **Climate Modeling:** AI-enhanced climate models improve the accuracy of long-term climate predictions, helping policymakers make informed decisions and allocate resources for climate adaptation.
- **Renewable Energy Optimization:** AI optimizes the generation and distribution of renewable energy, making wind, solar, and hydropower sources more efficient and cost-effective.
- **Carbon Capture and Storage:** AI assists in carbon capture and storage processes by optimizing the injection and storage of captured carbon dioxide, reducing emissions from industrial processes.

### 4.3 Sustainable Resource Management

The responsible management of natural resources is essential for sustainability. AI contributes to this by:

- **Precision Agriculture:** AI enables precision agriculture techniques, optimizing crop yields while minimizing resource use. AI-powered machinery, sensors, and data analysis promote sustainable farming practices.
- **Smart Cities:** AI aids in managing urban resources, from optimizing energy consumption and waste management to traffic flow and public transportation, contributing to overall urban sustainability.
- **Water Resource Management:** AI-driven systems can predict water supply and demand, enabling efficient water resource allocation and management, particularly in regions facing water scarcity.

AI's contributions to environmental sustainability are driven by its capacity to process vast datasets and make predictions, allowing for more informed decision-making. These applications align with multiple United Nations Sustainable Development Goals (SDGs), such as SDG 13 (Climate Action), SDG 14 (Life Below Water), and SDG 15 (Life on Land), and exemplify the potential of technology to address the interconnected challenges of our time.

We will further explore the ethical implications and responsible development of AI in the context of environmental sustainability. The integration of AI into our environmental efforts represents a powerful tool for safeguarding the future of our planet, and it is essential that we navigate this path with wisdom and responsibility.

## 5 Transforming Healthcare with AI

Healthcare is an arena where precision, speed, and accessibility can significantly impact the well-being of individuals and communities. Artificial Intelligence (AI) has rapidly emerged as a transformative force in healthcare, enhancing diagnostics, expanding healthcare access, and offering innovative solutions to complex medical challenges. In this chapter, we explore how AI is revolutionizing the healthcare sector, with a focus on its applications in diagnostics and treatment, healthcare access and telemedicine, as well as the critical ethical considerations that accompany its integration.

### 5.1 AI in Diagnostics and Treatment

AI is increasingly being integrated into diagnostics and treatment processes, improving medical accuracy and patient care in several ways:

- **Medical Imaging:** AI's image recognition capabilities enhance medical imaging, enabling more precise detection and diagnosis of conditions such as cancer, heart disease, and neurological disorders.
- **Drug Discovery:** AI accelerates drug discovery by analyzing massive datasets to identify potential drug candidates, thereby expediting the development of new treatments for various diseases.
- **Genomic Medicine:** AI helps in interpreting vast genomic data, allowing for personalized treatment plans tailored to an individual's genetic makeup.
- **Predictive Analytics:** AI-driven predictive models assist healthcare providers in forecasting disease outbreaks, identifying at-risk patients, and optimizing patient care.

## *5.2 Healthcare Access and Telemedicine*

AI is a catalyst for expanding healthcare access, particularly in underserved regions and during health crises:

- **Telemedicine:** AI-powered telemedicine platforms facilitate remote consultations, enabling patients to receive medical care from a distance, which is crucial in scenarios like the COVID-19 pandemic.
- **Health Chatbots:** AI-driven chatbots provide immediate medical advice, helping individuals assess their symptoms, determine whether they need to consult a healthcare provider, and access information about medical conditions.
- **Remote Monitoring:** AI is used to remotely monitor patients' health, especially those with chronic illnesses. These systems can alert healthcare professionals to deviations from normal health parameters.
- **Medical Supply Chain Management:** AI optimizes the supply chain for healthcare equipment and medications, ensuring that vital resources reach where they are needed promptly.

## *5.3 Ethical Considerations in Healthcare AI*

While the benefits of AI in healthcare are substantial, ethical considerations are paramount:

- **Privacy:** The use of patient data in AI applications must be carried out with strict adherence to patient privacy regulations and a commitment to data security.
- **Bias and Fairness:** Bias in AI algorithms can perpetuate healthcare disparities. Ensuring fairness in AI healthcare solutions is crucial to prevent inequities in diagnosis and treatment.
- **Transparency:** AI algorithms should be transparent and explainable, so that healthcare providers and patients can understand and trust the recommendations made.

- **Accountability:** In cases where AI is used in clinical decision-making, defining accountability and liability is vital to ensure that the technology is used responsibly.

It will also emphasize the ethical imperatives that must guide the development and deployment of AI in healthcare, ensuring that the innovations in this domain benefit all individuals and communities while maintaining patient trust and privacy. The integration of AI into healthcare represents a revolution in patient care and medical practice, with the potential to enhance health outcomes and access to medical expertise worldwide.

# 6 Enhancing Education Through AI

Education is a cornerstone of human development and a key driver of societal progress. In this chapter, we delve into the ways in which Artificial Intelligence (AI) is revolutionizing education. AI is enhancing learning experiences, expanding access to quality education, and offering new opportunities to address long-standing challenges in the field of education.

## *6.1 Personalized Learning*

Personalized learning is a pedagogical approach that tailors education to individual students' needs, preferences, and progress. AI is instrumental in implementing personalized learning by:

- **Adaptive Learning Systems:** AI-powered platforms analyze students' performance and adapt the curriculum to their specific strengths and weaknesses, ensuring optimal learning.
- **Recommendation Systems:** AI can recommend educational resources, from textbooks to online courses, based on students' interests and past learning patterns.
- **Natural Language Processing (NLP):** NLP technologies facilitate intelligent tutoring systems that can interact with students in natural language, providing personalized assistance.
- **Learning Analytics:** AI analyzes student data to help educators make data-driven decisions about curriculum design and teaching strategies, improving learning outcomes.

### *6.2 Access to Quality Education*

AI extends the reach of quality education by:

- **Online Learning Platforms:** AI-powered platforms provide access to quality educational content globally, bridging geographical and economic barriers.
- **Language Translation:** AI enables real-time language translation, making educational content more accessible to diverse linguistic communities.
- **Virtual Classrooms:** AI-driven virtual classrooms facilitate remote learning, enabling students to access quality education regardless of their physical location.
- **Specialized Education:** AI aids in creating personalized education plans for students with special needs, ensuring they receive quality education tailored to their requirements.

### *6.3 Challenges and Opportunities*

While AI offers remarkable opportunities in education, it also presents challenges:

- **Privacy and Data Security:** The collection and use of student data raise concerns about privacy and data security, necessitating robust safeguards.
- **Equity:** Access to AI-driven education is not uniform, and the digital divide can exacerbate educational inequalities.
- **Teacher Training:** Educators require training to effectively leverage AI tools in the classroom, and their role may evolve in AI-enhanced learning environments.

## 7 Ethical Considerations:

- The use of AI in education raises ethical questions, from algorithmic bias to transparency in decision-making.

We will examine case studies showcasing the transformative impact of AI in education and consider the challenges and opportunities that this integration brings. By harnessing the potential of AI in education and addressing its ethical and equity considerations, we can foster a more inclusive, adaptable, and effective educational system.

## 8 Responsible AI and Ethical Implications

The integration of Artificial Intelligence (AI) into various domains brings forth a range of ethical considerations and responsibilities. This chapter explores the critical dimensions of responsible AI, including the development of ethical frameworks, the assurance of fairness and accountability, and the identification of potential risks and challenges in the application of AI technologies.

### *8.1 Ethical Frameworks*

The development and application of AI must be guided by ethical principles and frameworks to ensure that AI systems operate in a manner aligned with societal values and human rights. In this section, we delve into:

## 9 Ethical AI Principles:

- An exploration of key ethical principles that underpin responsible AI, such as transparency, accountability, fairness, and privacy.
- **Regulatory and Governance Frameworks:** A discussion of emerging AI regulations and governance initiatives at the international, national, and organizational levels.
- **Case Studies:** Examples of organizations and initiatives that have successfully implemented ethical AI frameworks in practice.

### *9.1 Ensuring Fairness and Accountability*

Fairness and accountability are pivotal in AI systems, as biases and unintended consequences can impact individuals and communities. In this section, we examine:

- **Algorithmic Bias:** The challenges and consequences of algorithmic bias, and strategies to mitigate bias in AI systems.
- **Explainability and Transparency:** The importance of making AI systems more transparent and explainable to ensure accountability.
- **Auditing and Compliance:** How organizations can implement auditing and compliance mechanisms to uphold ethical standards in AI.
- **Case Studies:** Real-world examples that highlight the importance of fairness and accountability in AI systems.

### *9.2 Risks and Challenges*

AI introduces a set of risks and challenges that require close attention and proactive mitigation. This section explores:

- **Privacy Concerns:** The implications of AI for data privacy and strategies to protect personal information.
- **Job Displacement:** The potential impact of automation on the job market and strategies for addressing workforce displacement.
- **Security and Malicious Use:** The risks associated with AI being used for malicious purposes and strategies for enhancing AI security.
- **Ethical Dilemmas:** Complex ethical dilemmas that AI technologies may present and approaches to navigate them.
- **Unintended Consequences:** The unexpected consequences of AI adoption and methods for identifying and addressing them.
- **Global Collaboration:** The importance of international collaboration to address the global nature of AI challenges.

Through a comprehensive exploration of these dimensions, this chapter aims to foster a deeper understanding of responsible AI and the ethical implications that accompany its integration. Responsible AI is not just a technological imperative but a societal one, ensuring that AI is developed and used to benefit all, while minimizing harm and ensuring accountability for its impact. As the application of AI continues to expand, the ethical considerations and responsible practices outlined in this chapter will remain essential in guiding the path forward.

## 10 Case Study

The Sustainable Development Goals (SDGs) represent a global commitment to address some of the most pressing challenges facing humanity, ranging from eradicating poverty and ensuring access to clean water and sanitation to building sustainable cities and combating climate change. These 17 interconnected goals, established as part of the United Nations 2030 Agenda for Sustainable Development, provide a comprehensive framework for achieving a better and more sustainable future for all. Monitoring progress toward the SDGs is critical to assess the effectiveness of global efforts and to make informed policy decisions. This process involves the collection, cleaning, and analysis of a wide array of indicators that reflect various aspects of economic, social, and environmental development. These indicators serve as vital signposts on the journey towards a more sustainable, equitable, and prosperous world. In this case study, we embark on a data-driven exploration of SDG indicators, showcasing the power of data analysis in tracking and understanding progress toward the SDGs. Our aim is to provide a practical example of how data can be harnessed to shed light on the status of specific SDGs, contributing to evidence-based decision-making and targeted interventions. We will take you through the data collection process, highlighting the challenges and considerations involved in sourcing data from diverse, often fragmented, and occasionally inconsistent data repositories. Subsequently, we will delve into the intricate process of data cleaning, where we tackle issues like missing data and discrepancies to ensure that our analysis is based on high-quality information.

The heart of this case study lies in the Exploratory Data Analysis (EDA), where we will use statistical tools and data visualization techniques to unearth insights and patterns from the SDG indicators. By selecting a subset of the SDGs, we will provide a focused analysis, shedding light on the progress made and the areas that require further attention. Our findings will be presented through compelling visualizations, which not only make the data more accessible but also highlight the disparities and successes in our selected SDGs. The implications of our findings will be discussed, emphasizing the tangible ways in which data analysis can inform decision-makers and stakeholders. Finally, we will offer recommendations for actions and future work, recognizing that this case study represents just a snapshot of a continuous journey towards achieving the SDGs. The importance of data analysis in this context cannot be overstated; it is a powerful tool in the quest to build a more sustainable, equitable, and prosperous future for all, in line with the principles of the United Nations 2030 Agenda for Sustainable Development.

The code **sns.distplot(df['Human Development Index - undp_hdi'], kde_kws = {"color": "red"})** is used to create a distribution plot (histogram) of the Human Development Index (HDI) values from the DataFrame **df**, with a red overlay for the kernel density estimate (KDE) of the distribution. This plot helps visualize the distribution of HDI values across different countries or regions as shown in Fig. 1.

The Fig. 2 is a PairGrid that visualizes relationships and distributions among three selected variables: 'Deaths - Unsafe water source,' 'Urban population (% of total population),' and 'Government Expenditure (IMF based on Mauro et al. (2015)).' In the upper triangular portion, scatterplots with red regression lines depict how each pair of variables is related, helping us understand the direction and strength of

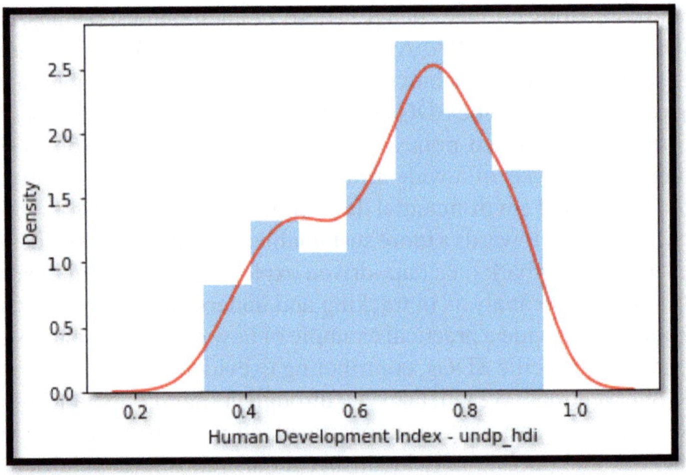


**Fig. 1** distribution plot (histogram) of the Human Development Index (HDI)

these relationships. The lower triangular section displays residual plots, enabling us to identify deviations from the regression lines and any patterns or outliers in the residuals. The diagonal exhibits histograms illustrating the individual distributions of each variable. This comprehensive visual analysis allows for the assessment of correlations, model assumptions, and variable distributions, aiding in gaining insights into the complex interplay of these factors in the dataset.

The Fig. 3 comprises a boxplot and a strip plot, both centered on the 'Urban population (% of total population)' variable from the dataset 'df.' The boxplot provides a summary of the variable's distribution, featuring the median, quartiles, and the presence of outliers. Concurrently, the strip plot displays individual data points along the x-axis as red dots, allowing for a granular visualization of data density and the identification of potential outliers. This combination offers a comprehensive view of the distribution characteristics of urban population percentages, illuminating central tendencies, variations, and anomalies within the dataset.

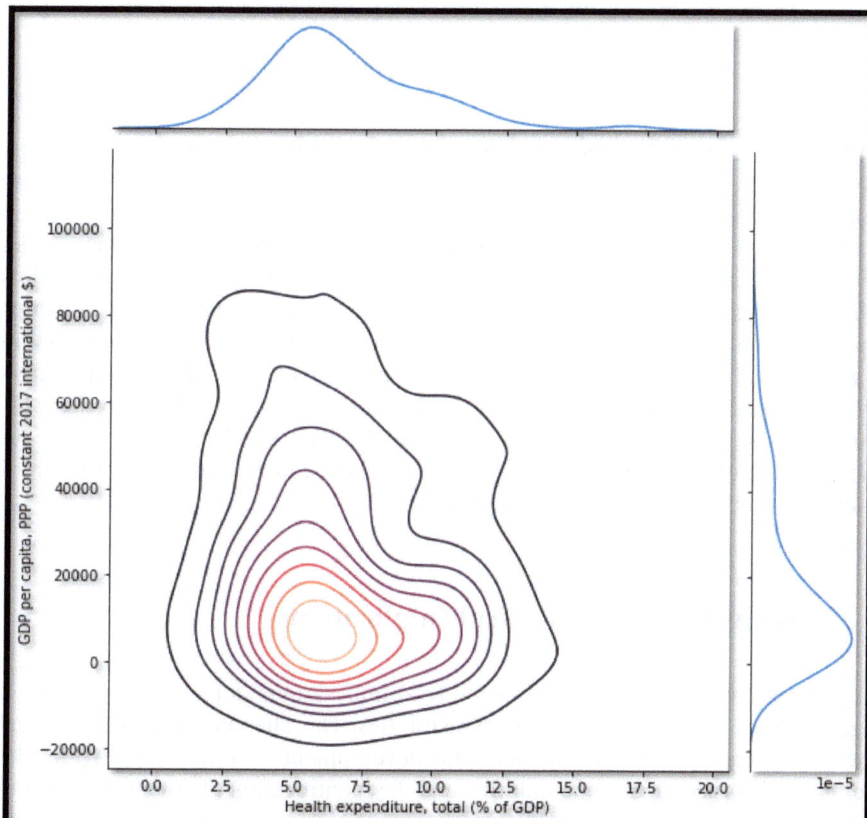


**Fig. 2** PairGrid that visualizes relationships and distributions

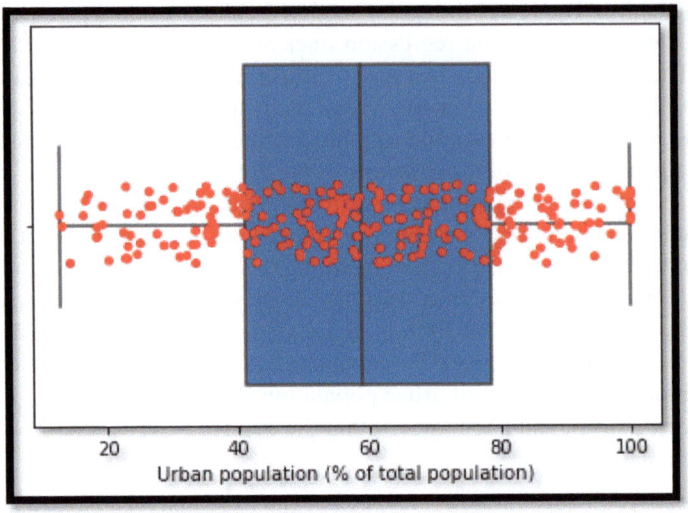


**Fig. 3** A boxplot and a strip plot, both centered on the 'Urban population'

The Fig. 4 is a PairGrid composed of scatterplots, regression lines, residual plots, and histograms, designed to comprehensively analyze relationships and distributions among three chosen variables ('Deaths - Unsafe water source,' 'Urban population (% of total population),' and 'Government Expenditure (IMF based on Mauro et al. (2015)).') from the DataFrame 'df.' The upper triangular section features scatterplots with red regression lines, illuminating the direction and strength of relationships between variable pairs. In the lower triangular part, residual plots aid in assessing the goodness of fit of the regression models, revealing patterns and potential outliers in the residuals. The diagonal exhibits histograms, offering insights into the individual variable distributions. This multi-layered visual exploration enhances our understanding of associations and data characteristics, facilitating the detection of correlations and anomalies within the dataset.

## 11 Responsible AI and Ethical Implications

The integration of Artificial Intelligence (AI) into various domains brings forth a range of ethical considerations and responsibilities. This chapter explores the critical dimensions of responsible AI, including the development of ethical frameworks, the assurance of fairness and accountability, and the identification of potential risks and challenges in the application of AI technologies.

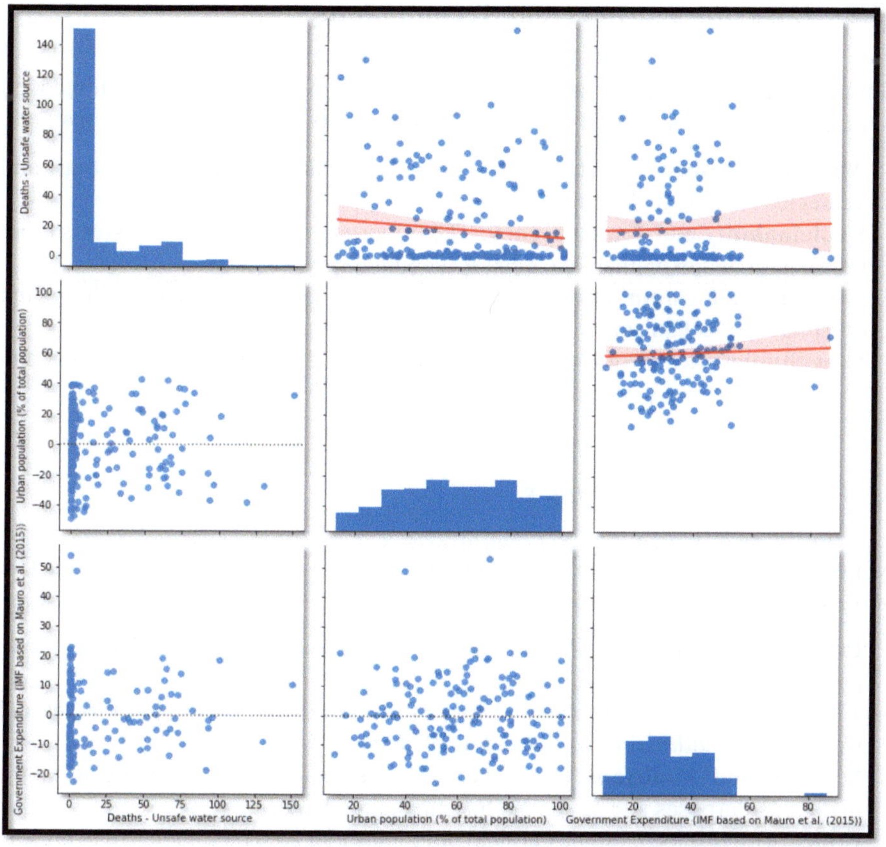


**Fig. 4** PairGrid composed of scatterplots, regression lines, residual plots, and histograms

## *11.1 Ethical Frameworks*

The development and application of AI must be guided by ethical principles and frameworks to ensure that AI systems operate in a manner aligned with societal values and human rights. In this section, we delve into:

- **Ethical AI Principles:** An exploration of key ethical principles that underpin responsible AI, such as transparency, accountability, fairness, and privacy.
- **Regulatory and Governance Frameworks:** A discussion of emerging AI regulations and governance initiatives at the international, national, and organizational levels.
- **Case Studies:** Examples of organizations and initiatives that have successfully implemented ethical AI frameworks in practice.

## 11.2 Ensuring Fairness and Accountability

Fairness and accountability are pivotal in AI systems, as biases and unintended consequences can impact individuals and communities. In this section, we examine:

- **Algorithmic Bias:** The challenges and consequences of algorithmic bias, and strategies to mitigate bias in AI systems.
- **Explainability and Transparency:** The importance of making AI systems more transparent and explainable to ensure accountability.
- **Auditing and Compliance:** How organizations can implement auditing and compliance mechanisms to uphold ethical standards in AI.
- **Case Studies:** Real-world examples that highlight the importance of fairness and accountability in AI systems.

## 11.3 Risks and Challenges

AI introduces a set of risks and challenges that require close attention and proactive mitigation. This section explores:

- **Privacy Concerns:** The implications of AI for data privacy and strategies to protect personal information.
- **Job Displacement:** The potential impact of automation on the job market and strategies for addressing workforce displacement.
- **Security and Malicious Use:** The risks associated with AI being used for malicious purposes and strategies for enhancing AI security.
- **Ethical Dilemmas:** Complex ethical dilemmas that AI technologies may present and approaches to navigate them.
- **Unintended Consequences:** The unexpected consequences of AI adoption and methods for identifying and addressing them.
- **Global Collaboration:** The importance of international collaboration to address the global nature of AI challenges.

Through a comprehensive exploration of these dimensions, this chapter aims to foster a deeper understanding of responsible AI and the ethical implications that accompany its integration. Responsible AI is not just a technological imperative but a societal one, ensuring that AI is developed and used to benefit all, while minimizing harm and ensuring accountability for its impact. As the application of AI continues to expand, the ethical considerations and responsible practices outlined in this chapter will remain essential in guiding the path forward.

# 12 Conclusion

The journey through the integration of Artificial Intelligence (AI) in advancing the United Nations Sustainable Development Goals (SDGs) has been one of exploration, inspiration, and a call to action. In this concluding chapter, we recap the key findings, issue a call to action for all stakeholders, and glimpse into the future prospects of AI-SDG integration.

## *12.1 Recap of Key Findings*

Throughout this book, we have uncovered critical insights and findings regarding AI's role in advancing the SDGs:

- **AI's Transformative Potential:** AI offers unparalleled potential to address the complex and interconnected global challenges embedded within the SDGs. Its data processing capabilities and predictive power enable more informed decision-making across various domains.
- **Real-World Impact:** Real-world case studies have illustrated the tangible outcomes of AI-SDG integration, from improving healthcare access and diagnostics to conserving the environment, advancing education, and eradicating poverty.
- **Ethical Imperatives:** Responsible AI development is central to AI-SDG integration. Ethical considerations, including fairness, transparency, accountability, and data privacy, must guide AI applications to ensure equitable and ethical outcomes.
- **Challenges and Opportunities:** AI brings both challenges and opportunities, from job displacement to security concerns. Addressing these challenges requires proactive mitigation and international collaboration.

## *12.2 Call to Action*

The journey doesn't end with these findings; it serves as a catalyst for collective action:

- **Collaboration:** The global community must foster collaboration among governments, organizations, and communities. The complex nature of the SDGs necessitates joint efforts to address the challenges and seize the opportunities presented by AI.
- **Ethical Commitment:** Ethical AI must be at the forefront of AI-SDG initiatives. Commit to responsible AI development, ensuring that AI technologies align with the values of the SDGs, respect human rights, and consider the well-being of all.

- **Inclusivity:** The principle of "Leave No One Behind" is central to the SDGs. AI-SDG projects must prioritize inclusivity, ensuring that marginalized and underserved populations benefit from these technologies.
- **Sustainability:** Sustainable development requires sustainable solutions. AI-SDG initiatives should incorporate long-term planning and practices that consider the environmental and societal impact.
- **Education and Training:** Preparing the workforce and the community to leverage AI technologies is essential. Invest in education and training programs to equip individuals with the skills and knowledge needed to maximize the benefits of AI.

## *12.3 Future Prospects*

The future prospects of AI-SDG integration are promising:

- **Innovation:** AI is an evolving field, with continuous innovation and advancements. We can expect new AI technologies and applications to emerge, further expanding our capacity to address the SDGs.
- **Global Commitment:** As the global community recognizes the potential of AI for sustainable development, we anticipate increased commitment to harness AI's power to advance the SDGs.
- **Cross-Domain Integration:** AI will continue to integrate across different SDG domains, offering cross-cutting solutions to complex global challenges.
- **Responsible AI**: The focus on responsible AI development and ethics will persist, guiding the path forward to ensure that AI serves the greater good.

The integration of AI into the United Nations Sustainable Development Goals signifies a transformative opportunity to reshape the future of our planet and its inhabitants. The journey may be complex, but the collective will, collaboration, and ethical commitment can steer us toward a more equitable, sustainable, and prosperous future. The responsibility rests with us to turn the promise of AI into reality, and this call to action inspires us to embark on that collective endeavor.

## References

Astobiza AM, Toboso M, Aparicio M, López D (2021) AI ethics for sustainable development goals. IEEE Technol Soc Mag 40(2):66–71

Bachmann N, Tripathi S, Brunner M, Jodlbauer H (2022) The contribution of data-driven technologies in achieving the sustainable development goals. Sustainability 14(5):2497

Banifatemi A, Miailhe N, BuseÇetin R, Cadain A, Lannquist Y, Hodes C (2021) Democratizing AI for humanity: A common goal. Reflections on Artificial Intelligence for Humanity. Springer, Cham, pp 228–236

George I (2023). The use of artificial intelligence to improve the Sustainable Development Goals (SDG).

Goralski MA, Tan TK (2023) Artificial Intelligence: Poverty Alleviation, Healthcare, Education, and Reduced Inequalities in a Post-COVID World. The Ethics of Artificial Intelligence for the Sustainable Development Goals. Springer International Publishing, Cham, pp 97–113

Isabelle DA, Westerlund M (2022) A review and categorization of artificial intelligence-based opportunities in wildlife, ocean and land conservation. Sustainability 14(4):1979

Jose A, Nandagopalan S, Akana CMVS (2021). Artificial Intelligence techniques for agriculture revolution: a survey. Annals of the Romanian Society for Cell Biology, 2580–2597.

Jungwirth D, Haluza D (2023). Artificial Intelligence and the Sustainable Development Goals: GPT-3s Reflections on the Society Domain.

Kasinathan P, Pugazhendhi R, Elavarasan RM, Ramachandaramurthy VK, Ramanathan V, Subramanian S, Alsharif MH (2022) Realization of sustainable development goals with disruptive technologies by integrating industry 5.0, society 5.0, smart cities and villages. Sustainability 14(22):15258

Khakurel J, Penzenstadler B, Porras J, Knutas A, Zhang W (2018) The rise of artificial intelligence under the lens of sustainability. Technologies 6(4):100

Leal Filho W, Yang P, Eustachio JHPP, Azul AM, Gellers JC, Gielczyk A, Kozlova V (2023) Deploying digitalisation and artificial intelligence in sustainable development research. Environ Dev Sustain 25(6):4957–4988

Mehmood H, Liao D, Mahadeo K (2020). A review of artificial intelligence applications to achieve water-related sustainable development goals. In 2020 IEEE/ITU International Conference on Artificial Intelligence for Good (AI4G) (pp. 135–141). IEEE.

Mercier-Laurent E (2021). Can Artificial Intelligence Effectively Support Sustainable Development? In Artificial Intelligence for Knowledge Management: 8th IFIP WG 12.6 International Workshop, AI4KM 2021, Held at IJCAI 2020, Yokohama, Japan, January 7–8, 2021, Revised Selected Papers 8 (pp. 144–159). Springer International Publishing.

Mhlanga D (2021) Artificial intelligence in the industry 4.0, and its impact on poverty, innovation, infrastructure development, and the sustainable development goals: Lessons from emerging economies? Sustainability 13(11):5788

Mhlanga D (2022) The role of artificial intelligence and machine learning amid the COVID-19 pandemic: What lessons are we learning on 4IR and the sustainable development goals. Int J Environ Res Public Health 19(3):1879

Neuerburg K (2022) Applications of artificial intelligence in achieving development goals. Int J Technol Knowledge Society 18(1):35

Pedro F, Subosa M, Rivas A, Valverde P (2019). Artificial intelligence in education: Challenges and opportunities for sustainable development.

Rakha NA (2023) Artificial Intelligence and Sustainability. Int J Cyber Law 1:3

Razak A, Nayak MP, Manoharan G, Durai S, Rajesh GA, Rao CB, Ashtikar SP (2023) Reigniting the power of artificial intelligence in education sector for the educators and students competence. Artificial Intelligence and Machine Learning in Smart City Planning. Elsevier, Hoboken, pp 103–116

Sætra HS (2021) AI in context and the sustainable development goals: Factoring in the unsustainability of the sociotechnical system. Sustainability 13(4):1738

Sætra HS (2022) AI for the sustainable development goals. CRC Press

Truby J (2020) Governing artificial intelligence to benefit the UN sustainable development goals. Sustain Dev 28(4):946–959

Whig P, Velu A, Nadikattu RR, Sharma P (2024a) Reinforcement Learning for Automated Medical Diagnosis and Dynamic Clinical Regimes. Handbook of Research on Artificial Intelligence and Soft Computing Techniques in Personalized Healthcare Services. Apple Academic Press, New Jersey, pp 169–187

Whig P, Velu A, Nadikattu RR, Alkali YJ (2024b) Role of AI and IoT in Intelligent Transportation. Artificial Intelligence for Future Intelligent Transportation. Apple Academic Press, New Jersey, pp 199–220

# Towards Implementing SDG-4: Leveraging AI in Today's Educational Landscape

**Anadi Trikha**

**Abstract** This chapter explores the influence of Artificial intelligence (AI) in the field of education, focusing on its potential benefits, challenges, and ethical consideration. The chapter attempts to correlate the UN Agenda 2030 underlining SDG 4 as equitable quality education and promoting lifelong learning for all with AI technology.

The chapter further discussed the initiatives taken by the government of India to leverage the best possible use of AI to educate its youth and the coming generations. This chapter reveals an important aspect of combining Artificial intelligence (AI) with education and illustrates emerging practices that can reduce the barriers to education and improve lifelong learning for all.

**Keywords** Artificial Intelligence (AI) · Digitalization · Digital Transformation · Sustainable Development Goals · SDG-4 · Life-long Learning

## 1 Introduction

The world is facing an education crisis and more than 200 million children globally are out of school and only education can help harness sustainable goals (Wayne Holmes, et al., 2019). Uneven global development of the technological infrastructure has created a digital divide among the continents. This can only be overcome if AI-enabled education is provided to all irrespective of nationality, culture, gender, or age. Also, COVID-19 has exposed the long-existing challenges in the education sector. It has intensified the challenge of creating a society where quality education can be accessed easily by all (Education 4.0 India, 2022).

Digital learning can address these difficulties effectively and can ensure no one is left behind. The emergence of Artificial Intelligence (AI) has rapidly become a crucial tool for optimizing the digital learning process and enhancing the user experience

A. Trikha (✉)
Centre for Distance and Online Education, Manipal University Jaipur, Jaipur, Rajasthan, India
e-mail: anadi.trikha@jaipur.manipal.edu



W. Leal Filho et al. (eds.), *Digital Technologies to Implement the UN Sustainable Development Goals*, World Sustainability Series,
https://doi.org/10.1007/978-3-031-68427-2_4

(Ray, 2023). Intelligent systems powered by AI technology have enabled personalized learning experiences and automated assessments and feedback for students. These AI-powered e-learning systems can analyse and monitor a student's learning activity, providing recommendations on how to improve their performance through continuous assessment and predictive analysis (Panjaburee, 2023).

AI has become one of the most prevailing forces in today's technological world because of its ability to handle large amounts of data quickly and accurately. AI has the potential to accelerate the process of achieving Sustainable Development Goals (SDGs). In developing countries, where educational equity and quality can be challenging to achieve, AI can contribute tremendously through its ability to exploit data (Tlili et al., 2023). Figure 1 depicts the sectors in which AI can be useful in the educational landscape.

Globally and locally, AI is creating a makeshift in the field of education. It has transformed the modes of learning and interaction for learners across disciplines. The unique benefits of AI have been observed in healthcare education with the introduction of AI in the learning system. This technology has enabled students to access information at their fingertips for quick learning.

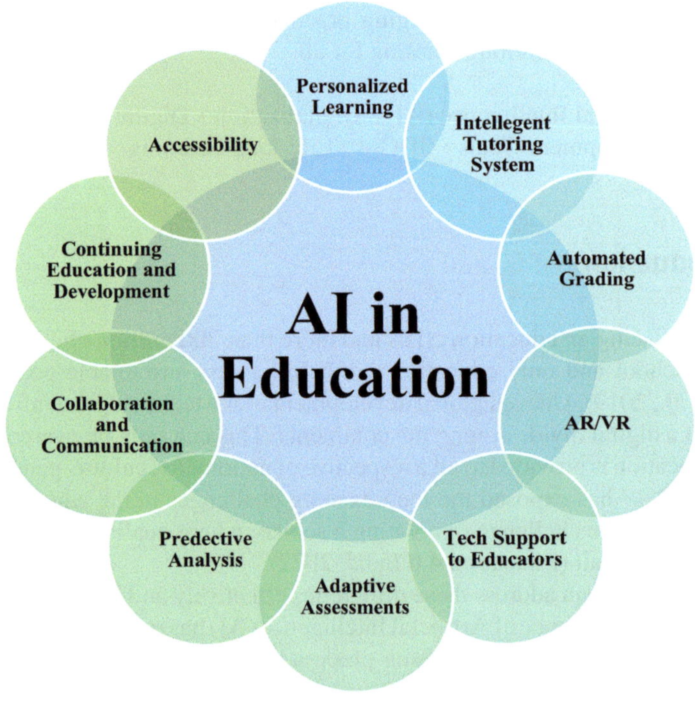


**Fig. 1** How AI Can Contribute to Education Source: Author-Creation

## 1.1 AI and SDG-4

The United Nations has identified AI as a tool to help achieve development goals. Intelligent, adaptive learning systems are being developed for schools and universities around the globe. AI technologies are used in facilitating the delivery of education. Augmented reality (AR) and Virtual Reality (VR) are two innovations that are now used in educational contexts to make learning more engaging (Walia, n.d.).

To increase awareness and assist members in tackling complex challenges, UNESCO has been actively encouraging discourse and information on AI in education. To achieve SDG 4 (Education 2030), the 2019 Beijing Consensus includes guidelines for making use of AI technologies for global audiences. UNESCO urges the provision of guidelines and instruments for capacity-building and incorporating AI skills into ICT competency frameworks. UNESCO's intellectual work on artificial intelligence in education aims to provide a shared understanding of the possibilities and hazards involved.

In another conference on AI for sustainable development, UNESCO emphasised on human-centric approach to AI. The conference addressed current inequalities regarding access to knowledge, research, and the diversity of cultural expressions (Wayne Holmes, et al., 2019). Transforming the education and training systems for the common good and sustainable future of society gives a benefit-risk evaluation, aggregates national policies, and acts as a handbook for policy formation (Miao, 2021). Table 1 mentions how AI can contribute to seven targets under SDG-4.

The advent of AI has paved the way for exciting new possibilities in the field of distance education. However, to guarantee the security and success of AI-powered online learning tools, several ethical considerations must be taken, just as they would be with any other technology. The study explored the moral dilemmas presented by AI in e-learning and provided guidelines for creating ethical AI-driven e-learning platforms. It focused on various forms of instructional technology in Namibia's education and training sector and explored how ethical principles may be used there. Ethical and moral considerations were explored regarding the use of Learning Management Systems (LMSs), Interactive White-Boards, digital games, AR/VR, gamification, mobile tech, and Virtual Learning environments (VLEs). To be ready for the rise of AI and instructional technologies in online education, both students and teachers need to be familiar with the inner workings of responsible technology and the process by which ethical decisions are made (Uunona, 2023).

By using machine learning and big data technology, educators have access to large datasets that can help in understanding student needs. Machine learning algorithms are invaluable in identifying patterns and trends that may otherwise go unnoticed. AI-based analysis can flag students who are at risk of dropping out or failing their classes. Moreover, assessing different teaching strategies for effectiveness using AI-like tools is an effective way to provide tailored teaching methods that will meet each student's individual needs (Alam, 2021, 2022).

A study investigation related to AI's influence on the classroom focuses on the tech's ability to solve students' most perplexing problems, boost their imagination,

**Table 1** SDG-4 Targets Source: https://en.unesco.org/education2030-sdg4/targets

| | |
|---|---|
| **SDG 4.1** | **Universal primary and secondary education**<br>AI can help enhance access to quality education by providing personalized learning experiences. Through adaptive learning platforms powered by AI can assess individual students' strengths and weaknesses and can provide tailor-made content |
| **SDG 4.2** | **Early childhood development and Universal pre-primary Education**<br>AI can assist in early childhood development by providing interactive learning tools and educational games. These tools can stimulate cognitive and social skills development in young children |
| **SDG 4.3** | **Equal access to technical/vocational and higher education**<br>Through online platforms, individuals from remote areas can access vocational and higher education courses. AI-powered language translation and voice-recognizing technologies can help overcome language barriers and enable diverse participation in education |
| **SDG 4.4** | **Relevant skills for decent work**<br>AI can help identify and address skill gaps by analysing labour market trends and predicting future skills demand. This information can help educators and policymakers to design a curriculum that aligns with the needs of the job market |
| **SDG 4.5** | **Gender Equality and Inclusion**<br>AI-powered content moderation tools can help identify and eliminate gender-based discrimination in textbooks, online resources, and assessments. Additionally, AI can support initiatives that encourage girls' participation in STEM fields by providing mentorship, online communities, and tailored educational content |
| **SDG 4.6** | **Universal youth literacy**<br>AI can assist in promoting literacy by providing personalized reading recommendations and language learning tools. Natural Language Processing (NLP) algorithms can analyse individuals' reading levels and offer appropriate reading materials, fostering a love for reading and improving literacy skills |
| **SDG 4.7** | **Education for sustainable development and global citizenship**<br>AI can facilitate education for sustainable development by providing real-time data on environmental issues and promoting awareness. It can support simulations and virtual experiences that allow students to understand complex sustainability challenges |
| Three means of implementation | |
| **4. a** | **Effective learning environments**<br>AI can contribute to creating effective learning environments by providing intelligent tutoring systems, virtual reality (VR) simulations, and interactive educational resources. These technologies enhance engagement, motivation, and knowledge retention among students |
| **4. b** | **Scholarships**<br>AI can assist in identifying and matching eligible students with scholarships based on their academic achievements, interests, and financial needs. AI-powered systems can streamline the scholarship application process, ensuring transparency and efficiency |
| **4. c** | **Teachers and educators**<br>AI can support teachers and educators by automating administrative tasks, such as grading and lesson planning, allowing them to focus more on personalized instruction and student support. AI can also provide professional development resources and access to a vast repository of educational materials, facilitating continuous learning for teachers |

and create innovative learning environments (Chassignol, 2018). While AI will not be able to replace everything, it can and should be used to enhance other learning methods like gaming and virtual and augmented reality. The development of AI has opened new possibilities for online testing, evaluation, and tracking of student progress in Massive Open Online Courses (MOOCs). However, without human mentors, AI assessment algorithms might not always be reliable. Artificial Intelligence (AI) tools are already being used in several educational contexts, all of which promote the cause of tech-enhanced education.

AI technology could improve education by speeding up information access, personalising learning for each pupil, automating administrative tasks, and reducing teacher workload. However, introducing AI to the education system required mindful administration to reduce inequalities, and contribute to the progress of education-related SDGs.

In addition to evaluating various teaching techniques' effectiveness, personalization is another area where AI systems excel at improving educational quality in developing nations. With intelligent recommendations from e-learning platforms like these, educators get updated insights based on individual students' interaction with the curriculum. The holistic approach taken by AI-driven algorithms often rewards creativity while also empowering learners who struggle with traditional classroom formats. Overall, developing countries should hope to leverage the transformative power of AI as it provides educators with the tools necessary for delivering a high-quality education equitable to the population regardless of geographical location (Pedro, 2019).

## 2 Government-Led Initiatives for Creating an Ecosystem of AI in India

Figure 2 shows the numerous initiatives that India has displayed that are attracting the attention of stakeholders from all over the world who thrive in the field of artificial intelligence. On a recent trip to India, Mr Sundar Pichai, the CEO of Google, put down his discoveries from his trip to India about the incredible things that individuals are doing with AI in India (Pichai, 2022).

**Fig. 2** Government of India AI Initiatives Source: Author-Creation

According to a recent report from Stanford University, India ranks fifth in terms of investments received by startups providing AI-based goods and services in 2022 (Ahaskar, 2023). The government of India has understood that AI is an emerging field and that its impact is not fully realized.

## 2.1 AI for All: NITI Aayog

*National Strategy for AI: Discussion Paper Part I* (2018) focused on five sectors that would benefit the most from AI in solving real-time issues: healthcare, agriculture, education, smart cities and infrastructure and smart mobility & transportation. The report also quoted the potential of AI to add $1 trillion to India's economy in 2035. Education is one of the primary focus areas. The Government of India has made significant efforts to encourage retraining and further education in the area of artificial intelligence (Aayog, National Strategy for Artificial Intelligence, 2018).

*Responsible AI: Approach Document for India Part II* was published in 2021 in collaboration with the World Economic Forum as a Knowledge Partner. Chapter 4 has emphasized how Indian educational institutions like NCERT, and AICTE have introduced foundation pedagogies related to AI for senior secondary and HEIs (Aayog, RESPONSIBLE AI #AIForAll, 2021).

## 2.2 FutureSkills PRIME

It is a B2C system for the upskilling and reskilling of IT professionals in emerging areas like AI. Under this scheme, 500 + trainers, and 4000 + government officials were trained on these technologies. This program employs an 'aggregator of aggregators' approach for digital skills training on a national level and is hosted as an online B2C e-commerce platform so that Indian citizens in Tier 2 and Tier 3 cities can participate (Technology, n.d.).

- 40 C-DAC/NIELIT Centres across the country are institutionalizing blended-learning programs, in a hub-and-spoke model, as lead and co-lead resource centres.
- The Career Portal offers details on a variety of IT-related employment opportunities, including hackathons, internships, and apprenticeships. Incentives are made available to trainees, particularly those who come from economically disadvantaged backgrounds, following successful evaluation and certification of the learner. The purpose of these incentives is to encourage more trainees to finish the online up-skilling and re-skilling programs.
- To promote capacity building in Artificial Intelligence, the Government of India has taken several initiatives, including the 'Visvesvaraya PhD Scheme' which focuses on research areas such as AI and machine learning.

## 2.3 Centre of Excellence

Setting-up of the Technologies Division by the Ministry of Electronics and Information and Technology, Government of India. This division of MeitY will be responsible for promoting cutting-edge technology in the country. Under this initiative, they have already established a Centre of Excellence for IoT at Gandhi Nagar, Bengaluru, Gurugram and Vizag. They have also established the Centre for Excellence in Virtual and Augmented Reality (VARCoE) at IIT Bhubaneswar. Another CoE for Gaming, VFX, Computer Vision & AI has been set up in Hyderabad. A CoE for Blockchain Technology has been set up in Gurugram.

The "National Programme on Responsible Use of AI for Youth" helps schoolchildren become AI-ready and closes the skills gap. To promote innovation through study, the Government has established "Centres of Excellence" on emerging technologies, including AI.

The Department of Science & Technology's "National Mission on Interdisciplinary Cyber-Physical Systems" promotes R&D, HRD, Technology Development, Entrepreneurship Development, and International Collaboration. 25 technology innovation hubs in advanced technologies like Machine Learning and AI have been created in prestigious schools nationwide as part of this mission.

The 'Global Partnership on Artificial Intelligence' is a multi-stakeholder, multinational effort to promote ethical AI research, development, and application; India is one of the organization's founding members. In 2020, the government organized the 'Responsible AI for Social Empowerment conference, which drew 79,000 + attendees from 147 nations and 320 presenters from 21 countries (PIB, Press Release, 2022).

The recent budget announcement in February 2023 Finance Minister Nirmala Sitharaman announced three centres of excellence in AI as part of GOI's "Make AI for India" and "Make AI Work for India" initiatives (Jain, 2023).

## 2.4 AI Research Analytics and Knowledge Dissemination Platform (AIRAWAT)

This is yet another initiative from the Government of India (GOI) to establish a shared infrastructure for AI research and data aggregation. This artificial intelligence supercomputer, named "AIRAWAT" (which, in Sanskrit, translates to "great white elephant"), recently placed 75th in the World 500 supercomputing list (Delhi, 2023). This lives up to its name by conforming to the goals of India's honourable prime minister, Shri Narendra Modi, who has advocated for "AI for All." The centre has begun accepting AI-related startups and MSMEs.

## 2.5 *INDIAai*

Launched in 2022, INDIAai is a hub for information, a laboratory for study, and a foundation for ecosystems. It has the potential to bring together and encourage cooperation among the many players in India's AI ecosystem. The portal features in-depth information on the major actors in the AI ecosystem as well as news and commentary on the state of AI both internationally and in India (Meit, n.d.).

INDIAai also provides students, business owners, professionals, and C-suite executives with a wealth of additional tools, including upskilling and career guidance. Regular partnerships with a wide range of stakeholders and international organizations are a vital aspect of INDIAai's goal.
INDIAai has collaborated with companies, start-ups and HEIs like IBM, Microsoft, NVIDIA, Meta, Fractal, Tech Mahindra, IIT Madras, IIT Jodhpur, and IIT Ropar, CoRover, Niramai, Haptik, Oncostem, Yellow AI, and TagHive.
INDIAai also plays a pivotal role as a content repository for the National Program on AI.

The government's approach to education policy should be placed a high priority on fostering a culture of lifelong learning as SDG 4 suggests, providing opportunities for educator professional development, and safeguarding students' private information. Partnerships based on mutual respect and cooperation can strengthen university-business collaboration. The guest editorial proposed an in-depth discussion between advocates of "cold" technology and "warm" humanity to help teachers and students better comprehend the opportunities and challenges presented by recent technological advancements, such as the big data explosion and the AI revolution (Yang, 2021).

# 3 Challenges and Risks of AI

Every revolution comes with challenges and risks. Despite AI's many potential benefits, the technology still faces several obstacles and has certain constraints. Misconceptions regarding AI's capabilities, both methodological and epistemological, have yet to reach their full potential. These are a direct product of the constraints imposed by the social setting in which ML applications are developed. Finally, they are the result of the current state of artificial intelligence technology (Hagendorff, 2019).

To begin, the structure, methods, teaching style, and learning content were all altered as an outcome of changes made by AI. Promoting digital literacy to build the capacity to achieve sustainable development goals is a huge challenge because doing so requires aligning and harmonizing several distinct components. The use of AI-powered technology has the potential to exacerbate existing disparities in one's odds of achieving success (Digest, n.d.).

India, which has the world's largest population, confronts a second challenge in providing its citizens with cost-effective access to technology, smart devices, and the internet, which provide access to educational resources that are not restricted by geographical boundaries. On the one hand, having access to education that is of high quality is essential for the accomplishment of SDG 4. On the other hand, the skills, and competencies necessary to make successful use of digital technology in educational contexts are also essential and need to be inclusive and equitable. This is because educational settings are becoming increasingly reliant on the use of digital technology.

Third, there is a risk that AI-driven technologies will lead to an overly mechanized classroom environment. AI systems are typically trained to draw conclusions based on data, so they may stifle student creativity and flexibility in the classroom. If AI systems in schools do not adapt to the needs of individual students, it could lead to a very inefficient and unresponsive system.

Fourth, there is a chance that additional scrutiny may be put on children as a result of AI-driven surveillance tools. Tracking students' movements and activities with AI technologies could endanger their privacy and safety. Finally, individuals may become less analytical as a result of AI-driven systems. Students may develop an unhealthy reliance on AI systems because of the possibility that they will be used repeatedly to complete mundane tasks. This could lead to a decrease in pupils' ability to think critically, which is crucial for success in school.

## 4 Conclusion

The globe is undergoing a dynamic digital shift, and for its residents to remain on the same page during this time of transition, it is imperative that they adopt and adapts to the wave of new technology. The ability to recognize a person's voice and face is rapidly becoming more commonplace. Voice assistants like Siri and Alexa are increasingly becoming available in homes around the country. Chatbots are assisting multinational corporations in managing their relationships with their customers. It is now time for developing countries to begin recognizing the significance of artificial intelligence (AI) in our everyday lives. To alleviate the current crisis confronting the world's educational system, artificial intelligence (AI) must be incorporated into both elementary and secondary schools.

AI has proved its mettle and has become useful in several fields like health, environment, sanitation, etc. By leveraging AI technology in the field of education various stakeholders of society can work together to create a future where quality education will be accessible to all regardless of geographical location or socio-economic background which in turn will help in realizing the SDGs. It is also essential to ensure responsible and ethical use of AI by prioritising human collaboration, and continuous adaptation of technology to meet the evolving needs of the learners. With these priorities in mind AI in education can contribute significantly to achieving sustainable development goals shaping a brighter future for global education.

India is slowly becoming a torch bearer in the incorporation of AI technology in a variety of industries, even though it is experiencing several obstacles. The vision of India to become a powerhouse in the future decade is highlighted by the initiatives, plans, and provisions implemented by the government of India to incorporate AI into the heart of the education system. According to Sustainable Development Goal 4, the collaborations that the government has formed with public and private organizations and HEIs will encourage young people to work together and learn from one another in an educational setting that is more equitable. These new capabilities earned out of the new AI curriculum will boost new enterprises and start-ups in India that can act as sustainable green businesses for future generations (Trikha & Kaushik, 2023).

The future of AI in education requires policymakers to develop an enabling, inclusive, and multidimensional framework to monitor and promote sustainable development for future generations. In addition, technologies must be governed by sound theoretical frameworks and subjected to extensive empirical testing. The future of AI in education is promising and has the potential to significantly transform how we learn and teach. With the growing prevalence of Metaverse as an AI tool for immersive learning, along with numerous other AI tools (McFarland, 2023), the distant future is nothing short of these tools becoming learners' first choice.

## References


Aayog N (2018). *National Strategy for Artificial Intelligence.* NITI Aayog. Retrieved from https://niti.gov.in/sites/default/files/2019-01/NationalStrategy-for-AI-Discussion-Paper.pdf

Aayog N (2021). *RESPONSIBLE AI #AIForAll.* NITI Aayog. Retrieved from https://www.niti.gov.in/sites/default/files/2021-08/Part2-Responsible-AI-12082021.pdf

Ahaskar A (2023). Retrieved May 12, 2023, from https://www.techcircle.in/2023/04/12/india-ranks-5th-in-countries-with-most-ai-investments

Alam A (2021) Possibilities and Apprehensions in the Landscape of Artificial Intelligence in Education. *International Conference on Computational Intelligence and Computing Applications (ICCICA).* https://doi.org/10.1109/ICCICA52458.2021.9697272

Alam A (2022) Employing Adaptive Learning and Intelligent Tutoring Robots for Virtual Classrooms and Smart Campuses: Reforming Education in the Age of Artificial Intelligence. Advanced Computing and Intelligent Technologies. Springer, Singapore, pp 395–406

Chassignol H (2018) Artificial Intelligence trends in education: a narrative overview. Procedia Computer Science 136:16–24. https://doi.org/10.1016/j.procs.2018.08.233

Delhi P (2023). Retrieved May 29, 2023, from Press Information Bureau: https://pib.gov.in/Pressreleaseshare.aspx?PRID=1926942

Digest DT (n.d.). Retrieved from https://www.bonnalliance-icb.de/wp-content/uploads/Thinkathon-digest_SDG-4.pdf

*Education 4.0 India.* (2022). UNICEF. Switzerland: World Economic Forum. Retrieved June 10, 2023, from www3.weforum.org/docs/WEF_Education_4.0_India_Report_2022.pdf

Hagendorff EA (2019) 15 challenges for AI: or what AI (currently) can't do. *AI & Society,* 355–365 https://doi.org/10.1007/s00146-019-00886-y

Jain S (2023). Retrieved from Business Insider: https://www.businessinsider.in/budget/news/government-to-set-up-three-centres-of-excellence-in-artificial-intelligence/articleshow/97520242.cms

McFarland A (2023). *unite.ai*. Retrieved June 14, 2023, from https://www.unite.ai/10-best-ai-tools-for-education/

Meit Y (n.d.). https://indiaai.gov.in. Retrieved May 20, 2023, from https://indiaai.gov.in

Miao EA (2021) *AI and education: guidance for policy-makers*. UNESCO https://doi.org/10.54675/PCSP7350

Panjaburee NK (2023) Trends and development of technology-enhanced personalized learning in science education: a systematic review of publications from 2010 to 2022. J Comput Edu. https://doi.org/10.1007/s40692-023-00276-w

Pedro FE (2019). Retrieved May 3, 2023, from https://www.edu-links.org/resources/artificial-intelligence-education

Pichai S (2022). Optimism for India's digital future. Retrieved May 9, 2023, from https://blog.google/inside-google/message-ceo/optimism-india-digital-future/

Ray PP (2023) ChatGPT: A comprehensive review on background, applications, key challenges, bias, ethics, limitations and future scope. Internet of Things and Cyber-Physical Systems 3:121–154. https://doi.org/10.1016/j.iotcps.2023.04.003

Technology M. o. (n.d.). Retrieved from https://futureskillsprime.in/about-us

Tlili A et al (2023) Smart Learning Environments. https://doi.org/10.1186/s40561-023-00237-x

Trikha A, Kaushik MB (2023) Green Entrepreneurship: A Way Ahead to Sustainable Development. Futuristic Technology Perspectives on Entrepreneurship and Sustainable Innovation. IGI Global, Pennsylvania, pp 91–103

Uunona GN (2023) Leveraging Ethical Standards in Artificial Intelligence Technologies: A Guideline for Responsible Teaching and Learning Applications. Handbook of Research on Instructional Technologies in Health Education and Allied Disciplines. IGI Global, Pennsylvania, p 21

Walia, P. (n.d.). *turnthebus.org*. Retrieved from, June 5, 2023 https://www.turnthebus.org/blog/how-can-artificial-intelligence-help-reach-the-sdg-for-education

Wayne Holmes, et al. (2019). *Artificial Intelligence for Sustainable Development Synthesis Report. Mobile Learning Week 2019*. Paris: United Nations Educational, Scientific and Cultural Organization (UNESCO). Retrieved June 2, 2023, from https://www.researchgate.net/publication/361653445_Artificial_Intelligence_for_Sustainable_Development_Synthesis_Report_Mobile_Learning_Week_2019

Yang SJ (2021) Precision Education - A New Challenge for AI in Education. J Educ Technol Soc 24(1):105–108

# Biotechnology and Digital Health Advancements in the 21st Century: A Case Study on the Adoption of Telemedicine for Managing Chronic Diseases

**Jhansi Bharathi Madavarapu, Pawan Whig, Balaram Yadav Kasula, and Sandeep Kautish**

**Abstract** Biotechnology and digital health represent dynamic and transformative domains with the potential to reshape the healthcare landscape. Biotechnology harnesses the power of living organisms and their components to yield valuable products and processes, while digital health leverages digital technologies to enhance health and healthcare delivery. This chapter delves into the confluence of biotechnology and digital health, shedding light on the profound advancements and challenges in the twenty-first century. In addition to this exploration, we will present a compelling case study illustrating the adoption of telemedicine for managing chronic diseases, facilitated by synthetic patient data generated through Synthea™. The narrative will encompass the role of biotechnology in digital health, encompassing genetic engineering, gene editing, and synthetic biology. We will also explore how digital technologies, including mobile health applications, telemedicine, and wearable devices, are seamlessly integrated to elevate patient care. Furthermore, this chapter will take a comprehensive look at the ethical and regulatory dimensions that surround the use of biotechnology and digital health technologies within the healthcare sector. Through the inclusion of the case study, we aim to provide a tangible, real-world illustration of the profound impact and potential ethical considerations inherent to this intersection of innovation.

J. B. Madavarapu · B. Y. Kasula
Department of Information Technology, University of the Cumberlands, Williamsburg, KY, USA
e-mail: jmadavarapu0693@ucumberlands.edu

P. Whig (✉)
Dean Research, Vivekananda Institute of Professional Studies-TC, New Delhi, India
e-mail: pawanwhig@gmail.com

S. Kautish
LBEF Campus, Kathmandu, Nepal



W. Leal Filho et al. (eds.), *Digital Technologies to Implement the UN Sustainable Development Goals*, World Sustainability Series,
https://doi.org/10.1007/978-3-031-68427-2_5

**Keywords** Biotechnology · Digital health · Genetic engineering · Gene editing · Synthetic biology · Mobile health applications · Telemedicine · Wearable devices · Ethics · Regulation

## 1 Introduction

The twenty-first century has witnessed remarkable progress in two transformative fields that are shaping the landscape of healthcare and medicine: biotechnology and digital health (Assidi et al., 2022). These twin pillars of innovation have not only expanded our understanding of the human body and its intricacies but have also revolutionized the way we diagnose, treat, and manage diseases (Bhatti et al., 2022). The synergy between biotechnology and digital health has ushered in an era of unprecedented advancements, offering new hope and opportunities for patients and healthcare providers alike (Hird et al., 2016; Petersen, 2018; Raijada et al., 2021). Biotechnology, with its roots in the manipulation of biological processes and organisms, has paved the way for groundbreaking discoveries in genomics, personalized medicine, and regenerative therapies (Crawford and Serhal, 2020). As we delve deeper into the genetic code of living organisms, we are gaining a more profound understanding of the underlying mechanisms of diseases and the potential for targeted interventions (Tu et al., 2020). This knowledge has spurred the development of innovative biopharmaceuticals, gene therapies, and precision medicine approaches, all of which hold the promise of more effective and tailored healthcare solutions (Filkins et al., 2016).

In parallel, the digital health revolution has harnessed the power of technology, data, and connectivity to enhance the delivery and management of healthcare (Desveaux et al., 2019). Smart devices, health apps, telemedicine, and wearable sensors have become integral components of modern healthcare, allowing for real-time monitoring, remote consultations, and improved patient engagement (Gordon and Catalini, 2018). Furthermore, the integration of artificial intelligence and machine learning algorithms has given rise to predictive analytics and decision support systems, enabling healthcare professionals to make more informed and timely decisions (Pawelek et al., 2022). However, as we stand on the precipice of a new era in healthcare, these advancements also bring forth a host of challenges (New, 2018). Ethical concerns surrounding privacy and data security in the digital health sphere, the cost and accessibility of cutting-edge biotechnology treatments, and the regulatory frameworks governing these rapidly evolving fields are among the issues that demand careful consideration (Kavakiotis et al., 2017). In this comprehensive exploration, we will delve into the remarkable advancements and transformative potential of biotechnology and digital health in the twenty-first century (Lall et al., 2019). We will also critically examine the challenges that must be addressed to ensure that these innovations are harnessed for the benefit of all, leaving no one behind in the quest for improved healthcare (Bettinger, 2015). The biotechnology and digital health nexus is poised to define the healthcare landscape in the years to come, and our understanding of its accomplishments and constraints is crucial for shaping a healthier and more equitable future (Altman et al., 2011).

## 2 Biotechnology: Definitions and Overview

Biotechnology is a multidisciplinary field of science and technology that harnesses biological systems, living organisms, or their derivatives to develop products and processes that improve human life and address various challenges (Banner, 2017). It encompasses a wide range of applications, from agriculture and environmental conservation to healthcare and pharmaceuticals. At its core, biotechnology involves the manipulation and modification of biological materials to create new and innovative solutions.

The field of biotechnology can be categorized into several key areas:

1. **Medical Biotechnology:** This branch focuses on using biological systems to develop diagnostics, pharmaceuticals, and therapies. It includes areas such as genomics, personalized medicine, gene therapy, and the production of biopharmaceuticals (Dunleavy et al., 2019).
2. **Agricultural Biotechnology:** Agricultural biotechnology aims to enhance crop and livestock production, making agriculture more sustainable and resilient. It includes genetically modified organisms (GMOs), crop breeding techniques, and pest-resistant plants.
3. **Environmental Biotechnology:** This sector concentrates on addressing environmental challenges through biological means. It includes wastewater treatment, bioremediation, and the development of eco-friendly technologies (Greaves et al., 2019).
4. **Industrial Biotechnology:** Industrial biotechnology involves using biological processes to manufacture chemicals, materials, and fuels more sustainably. It includes the production of biofuels, bio-based materials, and enzymes for various industries (Wickramasinghe et al., 2022).
5. **Bioinformatics:** Bioinformatics is the application of computational techniques to analyze and interpret biological data, particularly in genomics and proteomics research (Seçkin et al., 2016).

## 3 Digital Health: Definitions and Overview

Digital health, also known as eHealth or HealthTech, represents the intersection of healthcare and technology (Colombo et al., 2020). It encompasses a broad range of digital solutions designed to improve healthcare delivery, patient care, and health outcomes. Digital health leverages information technology, connectivity, and data analytics to provide innovative and more efficient healthcare services (Hood and Flores, 2012).

Key components of digital health include:

1. **Telemedicine and Telehealth:** These technologies enable remote patient consultations, diagnosis, and treatment through videoconferencing, virtual clinics, and mobile health apps. Telemedicine has played a vital role in extending healthcare access, especially in remote areas.
2. **Wearable Devices:** Wearable health technologies, such as fitness trackers and smartwatches, monitor various health metrics like heart rate, sleep patterns, and physical activity. They offer individuals real-time insights into their well-being.
3. **Health Information Systems:** Electronic Health Records (EHRs) and Health Information Exchanges (HIEs) streamline the storage and sharing of patient data among healthcare providers, improving care coordination and reducing errors.
4. **Remote Monitoring:** Remote patient monitoring devices collect data from patients' homes and transmit it to healthcare providers, allowing for continuous tracking of chronic conditions and early intervention.
5. **Artificial Intelligence and Data Analytics:** Machine learning and AI algorithms analyze vast datasets to make predictions, aid in diagnostics, and provide treatment recommendations. These tools have the potential to revolutionize healthcare decision-making.
6. **Mobile Health Apps:** Mobile applications provide users with health and wellness information, symptom tracking, medication reminders, and access to telehealth services. They empower individuals to take a more active role in managing their health.

Biotechnology and digital health often intersect, with biotechnological advancements contributing to the development of new medical treatments and diagnostics, and digital health technologies enhancing the accessibility and efficiency of healthcare delivery (Hood and Flores, 2012, c, S., 2016). This synergy between biotechnology and digital health holds significant promise for the future of medicine and patient care (Whig et al., 2024a, 2024b).

## 4 Role of Biotechnology in Digital Health

Biotechnology plays a pivotal role in advancing digital health by contributing to the development of cutting-edge tools, techniques, and therapies as shown in Fig. 1. Below, we explore how biotechnology, specifically genetic engineering, gene editing, and synthetic biology, intersects with and supports the digital health landscape:

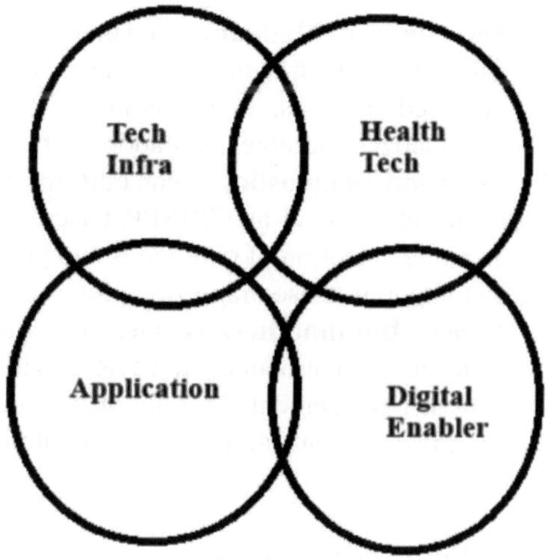


**Fig. 1** Role of Biotechnology in Digital Health

## *4.1 Genetic Engineering*

Genetic engineering is a fundamental biotechnological technique that involves the modification of an organism's DNA. In the context of digital health, genetic engineering is instrumental in several ways:

1. **Genomic Data Analysis:** Genetic engineering techniques enable the sequencing and profiling of an individual's genetic code. This genomic information can be used for disease risk assessment, personalized medicine, and pharmacogenomics. Digital health platforms use this genetic data to provide personalized health recommendations and medication guidance.
2. **Biomarker Discovery:** Genetic engineering is vital for identifying and creating biomarkers associated with various diseases. These biomarkers serve as diagnostic tools, and digital health applications can integrate them for early disease detection and monitoring.
3. **Therapeutic Development:** Genetic engineering is crucial for developing gene therapies, which are becoming increasingly important in treating genetic disorders. Digital health platforms support the delivery and monitoring of these therapies, ensuring patients receive the right treatment and experience optimal outcomes.

## *4.2 Gene Editing*

Gene editing is a more precise genetic manipulation technique, allowing for the addition, deletion, or modification of specific DNA sequences. It has significant implications for digital health:

1. **Personalized Medicine:** Gene editing, particularly through techniques like CRISPR-Cas9, has the potential to correct genetic mutations responsible for inherited diseases. In the context of digital health, this translates to highly personalized therapies that address the root causes of genetic disorders.
2. **Precision Diagnostics:** Gene editing can be used to create highly specific diagnostic tools, such as CRISPR-based diagnostic tests, which enable rapid and accurate detection of pathogens and genetic markers. Digital health applications can integrate these diagnostics for convenient testing and result reporting.
3. **Cancer Immunotherapy:** Gene editing is advancing the development of precision cancer treatments, like CAR-T cell therapy. Digital health platforms support patient management throughout the course of treatment, including monitoring for adverse events and assessing treatment responses.

## *4.3 Synthetic Biology*

Synthetic biology involves designing and constructing biological components or systems for specific purposes. In digital health, synthetic biology contributes in the following ways:

1. **Biosensors:** Synthetic biology allows for the creation of biosensors that can detect specific molecules or changes in biological conditions. These biosensors can be integrated into wearable and implantable digital health devices to monitor various health parameters in real time.
2. **Bioinformatics and Data Processing:** Synthetic biology techniques can be employed to design biological circuits and systems for processing biological data. This is vital in the development of advanced algorithms and analytical tools for managing large-scale health data, as seen in genomics and proteomics.
3. **Drug Development:** Synthetic biology is used to engineer microorganisms for the production of bio-based pharmaceuticals and therapies. Digital health platforms support the development and monitoring of these biopharmaceuticals and their administration to patients.

The synergy between biotechnology and digital health is a dynamic force, enhancing healthcare by providing more personalized, efficient, and effective solutions. These biotechnological approaches are integral to the digital health ecosystem, enabling precise diagnostics, targeted therapies, and real-time health monitoring, ultimately leading to improved patient outcomes and well-being.

# 5 Role of Digital Technologies in Biotechnology and Healthcare

Digital technologies have revolutionized the fields of biotechnology and healthcare by providing innovative solutions that enhance patient care, diagnostics, and research [23]. Here, we explore the roles of mobile health applications, telemedicine, and wearable devices in these domains:

## *5.1 Mobile Health Applications*

Mobile health applications, commonly known as mHealth apps, have become indispensable tools in the healthcare industry. They offer a wide range of benefits in biotechnology and healthcare:

1. **Remote Monitoring:** mHealth apps enable patients to monitor various health parameters, such as blood pressure, glucose levels, and heart rate, using connected devices. This data is crucial for managing chronic conditions and ensuring timely interventions.
2. **Medication Management:** These apps provide medication reminders and dosage tracking, promoting medication adherence and reducing the risk of errors. They also allow healthcare providers to monitor patients' medication compliance remotely.
3. **Telehealth:** Many mHealth apps incorporate telehealth features, allowing patients to schedule virtual consultations with healthcare professionals. This enhances access to healthcare services, especially for those in remote areas or with mobility challenges.
4. **Health Records:** Patients can maintain electronic health records on these apps, providing a convenient and secure way to store and access their medical history, prescriptions, and test results. This simplifies information sharing between patients and healthcare providers.
5. **Wellness and Lifestyle Management:** mHealth apps provide information and tools for maintaining a healthy lifestyle, including fitness tracking, nutrition advice, and mental health support. They empower individuals to take proactive steps to improve their well-being.

## *5.2 Telemedicine*

Telemedicine involves the remote provision of healthcare services through digital communication technologies. Its role in biotechnology and healthcare is profound:

1. **Expanded Access:** Telemedicine extends healthcare access to underserved populations and remote areas, reducing healthcare disparities. Patients can consult specialists without the need for long-distance travel.
2. **Real-Time Consultations:** Telemedicine facilitates real-time interactions between patients and healthcare providers, making it possible to diagnose and manage acute conditions promptly.
3. **Chronic Disease Management:** Patients with chronic conditions can benefit from continuous monitoring and remote consultations. Telemedicine platforms integrate with wearable devices and mHealth apps to provide comprehensive care.
4. **Elderly Care:** Telemedicine is instrumental in providing healthcare services to the elderly, who may have difficulty accessing physical healthcare facilities. Remote monitoring and virtual consultations ensure their health and well-being.
5. **Cost Savings:** Telemedicine reduces the cost of healthcare delivery by minimizing the need for physical infrastructure and streamlining administrative processes. This benefits both patients and healthcare systems.

### *5.3 Wearable Devices*

Wearable devices are becoming increasingly popular for monitoring and improving health. Their roles in biotechnology and healthcare include:

1. **Continuous Monitoring:** Wearable devices, such as fitness trackers and smartwatches, continuously monitor vital signs, physical activity, and sleep patterns. This data is valuable for tracking overall health and well-being.
2. **Chronic Disease Management:** Wearables can be tailored to monitor specific health parameters, such as glucose levels for diabetics or heart rate for cardiac patients. This data aids in the management of chronic diseases.
3. **Remote Patient Monitoring:** Healthcare providers can remotely monitor patients with chronic conditions, ensuring early intervention and reducing hospital readmissions. Wearables transmit data in real time.
4. **Preventive Healthcare:** Wearables encourage users to adopt healthier lifestyles by providing insights into their daily habits and activities, promoting physical fitness and mental well-being.
5. **Clinical Research:** Wearable Devices Are Increasingly Used in Clinical Trials and Medical Research to Collect Objective Data, Enhancing the Accuracy of Studies.

Digital technologies like mobile health apps, telemedicine, and wearable devices have fundamentally transformed healthcare by improving access, patient engagement, and the quality of care [24]. These tools have also facilitated the integration of biotechnology innovations into everyday healthcare practices, creating a more interconnected and patient-centered approach to medicine.

## 6 Advancements and Applications of Biotechnology and Digital Health in Healthcare

The dynamic convergence of biotechnology and digital health has yielded a host of advancements and applications that are shaping the future of healthcare. These innovations offer new ways to diagnose, treat, and manage diseases, improve patient outcomes, and enhance the overall healthcare experience. Here, we explore some of the key advancements and applications in this realm:

### *6.1 Genomic Medicine and Precision Healthcare:*

Advancement: The sequencing of the human genome and the development of high-throughput DNA sequencing technologies have ushered in the era of precision medicine.

Applications:

- **Genomic Profiling:** Genetic testing and sequencing enable the identification of disease susceptibility and personalized treatment strategies.
- **Pharmacogenomics:** Tailoring drug therapies to an individual's genetic makeup, optimizing drug selection and dosages.
- **Cancer Genomics:** Identifying specific genetic mutations in cancer cells to develop targeted therapies and predict disease progression.

### *6.2 Gene Editing and Gene Therapy*

Advancement: The emergence of precise gene editing techniques, including CRISPR-Cas9, has revolutionized the ability to modify and correct genes.

Applications:

- **Treatment of Genetic Disorders:** Gene therapy is being used to treat genetic diseases like sickle cell anemia and cystic fibrosis.
- **Cancer Immunotherapy:** Genetically engineered immune cells, such as CAR-T cells, are employed in cancer treatment.
- **Gene Editing in Drug Discovery:** Modifying genes in model organisms to develop and test potential drug targets.

### *6.3 Telemedicine and Telehealth*

Advancement: Telemedicine has seen widespread adoption, driven by advancements in communication technology and healthcare regulations.

Applications:

- **Virtual Consultations:** Patients can consult with healthcare providers via video calls, expanding access to healthcare services.
- **Remote Monitoring:** Real-time monitoring of patients with chronic diseases, allowing for early intervention.
- **Mental Health Services:** Telehealth platforms offer mental health counseling and support.
- **Tele-ICU:** Remote monitoring of intensive care patients by critical care specialists.

## *6.4 Wearable Devices and Remote Monitoring*

Advancement: Wearable technology has become increasingly sophisticated, offering comprehensive health data monitoring.

Applications:

- **Fitness Tracking:** Wearables monitor physical activity, sleep patterns, and heart rate, motivating individuals to adopt healthier lifestyles.
- **Chronic Disease Management:** Remote monitoring of vital signs and glucose levels for patients with chronic conditions like diabetes.
- **Fall Detection:** Wearables can alert caregivers or emergency services in the event of a fall, especially beneficial for the elderly.

## *6.5 AI and Machine Learning in Diagnostics*

Advancement: Artificial intelligence and machine learning algorithms are being used to analyze large datasets and make predictions.

Applications:

- **Radiology:** AI assists radiologists in interpreting medical images, increasing diagnostic accuracy.
- **Pathology:** Machine learning models are used for pathology image analysis to detect diseases like cancer.
- **Drug Discovery:** AI accelerates drug development by identifying potential drug candidates and predicting their effectiveness.

## *6.6 Digital Health Records and Interoperability*

Advancement: Electronic health records (EHRs) and health information exchange (HIE) systems are becoming more comprehensive and interoperable.

Applications:

- **Improved Patient Care:** EHRs provide quick access to patient information, reducing medical errors and improving care coordination.
- **Data Sharing:** Interoperable systems allow for the seamless exchange of patient data between healthcare providers, enhancing patient care continuity.

### *6.7 3D Printing in Healthcare*

Advancement: 3D printing technology has evolved to create custom implants, prosthetics, and medical models.

Applications:

- **Custom Implants:** 3D printing is used to create patient-specific implants for orthopedic and dental surgeries.
- **Prosthetics:** Customized 3D-printed prosthetic limbs offer better comfort and functionality.
- **Surgical Planning:** 3D-printed anatomical models aid surgeons in planning complex procedures.

These advancements and applications of biotechnology and digital health are fundamentally transforming healthcare, making it more personalized, accessible, and effective. As technology continues to evolve, the potential for further innovations in healthcare remains vast, promising better patient outcomes and the continued advancement of medical science.

## 7 Challenges and Limitations of Biotechnology and Digital Health

The rapid evolution of biotechnology and digital health brings with it several challenges and limitations that need to be addressed for these innovations to be harnessed safely and effectively. Here, we discuss the technical challenges, ethical considerations, and regulatory challenges associated with these fields:

### *7.1 Technical Challenges*

1. **Data Security and Privacy:** The vast amounts of health data generated by digital health technologies raise concerns about data security and patient privacy. Data breaches and cyberattacks can compromise sensitive medical information, and maintaining strong security measures is a constant challenge.

2. **Interoperability:** Ensuring that different digital health systems can seamlessly exchange data remains a significant technical challenge. Lack of interoperability can impede the efficient sharing of patient information across healthcare providers and systems.
3. **Standardization:** The absence of universal standards for digital health data and technologies can hinder their widespread adoption. Establishing consistent standards is essential for compatibility and data integrity.
4. **Data Accuracy and Quality:** The accuracy and reliability of data generated by wearables, sensors, and health apps can vary. Ensuring data quality and consistency is a technical challenge in digital health.
5. **Technical Infrastructure:** Access to high-quality internet and digital devices is not uniform worldwide, limiting the reach of digital health technologies, particularly in remote or underserved areas.
6. **Complexity and Cost of Implementation:** Some digital health solutions, especially those involving advanced technologies like AI and big data analytics, can be complex to implement and costly for healthcare providers and institutions.

## *7.2 Ethical Considerations*

1. **Data Ownership and Consent:** Questions about who owns and controls health data and how informed consent is obtained from patients for data collection and sharing raise ethical concerns. Patients should have agency over their health information.
2. **Patient Autonomy:** The use of AI and algorithms in healthcare decision-making may raise concerns about patient autonomy. Decisions driven by algorithms may limit the role of healthcare professionals and patients in medical choices.
3. **Bias and Discrimination:** The algorithms used in digital health technologies can inherit biases from the data they are trained on. This can lead to discrimination in diagnosis and treatment recommendations, particularly against marginalized or underrepresented groups.
4. **Informed Decision-Making:** Patients may not fully understand the implications of using digital health tools or the limitations of AI-driven diagnoses. Ensuring that individuals make informed decisions is an ethical concern.
5. **End-of-Life and Palliative Care:** Ethical dilemmas arise in end-of-life and palliative care settings where technology may prolong life against a patient's wishes, raising questions about the balance between preserving life and respecting patient autonomy.

## *7.3 Regulatory Challenges*

1. **Regulatory Frameworks:** Developing and adapting regulatory frameworks to keep pace with rapidly evolving biotechnology and digital health innovations is

a significant challenge. Regulators must balance the need for safety and efficacy with the drive for innovation.
2. **Data Protection Laws:** Ensuring compliance with data protection and privacy laws, such as GDPR (General Data Protection Regulation) in the European Union, is a regulatory challenge for companies and healthcare organizations operating globally.
3. **Lack of Global Standards:** The absence of uniform international standards complicates the regulatory environment for digital health, especially when devices and services are marketed globally.
4. **Medical Device Classification:** Many digital health technologies, such as mobile apps and wearable devices, may fall into regulatory gray areas, making it challenging to determine the appropriate level of oversight.
5. **Regulatory Delays:** The time required to obtain regulatory approvals can slow down the introduction of new biotechnology and digital health products to the market, potentially limiting timely access to innovative solutions.

Addressing these challenges and ethical considerations while developing and implementing regulatory frameworks that strike the right balance is crucial for realizing the full potential of biotechnology and digital health while safeguarding patient rights, privacy, and health outcomes.

## 8 Case Study

Synthea is an open-source tool designed to generate synthetic patient data and simulate the medical history of these "virtual" patients. This tool plays a crucial role in healthcare research and case studies, as it provides researchers and professionals with realistic yet entirely synthetic patient data and health records. Here are some key points about Synthea:

1. Synthetic Patient Data Generation: Synthea creates synthetic patients with diverse characteristics, including age, gender, medical conditions, and medical history. These simulated patients can mimic real-world individuals, making them valuable for a wide range of healthcare research and case study applications.
2. Realistic, But Not Real: The patient data generated by Synthea is designed to be highly realistic, reflecting the demographics and health conditions of a diverse population. However, it's essential to emphasize that this data is entirely synthetic and does not represent real individuals. This aspect ensures privacy and compliance with healthcare data regulations.
3. Use in Case Studies: Synthea is a valuable resource for conducting case studies, clinical trials, software testing, and other healthcare research initiatives. It provides researchers with standardized, consistent, and customizable patient data to evaluate and test healthcare systems, interventions, and technologies.

4. Comprehensive Health Records: The synthetic patient data created by Synthea includes comprehensive health records, encompassing medical history, treatments, medications, laboratory results, and more. This breadth of information allows for in-depth analysis in various healthcare scenarios.
5. Customization: Synthea offers customization options, allowing researchers to define parameters for the generated synthetic patients. This flexibility ensures that the generated data aligns with specific research objectives and scenarios.
6. Privacy and Compliance: The use of synthetic patient data from Synthea helps protect the privacy and confidentiality of real patients while complying with healthcare data regulations such as HIPAA (Health Insurance Portability and Accountability Act) in the United States.
7. Open Source: Synthea is open-source, meaning that its source code is available for anyone to view, modify, and use. This fosters collaboration and transparency in the healthcare research community.

Synthea is a valuable tool for generating synthetic patient data and health records that closely resemble real-world scenarios. It serves as a crucial resource for healthcare researchers, professionals, and organizations looking to conduct case studies, test healthcare systems, and perform various analyses without compromising patient privacy or data security.

The case study aims to assess the effects of the adoption of telemedicine on the management of chronic diseases using synthetic patient data generated by Synthea. Telemedicine has gained popularity as a remote healthcare delivery method, but its impact on chronic disease management remains an area of interest and study.

## *8.1 Methodology*

1. **Data Source:** Synthetic patient data was created using Synthea™ to simulate a diverse patient population with chronic conditions such as diabetes, hypertension, and chronic obstructive pulmonary disease (COPD).
2. **Patient Selection:** A representative sample of synthetic patients with chronic diseases was selected from the Synthea dataset.
3. **Intervention:** A virtual telemedicine program was introduced for selected synthetic patients, allowing them to schedule remote consultations with healthcare providers and monitor their chronic conditions using digital health tools.

## *8.2 Key Findings*

1. **Improved Access to Care:** The introduction of telemedicine resulted in increased access to care, particularly for patients in remote or underserved areas. Synthetic patients reported reduced travel time and increased convenience.

2. **Enhanced Chronic Disease Monitoring:** Patients using telemedicine showed improved adherence to treatment plans and better control of their chronic conditions. Real-time monitoring through wearable devices contributed to early intervention and improved outcomes.
3. **Cost Savings:** The case study found that the use of telemedicine led to cost savings for patients in terms of reduced transportation costs and time away from work. Additionally, it lowered healthcare expenses by reducing emergency room visits.
4. **Patient Satisfaction:** Synthetic patients participating in telemedicine reported high levels of satisfaction with the convenience and quality of care received. They appreciated the ability to consult with healthcare providers from the comfort of their homes.

**Discussion:** The case study demonstrates that the adoption of telemedicine has a positive impact on chronic disease management, as evidenced by improved patient outcomes, cost savings, and patient satisfaction. The use of synthetic patient data from Synthea™ allowed for the analysis of a diverse patient population without compromising patient privacy or data security.

**Conclusion:** This case study illustrates the value of telemedicine in improving access to care and enhancing chronic disease management. The use of synthetic patient data, as generated by Synthea™, provides a powerful tool for healthcare research, enabling the assessment of healthcare interventions, technologies, and strategies in a controlled and privacy-compliant manner. The findings of this case study may inform healthcare policy and decision-making related to telemedicine adoption for chronic disease management.

The code is creating a count plot that shows the distribution of patient genders in the patient_df dataset. Each gender category will be represented on the x-axis, and the number of occurrences of each gender will be represented on the y-axis, with "Patient gender" as the label for the y-axis as shown in Fig. 2. This type of plot provides a visual summary of the gender distribution within the patient dataset.

The resulting chart will show two lines, one in blue and one in orange, with points representing the count of male and female patient births over different birth years as shown in Fig. 3. The legend will indicate which line corresponds to which gender. Researchers and analysts can use this visualization to observe trends in patient births over the years and compare gender-based differences in these trends.

The function is expected to generate a visual representation of the distribution of encounter class codes in the DataFrame, helping users to better understand the data related to healthcare encounters as shown in Fig. 4.

The provided code generates a $4 \times 5$ grid of subplots, each displaying a distribution plot (using Seaborn's **sns.distplot**) for a unique set of resource codes and their corresponding values from an "observation_df" DataFrame. Resource code labels are extracted and displayed on the X-axes of each subplot. Different colors are used to distinguish the plots, and there are 20 such plots in total as shown in Fig. 5. The vertical distribution plots show the frequency of resource values, offering insights into the distribution patterns of these resources in the dataset. This visualization provides

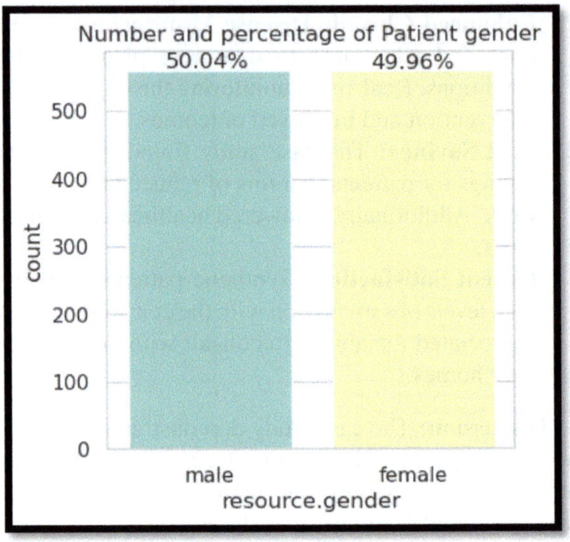


**Fig. 2** Number and percentage of patient gender

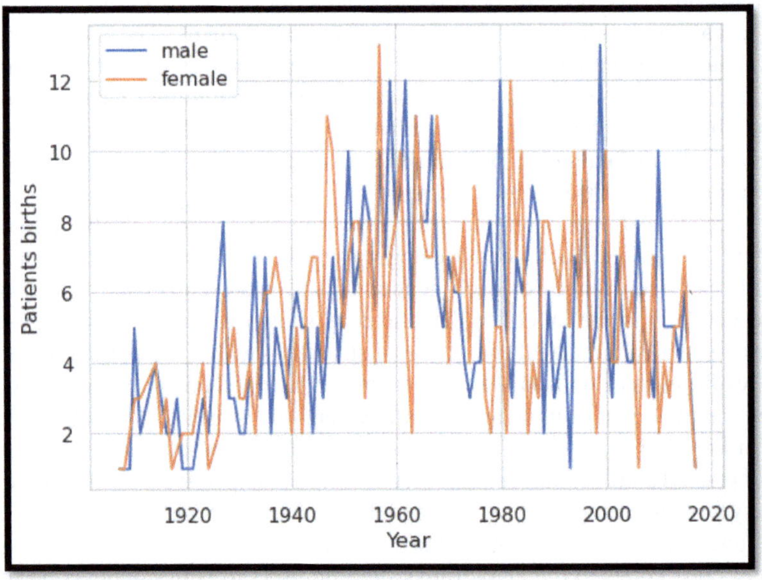


**Fig. 3** Patients Birth data

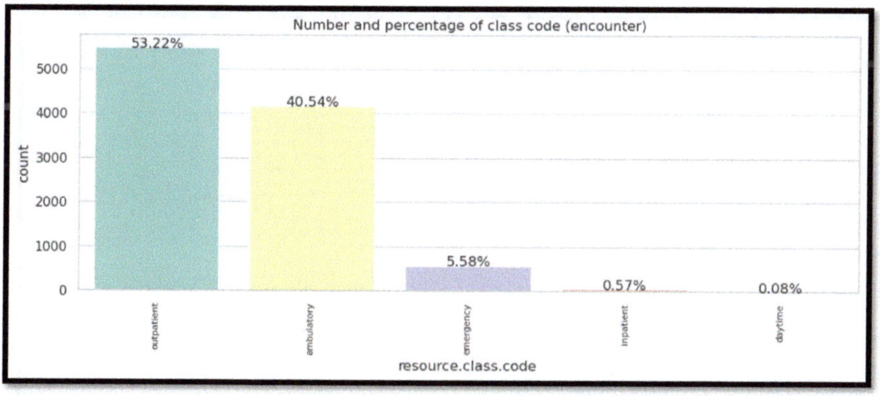


**Fig. 4** Number and percentage of class code encounter

a compact overview of the data, making it easier to compare the distributions of various resource codes, potentially revealing trends or variations in the dataset.

The provided code generates a series of 4 × 5 grid subplots for each of the top four patients in terms of the number of observations they have, as indicated by the "resource.subject.reference" column. For each selected patient, the code plots the time series of various resource values over multiple dates. The resource values are categorized by the "resource.code.display," and each subplot in the grid represents a specific resource code with its corresponding values over time as shown in Fig. 6. Different colors are used to distinguish the resource codes, and labels within the plots indicate the resource code and the associated quantity code. These time series plots provide a visual representation of how resource values change over time for each of the top four patients, potentially revealing trends or patterns in their healthcare observations.

## 9 Future Directions and Conclusion

The dynamic synergy of biotechnology and digital health has the potential to reshape the landscape of healthcare in profound ways. As we look to the future, several exciting developments and directions in these fields can be anticipated:

1. **Precision Medicine and Personalized Healthcare:** The integration of genomics, gene editing, and data analytics will continue to advance, enabling the delivery of highly individualized medical treatments and interventions.
2. **Telemedicine Evolution:** Telemedicine will evolve further, with improved virtual reality (VR) and augmented reality (AR) applications, offering more immersive remote healthcare experiences.

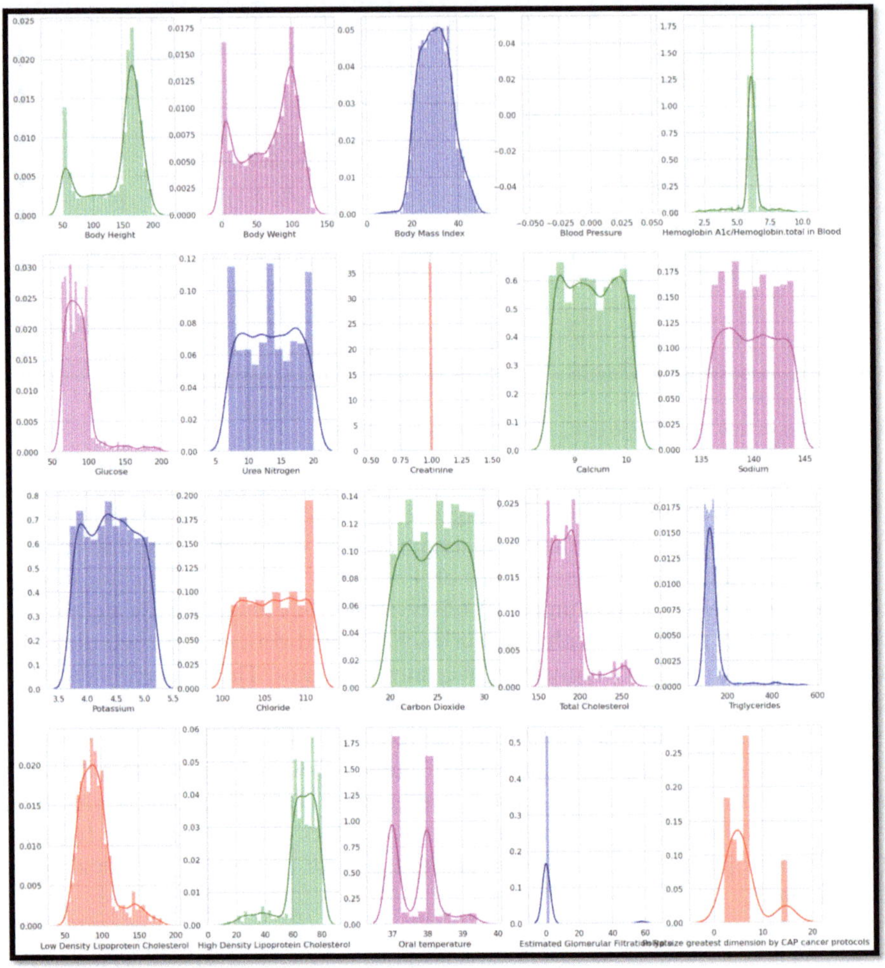


**Fig. 5** 4 × 5 grid of subplots displaying a distribution plot

3. **AI-Driven Diagnostics:** Artificial intelligence and machine learning will play an increasingly prominent role in diagnosing and treating diseases, as algorithms become more accurate and capable.
4. **Digital Therapeutics:** The development of prescription-based digital therapeutics, which use software to treat medical conditions, is on the horizon, offering drug-free alternatives for certain ailments.
5. **Remote Monitoring and Wearables:** Wearable devices will become even more advanced, offering more comprehensive health monitoring and real-time feedback for individuals and healthcare providers.

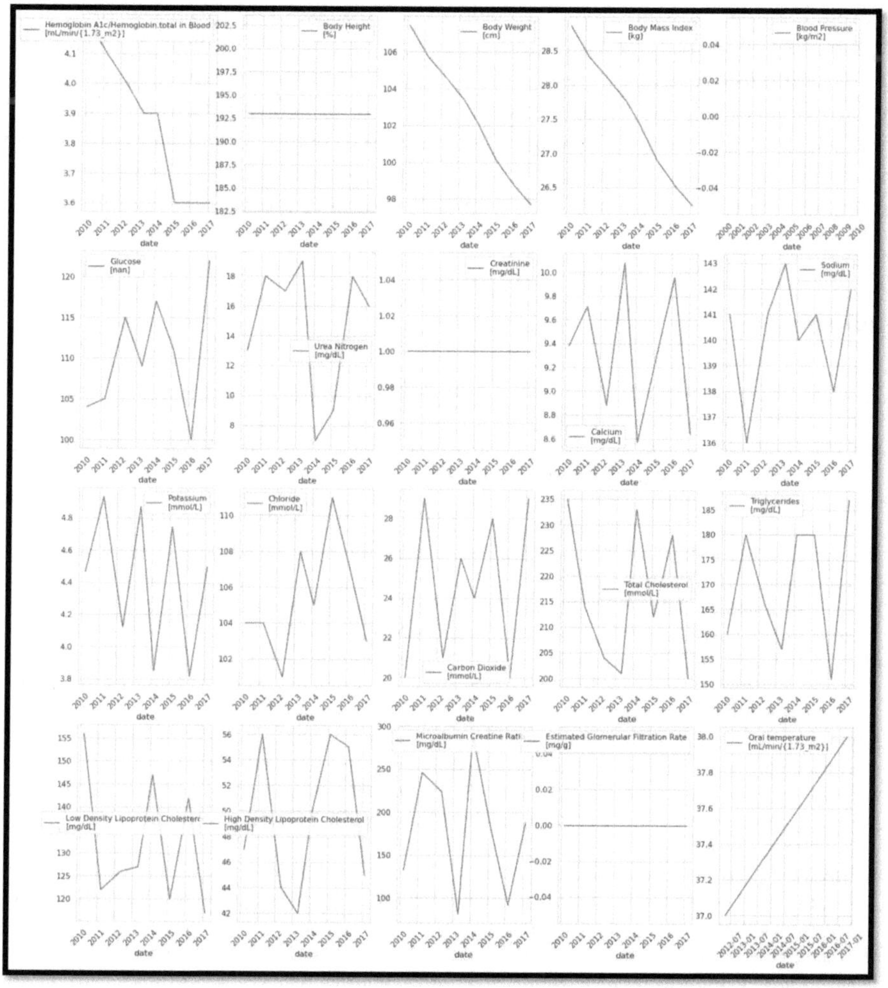


**Fig. 6** 4 × 5 grid subplots for each of the top four patients

6. **Regulatory Adaptation:** Regulatory bodies will work to adapt to the rapidly changing landscape, creating clear guidelines for the approval and monitoring of biotechnology and digital health solutions.
7. **Patient Empowerment:** Patients will have greater access to their own health data, fostering more active participation in their care and promoting a shift toward wellness and preventive healthcare.

In conclusion, the fusion of biotechnology and digital health is reshaping healthcare paradigms, making it more personalized, accessible, and efficient. As we navigate the challenges of data security, ethics, and regulation, we must remain vigilant and adaptable to ensure these innovations benefit patients and society as a whole. The future of healthcare is not just about addressing diseases but also about promoting well-being and early interventions that prevent illnesses. With biotechnology and digital health as our allies, we stand on the cusp of a transformative era that holds the promise of a healthier, more connected, and more informed world. As we continue to innovate and refine these technologies, we must do so with a commitment to ensuring that healthcare remains a human-centric, compassionate, and equitable endeavor, embracing the full spectrum of care from the cellular to the digital.

## References

Altman A, Hasegawa PM (eds) (2011) Plant biotechnology and agriculture: prospects for the 21st century. Academic press, Cambdrige

Assidi M, Buhmeida A, Budowle B (2022) Medicine and health of 21st Century: Not just a high biotech-driven solution. NPJ Genom Med 7(1):67

Banner O (2017) Communicative biocapitalism: The voice of the patient in digital health and the health humanities. University of Michigan Press, Ann Arbor, MI

Bettinger CJ (2015) Materials advances for next-generation ingestible electronic medical devices. Trends Biotechnol 33(10):575–585

Bhatti A, Rehman A, John P (2022) Challenges and opportunities in healthcare biotechnology. Biotechnology in Healthcare. Elsevier, Amsterdam, pp 321–342

Colombo F, Oderkirk J, Slawomirski L (2020). Health information systems, electronic medical records, and big data in global healthcare: Progress and challenges in OECD countries. Handbook of global health, 1–31.

Crawford A, Serhal E (2020) Digital health equity and COVID-19: the innovation curve cannot reinforce the social gradient of health. J Med Internet Res 22(6):e19361

Desveaux L, Soobiah C, Bhatia RS, Shaw J (2019) Identifying and overcoming policy-level barriers to the implementation of digital health innovation: Qualitative study. J Med Internet Res 21(12):e14994

Dunleavy G, Nikolaou CK, Nifakos S, Atun R, Law GCY, Tudor Car L (2019) Mobile digital education for health professions: systematic review and meta-analysis by the digital health education collaboration. J Med Internet Res 21(2):e12937

Filkins BL, Kim JY, Roberts B, Armstrong W, Miller MA, Hultner ML, Steinhubl SR (2016) Privacy and security in the era of digital health: what should translational researchers know and do about it? American J Trans Res 8(3):1560

Gordon WJ, Catalini C (2018) Blockchain technology for healthcare: facilitating the transition to patient-driven interoperability. Comput Struct Biotechnol J 16:224–230

Greaves RF, Bernardini S, Ferrari M, Fortina P, Gouget B, Gruson D, Kricka LJ (2019) Key questions about the future of laboratory medicine in the next decade of the 21st century: a report from the IFCC-Emerging Technologies Division. Clin Chim Acta 495:570–589

Hird N, Ghosh S, Kitano H (2016) Digital health revolution: perfect storm or perfect opportunity for pharmaceutical R&D? Drug Discovery Today 21(6):900–911

Hoffman S (2016) Electronic health records and medical big data. Cambridge University Press, Cambdrige, p 32

Hood L, Flores M (2012) A personal view on systems medicine and the emergence of proactive P4 medicine: predictive, preventive, personalized and participatory. New Biotechnol 29(6):613–624

Kavakiotis I, Tsave O, Salifoglou A, Maglaveras N, Vlahavas I, Chouvarda I (2017) Machine learning and data mining methods in diabetes research. Comput Struct Biotechnol J 15:104–116

Lall P, Rees R, Law GCY, Dunleavy G, Cotič Ž, Car J (2019) Influences on the implementation of mobile learning for medical and nursing education: qualitative systematic review by the digital health education collaboration. J Med Internet Res 21(2):e12895
New CA (2018). Digital health: creating a new growth industry for Australia.
Pawelek J, Baca-Motes K, Pandit JA, Berk BB, Ramos E (2022) The power of patient engagement with electronic health records as research participants. JMIR Med Inform 10(7):e39145
Petersen A (2018) Digital health and technological promise: A sociological inquiry. Routledge, UK
Raijada D, Wac K, Greisen E, Rantanen J, Genina N (2021) Integration of personalized drug delivery systems into digital health. Adv Drug Deliv Rev 176:113857
Seçkin G, Yeatts D, Hughes S, Hudson C, Bell V (2016) Being an informed consumer of health information and assessment of electronic health literacy in a national sample of internet users: validity and reliability of the e-HLS instrument. J Med Internet Res 18(7):e161
Tu J, Torrente-Rodríguez RM, Wang M, Gao W (2020) The era of digital health: A review of portable and wearable affinity biosensors. Adv Func Mater 30(29):1906713
Whig P, Velu A, Nadikattu RR, Sharma P (2024a) Reinforcement Learning for Automated Medical Diagnosis and Dynamic Clinical Regimes. Handbook of Research on Artificial Intelligence and Soft Computing Techniques in Personalized Healthcare Services. Apple Academic Press, NJ, pp 169–187
Whig P, Velu A, Nadikattu RR, Alkali YJ (2024b) Role of AI and IoT in Intelligent Transportation. Artificial Intelligence for Future Intelligent Transportation. Apple Academic Press, NJ, pp 199–220
Wickramasinghe N, Thompson BR, Xiao J (2022) The opportunities and challenges of digital anatomy for medical sciences: Narrative review. JMIR Med Edu 8(2):e34687

# Innovative Technology Adoption in Food Retail: Identifying Drivers for Embracing Vertical Farming to Achieve Sustainable Food Supply Chains

**Marie Asma Ben-Othmen, Cyprien Julienne, and Fahim Shaikh**

**Abstract** Vertical farming is a new and innovative technology that is gaining popularity worldwide. This method uses advanced technologies to increase the production of fresh food in urban areas. It is recognized as a way to achieve Sustainable Development Goals (SDGs), particularly SDG 11 (Sustainable Cities and Communities) and SDG 12 (Responsible Consumption and Production). This chapter overviews vertical farming and its various forms, ranging from hydroponic greenhouses to aeroponics, aquaponics, and plant factories. It also discusses how technology can optimize these production systems to create sustainable food supply chains.

In addition, the chapter presents the results of a survey conducted with 98 food retail managers operating in Rouen City, France. The survey aimed to understand the reasons behind their decision to implement vertical farming technology in their stores. The findings indicate that performance expectancy and facilitating conditions are critical factors that influence the behavioral intention of retail store managers to adopt vertical farming technologies. However, perceived risks, such as high energy consumption and costs, have a negative impact on their intention.

**Keywords** Sustainable food supply chains · Technology adoption · SDG11 · SDG12 · Vertical farming · Unified Theory of Acceptance and Use of Technology (UTAUT)

M. A. Ben-Othmen (✉)
INTERACT Research Unit–Innovation, Land Management, Agriculture, Agro-Industries, Knowledge, and Technology, UniLaSalle-France, Mont-Saint-Aignan, France
e-mail: marie-asma.benothmen@unilasalle.fr

C. Julienne · F. Shaikh
MSc Urban Agriculture & Green Cities, UniLaSalle-France, Mont-Saint-Aignan, France
e-mail: Cyprien.JULIENNE@etu.unilasalle.fr

F. Shaikh
e-mail: Fahim.SHAIKH@etu.unilasalle.fr



W. Leal Filho et al. (eds.), *Digital Technologies to Implement the UN Sustainable Development Goals*, World Sustainability Series,
https://doi.org/10.1007/978-3-031-68427-2_6

# 1 Introduction

With the burgeoning and densely populated urban landscape transforming cities into megacities globally, designers are striving to promote the idea that cities possess the potential to serve as prolific hubs for cultivating the substantial food needs of communities (Zareba et al., 2021). Mainly, vertical farming, as one of the digital technology applications in agriculture, is experiencing a surge in popularity with regard to its potential to optimize land utilization, leverage advanced technologies and smart devices, and enhance local food production within urban areas, aligning with the goals of efficiency and sustainability while securing access to fresh and local food production within city boundaries (Graamans et al., 2018).

Vertical farms are beneficial not only to urban farmers but also to food outlets and supermarkets, as they bridge the gap between city dwellers and the food they buy, reducing physical and emotional disconnection (Trivette, 2019). However, there are significant questions about the role of retail actors in expanding local food activities as they are increasingly recognized as essential gatekeepers between the food industry and final consumers (Burt & Sparks, 2002).

In France, most supermarkets depend on global supply chains that cater to seasonal demand and provide high-quality products at affordable prices. They achieve this by utilizing production facilities and distribution channels from diverse suppliers across the globe. However, as concerns over climate change have risen, so have concerns about global supply chains greenhouse gas emissions (Galli et al., 2017) along with the waste of resources due to global sourcing and shipping practices that translate into environmental and social costs (Albizzati et al., 2019). Many retailers have attempted to address sustainability challenges by implementing sustainability strategies. Despite the gap between the theory of sustainable supply chains and their practical implementation (Govindan, 2018), some food outlets are already adopting innovative technologies like indoor vertical farming to cater to consumers who prioritize locally sourced food. This shift aims to address environmental concerns and attract the segment of customers interested in supporting local food production. Despite this, the adoption rate in the retail industry remains low, and there is insufficient literature to explain this trend.

This chapter explores the potential of vertical farming as a sustainable solution for urban food supply chain strategies. The idea is that vertical farming can provide a way to move away from reliance on globalized supply chains by shifting the focus and operations. The methodology is grounded in a survey conducted with food retailers in France, including supermarkets, grocery stores, specialized food shops, and food cooperatives. The first section presents a brief overview of vertical farming technologies and their potential to achieve sustainability in food supply chains. Section 2 presents the theoretical background of the Unified Theory of Acceptance and Use of Technology that allows investigating variables that directly predict behavioral intention of technology adoption, including social influence, facilitating conditions, performance expectancy, and effort expectancy. The section also presents the hypothesis, and the survey design. The results are described in Sect. 4, followed by a discussion and the chapter's conclusion.

## 2 Setting the Stage

### *2.1 Defining Vertical Farming*

Vertical farming is a modern form of agriculture that involves growing crops indoors in a stacked arrangement within a controlled and automated environment (Petrovics & Giezen, 2022). Its primary aim is to produce food sustainably and efficiently, utilizing limited space and resources (Gan et al., 2023). This is achieved through controlled environment agriculture (CEA) techniques, which carefully monitor and control factors such as light, humidity, temperature, and nutrient levels. By doing so, the production system is more precise, resulting in improved plant growth and crop yield consistency while also reducing the need for water resources and pesticides. Numerous studies have highlighted the significant role that vertical farming could play in tackling environmental challenges such as land degradation and biodiversity loss, as well as addressing issues of food insecurity (Gibbons et al., 2020).

Historically, Patrick Blanc – a French scientist and botanist – is credited as the father of vertical gardens. In 1988, he patented a method for growing plants upright without soil, using a felt fabric that absorbs and retains water instead. The felt layer is attached to a PVC board with a special mesh to support the plant structure, and an irrigation system is used to deliver water to the plants (Fig. 1).

The vertical Garden by Patrick Blanc

Modern vertical farm

**Fig. 1** The vertical Garden by Patrick Blanc *versus* Modern Vertical Farm

In the early 2000s, a public health professor and ecologist named Dickson Despommier popularized the concept of vertical farming. According to Despommier's book "The Vertical Farm: Feeding the World in the twenty-first Century," vertical farming refers to using advanced greenhouse technologies to cultivate plants and animals for commercial purposes in skyscrapers (Despommier, 2010). This method can produce a variety of food items such as fish, poultry, fruits, and vegetables. Since Despommier's initial concept, modern vertical farming has progressed to include various techniques that can be adapted to different scales, such as hydroponic greenhouses, plant factories, container farms, and in-home systems.

### 2.1.1 Hydroponic Greenhouses

Hydroponic greenhouses are designed to maintain optimal growing conditions, including temperature, humidity, and lighting, while providing a stable and nutrient-rich growing medium for plants (Khan et al., 2021). In a hydroponic greenhouse, plants are typically grown in containers or channels filled with an inert growing substrate, such as rock wool or coconut coir, supporting the roots. The term "hydroponics" originates from the Greek words "hydro" and "ponos," which mean "water doing the work" or "water works." This system uses water to deliver nutrient-rich water to crops through pipes or tubing. The plants then absorb the necessary nutrients directly from the water.

One of the significant advantages of hydroponic greenhouses is their ability to produce high yields of crops in a relatively small space, using less water and fertilizer than traditional farming methods. They offer a solution to cultivate plants throughout the year, regardless of the weather conditions or climate outdoors (Despommier, 2011). Even though these systems require a less-intensive way to manage large areas of production, they still require careful monitoring and management to ensure optimal growing conditions, including controlling pests and diseases, maintaining proper pH levels and nutrient concentrations, and ensuring adequate ventilation and air circulation (Rufí-Salís et al., 2020) (Fig. 2).

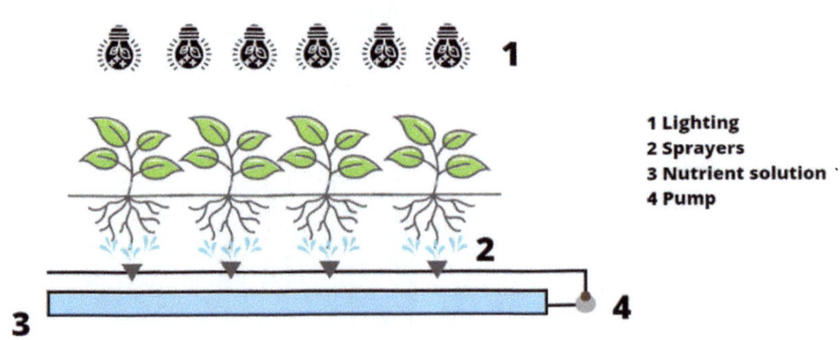


**Fig. 2** A schematic representation of a hydroponic system adapted from (Son et al., 2020)

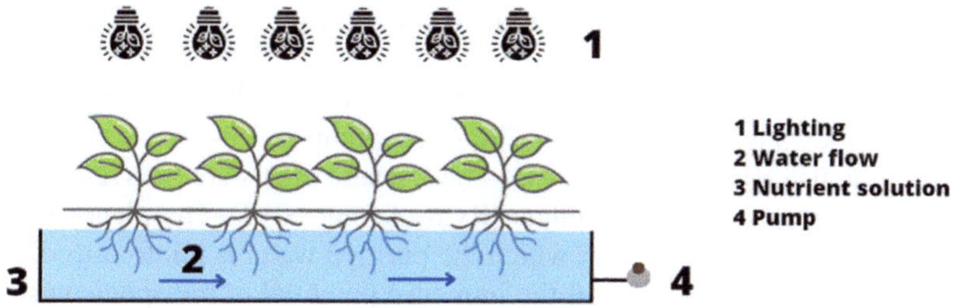


**Fig. 3** A schematic representation of an aeroponic system adapted from (Son et al., 2020)

### 2.1.2 Aeroponics

Aeroponics is an advanced technology that was developed by NASA in the 1990s to find ways to grow plants in space. It is a technique that involves growing plants without soil in a misty air environment and is gaining popularity due to its superior efficiency. According to Al-Al-Kodmany (2018), aeroponics surpasses traditional hydroponics and uses up to 90% less water than even the most proficient hydroponic systems.

In an aeroponic system, plants are placed in containers or chambers with their roots suspended in the air. Nutrient-rich water is delivered to the roots by a network of misting nozzles or foggers that disperse a fine mist of water and nutrients. This technique allows for very little waste of water resources, which is a significant concern in traditional farming methods. As a result, faster plant growth and higher yields are achieved than with other soilless growing methods, as mentioned by Lakhiar et al. (2018).

However, aeroponic systems can be more complex and expensive to set up and maintain than other soilless growing methods. They require careful monitoring and management to ensure that the misting nozzles or foggers deliver the correct amount of water and nutrients to the roots and that the growing environment is optimal for plant growth, as noted by Eldridge et al. (2020).

When combined with vertical farming techniques, aeroponic systems allow for high-density planting and efficient use of space. Plants can be stacked in layers on top of each other, making it particularly useful in urban areas where space is limited. Plant boxes can be stacked vertically in any environment, including basements or warehouses (Fig. 3).

### 2.1.3 Aquaponics

Aquaponics is a bio-system that combines aquaculture, which involves raising aquatic animals like fish, with hydroponics, which involves growing plants in water without soil. This system is designed as a closed loop, where the waste produced by fish is used as nutrients for the plants that, in turn, filter and purify the water for the fish, thereby decreasing the requirement for extra fertilizers as highlighted by Goddek et al., (2015).

The basic setup of an aquaponic system involves a fish tank, a growing (or production) bed, and a water pump (Fig. 4). The water pump facilitates the circulation of water from the fish tank to the production bed, where plants are cultivated in a soilless medium like gravel or clay pellets (Kyaw & Ng, 2017). The plants assimilate nutrients from the fish waste, after which the purified water is reintroduced into the fish tank. Experts have highlighted the immense potential of aquaponic systems in sustainable agriculture by achieving the 3Rs (reduce, reuse, recycle) (Okomoda et al., 2022). These systems offer many advantages, such as utilizing water and nutrients more efficiently, utilizing fish waste as plant fertilizer, and designing the system based on a symbiotic relationship between plants, microorganisms, and fish.

Although aquaponics is a versatile method of farming that can be used in urban areas where space for traditional farming is limited, setting up and maintaining such a system can be challenging, as it demands meticulous monitoring to ensure good water quality, nutrient balance, and animal and plant health. Scientists are exploring technological advancements to improve the performance of aquaponic systems, making them more intelligent, accurate, and efficient (Wei et al., 2019). Recently, these systems have gained greater attention from the public due to their integration within value chains, increased productivity, and less harmful environmental impact than other production methods (Palm et al., 2018).

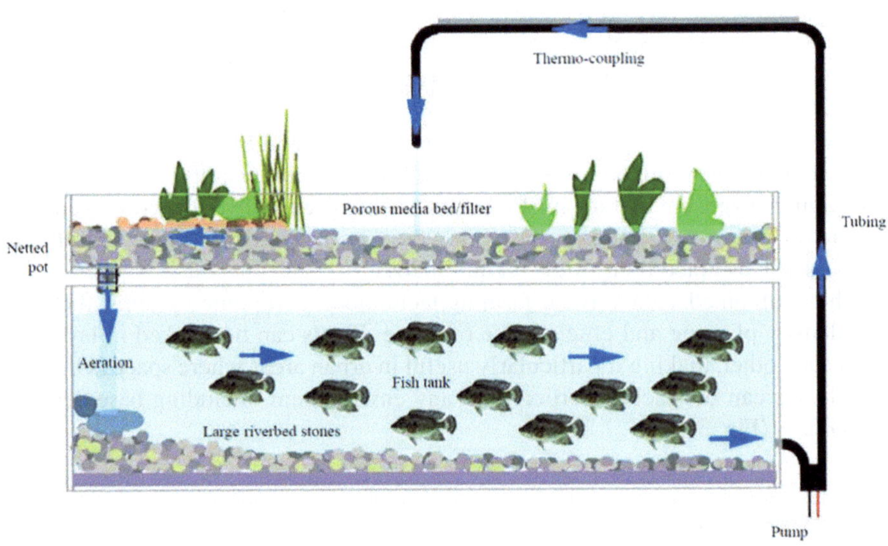


**Fig. 4** Basics of an aquaponic system (Maurya et al., 2023)

### 2.1.4 Plant Factories

Plant factories are innovative indoor farming systems that use vertical racks or shelves to grow plants. These vertical farms typically have multiple levels of plants, and LED lights are crucial for their growth as they provide the necessary spectrum and intensity of light. However, advancements in LED technology are expected to significantly improve efficiency, resulting in reduced operational costs and a smaller environmental footprint. Dutch-based group, PlantLab, recently invented new "induction" lighting technology that produces optimal wavelengths of light, including blue, red, and infrared light that plants require for photosynthesis (den Besten, 2019). This lighting technology emulates specific color spectrums of sunlight to promote the development of vegetables and fruits, creating an ideal microenvironment for growing plants without wasting energy on heating the entire building. It generates more warmth than LED lighting but less than conventional bulbs.

In addition to using LED lighting, hydroponics, aeroponics, and aquaponics can help achieve economies of scale in plant factories. However, optimizing Heating, Ventilation, and Air Conditioning (HVAC) is crucial for crop health and growth in indoor farming environments. The HVAC system is designed to create a controlled environment that stimulates ideal growing conditions for crops. This system typically includes heating and ventilation components to control temperature, humidity levels, and air circulation to distribute carbon dioxide and provide fresh air to the plants. Integrating the HVAC system with other critical components of the vertical farm, such as water, nutrient provision, and lighting, helps plants get the optimal growth and health conditions they require to thrive (Cheng & Lee, 2019).

Experts anticipate that plant factories and vertical farms will soon be highly automated and even benefit from the advances of artificial intelligence. This means that sensors will be placed near each plant bed to detect its water and nutrient needs, as well as any harmful pathogens or microorganisms that may cause diseases. Gas chromatography technology will also be utilized to analyze flavonoid levels, recommending optimal harvesting times accurately. Therefore, farming operations in these facilities are expected to become increasingly efficient and effective soon (Halgamuge et al., 2021).

Overall, plant factories are innovative indoor farming systems that offer a promising solution to feeding the growing global population. With the use of advanced lighting technology, HVAC systems, and automation, plant factories can provide an efficient and effective way of growing crops. These technological advancements could significantly reduce the environmental impact of traditional agriculture while providing fresh, healthy, and sustainable food (Fig. 5).

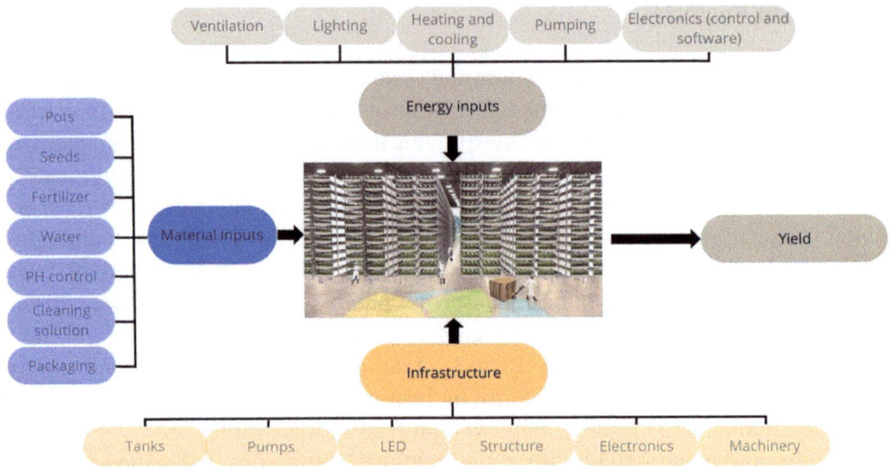


**Fig. 5** Schematic representation of vertical farming inputs, including material, energy, and infrastructure (adapted from Martin et al. (2022)

## 3 The Potential of Vertical Farms in Crafting Sustainable Food Supply Chains

The sustainability of food supply chains is becoming more crucial, especially since consumers are now more aware and prefer products that cause minimal environmental harm and have a positive social impact (Erez, 2019). At the same time, companies in the retail industry are increasingly recognizing the benefits of implementing sustainable practices, such as improving reputation and boosting employee engagement. These outcomes can result in increased customer loyalty and overall business success. Additionally, sustainability can help retailers save money by reducing waste, conserving energy, and lowering packaging and transportation costs. These cost-cutting measures can ultimately contribute to a wider profit margin for the business (Vadakkepatt et al., 2021). Moreover, according to (Elkington, 1994), it is crucial to recognize that businesses aiming for sustainability are often measured by what is known as the triple bottom line or the 3P's. These entail that companies must consider not only their economic performance (profit) but also their impact on the environment (planet) and society (people). In this context, retailers have a crucial role to play in promoting sustainability since they serve as a vital link between the various upstream (*i.e.,* natural resources, resources producers, suppliers) and downstream (*i.e.,* consumers, households and communities) actors involved in the food supply chain (Grewal & Levy, 2007).

One way for retailers to promote sustainability in the food supply chain is by adopting vertical farming. This allows them to produce more local food and address sustainability challenges in their operational models by promoting the "*selling where you produce*" model. In fact, vertical farms are considered "*ecosystemic*" in some urban regions because they function as a single company, producing and selling their produce within the same building. This approach avoids carbon emissions from transportation and ensures better traceability of food items. Additionally, vertical farming provides retail companies with a unique approach to product offerings, utilizing innovative production methods that incorporate diverse multilayers, techniques, and integrated organizational interactions. This streamlines the food production process from seed to retail and offers an organizational interface that holds great promise for sustainability as it reduces the ecological footprint of food production while ensuring that densely populated urban areas have continuous access to fresh food of high quality and freshness.

The advantages and challenges of vertical farming adoption in the food retail sector have not been extensively researched yet, but past studies have shed light on the factors that influence innovation adoption by food retailers. According to Esbjerg et al. (2016), retailers prioritize consumers' interests even if it could hinder innovation adoption. This means that adopting new technologies like vertical farming must demonstrate clear consumer benefits and pose minimal risk for retailers. Furthermore, Jung et al. (2022) emphasize the importance of communicating these benefits in terms of business instead of technical jargon, so both retailers and consumers can comprehend them better.

Studies have researched the factors that drive the adoption of vertical farming by catering industry actors. For example, according to Clark et al. (2022), although chefs believe that vertical farms can significantly reduce food miles and produce high-quality, fresh, and visually appealing produce, they have concerns about whether the taste of vertical farm produce can match that of conventional crops. The quality and taste perceptions of vertical farm produce are the main factors that influence chefs' willingness to use or reject these production techniques.

Other studies have investigated consumers' attitudes toward produce from vertical farms. One such study by Ares et al. (2021) examined such attitudes in the USA, UK, Singapore, and China. The findings indicated that consumers in these countries mainly had positive attitudes towards vertical farming, with the participants from China and Singapore being the most positive overall. Moreover, the study put forth that the characteristics of vertical farms aligned with the United Nations Sustainable Development Goals, including higher yield, reduction of carbon emissions, and securing access to food, were identified as critical drivers of positive attitudes. Conversely, negative attitudes towards vertical farms were influenced by factors such as high energy consumption and elevated pricing.

In line with this stream of research, in this study, we argue that food retailers in urban areas are the key to the spread of vertical farming and analyze the factors motivating them to embrace it.

# 4 Material and Methods

## *4.1 Conceptual Framework and Hypothesis*

The Unified Theory of Acceptance and Use of Technology (UTAUT) is a model proposed by Venkatesh et al. in 2003. Its purpose is to provide a more accurate explanation and prediction of technology acceptance behavior. It is one of the most recent models used for researching technology acceptance and has been applied in various fields, including healthcare (Ben Arfi et al., 2021), education (El-Masri & Tarhini, 2017), and agriculture (Li et al., 2020). However, the use of the UTAUT model in food retail research is still limited. It has been primarily utilized in previous studies to explore the adoption of blockchain technology for improving supply chain sustainability and to investigate innovative strategies for e-grocery (Dirsehan & Cankat, 2021). To gain a better understanding of retailers' behavioral intention to use vertical farming, the following paragraphs examine the available literature and describes the theoretical definitions of the different UTAUT components. Our research add to this model a component to measure risk perception. Additionally, the underlying hypothesis of this research are presented.

**Performance expectancy (PE):** refers to the extent to which an individual thinks that technology use could be beneficial to him or her to reach performance (Venkatesh et al., 2003). In this study, PE is described as the retailers' understanding of the usefulness and advantages of vertical farms technology. Hence, the following hypothesis was formulated:

**H1.** Performance expectancy directly, positively, and significantly affects retailers' behavioral intention to adopt vertical farming technologies.

**Effort expectancy (EE):** refers to the users' perception of how difficult or easy it is for them to use new technology (Venkatesh et al., 2003). Many studies consider effort expectancy the most influential factor in determining users' behavioral intention to adopt new technologies (see Sun et al., (2013) for a review). In this study, EE refers to the extent to which retailers believe it is easy to learn and use vertical farming technologies.

**H2.** Effort expectancy directly, positively, and significantly affects retailers' behavioral intention to adopt vertical farming technologies.

**Social influence (SE):** refers to the impact of individuals close to or important to a person, who can persuade him or her to adopt new technologies. Studies put forward the importance of SE in explaining individuals' intention to adopt as well as their effective adoption of new technologies (Sun et al., 2013).

**H3:** Social influence directly, positively, and significantly affects retailers' behavioral intention to adopt vertical farming technologies.

**Facilitating conditions (FC)** include "the users' attitudes toward the organizational infrastructure and financial and technical support after accepting the new technology" (Venkatesh et al., 2003). Chen & Chang (2013) have shown that facilitating conditions are significant factors in determining behavioral intention in adopting

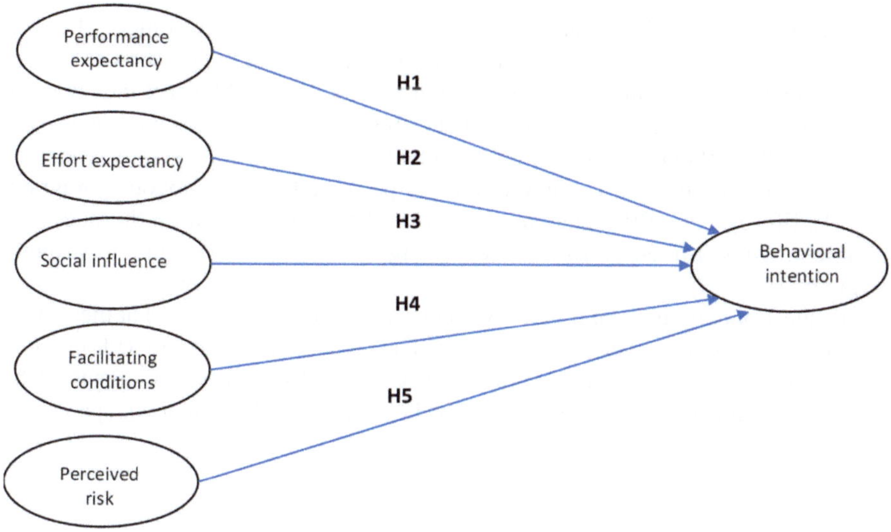


**Fig. 6** conceptual model and hypothesis

technologies. In this study, FC refers to resources and facilities that make the vertical farming technologies adoption process more accessible.

**H4:** Facilitating conditions (FC) directly, positively, and significantly affects retailers' behavioral intention to adopt vertical farming technologies.

**Perceived risk (PR):** perceived risk refers to the users' attitude toward potential risks that may arise in various areas (financial, technical, production, management-related, etc.) because of innovative technologies adoptions (Venkatesh et al., 2003). In this study, perceived risk refers to the risk that retailers assume by adopting vertical farming technologies and its impact on them and their businesses. Thus, the hypothesis that follows is:

**H5**: perceived risk negatively affects retailers' intention to adopt vertical farming techniques.

**Behavioral intention (BI)** refers to "the possibility that a person engages in certain behaviors in the future under certain conditions and to do something" (Venkatesh et al., 2003) and has been put forward by several researchers as a central in determining an individual's acceptance of new technologies (Ajzen, 1991). In this study, BI refers to the intention of retailers to adopt vertical farming technologies, which is a central concept of this study's model (Fig. 6).

## 4.2 *Survey Design and Measurement*

The survey conducted in this study was based on a conceptual model (Fig. 6) and consisted of two main sections. The initial section focused on the characteristics of the retail business, such as total areas, the number of annual customers, the year of opening, and the primary fresh products sold in the store. The second section aimed to explore the attitudes and behavioral factors that could affect the adoption of vertical farming technologies by retailers. Respondents were shown images and provided with explanations of vertical farming technologies, followed by questions that aimed to measure the latent variables as presented in the previous paragraph (i.e., PE, EE, SE, FC, PR). The variables were determined based on the questions in Table 1, using a five-point Likert scale ranging from strongly agree to strongly disagree.

The questionnaire used in this study comprised four sections aimed at measuring the constructs of perceived ease of use (PE), perceived usefulness (PU), social influence (SI), and facilitating conditions (FC), based on the scale suggested by Venkatesh & Zhang (2010) and supported by Foon & Fah (2011). To assess perceived

**Table 1** Measurement items included in the survey

| Variables | Measurement items |
|---|---|
| **Performance expectancy [PE]** | **[PE1]** Using vertical farming technology would improve my business performance<br>**[PE2]** Using vertical farming technology would save me time<br>**[PE3]** I find vertical farming technology useful |
| **Effort expectancy [EE]** | **[EE1]** It is easy for me to learn vertical farming technologies<br>**[EE2]** I can easily become skillful in using vertical farming technologies<br>**[EE3]** I can easily interact with vertical farming technologies<br>**[EE4]** I would find vertical farming technologies easy to use |
| **Social Influence [SI]** | **[SI1]** People I consider important think I should use vertical farming technologies in my store<br>**[SI2]** People I am familiar with think I should use vertical farming technologies as part of my business strategy<br>**[SI3]** People who influence my behavior think I should use vertical farming technologies as part of my business strategy<br>**[SI4]** Most people surrounding me think I should use vertical farming technologies in my business strategy |
| **Perceived risk [PR]** | **[PR1]** Using vertical farming in my store is risky<br>**[PR2]** I feel that using vertical farming technologies would cause me much trouble if something went wrong<br>**[PR3]** I am sure I would make a mistake if I used vertical farming technologies in my store |
| **Facilitating Condition [FC]** | **[FC1]** My living environment encourage me to use vertical farming technologies<br>**[FC2]** My business environment encourages me to use vertical farming technologies<br>**[FC3]** Using vertical farming technologies is compatible with my business strategy<br>**[FC4]** Help is available if I encounter any problems related to my use of vertical farming technologies |
| Behavioral Intention [BI] | **[BI]** I intend to use vertical farming technologies in my store |

risk, questions were developed using the scale proposed by Stone & Grønhaug (1993). Furthermore, behavioral intention was measured using a single item adapted from Venkatesh and Zhang (2010). The third section of the questionnaire aimed to collect sociodemographic information about farmers, including their age, education level, and any prior agriculture-related education.

To ensure maximum participation from retailers, both in-person and online surveys were conducted. The target retailers were based in Rouen, a city in north-western France with a population of 111,000 inhabitants (Ben-Othmen et al., 2022). The survey was carried out from February to July 2022 and resulted in 98 completed questionnaires. However, only 90 of these responses were deemed usable and included in the subsequent analysis.

### *4.3 Data Analysis Procedure*

The collected data was analyzed using STATA software version 16. The results showed that the majority of retail managers who participated in the survey were male, constituting 75% of the sample. On average, they were 33 years old and had been operating their retail stores for six years. Interestingly, only 15% of the respondents had earned a business degree, and just 7% had been trained in food or agricultural science during their studies. The survey also revealed that the average store size was 645 square meters, with a significant disparity in the sample as indicated by the standard deviation of 115.05. Food supermarkets were the most represented category in the sample, accounting for 45%, followed by food coops (28%), greengrocers (23%), and hard discounts (10%). Regarding vertical farming, the data show that while 85% of the respondents had some knowledge about this technology, only 20% regularly sourced fresh produce from vertical farms. However, there was a consensus among the respondents about the potential of vertical farming technology in ensuring access to fresh food for city dwellers (80% totally agreed) and preserving water resources (60% totally agreed). Nonetheless, there was less agreement concerning the potential of vertical farming in addressing the lack of arable land in and around cities for growing fresh food (25% totally disagreed). These findings are presented in Fig. 7.

In order to comprehensively understand the relationship between different drivers of behavioral intetion to adopt vertical farming technologies, we employed the Structural Equation Modeling (SEM) approach. This approach is based on mathematical statistics and enables us to accurately measure the complex correlations between various factors (Hair et al., 2012). For this study, we used the Maximum Likelihood method of the STATA command language to conduct the SEM. We utilized the STATA builder to create both the path analysis and model parameters.

The SEM analysis consisted of two steps. The first step was to evaluate and analyze the measurement model used in this research. We used the non-parametric indicator Cronbach's alpha (>0.7) and analyzed it to determine if it was in the desired range. In such a case, we included it in the structural model (Bagozzi et al., 1992). The next

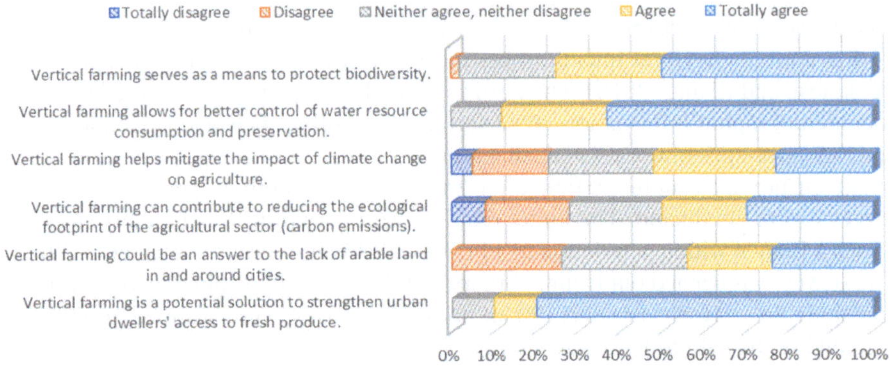


**Fig. 7** Retailers' perceptions of sustainability of vertical farming technologies

step involved performing confirmatory factor analysis and examining the results of the structural models.[1]

# 5 Results

## 5.1 Exploratory Factor Analysis

In order to examine the relationship between the questionnaire items and the behavioral intention of retailers towards adopting vertical farming technologies, an exploratory factor analysis was conducted. The correlation matrix between the items, provided in Table 2, shows that the data are not highly correlated. This is a positive indication as highly correlated data may not provide unique information. Additionally, the Barlet test was performed, which showed significant results ($Chi^2 = 4051.31$, $p$-value = 0.000) indicating the presence of sufficient adequate correlations. Moreover, the Kaiser Meyer Auckland measure of sampling adequacy (KMO $= 0.906 > 0.50$) confirmed that there are significant intercorrelations between the data, underlining the importance of conducting the factor analysis.

The measurement scale underwent an exploratory factor analysis, and the results are presented in Table 3. Reliability and validity of the items were verified using Cronbach's alpha and Dillon-Goldstein's rho, respectively, following the guidelines of Hair et al. (2020). These coefficients were found to be higher than 0.7 and 0.8, respectively. However, items with a factor loading below 70% were removed from

[1] This helped us analyze the internal consistency of the measures and improve the model fit. We mainly applied the Satorra-Bentler Estimator to address any potential non-normality of the measures. Additionally, we tested the goodness of fit using the Root Mean Square Error of Approximation (RMSEA), the Comparative (CFI), the Tucker-Lewis index (TLI), and the Standardized root mean squared residual (SRMR).

**Table 2** Measures correlation matrix

| | Performance Expectancy | Facilitating Conditions | Social Influences | Risk Perception | Effort Expectancy | Behavioral Intention |
|---|---|---|---|---|---|---|
| PE_1 | **0.913** | 0.578 | 0.513 | –0.115 | 0.546 | 0.664 |
| PE_2 | **0.911** | 0.432 | 0.499 | 0.027 | 0.465 | 0.596 |
| PE_4 | **0.890** | 0.595 | 0.537 | 0.026 | 0.645 | 0.679 |
| EE_1 | 0.476 | 0.597 | 0.382 | –0.082 | **0.872** | 0.513 |
| EE_2 | 0.515 | 0.628 | 0.487 | –0.112 | **0.919** | 0.507 |
| EE_4 | 0.410 | 0.506 | 0.476 | –0.049 | **0.898** | 0.582 |
| SI_1 | 0.473 | 0.517 | **0.924** | 0.070 | 0.565 | 0.521 |
| SI_2 | 0.510 | 0.531 | **0.910** | 0.034 | 0.494 | 0.473 |
| SI_3 | 0.594 | 0.477 | **0.912** | –0.019 | 0.480 | 0.459 |
| SI_4 | 0.377 | 0.410 | 0.421 | –0.044 | 0.469 | 0.353 |
| PR_1 | –0.046 | –0.111 | –0.016 | **0.831** | –0.085 | –0.187 |
| PR_2 | 0.174 | 0.078 | 0.113 | **0.896** | 0.045 | 0.075 |
| PR_3 | –0.157 | –0.114 | –0.097 | **0.892** | –0.113 | –0.276 |
| FC_1 | 0.471 | **0.887** | 0.578 | –0.023 | 0.640 | 0.596 |
| FC_2 | 0.422 | **0.888** | 0.433 | –0.034 | 0.641 | 0.588 |
| FC_3 | 0.637 | **0.857** | 0.439 | –0.087 | 0.640 | 0.675 |
| FC_4 | 0.510 | **0.630** | 0.418 | 0.035 | 0.742 | 0.672 |
| BI | 0.612 | –0.110 | 0.608 | 0.531 | 0.687 | 1.000 |

the scale. Specifically, the item related to social influence (SI_4) and performance expectancy (PE_3) were deleted due to their low factor loadings.

## 5.2 *Confirmatory Factor Analysis*

A Confirmatory Factor Analysis (CFA) was carried out to assess the internal consistency of the measures used in the structural model, following the Exploratory Factor Analysis (EFA) that was previously conducted. While the factorial structure of the constructs (PR, EE, PE, SI, FC) can be considered established based on extensive literature on the use of technology in different contexts (UTAUT), the EFA helped to redefine and tailor the measurement items that would be most suitable for the retail sector context under study. As such, a CFA was performed for all latent variables of both the UTAUT model and the EFA. The goodness of fit indices, such as the Root Mean Squared Error of Approximation (RMSEA), the Comparative Fit Index (CFI), the Tucker-Lewis Index (TLI), the Standardized Root Mean Squared Residual (SRMR), and the Coefficients of Determination (CD) are presented in Table 4. According to Pituch and Stevens (2016), RMSEA values below 0.05 indicate a good

**Table 3** Measurement scales psychometric properties

| Latent variable | Dimensions | Factor loadings | Cronbach's alpha | AVE |
|---|---|---|---|---|
| Performance Expectancy | PE_1 | 0.718 | 0.896 | 0.795 |
| | PE_2 | 0.759 | | 0.829 |
| | PE_4 | 0.753 | | 0.782 |
| Effort Expectancy | EE_1 | 0.786 | 0.887 | 0.792 |
| | EE_2 | 0.832 | | 0.852 |
| | EE_4 | 0.838 | | 0.775 |
| | FC_4 | 0.486 | | 0.875 |
| Social Influence | SI_1 | 0.866 | 0.862 | 0.890 |
| | SI_2 | 0.889 | | 0.853 |
| | SI_3 | 0.844 | | 0.822 |
| Perceived Risk | PR_1 | 0.828 | 0.783 | 0.777 |
| | PR_2 | 0.782 | | 0.798 |
| | PR_3 | 0.824 | | 0.737 |
| Facilitating Conditions | FC_1 | 0.734 | 0.892 | 0.734 |
| | FC_2 | 0.648 | | 0.776 |
| | FC_3 | 0.747 | | 0.799 |

fit, whereas values above 0.95 for both CFI and TLI are indicative of a very good fit. Regarding SRMR, a value higher than 0.10 is considered acceptable. Results indicate that the social influence variable's goodness of fit was not satisfactory and was therefore excluded from the model. Consequently, only four latent variables, namely PE, EE, FC, and PR, were included in the baseline Structural Equation Model.

**Table 4** Confirmatory factor analysis results

| Latent variable | Number of items | RMSEA | CFI | TLI | SRMR | CD | Baseline model |
|---|---|---|---|---|---|---|---|
| SI | 4 | 0.294 | 0.987 | 0.911 | 0.025 | 0.938 | Excluded from the model |
| EE | 4 | 0.103 | 0.946 | 0.973 | 0.313 | 0.991 | Included in the model |
| PE | 3 | 0.047 | 0.907 | 0.962 | 0.033 | 0.914 | Included in the model |
| FC | 3 | 0.202 | 0.991 | 0.829 | 0.065 | 0.801 | Included in the model |
| PR | 3 | 0.212 | 0.888 | 0.865 | 0.082 | 0.771 | Included in the model |

## 5.3 The Structural Equation Model

The Structural Equation Model was utilized to evaluate the association among Performance expectancy, Effort expectancy, Facilitating conditions, Perceived risk, and Behavioral intention to adopt vertical farming technologies. After conducting the Exploratory Factor Analysis, one item related to the Effort expectancy construct was replaced by the item FC4 linked to the Facilitating conditions latent variable. Moreover, one item PE_3 was eliminated from the Performance expectancy construct due to its low associated factor loading (0.53 < 0.7).

The initial baseline model fit was unsatisfactory (RMSEA_SB = 0.137, CFI_SB = 0.798, TLI_SB = 0.751, SRMR = 0.308, CD = 1.000), indicating a potential specification error. Consequently, following MacCallum's (1986) suggestion, we improved the model specification by allowing covariance between the omitted latent variables. The modification indices for path coefficients and covariances suggested that including three covariances between PE and EE, PE and FC, and EE and FC could enhance the model fit (RMSEA_SB = 0.084, CFI_SB = 0.927, TLI_SB = 0.906, SRMR = 0.081, CD = 0.999). The factor loadings and path coefficients for this model, are illustrated in Fig. 8.

Based on our findings, the factors that are most crucial in influencing retail store managers to adopt vertical farming technologies are performance expectancy and facilitating conditions. Although effort expectancy also plays a role, its significance in explaining the intention to adopt these technologies is relatively low. On the other hand, the perceived risk associated with using vertical farming technology in retail stores has a negative and significant impact on the intention to use it. Furthermore, our

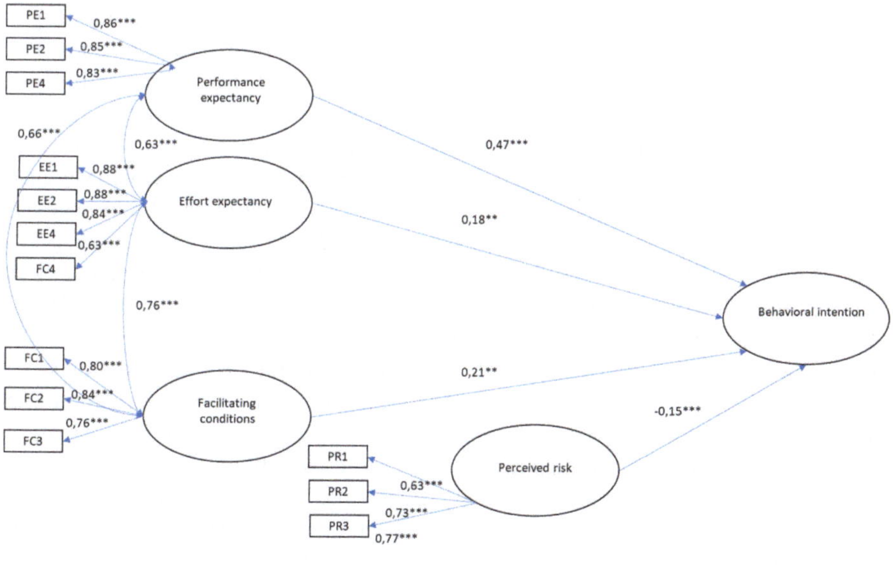


**Fig. 8** Path diagram of the structural model

results support the hypothesis that both facilitating conditions and effort expectancy have a positive and significant impact on the intention to use vertical farming technology in retail stores. It also highlights a clear and significant correlation between performance and effort expectancy, as well as a positive and significant relationship between effort expectancy and facilitating conditions.

Overall, these findings suggest that in order to encourage the adoption of vertical farming technologies in retail stores, it is important to focus on improving performance expectancy and facilitating conditions while also addressing concerns around perceived risk.

## 6 Discussion

In today's ever-changing society, sustainability has become a crucial aspect of food retail businesses' core strategies. As customer-facing entities, food retailers have a unique position in the supply chain, enabling them to influence both upstream suppliers and downstream customers. This chapter proposes vertical farming technologies as a potential solution to address sustainability in food supply chains. The study focuses on understanding the key drivers of vertical farming technology adoption by food retailers in Rouen City, France. This research is the first, to the best of the authors' knowledge, to utilize the Unified Theory of Acceptance and Use of Technology to comprehend the five primary drivers of vertical farming technology adoption, which include performance expectancy, effort expectancy, facilitating conditions, perceived risk, and social influence.

The results showed that performance expectancy had a positive and the most significant effect on food retailers' behavioral intention to adopt vertical farming in their stores. In other words, the more retailers perceive the suitable performance of vertical farming technologies, the more their motivation to use them. Similarly to Kshetri (2018), we argue that we should consider performance expectancy as a set of dimensions that cover quality, cost and risk reduction, and sustainability. Our findings are consistent with prior research on the food and agriculture sectors, which suggests that, for example, farmers are inclined to embrace new technologies if they perceive potential benefits in their performance (Beza et al., 2018; Faridi et al., 2020). The same applies to the retail supply chain sector, where better performance expectations lead to a higher intention to adopt new technologies according to Wong et al. (2020). It is noteworthy that getting fresh food products made and delivered in time to city dwellers at the lowest cost possible while ensuring that the product remains intact during transport is an ongoing challenge for urban fresh food retailers, which may explain why performance expectancy proved to have the most potent direct effect on retailers' behavioral intention to adopt vertical farming technologies.

Our results highlight that facilitating conditions are crucial in determining retailers' behavioral intention to adopt vertical farming. Although this finding contradicts the results of a recent study by Mukherjee et al. (2023), putting forth that the facilitating conditions did not promote the adoption of blockchain technology in food retail sector, it is consistent with the results of a survey conducted by Shao et al. (2022) on the factors influencing public intention to use micro-vertical farming and putting forth that the availability of resources needed for vertical farming significantly determined people's intention to adopt it.

The results show that the ease of putting effort into vertical farming (i.e., effort expectancy) positively impacts retail store managers' intention to participate. However, this influence is not as strong as the impact of facilitating conditions, such as having the necessary resources, knowledge, and investments and ensuring that vertical farming aligns with the business strategy of the retail store. This suggests that facilitating conditions significantly influence behavioral intention more than the ease of use of vertical farming as a new technology.

This study found that risk perception negatively affected food retailers' intention to adopt vertical farming. The importance of risk perception in determining technology acceptance is well-documented in numerous studies (Bearth & Siegrist, 2016). Specifically, in the food industry, perceived risks and benefits play a crucial role in shaping public opinion and acceptance of digital technology (Pfeiffer et al., 2021). Gaskell (2000) states that the fundamental need for innovative technology is questioned if the public associates too few benefits with technology. According to our survey results, retailers have raised concerns about the high operational costs associated with running vertical farms. One of the primary risks identified by respondents was the energy dependency of vertical farms, which require a significant amount of power to operate continuously throughout the year. Respondents have noted that the lighting and climate control systems necessary for vertical farming would consume a tremendous amount of energy.

The final structural model of the study did not include social influence. This is in contrast to previous studies that suggested social influence as a factor that could affect but not determine people's use of vertical farms to grow their food, as demonstrated by Shao et al. (2022). However, research in the agricultural field has shown that social influence has little impact on farmers' willingness to adopt innovative technologies. According to Li et al. (2020), while social impact may motivate farmers to adopt new technologies, actual adoption is more likely when there are no barriers to accessing necessary resources and knowledge, and perceived financial risks are low.

## 7 Conclusion

Urban areas are seeing a rising demand for fresh, locally sourced food. To ensure sustainability and success in the local food system and overall supply chain, all players in the food supply chain need to come together. In this article, we propose that food retailers can help achieve these goals by implementing vertical farming technology. Vertical farming involves using soil-less cultivation technology and indoor vertical space to grow fresh food directly in-store and has the potential to promote sustainable cities and communities, as well as sustainable consumption and production.

Compared to farmers, retailers have a greater interconnectivity in the local food system. Hence, their motivations and drivers for adopting vertical farming technologies may differ. Our research shows that in Rouen City, France, while retailers recognize the benefits of integrating vertical farming technologies into their business strategies, they are hesitant to take the risk. This is not surprising as most people are risk-averse, leading to extensive research on the impact of risk and risk-aversion on technology adoption.

Although vertical farming has been studied extensively in terms of improving its technological and agricultural aspects, its economic evaluation and risk assessment are relatively new topics. Currently, there is a lack of publicly available data on vertical farming business models in France, with these businesses still considered emerging. They face three main challenges: standardization, environmental sustainability, and profitability, with a high risk of failure that could be costly. This is mainly due to premature business decisions about location choice, pricing strategy, system design, and other critical issues. Therefore, to support those who wish to adopt vertical farming technologies and develop a more local and sustainable supply chain, it is necessary to consider retailers' risk aversion by improving knowledge transfer and developing adaptable business plans.

**Acknowledgements** The authors would like to thank all the retail store managers in Rouen City (France) who have agreed to answer the questionnaire; the authors also thank students of the fourth year of engineering studies Farming for the Future at UniLaSalle Rouen Campus: Rose Poilleux, Pauline Triboulet, and Cécile Millet who have contributed to the data collection of this study.

## References

Ajzen I (1991) The theory of planned behavior organizational behavior and human decision processes. Organ Behav Hum Decis Process 50(2):179–211

Albizzati PF, Tonini D, Chammard CB, Astrup TF (2019) Valorisation of surplus food in the French retail sector: Environmental and economic impacts. Waste Manag 90:141–151. https://doi.org/10.1016/j.wasman.2019.04.034

Al-Kodmany K (2018) The vertical farm: A review of developments and implications for the vertical city. Buildings 8(2):24. https://doi.org/10.3390/buildings8020024

Ares G, Ha B, Jaeger SR (2021) Consumer attitudes to vertical farming (indoor plant factory with artificial lighting) in China, Singapore, UK, and USA: A multi-method study. Food Res Int 150(November):110811. https://doi.org/10.1016/j.foodres.2021.110811

Bagozzi RP, Yi Y, Fornell C, Larcker DF, Bagozzi RP, Yi Y, Phillips LW, Henseler J, Ringle CM, Sarstedt M, Schuenemeyer JH, Murtagh F, Heck A (1992) Assessing construct validity in organizational research. Adm Sci Q 36(3):421–458

Bearth A, Siegrist M (2016) Are risk or benefit perceptions more important for public acceptance of innovative food technologies: A meta-analysis. Trends Food Sci Technol 49:14–23. https://doi.org/10.1016/j.tifs.2016.01.003

Ben Arfi W, Nasr IB, Kondrateva G, Hikkerova L (2021) The role of trust in intention to use the IoT in eHealth: Application of the modified UTAUT in a consumer context. Technol Forecast Soc Chang 167(February):120688. https://doi.org/10.1016/j.techfore.2021.120688

Ben-Othmen MA, Canchel J, Devillers L, Hennart A, Rouyer L, Ostapchuk M (2022) Peri-urban Farmers' Perception of Climate Change: Values and Perspectives – A French Case Study BT. In: Chatterjee U, Akanwa AO, Kumar S, Singh SK, Dutta Roy A (eds) Ecological Footprints of Climate Change : Adaptive Approaches and Sustainability. Springer International Publishing, Cham, pp 349–372

den Besten J (2019). *Chapter 8.2 - Vertical Farming Development; the Dutch Approach* (M. Anpo, H. Fukuda, & T. B. T.-P. F. U. A. L. Wada (eds.); pp. 307–317). Elsevier. https://doi.org/10.1016/B978-0-12-813973-8.00027-0

Beza E, Reidsma P, Poortvliet PM, Belay MM, Bijen BS, Kooistra L (2018) Exploring farmers' intentions to adopt mobile Short Message Service (SMS) for citizen science in agriculture. Comput Electr Agric 151:295–310. https://doi.org/10.1016/j.compag.2018.06.015

Burt SL, Sparks L (2002) Corporate Branding, Retailing, and Retail Internationalization. Corp Reput Rev 5(2):194–212. https://doi.org/10.1057/palgrave.crr.1540174

Chen K-Y, Chang M-L (2013) User acceptance of 'near field communication' mobile phone service: an investigation based on the 'unified theory of acceptance and use of technology' model. Serv Ind J 33(6):609–623. https://doi.org/10.1080/02642069.2011.622369

Cheng CC, Lee D (2019) Artificial intelligence-assisted heating ventilation and air conditioning control and the unmet demand for sensors: Part 1. Problem formulation and the hypothesis. Sensors (switzerland) 19(5):1131. https://doi.org/10.3390/s19051131

Clark A, Ellison M, Robinson G, Hill TR (2022) Consumer perceptions of vertical farming with a focus on the catering industry. Proc Nutr Soc 81(OCE5):2022. https://doi.org/10.1017/s0029665122002506

Despommier D (2011) The vertical farm: controlled environment agriculture carried out in tall buildings would create greater food safety and security for large urban populations. J Verbr Lebensm 6(2):233–236. https://doi.org/10.1007/s00003-010-0654-3

Despommier D (2010). *The Vertical Farm: Feeding the World in the 21st Century* (T. D. Books (ed.)). Saint Martin's Press.

Dirsehan T, Cankat E (2021) Role of mobile food-ordering applications in developing restaurants' brand satisfaction and loyalty in the pandemic period. J Retail Consum Serv 62:102608

Eldridge BM, Manzoni LR, Graham CA, Rodgers B, Farmer JR, Dodd AN (2020) Getting to the roots of aeroponic indoor farming. New Phytol 228(4):1183–1192. https://doi.org/10.1111/nph.16780

Elkington J (1994) Towards the sustainable corporation: Win-Win-Win business strategies for sustainable development. Calif Manage Rev 36(2):90–100. https://doi.org/10.2307/41165746

El-Masri M, Tarhini A (2017) Factors affecting the adoption of e-learning systems in Qatar and USA: Extending the Unified Theory of Acceptance and Use of Technology 2 (UTAUT2). Education Tech Research Dev 65(3):743–763. https://doi.org/10.1007/s11423-016-9508-8

Erez R (2019). *Sustainability In Retail: Good For Business, Great For Humanity.* Forbes. https://www.forbes.com/sites/royerez/2019/10/01/sustainability-in-retail-good-for-business-great-for-humanity/?sh=7912255f6dec

Esbjerg L, Burt S, Pearse H, Glanz-Chanos V (2016) Retailers and technology-driven innovation in the food sector. British Food J 118(6):1370–1383. https://doi.org/10.1108/BFJ-10-2015-0367

Faridi AA, Kavoosi-Kalashami M, Bilali HE (2020) Attitude components affecting adoption of soil and water conservation measures by paddy farmers in Rasht County, Northern Iran. Land Use Policy 99(April):104885. https://doi.org/10.1016/j.landusepol.2020.104885

Foon YS, Fah BCY (2011) Internet banking adoption in kuala lumpur: an application of UTAUT model. Int J Bus Manage 6(4):161–167. https://doi.org/10.5539/ijbm.v6n4p161

Galli A, Iha K, Halle M, El Bilali H, Grunewald N, Eaton D, Capone R, Debs P, Bottalico F (2017) Mediterranean countries' food consumption and sourcing patterns: An Ecological Footprint viewpoint. Sci Total Environ 578:383–391. https://doi.org/10.1016/j.scitotenv.2016.10.191

Gan CI, Soukoutou R, Conroy DM (2023) Sustainability framing of controlled environment agriculture and consumer perceptions: A review. Sustainability (switzerland) 15(1):304. https://doi.org/10.3390/su15010304

Gaskell G, Allum N, Bauer M, Durant J, Allansdottir A, Bonfadelli H, ..., Wagner W (2000) Biotechnology and the European public. Nat Biotechnol 18(9):935–938

Gibbons C, Morgan B, Kavouras JH, Ben-Othmen M (2020) Sustainability in Agriculture and Local Food Systems: A Solution to a Global Crisis BT. In: LealFilho W, Azul AM, Brandli L, Özuyar PG, Wall T (eds) Zero Hunger. Springer International Publishing, Cham, pp 1–12

Goddek S, Delaide B, Mankasingh U, Ragnarsdottir KV, Jijakli H, Thorarinsdottir R (2015) Challenges of sustainable and commercial aquaponics. Sustainability (switzerland) 7(4):4199–4224. https://doi.org/10.3390/su7044199

Govindan K (2018) Sustainable consumption and production in the food supply chain: A conceptual framework. Int J Prod Econom 195(March 2017):419–431. https://doi.org/10.1016/j.ijpe.2017.03.003

Graamans L, Baeza E, van den Dobbelsteen A, Tsafaras I, Stanghellini C (2018) Plant factories versus greenhouses: Comparison of resource use efficiency. Agric Syst 160(November 2017):31–43. https://doi.org/10.1016/j.agsy.2017.11.003

Grewal D, Levy M (2007) Retailing research: Past, present, and future. J Retail 83(4):447–464. https://doi.org/10.1016/j.jretai.2007.09.003

Hair JF, Howard MC, Nitzl C (2020) Assessing measurement model quality in PLS-SEM using confirmatory composite analysis. J Bus Res 109:101–110. https://doi.org/10.1016/j.jbusres.2019.11.069

Hair JF, Ringle CM, Sarstedt M (2012) Partial least squares: the better approach to structural equation modeling? Long Range Plan 45(5–6):312–319

Halgamuge MN, Bojovschi A, Fisher PMJ, Le TC, Adeloju S, Murphy S (2021) Internet of Things and autonomous control for vertical cultivation walls towards smart food growing: A review. Urban for Urban Green 61(June 2020):127094. https://doi.org/10.1016/j.ufug.2021.127094

Jung IN, Sharma A, Mattila AS (2022) The impact of supermarket credibility on purchase intention of novel food. J Retail Consum Serv 64:102754. https://doi.org/10.1016/j.jretconser.2021.102754

Khan S, Purohit A, Vadsaria N (2021) Hydroponics: current and future state of the art in farming. J Plant Nutr 44(10):1515–1538. https://doi.org/10.1080/01904167.2020.1860217

Kshetri N (2018) 1 Blockchain's roles in meeting key supply chain management objectives. Int J Inform Manag 39(December 2017):80–89. https://doi.org/10.1016/j.ijinfomgt.2017.12.005

Kyaw TY, Ng AK (2017) Smart Aquaponics System for Urban Farming. Energy Procedia 143:342–347. https://doi.org/10.1016/j.egypro.2017.12.694

Lakhiar IA, Gao J, Syed TN, Chandio FA, Buttar NA (2018) Modern plant cultivation technologies in agriculture under controlled environment: A review on aeroponics. Journal of Plant Interactions 13(1):338–352. https://doi.org/10.1080/17429145.2018.1472308

Li W, Clark B, Taylor JA, Kendall H, Jones G, Li Z, Jin S, Zhao C, Yang G, Shuai C, Cheng X, Chen J, Yang H, Frewer LJ (2020) A hybrid modelling approach to understanding adoption of precision agriculture technologies in Chinese cropping systems. Comput Electron Agric 172(February):105305. https://doi.org/10.1016/j.compag.2020.105305

MacCallum R (1986) Specification searches in covariance structure modeling. Psychol Bull 100(1):107–120. https://doi.org/10.1037/0033-2909.100.1.107

Martin M, Weidner T, Gullström C (2022) Estimating the potential of building integration and regional synergies to improve the environmental performance of urban vertical farming. Frontiers in Sustainable Food Systems 6(March):1–18. https://doi.org/10.3389/fsufs.2022.849304

Maurya P, Jain S, Kumar V, Kiran B (2023) Vertical farming: an option in modern food production: a review. IJECC 13:883–893. https://doi.org/10.9734/ijecc/2023/v13i92308

Mukherjee S, Baral MM, Lavanya BL, Nagariya R, Singh Patel B, Chittipaka V (2023) Intentions to adopt the blockchain: investigation of the retail supply chain. Manag Decis 61(5):1320–1351. https://doi.org/10.1108/MD-03-2022-0369

Okomoda VT, Oladimeji SA, Solomon SG, Olufeagba SO, Ogah SI, Ikhwanuddin M (2022) Aquaponics production system: A review of historical perspective, opportunities, and challenges of its adoption. Food Sci Nutr 11:1157–1165. https://doi.org/10.1002/fsn3.3154

Palm HW, Knaus U, Appelbaum S, Goddek S, Strauch SM, Vermeulen T, Haïssam Jijakli M, Kotzen B (2018) Towards commercial aquaponics: a review of systems, designs, scales and nomenclature. Aquacult Int 26(3):813–842. https://doi.org/10.1007/s10499-018-0249-z

Petrovics D, Giezen M (2022) Planning for sustainable urban food systems: an analysis of the up-scaling potential of vertical farming. J Environ Planning Manage 65(5):785–808. https://doi.org/10.1080/09640568.2021.1903404

Pfeiffer J, Gabriel A, Gandorfer M (2021) Understanding the public attitudinal acceptance of digital farming technologies: a nationwide survey in Germany. Agric Hum Values 38(1):107–128. https://doi.org/10.1007/s10460-020-10145-2

Pituch KA, Stevens JP (2016) *Applied Multivariate Statistics for the Social Sciences* (6th edn). Routledge/Taylor & Francis Group

Rufí-Salís M, Calvo MJ, Petit-Boix A, Villalba G, Gabarrell X (2020) Exploring nutrient recovery from hydroponics in urban agriculture: An environmental assessment. Res Conserv Recycl 155:104683. https://doi.org/10.1016/j.resconrec.2020.104683

Shao Y, Wang Z, Zhou Z, Chen H, Cui Y, Zhou Z (2022) Determinants affecting public intention to use micro-vertical farming: a survey investigation. Sustainability (switzerland) 14(15):1–26. https://doi.org/10.3390/su14159114

Son JE, Kim HJ, Ahn TI (2020). *Chapter 20 - Hydroponic systems* (T. Kozai, G. Niu, & M. B. T.-P. F. (Second E. Takagaki (eds.); pp. 273–283). Academic Press. https://doi.org/10.1016/B978-0-12-816691-8.00020-0

Stone RN, Grønhaug K (1993) Perceived risk: Further considerations for the marketing discipline. Eur J Mark 27(3):39–50. https://doi.org/10.1108/03090569310026637

Sun Y, Wang N, Xitong G, Zeyu P (2013) Understanding the acceptance of mobile health services: a comparison and integration of alternative models. J Electron Comm 9(1):179–185

Trivette SA (2019) The importance of food retailers: applying network analysis techniques to the study of local food systems. Agric Hum Values 36(1):77–90. https://doi.org/10.1007/s10460-018-9885-1

Vadakkepatt GG, Winterich KP, Mittal V, Zinn W, Beitelspacher L, Aloysius J, Ginger J, Reilman J (2021) Sustainable Retailing. J Retail 97(1):62–80. https://doi.org/10.1016/j.jretai.2020.10.008

Venkatesh V, Morris MG, Davis GB, Davis FD (2003) User Acceptance of Information Technology: Toward a Unified View. MIS Quart 27(3):425–478. https://doi.org/10.2307/30036540

Venkatesh V, Zhang X (2010) Unified theory of acceptance and use of technology: US vs. China. J Glob Inf Technol Manag 13(1):5–27

Wei Y, Li W, An D, Li D, Jiao Y, Wei Q (2019) Equipment and intelligent control system in aquaponics: A review. IEEE Access 7:169306–169326. https://doi.org/10.1109/ACCESS.2019.2953491

Wong LW, Tan GWH, Lee VH, Ooi KB, Sohal A (2020) Unearthing the determinants of Blockchain adoption in supply chain management. Int J Prod Res 58(7):2100–2123. https://doi.org/10.1080/00207543.2020.1730463

Zareba A, Krzeminska A, Kozik R (2021) Urban vertical farming as an example of nature-based solutions supporting a healthy society living in the urban environment. Resources 10(11):109. https://doi.org/10.3390/resources10110109

# Quality Education: A Key Component of the United Nations Sustainable Development Goals (SDGs)

**Apurva Goyal, Puneet Kumar, and Shalini**

**Abstract** Education is seen as a basic human right and as a key component of progress of any country as India. Quality education is commentative for achieving the United Nations Sustainable Development Goals (SDGs) adopted in 2015, particularly Goal 4, which aims to confirm inclusive and unbiased quality education and promote permanent learning opportunities for all and encouraginginter-national bonds, cultural legacy, and the equivalence of all civilizations, it seeks to assurance that all child and citizen has contact to a decent education. The United Nations (UN) developed the Sustainable Development Goals (SDGs), also known as the Worldwide Goals, in 2015 as a worldwide call to action to end poverty, keep the environment, and confirm that everyone lives in liberty and peace by 2030 (de Villiers et al., 2021).Therefore we can say that the Sustainable growth Goals (SDGs) are an essential aspect of the nation's rich culture and history and essentially reflect India's aspirations for growth.

This chapter provides an overview of the importance of quality education in the perspective of the SDGs, highlighting the challenges and opportunities that exist in achieving this goal. By including SDGs into the formation of a new education strategy, we hope to present and discuss India's current tactic or initiatives with an eye towards reaching or contextualising the SDG for education, as well as their hopeful impression on the outcomes of superior training, as projected by 2030.

**Keywords** SDGs · Lifelong Learning · Inclusive Education · e-learning · RTE · SWYAM · NEP

A. Goyal (✉) · P. Kumar · Shalini
Department of Commerce, IIMT University, Meerut, UP, India
e-mail: aapurvab@gmail.com

 
W. Leal Filho et al. (eds.), *Digital Technologies to Implement the UN Sustainable Development Goals*, World Sustainability Series,
https://doi.org/10.1007/978-3-031-68427-2_7

## 1 Introduction

The idea of sustainable development aims to establish a stability between economic growth, social advancement, & environmental conservation in order to satisfy present requirements without negotiating the potential of upcoming groups to satisfy their individual requirements. It involves encouraging financially viable expansion, social equity, and ecological preservation through responsible policymaking and actions. Sustainable development seeks to address the interconnected challenges of poverty reduction, environmental degradation, and social inequality in a way that is mutually reinforcing and integrated. It is a global goal and is enshrined in the United Nations Sustainable Development Goals adopted in 2015 (Kawaguchi and Kuroda, 2013). Education is important human right and a high-powered tool for promoting financial development, social development, and individual empowerment (Mohanty, 2018).

Quality education is crucial for building prosperous and sustainable societies, reducing poverty and inequality, and promoting democratic governance. However, despite significant progress over the past limited decades, many countries are still struggling for offering reasonable admittance to admirable learning for all. In response, the United Nations (UN) adopted the Sustainable Development Goals in 2015, (UNESCO 2015) it include a specific target on education (SDG 4). Education is a powerful tool that can transform individuals, societies, and nations. Quality education is essential for achieving sustainable development and building a better world. The United Nations' Sustainable Development Goals recognize the importance of education in creating a sustainable future for all. Quality education is one of the 17 SDGs accepted by the United Nations in 2015, with the goal of ensuring inclusive and unbiased quality education and helping lifetime learning chances for all (Bajpai and Goyal, 2004). The SDGs acknowledge that access to high-quality education is aimportant human right and is necessary for achieving sustainable development and decreasing poverty. Goal 4 of the SDGs is set In order to make sure wide-ranging and balanced excellence education and encourage permanent education probability for each and every one by the year 2030. (Pandey, 2018).

Several goals have been established in order to accomplish the SDG on excellent education. These include attaining universal primary and secondary education, offering equivalent right to use to inexpensive as well as high-value scientific, professional, and higher education, & guaranteeing that totally kids have entry to eminence premature infancy improvement, prevention, and pre-primary schooling.

In addition to access, the SDG on quality education also focuses on improving the superiority and worth of learning. This includes ensuring that all students receive education for sustainable growth and worldwide nationality learning, as well as the information and abilities required to support sustainable development. It also includes enhancing teacher training and support, improving learning environments, and increasing the availability of educational resources and technologies.

The SDGs emphasize the need for international cooperation and partnerships to achieve their goals, and quality education is no exception. Association between governments, social culture organizations, & the private segment is essential for improving access to quality education, especially in low-income and conflict-affected nations. This chapter examines the purpose of excellence learning in achieving the SDGs and explores innovative approaches to get better way in learning furthermore enhance learning outcomes.

## 2 Related Work

The idea of sustainable expansion is an old concept. Numerous studies have dedicated their careers to protect the globe's life. Since its foundation, the United Nations has enthusiastic for building the globe in an advanced situation to survive. (Tomislav, 2018). Establishing recommendations that would help everyone move towards a more prosperous and sustainable future is the basic and the major objective stated under the in 2015, United Nations embrace the Sustainable Development Goals. All people on the earth are to have access to a sustainable learning organism, which is one of the seventeen SDGs. There are seven objectives of SDG 4 to deal with concerns including the inflated dropout figure in elementary, high, and higher learning along with the requirement to offer lucky chance for the less auspicious, such as marginalized communities, indigenous working class, and incapacitated people. There are three supplementary objectives which help in attaining targets of SDG 4.

The Government of India is focusing upon this appropriate aspect has launched several initiatives such as SWAYAM, National Education Policy (NEP), MOOCS, SARANSH and UDAAN etc. They are all build up to attain the SDG goal 4 exclusively. However all these efforts proved to be ineffective due to the COVID-19 Pandemic that stroked the World, and distorted the whole situation (Singh et al., 2021).

According to numerous academics, Gender gaps in education must be eliminated. Women are graduating with higher grades and at a higher rate than men. However contrary to this in India women's educational achievement continues to lag well behind that of men (Rammohan and Vu, 2018). By 2030, India has committed to provide all children with broad, unrestricted, and high-class learning. The Right to Education (RTE) and the Sarva Shiksha Abhiyan (SSA) have enhanced India's learning organism (Pandey, 2018). But still the government needs to address some of the major challenges that are unique in our Indian society due to our religious and cultural mindsets. They are somehow or the other proves to be a break towards the achievement of our goals.

In the framework of the SDGs, this work presents an overview of the significance of high-quality education while highlighting the opportunities and challenges that must be overcome to accomplish this objective. We hope to present and discuss India's current strategy or initiatives with a view to achieving the SDG for education along with their encouraging impact on the end result of superior preparation, as projected till 2030, by comprising SDGs addicted to the creation of a new education policy.

## 3 Importance of Quality Education as Per SDG in India

The cornerstone of social advancement, education empowers people and shapes nations. The quality of education is crucial for promoting equitable progress, innovation, and economic and social growth in India, a nation with a sizable and varied population. Nevertheless, despite admirable advancements in increasing access to education, maintaining a good level of education for all students continues to be a difficult task. Quality education is not just about obtaining understanding and expertise but also regarding upward critical thinking, problem-solving, creativity, and social skills. Quality education enables individuals to make informed decisions, full contribution in society, financial growth and sustainable expansion. Furthermore, quality education is essential for achieving other SDGs through reducing poverty and hunger (SDG 1 and 2), encouraging femininity parity (SDG 5), and securing healthy lives (SDG 3). Children in India's expulsion from the system for low academic performance leaves kids open to child labour, abuse, and violence. Many schools still practise discrimination, corporal punishment, and teacher-centered rote learning.

The 2009 Right to Education Act grants right to children to be free and essential education. While we have almost universal school enrollment thanks to the focus on increasing enrollment, the advantages of education can be seen not by the number of years spent in school but rather by the outcomes of learning.

A prime example of merit goods, as defined by renowned economist Richard Musgrave, is education. Unlike other common purchases, merit items have benefits that are not completely understood by the consumer at the stage of consumption. Children along with their parents are unable to understand the genuine personal benefits of studying hard in terms of potential careers, salaries, or social prestige; all they can perceive is the current sacrifice that is necessary. As a result, people could underinvest in education. (Saini et al., 2023). Even more crucially, the consumption of merit items has advantageous knock-on effects. In other words, education has value that extends beyond the individual and has positive effects on society as a whole. A well-educated and talented person increases economic production, which is essential for success. Given the advantages for both individuals and society as a whole, the majority of nations have established minimum educational requirements and offer free or heavily discounted basic education.

## 4 Major Targets to Be of SDG Goal 4

1. By 2030, make sure that every girl and boy receives a free, fair, and brilliance primary and secondary education, foremost to successful knowledge outcomes.
2. By 2030, guarantee access to quality premature babyhood growth, care, and pre-primary education for all girls and boys to get ready them for primary education.
3. By 2030, promote equal access to reasonable and excellence mechanical, professional, and tertiary education, including universities, for both women and men.

4. By 2030, significantly raise the quantity of young citizens and adults with the required skills for service, highly regarded service, and entrepreneurship.
5. By 2030, Ensure that all weak groups, including people with disabilities, indigenous peoples, and children in difficult situations, have equal access to all levels of education and vocational training.
6. By 2030, Ensure that All Young People and a Substantial Proportion of Adults, Regardless of Gender, Achieve Literacy and Numeracy.
7. By 2030, make sure that all students are equipped with the knowledge and abilities required to advance sustainable development, including instruction in such areas as global citizenship, human rights, gender equality, nonviolence, and education for sustainable development.
8. Construct and improve education facilities that are inclusive, safe, nonviolent, and gender-sensitive, providing effective learning environments for all, including children and individuals with disabilities.
9. By 2020, Increase the global availability of scholarships for students from rising nations, particularly those from the least developed, little land mass developing states, and African nations, to be present at advanced learning, including scientific, engineering, and technical programs in developed and other developing nations, as well as programmes for professional guidance and information and communications technology.
10. By 2030, enlarge the figure of capable educators significantly, together with worldwide alliance on educator teaching in developing nations, mainly the least developed and miniature island developing states.

## 5 Understanding of Inclusive and Equitable Quality Education

The Salamanca Statement, which was issued by the World Congress on Special Needs Education (WCSNE) in 1994, made the case for inclusive education on a global scale. There are many different definitions of inclusion. According to UNESCO (2003, p. 7), when it comes to education, inclusion is known as a process of addressing and responding to the variety of requirements of all learners with growing involvement in information, cultures, and communities, and dropping segregation from and within education. In this sense, inclusive education is viewed as a method of instruction that realises the idea of inclusive education that emphasizes its approach and ability to meet a variety of requirements. The integrated education ideology, which called for the inclusion of children with special needs prior to the establishment of inclusive education, required that students adapt to typical classroom environments. Contrary to this concept of integrated education, inclusive education advocates for educators to pay attention to students' needs (Kawaguchi and Kuroda 2013).

Thus we can say that more than just information dissemination is involved in quality education. It includes the building of character as well as the improvement of critical thinking, creativity, and problem-solving abilities. It equips people to think critically, analyze problems, and look for solutions in order to become active members of society. The various essential components that make up quality and inclusive education:

- Curriculum Design: Designing a strong curriculum that represents a thorough and balanced perspective on learning is a crucial aspect of high-quality education. The curriculum ought to promote interdisciplinary thinking, originality, and practical relevance. Adopting learner-centered pedagogical strategies that promote active engagement, inquiry-based education, and the cultivation of practical skills is equally vital.
- Faculty Development Programs: The main proponent of high-quality education is teachers. It is essential to ensure their professional growth and empowerment. This entails finding skilled and driven instructors, offering ongoing training opportunities, and fostering their development as knowledge facilitators. Education providers are better able to provide high-quality instruction when innovation, creativity, and critical thinking are encouraged.
- Assessment System: An effective assessment system and regular feedback ensures educators to identify strengths and weaknesses with customization of teaching strategies as per the requirements and continual enhancement in learning processes.
- Infrastructure: The foundation of a high-quality education is a supportive learning environment. A stimulating and rewarding learning environment may be fostered by adequate infrastructure, which includes well-stocked classrooms, libraries, labs, and accessibility to online resources. Fostering holistic growth requires a secure, welcoming atmosphere that values variety, cooperation, and respect.
- Inclusion along with equity: Bridging the gap between urban and rural communities, reducing gender disparities, and addressing the needs of disparaged communities is essential for achieving inclusive and equitable education. Focuses and effective attention needs to be paid towards providing adequate resources, scholarships, and support systems for underprivileged students. Quality education must reach all segments of society, leaving no child behind.

## 6 Issues Related to Inclusive and Equitable Quality Education in India

Numerous factors impede the efficacy and influence of India's educational system's quality of education. Some of the most significant problems include:

- **Mindless learning:** Majority of the schools place a heavy emphasis on rote memorization and learning strategies that are focused on exams. This method frequently prevents pupils from developing their critical thinking, problem-solving, and creative abilities.

- **Untrained Teachers:** In many schools, especially in rural regions, there is a lack of well-trained, competent instructors. This problem is exacerbated by insufficient professional development opportunities, recruiting, and retention strategies. (Education For All in India, 2023)
- **Ineffective Curriculum:** Many educational institutions' curricula are out-of-date and distant from practical applications. It might not effectively foster a comprehensive grasp of subjects or handle the changing requirements of the modern workforce.
- **Unavailability of faculty Development Programs**: The ability of educators to provide high-quality education is hampered by an inadequate focus on teacher training and professional development programmes. Their capacity to adjust to shifting educational demands is constrained by the lack of options for continual learning and upskilling.
- **Resources and Infrastructure Disparities:** Many schools, particularly those in rural areas, lack the most basic facilities, such functional classrooms, libraries, and labs. The lack of access to technology and educational materials makes the quality disparity even worse.
- **Gender Discrimination:** Access to education continues to be unequal between men and women, especially in rural regions, Mitra et al. (2023).The overall quality of education is impacted by lower enrollment and retention rates among females due to social and cultural issues, safety concerns, and discriminatory views.
- **Lack of focus towards Extra-curricular development:** Neglecting Students total growth and well-being might restrict their capacity to prosper in a variety of spheres of life. The emphasis on academic success frequently sidelines holistic development, including physical education, the arts, and extracurricular activities. (Education For All in India, 2023)
- **Outdated Education policies:** The efforts to raise the standard of education are hampered by inadequacies in the execution of policy, ineffective bureaucracy, and a lack of accountability in the educational system.

## 7 Current Status of Education in India as Per Unified District Information System for Education Plus (UDISE + ) Report 2022

After 2018, the number of schools in India began to decline, as depicted in Fig. 1. Before then, there had been a extensive enhance in the figure of schools, particularly after it soared to 15,58,903 in 2017 from 15,18,160 in 2014. However, the decreasing tendency does not point to a low level of education. In actuality, the consolidation of schools throughout India was the cause of the decline in the number of schools. However, the number of schools grew to 15,09,136 in 2021 from just 15,07,708 in 2020, putting a halt to the falling trend. Additionally, in 2022, there will be 14,08,115 schools, as a result of school mergers.

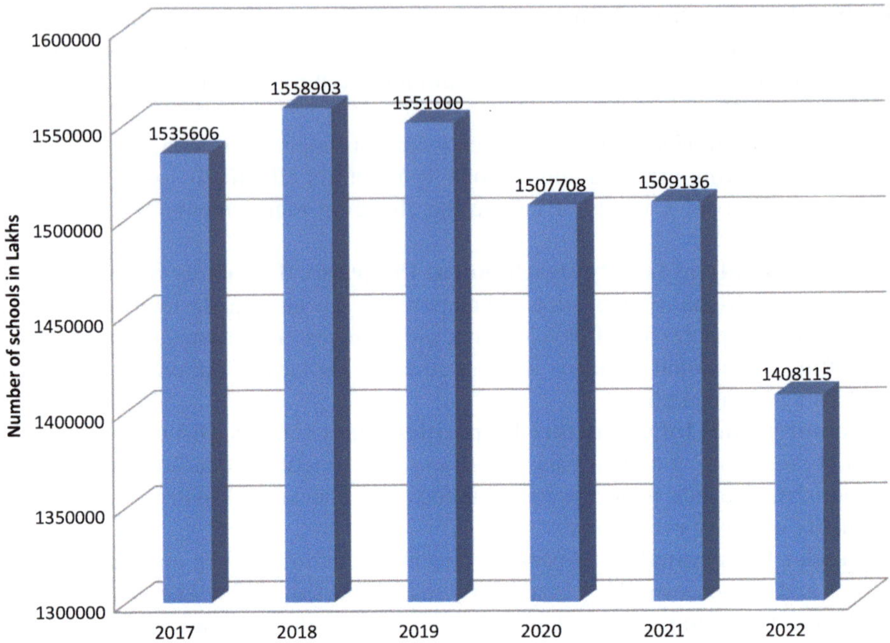


**Fig. 1** Number of Schools in India. *Source: Latest UDISE + Report 2022*

India is estimated to have 14, 89,115 schools as of 2022, which is higher than 14 lakhs. The level of education, such as elementary, secondary, etc., as well as masculinity distinctions, such as girls-only schools, boys-only schools, and co-educational schools, are used to categorise these schools into several groups and India has to maintain number of schools for achieving its SDG goal 2030.

Primary, secondary, and higher secondary are the three educational levels that schools in India as stated in Fig. 2. According to the most recent official statistics, 11,96,265 of India's schools are classified as elementary schools, compared to 1,50,452 secondary schools and 1,42,398 upper secondary schools. Since the 86th alteration to the Indian Constitution took effect in 2002, all children in India between the ages of 6 and 14 are now required to get free basic education as an effort towards achieving the SDG goals by 2030, therefore major focus in towards increasing the number of primary education as well the number of enrollments in primary education irrespective of gender and other disparities prevailing in the economy.

Government of India is making great efforts towards achieving United Nations sustainable development goals and therefore focusing towards enhancing government schools and education, so that it can be easily accessible to large number of students as depicted in Fig. 3. As per the latest official data for 2022, government schools are the most dominant, with the overall number of government schools in India being 10, 22,386, and the number of government-aided schools being 82,480. The government schools enhances our education system through its affordability

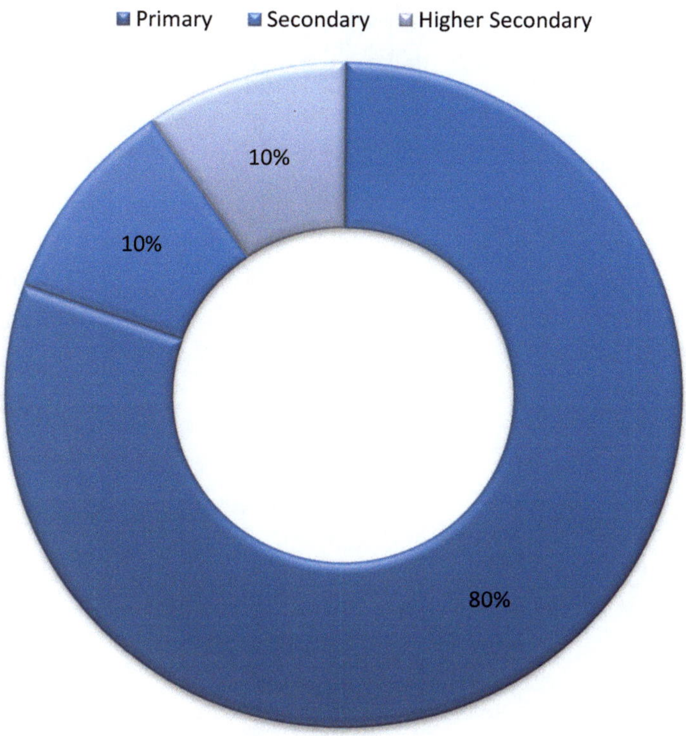


**Fig. 2** Number of Schools distributed as per level of education. *Source: Latest UDISE + Report 2022*

and various government aided initiatives towards providing whole without charge, impartial and excellence primary and secondary schooling leading to associated and Goal-4 efficient learning outcomes.

The Fig. 4 clearly shows that there has been a significant increase in amount of students enrollment in India. A large figure of students were enrolled in basic through higher secondary education in total in 2021–22. 13.28 crore of males and 12.28 crore of girls were enrolled. Over 2020–21, this represented a rise of more than 19 lakhs. Many primary schools inside the Pre-primary lessons were offered under UDISE + system. Pre-primary through higher secondary institutions had a combined enrollment of almost 26.52 billion students in 2021–22. School enrollment has grown at the upper primary, secondary, and higher secondary levels, representative an improvement in the system's capacity to grasp onto more kids in the educational system over time. Figures 5, 6 demonstrate the rise in the number of boys and girls enrolled in various educational levels.

According to the Indian government's latest UDISE + report for 2020–21 and Fig. 7, depicts that the total number of students in India attending school has increased significantly over the past few years, which has helped the country achieve the United

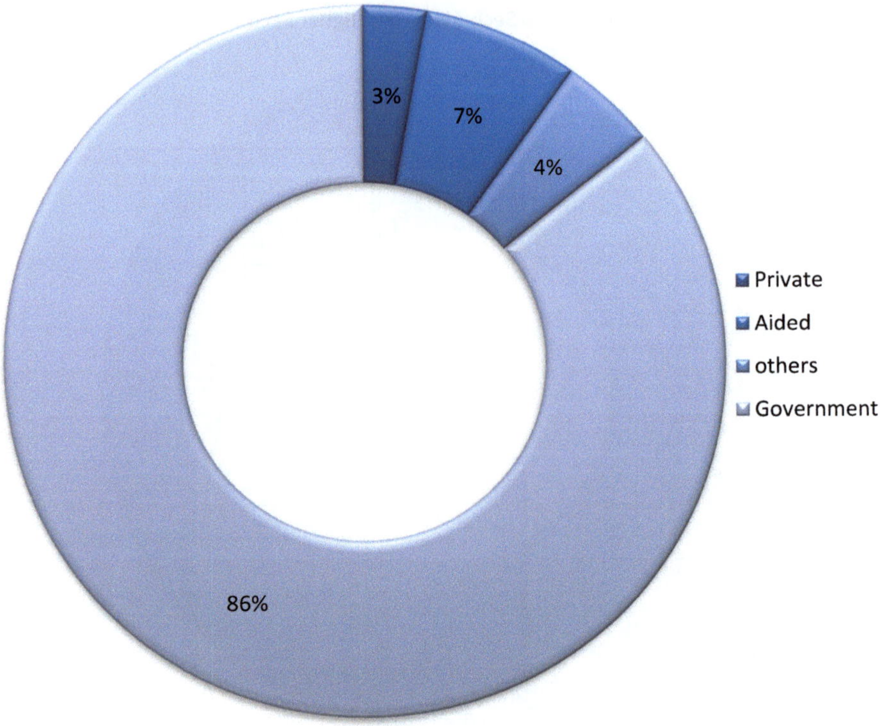


**Fig. 3** Number of Schools as per ownership in India. *Source: Latest UDISE + Report 2022*

Nations Sustainable Development Goals, particularly Goal 4, which calls for equivalent approach to all position of literacy for men and women and supports their attainment of literacy and numeracy.

Focusing on the first and foremost goal of SGD4 in India there has been an increase in the elementary schools, the Gross Enrollment Ratio from 101.3% in 2018–19 to 103.3% in 2020–21, and then to 103.39% in 2021–22. Similarly in upper primary schools there has been a noticeable increase from 92.2% in 2020–21 to 94.67% in 2021–22. According to the UDISE + report for 2020–21, the Ratio for secondary schools was 79.8%, however it is now 79.56 in 2021–22. In 2018–19, the GER for secondary schools was just 76.9%. In comparison to the other levels of schooling, the higher secondary level has the lowest gross enrollment ratio. In 2018–19, the GER was just 50.14%. The GER did, however, rise to 53.8% in 2020–21 and subsequently to 57.56 that depicts majestic growth curve in Fig. 8.

The Fig. 9 clearly shows that there has been a noticeable increase in number of female students in India as of 2022.

Figure 10, compares the availability of a few key infrastructural amenities in schools from 2018–19 to 2021–22. This includes access to the internet, a library, a ramp, a washroom, and water. According to the UDISE + 2021–22 statistics, all

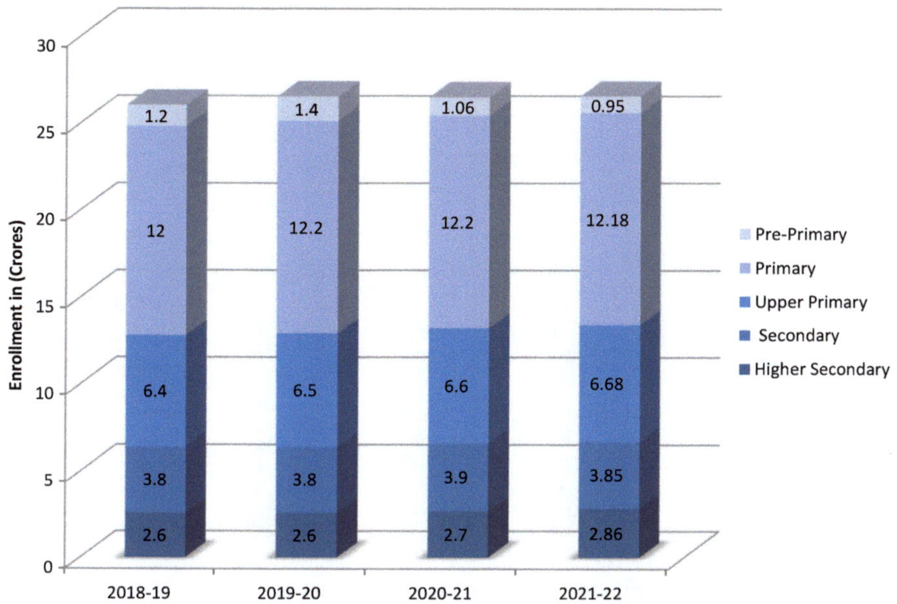


**Fig. 4** Total School Enrollment in India. *Source: Latest UDISE + Report 2022*

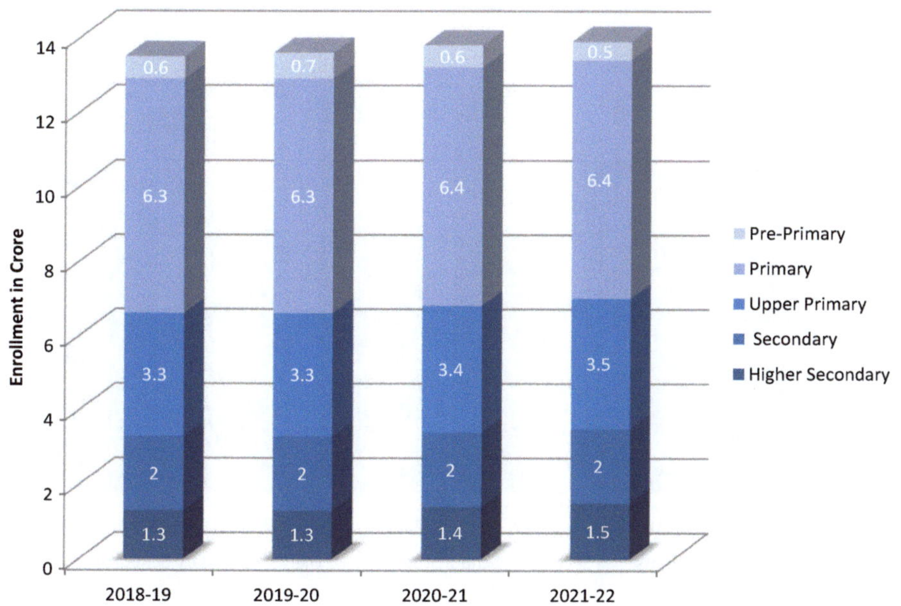


**Fig. 5** Enrollment of boys in Schools in India. *Source: Latest UDISE + Report 2022*

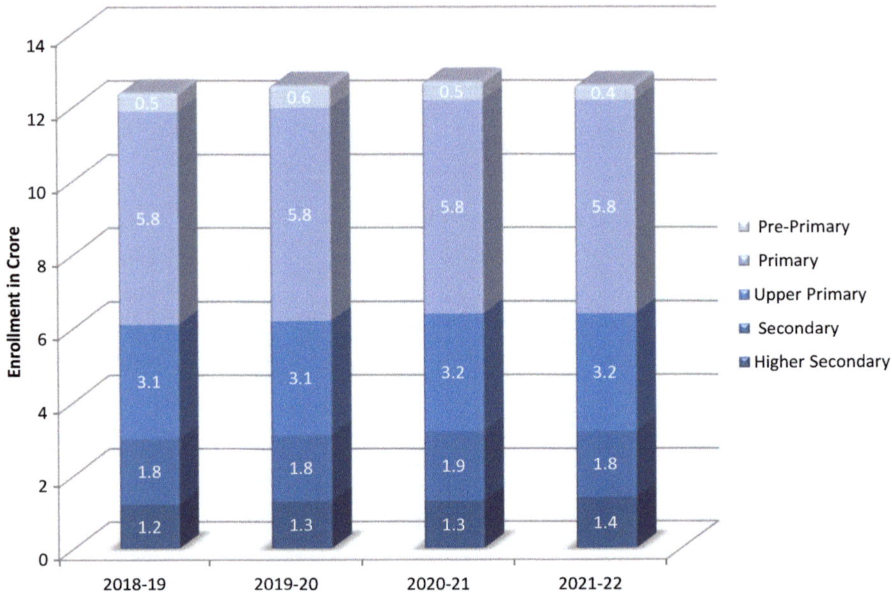


**Fig. 6** Enrollment of Girls in Schools in India. *Source: Latest UDISE + Report 2022*

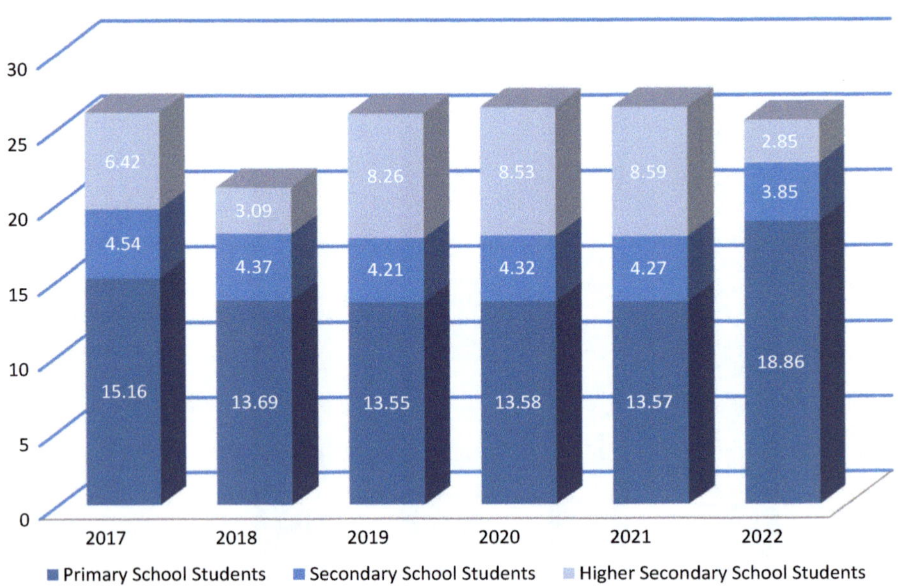


**Fig. 7** Number of School Students in India. *Source: Latest UDISE + Report 2022*

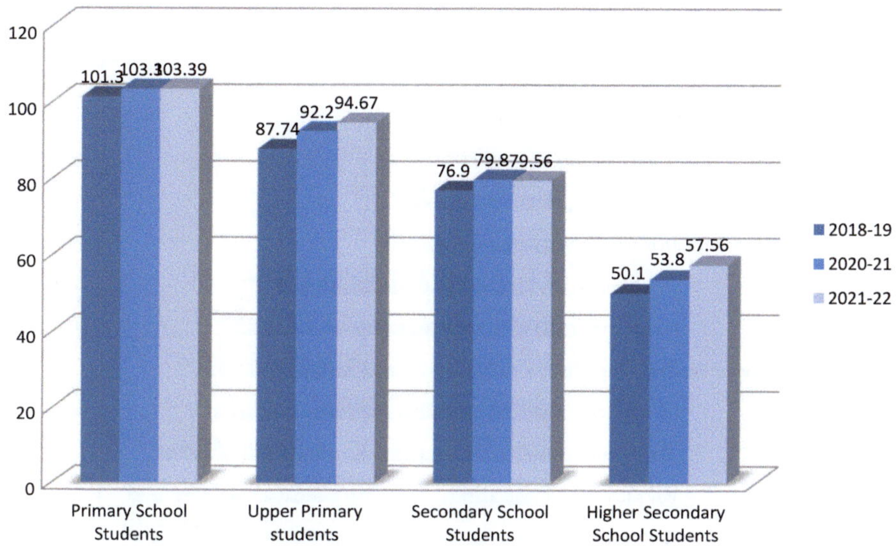


**Fig. 8** Gender Enrollment Ratio in India. *Source: Latest UDISE + Report 2022*

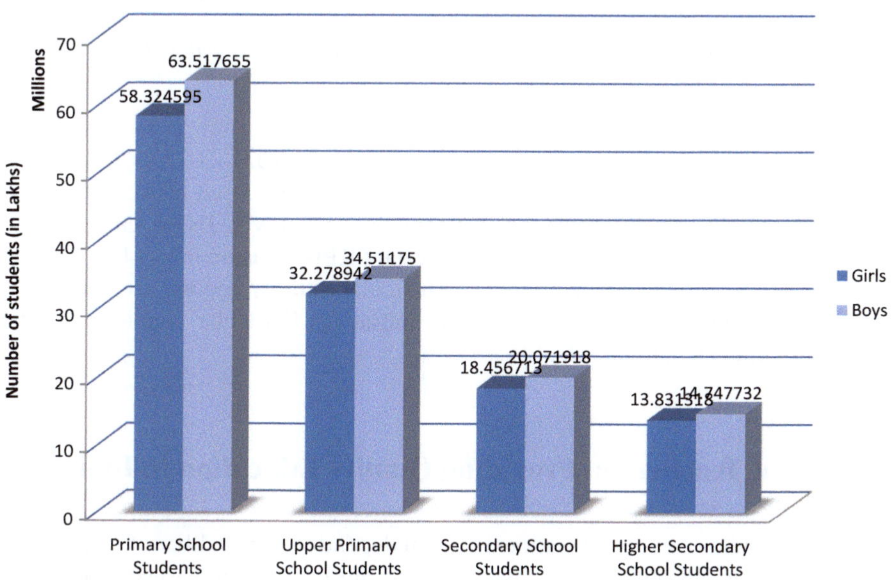


**Fig. 9** Distribution of students as per gender in India. *Source: Latest UDISE + Report 2022*

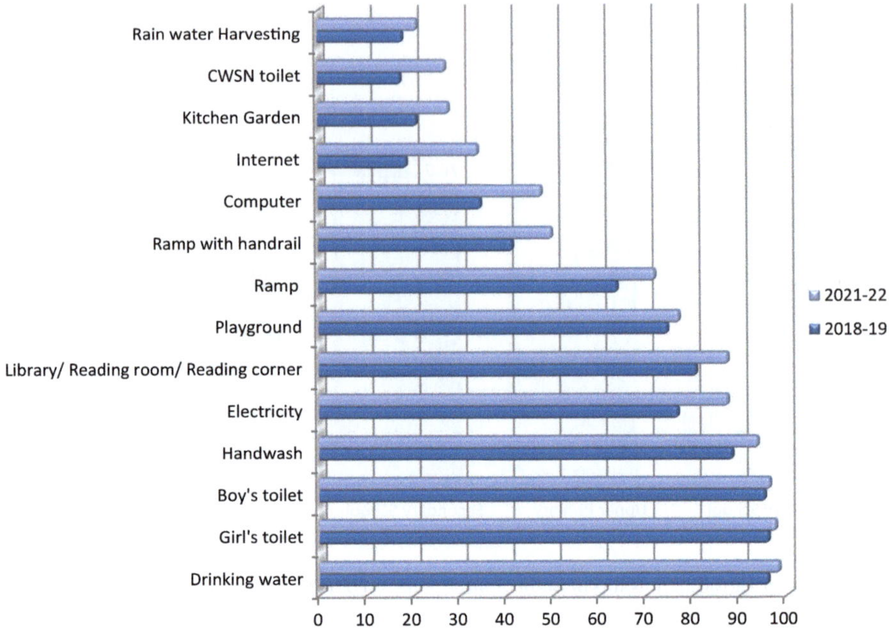


**Fig. 10** Status of Infrastructure and Basic Facilities in Indian Schools. *Source: Latest UDISE + Report 2022*

of the key infrastructure facilities saw an increase from 2021–2022 to 2021–2022. For the first time, additional facility-related statistics have been broken down and presented individually for State Government Schools, KVS, JNVs, Government-aided, Private, etc. Similar to this, some important KPIs measure things like whether a school has integrated science labs, co-curricular activity rooms, arts and crafts facilities, and environmental sustainability initiatives like solar panels, rainwater harvesting, kitchen gardens, etc.

## 8 Role of Teachers in Providing Quality Education in India

The effectiveness of every educational system depends on the caliber of its teachers, who in turn depend on how effectively lessons are taught and learned. The development of society and the necessary social reform depend on the role of teachers. As a result, the quality of higher education is determined by the quality of people who deliver it. The most crucial elements of any educational setting are the teachers. Teachers are essential to the development of the educational system as a whole, the sharing of knowledge, and the preservation of values in higher education.

The various roles played by a teacher are:

- **Motivating Factor:** It is the accountability of the educator to produce an atmosphere in which the students' wish and capability to learn can perform role most efficiently. A teacher should operate as a driving strength and be willing and able to establish an encouraging environment for learning in which learners are given the opportunity to think with awareness, logically, communicate their opinions, and decide on the situation and difficulties.
- **Skill Development:** The capacity of learners to develop their skills will determine how successful they are in the employment market. Making sure that students' skill development is similar to that of their counterparts overseas is a crucial part of enhancing the standard of higher education. In current era of economic liberalisation and globalisation, it is essential to produce highly qualified human resources. As a result, it is obvious that higher education must follow accepted global standards.
- **Designing of an effective curriculum:** By creating need-based curricula while taking into account the demands of the national and international employment market, higher education's quality may be improved.
- **Imparting expertise:** It has been demonstrated by numerous research studies that the quality of instructors has significant influence on the caliber of the pupils. Hence, a teacher, who is a constant learner has to refresh the subject knowledge continually and should be aware of recent advancement in their topic.
- **Academic Development:**The teachers are the most important components of the higher education system. Since instructors are the primary motivators and catalysts for the all-around development of students, the success of the higher education system rests on their academic progress.In order to impart superior knowledge to the students, instructors must have a high level of professional expertise; they are essential to maintaining and raising the bar for higher education (Jha and Kumar, 2012).This is known as continuing professional development of the educators, which is a vital guarantee of a high-quality education.

## *8.1 Availability of Teachers in India*

There are 16,20,001 secondary school teachers, 27,01,507 upper secondary school teachers, and 51,85,615 elementary school teachers in India. In other words, the total amount instructors in India may be further classified based on the type of education they provide (Figs. 11, 12, 13, 14).

Undoubtedly, there has been a major improvement in the student–teacher ratio during 2020–21.

There is no doubting that the Covid-19 epidemic had a extensive harmful influence on education. Along with all the other sectors, education was also halted, which led to a decline in student enrollment, especially for those in lower grades and those with special needs. In spite of all the challenges, the Indian educational system has succeeded in providing the best instruction possible, especially given the country's growing popularity of online learning.

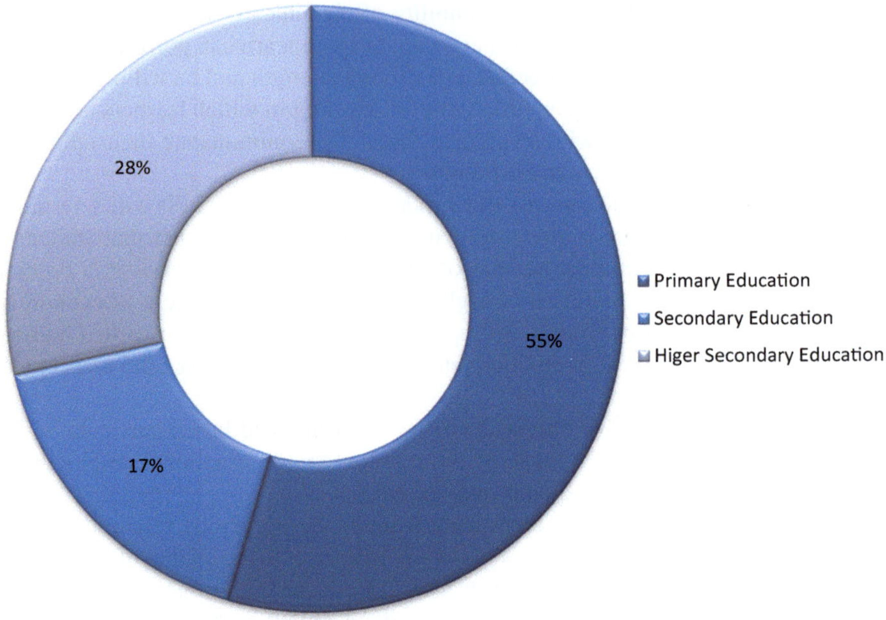


**Fig. 11** Number of teachers as per level of education. *Source: Latest UDISE + Report 2022*

## 9 Civil Society Organizations (CSOs) and Their Contribution Towards Providing Quality Education in India

CSOs are non-governmental, nonprofit groups that strive to improve society. By tackling diverse social, economic, and political challenges, they want to achieve social justice and empower individuals. Non-governmental organisations (NGOs), community-based organisations (CBOs), and advocacy groups are all examples of CSOs and are crucial to civil society. The various CSOs working in India includes The Akshay Patra Foundation, Teach-for-India and Pratham etc.

CSOs work in accordance with government of India to promote education. The government has created a number of partnerships and collaborations to strive towards this objective because it recognises the value of CSOs in advancing the educational system. Indian civil society organisations (CSOs) work in a variety of fields with the goal of enhancing the social, economic, and political climate of the nation. CSOs in the field of education seek to provide access to high-quality education, develop the abilities of teachers and school administrators, and support creative approaches to teaching and learning. CSOs in India mostly work in the following areas:

- Digital Education: CSOs utilise technology to expand educational opportunities and encourage digital literacy. For the purpose of educating students in far-off places, they create online courses and electronic education systems.

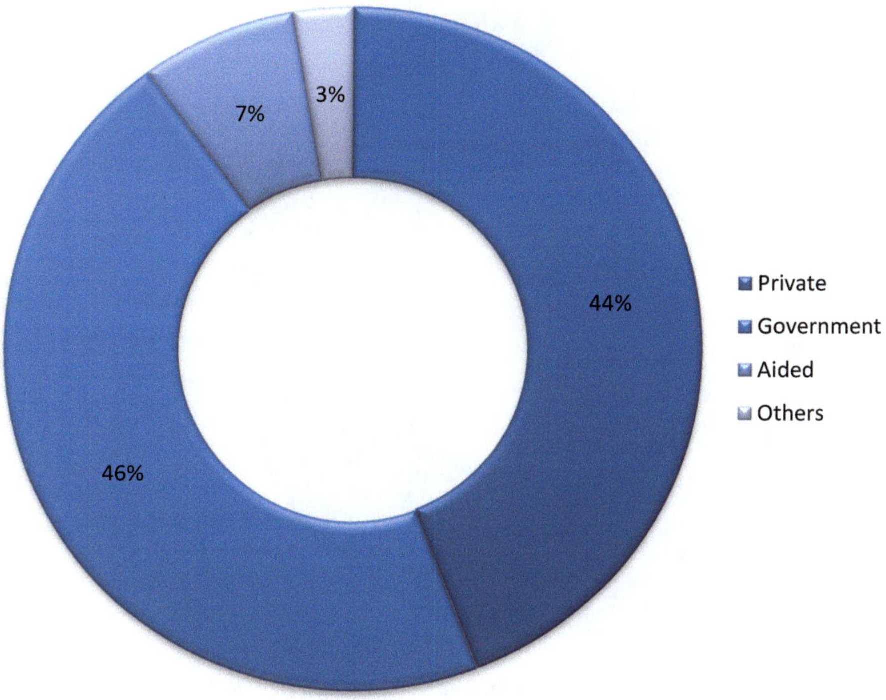


**Fig. 12** Number of Female Teachers in India. *Source: Latest UDISE + Report 2022*

- Primary Education: CSOs seek to increase children's access to high-quality education in both town and countryside regions. To improve the standard of education, they assist government schools, build new ones, and offer teacher preparation programmes.
- Special Education: CSOs seek to provide special needs and disabled children with education. In addition to offering special education and therapeutic services, they encourage inclusive education.
- Community Engagement: CSOs work with stakeholders and local communities to increase their involvement in the educational system. They strive to encourage neighborhood-based projects and inclusive decision-making procedures.
- Educational Policy: In order to impact education policy and advance systemic changes, CSOs participate in research, lobbying, and policy analysis. They want to increase expenditures on education and raise educational standards.

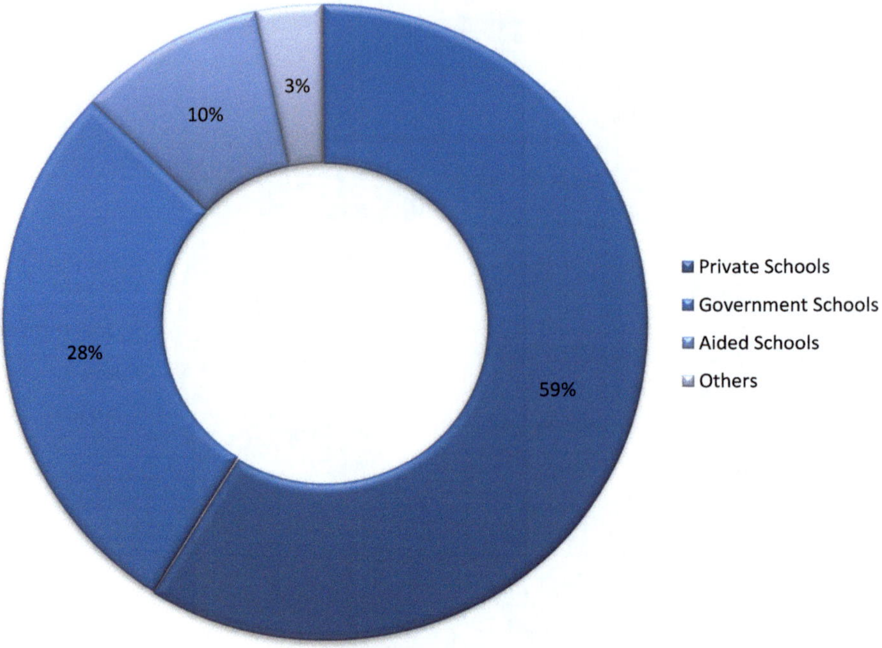


**Fig. 13** Number of Male Teachers in India. *Source: Latest UDISE + Report 2022*

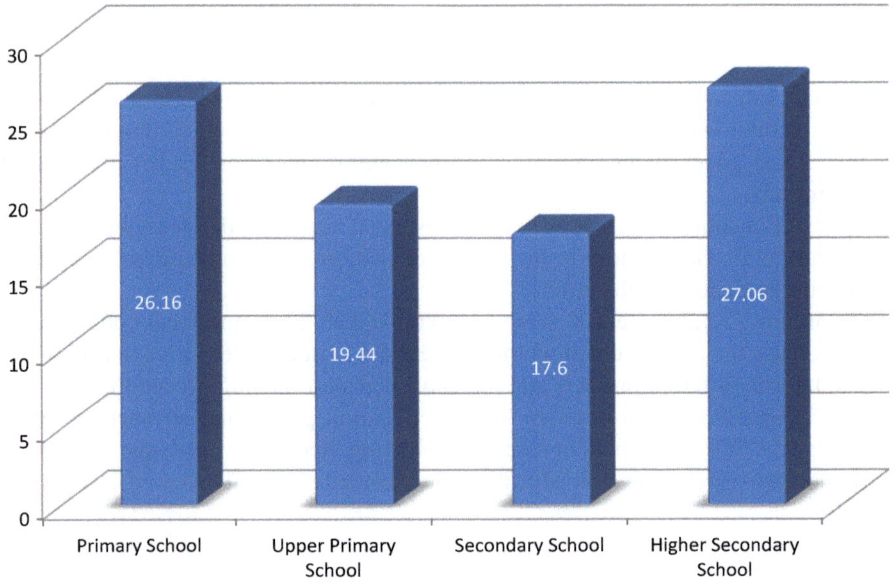


**Fig. 14** Student–Teacher Ratio in Schools in India. *Source: Latest UDISE + Report 2022*

## 10 Contribution of CSOs in promoting education in India

- Evaluation and Monitoring: CSOs are authorized for keeping a check on education system's performance and also provide improvement suggestions for any discrepancies. Therefore they may offer criticism and recommendations for improvement, as well as keep an eye on the effectiveness of government programmes and the quality of education.
- Providing of essential Resources: Particularly in isolated and underprivileged locations, CSOs assist in providing educational materials. In addition to offering textbooks and other educational resources, they can build up libraries. To students who deserve it, they can also provide scholarships and other forms of financial aid.
- Community Mobilization: CSOs encourage community support for education. They collaborate with educators, parents, and local authorities to foster a climate that is conducive to learning. The establishment of community-based organisations that may strive to improve education in their region can be aided by CSOs.
- Creates Awareness: CSOs increase public awareness and support for the value of education in society. They can strive to sway public opinion in favour of education and push the government to make more educational investments.

## 11 Measures Undertaken for Enhancement of Quality Education

- **Infrastructure Development:** To foster an atmosphere that is favourable to learning, infrastructure improvement has been a priority. To enhance school infrastructure, the government has launched initiatives like the Sarva Shiksha Abhiyan (SSA) and the Rashtriya Madhyamik Shiksha Abhiyan (RMSA), which include building classrooms, libraries, labs, and sanitary facilities.
- **Public–private partnerships:** The quality of education has enlarged recognition to coordinated efforts by the government, non-profits, and private sector initiatives. In order to improve educational outcomes, public–private partnerships have made it easier to embrace new teaching techniques, integrate technology, and share resources.
- **Learning Objectives & Evaluations:**Exams and rote memorising have lost importance since learning outcomes are now being measured. The Performance Grading Index (PGI) and the National Achievement Surveys (NAS) evaluate student learning levels, enabling educators to spot gaps and offer focused interventions. (Education for All in India). These evaluations support monitoring and raising the standard of education.

- **NEP 2020:** The National Education Policy: A broad strategy framework called NEP 2020 was created with the intention of reforming India's educational system. It places a strong emphasis on holistic and transdisciplinary education while encouraging creativity, critical thinking, and hands-on learning. (Kishore Kumar and Ajai Prakash 2020). The policy also emphasises teacher development, technological integration, and lessening the need on memorization.
- **Professional Development for Teachers:** Numerous efforts have been made to improve teachers' abilities and knowledge while keeping in mind their crucial position in society. The goal of initiatives like NISHTHA: National Initiative for School Heads' and Teachers' Holistic Advancement is to provide teachers with innovative pedagogical strategies, subject-matter expertise, and classroom management abilities (Alcántara-Ayala, 2020).
- **Integration of technology:** To improve student learning results, the government has promoted the employ of tools in education. Access to learning equipment, online courses, and teacher training resources is made possible by initiatives like the the Digital India initiative effort, the SWAYAM online platform, and the e-Pathshala digital resources. Interactive and individualised learning experiences are made possible by the use of digital tools and resources.
- **Vocational training and skill advancement:** In order to provide students with industry-relevant skills, skill development programmes have been created in recognition of the significance of vocational education. Vocational training programmes and initiatives like the National Skill Development Corporation (NSDC) can be crucial in helping students develop the practical skills they need for a variety of professional options.

## 12 Various Government Initiatives Towards Providing Quality Education:

- **NEP 2020**: NEP 2020 has been the most productive and significant to date. Goal 4 (SDG4) of the Sustainable Development Goals (SDGs) for 2030, which India accepted in 2015, replicate the universal learning accomplishment purpose and seeks to "make sure inclusive and impartial excellence learning and encourage lifetime education opportunities for all" by that year. To accomplish this challenging goal and all of the crucial targets and goals (SDGs) of the 2030 Agenda for Sustainable Development, the entire educational conglomerate must be reorganized to promote and encourage learning.

Major changes that fetch the supreme superiority, equity, and reliability into the organization, from premature infancy education and care through advanced education, are necessary to close the gap between the current level of learning outcomes and what is required. India must therefore create a top-notch educational system by the year 2040, ensuring that all pupils, regardless of their social or economic standing, have access to the greatest instruction. The NEP 2020 is the first education policy which focuses to deal with the many growing refining obligatory of our country. (Kumar and Prakash, 2020) To grow a new system that in keeping with the inspirational aims of 21st-century of education, together with SDG-4, this Policy advocates that all parts of the educational organization be altered and renovated, including its governance and regulation. It also incorporates the customs and morals of India.

The new education plan must support in reestablishing teachers as the most esteemed and significant members of our society. It must take all necessary measures to delegate educators and aid them in their work. In order to draw the finest and the dazzling people to the teaching occupation at the whole stages, the NEP must guarantee all teachers a living wage, respect, regality, and sovereignty as well as instill in the organism elementary dealings for quality assurance and accountability. Under the new educational strategy, all kids must have access to a top-notch educational system, with a focus on underrepresented, neglected, and criticized populations in past.

The most effective way to achieve inclusivity, equality, and monetary and societal mobility is through education since it levels the playing field. Programmes must be in place to ensure that, despite insurmountable obstacles, all students from these groups have access to a variety of targeted opportunities to enrol in and thrive in the educational system.

- **MOOCS**: A normal classroom can only hold so many students, however, there are millions of people worldwide who require and desire access to high-quality education. Massive open online courses, sometimes known as MOOCs, are the answer. Businesses came up with the concept as a means to give millions of students free access to online courses. The community interactions among students, professors, and teaching assistants (TAs) are supported by many MOOCs' interactive courses, which also include user forums or social media discussions. These courses also provide instant feedback for quick quizzes and assignments. Online courses designed for open access and limitless enrollment are known as massive open online courses (MOOCs) or open online courses. To promote resource reuse and remixing, early MOOCs (cMOOCs: Connectivist MOOCs) commonly prioritised open-access features, such as open licencing of content, structure, and learning objectives. Some later MOOCs (sometimes referred to as "xMOOCs" or "extended MOOCs") use closed licencing for their course materials while preserving student access without charge.
- **SWAYAM programme**: The goal of this initiative is to bring the finest resources for learning and instruction to all, which includes the most disadvantaged, while bridging the technology gap for students who have so far been left out of the digital revolution and are unable to participate in the knowledge economy. SWAYAM is a programme launched by the Government of India and intended to implement the three fundamental tenets of Education Policy, namely, access, equity, and quality. Study Webs of Active-Learning for Young Aspiring Minds (Kumar and Prabu, 2021) is the source of the program's name. As part of this plan, instructors from government-funded institutions including as IITs, IIMs, and central universities will offer online instruction to citizens of the country. There will be no charge

to take any of the courses. If the student wants an authenticated certificate, there will be a little cost. In the first phase, students will receive training in engineering education, social science, power, governance, and science courses at IIT Bombay, IIT Chennai, IIT Kanpur, IIT Guwahati, University of Delhi, Jawahar Lal Nehru University, IGNOU, IIM Bangalore, and IIM Calcutta, independently and with the assistance of academic staff from international schools.

- **SARANSH**: A decision support tool called "Saransh" was the initiative of the Central Board of Secondary Education (CBSE) with the purpose of "improving children's schooling by enhancing communication between schools and parents and providing data driven decision support system to support them in taking the most excellent decisions related to their children's upcoming time." By comparing the outcomes, this method enables schools to pinpoint areas for development in terms of students, instructors, and curriculum and to implement the required changes. At the National Conference on ICT on November 7, 2015, the Honourable HRM officially unveiled the mobile app for Saransh. This will make it possible for parents and kids to examine and contrast their performance with that of the school, state, and nation.
- **Samagra Shiksha**: A comprehensive approach to pre-kindergarten through class twelve education has been proposed in the Union Budget for 2018–19. With the extensive objective for making better school efficiency as measured by fair learning outcomes and equal access to education, Samagra Shiksha, a wide-ranging project for the school education segment that develops from pre-school to $12^{th}$ class. It includes the three programmes Teacher Education (TE), Rashtriya Madhyamik Shiksha Abhiyan (RMSA), and Sarva Shiksha Abhiyan (SSA).
- **UDAAN**: UDAAN is a Central Board of Secondary Education (CBSE) initiative that helps poor female students, including minority group's learners and defenseless situations, make the shift from secondary school to technical learning at postsecondary institutions, notably in the subjects of mathematics and science. UDAAN's first goal is to address the poor enrollment of girls in engineering programmes, which are now at 23% of girls vs 77% of boys. By concentrating on three dimensions—syllabus plan, operation, and measurement—it tries to close the excellence gap between post-secondary and engineering education admission systems. The Senior Secondary Mathematics and Science curricula will be improved and supplemented to achieve this goal. The CBSE will provide all students with free online materials along with additional incentives and support.

## 13 Conclusion

Quality education is a critical component of the United Nations Sustainable Development Goals (SDGs). All parties, including governments, civil society organisations, the commercial sector, etc., must work together to advance quality education. India is leading the way in the twenty-first century by advancing its creative flagship projects to quickly accomplish SDG objective 4 on education, which is a very broad topic.

As the SDG highlights, it is imperative to have an adequate physical infrastructure and secure, inclusive environments that support learning outcomes for all students, regardless of sex, background, or level of handicap.

Achieving SDG 4 requires a collective effort by all stakeholders.

- To advance quality education, governments should prioritize education funding, promote equity in access to education, and improve the quality of education.
- Civil society organizations should advocate for policy changes that support quality education and engage in education programs.
- The private sector should invest in education and partner with schools and universities to promote quality education. International organizations should support governments and other stakeholders in promoting quality education through technical assistance, funding, and advocacy.
- To significantly strengthen the education sector, the government must work with current NGOs and civil society in an extremely efficient manner.
- Additionally, NGOs that support government agencies' concentrated efforts to impart knowledge at all levels need to actively and effectively communicate how they contribute to the achievement of the SDGs across many targets.
- Similar efforts should be undertaken by NGOs and civic organisations to educate the public about the advantages and justification of high-quality education for their own good.
- In addition, the Indian educational system has adopted e-learning platforms like Massive Open Online Courses (MOOCs) to make higher education more accessible and inexpensive in light of the growing popularity of online learning.

Some of the determining factors that affect the demand for quality education include favourable educational outcomes, qualified teachers, proximity to school, relevant curriculum, provision of proper infrastructure, including chalk boards and boundaries walls, safe drinking water facilities, functional toilets, and MDMS. For this the Indian government has started a number of programmes to address these issues, including the RashtriyaUchchatarShikshaAbhiyan (RUSA) and the National Institutional Ranking Framework (NIRF). By providing financing to universities, supporting research and innovation, and fostering institutional collaboration, these efforts seek to raise the standard of higher education in India.

Despite India's RTE legislation, the fact that millions of kids are still not in school raises serious concerns. Until the parents are persuaded of the advantages that education may bring to their life, conditional monetary transfer arrangements can play a transforming role.

The SDGs agenda takes a fairly balanced stance towards good jobs and high-quality education. Along with the imperative to invest in high-quality learning and the significance of professional development in the education sector, there is a need to coordinate the plans of the State and the Centre to ensure parity in educational achievements.

For the Indian education system to improve in terms of accessibility, quality, and learning outcomes—not to mention, of course, the safety and security of all the children—adequate funding must be allocated. Standardisation, harmonisation, and regularity of data on education would aid in accurately measuring and quantifying the SDG 4. The provision of necessary resources to fund education, together with the strict implementation of these activities and efforts, must be done on a regular basis. Education is a cross-cutting problem and a goal in and of itself.

Despite all the initiatives, India's higher education system still has a long way to go before it can compete with the standards of the top educational establishments in the world. We must work together to create a system that offers high-quality education, promotes innovation, and equips our children with the tools they need to take on the world.

To stay competitive as the globe gets more cutthroat, it is essential to invest in higher education. To give the country's youth access to high-quality higher education and give them the tools they need to compete on a global scale, the Indian government, educational institutions, and stakeholders must continue to collaborate.

## References


Alcántara-Ayala I (2020) Education for Sustainable Development (ESD) in Teacher Education Programs: The Case of Mexico. In: Farooq M, Schöbel B (eds) Education for Sustainable Development: Addressing the Global Challenges of Our Time. Springer International Publishing, pp 95–108

Asif M, Singh KK (2022) Libraries @ national education policy (NEP) in India. IP Indian J Library Sci Inf Technol 7(1):18–21

Bajpai, Nirupam; Goyal, Sangeeta. (2004). Primary Education in India: Quality and Coverage Issues. Retrieved from https://academiccommons.columbia.edu/doi/https://doi.org/10.7916/D8G1604S

Brotherhood T, Yang L, Chankseliani M (2020). COVID-19 and higher education: Implications for equity, and a return to the common good. The post-pandemic university. https://postpandemicuniversity.net/2020/09/10/covid-19-and-higher-education-implications-for-equity-and-a-return-to-the-common-good//

Chankseliani M, McCowan T (2021) Higher education and the Sustainable Development Goals. High Educ 81:1–8. https://doi.org/10.1007/s10734-020-00652-w

Elmassah S, Biltagy M, Gamal D (2022) Framing the role of higher education in sustainable development: a case study analysis. Int J Sustain High Educ 23(2):320–355. https://doi.org/10.1108/IJSHE-05-2020-0164

Jha S, Kumar M. (2012) Management Education in India: Issues & Challenges. Available at SSRN: https://ssrn.com/abstract=2140807 or https://doi.org/10.2139/ssrn.2140807

Kawaguchi J, Kuroda K (2013). The Sustainable Development Goals Diffusion and Contestation in Asia and Europe.

Klieme E, Leutner D, Pelletier J (2018) Quality of education and learning outcomes: From a dimension-based to a competence-based approach. In: Townsend T, Schöne B, Wolter K (eds) Handbook of the Economics of Education, vol 5. Elsevier, Amsterdam, pp 139–201

Kumar K (2005) Quality of Education at the Beginning of the 21st Century – Lessons from India. Indian Edu Rev 41:3–28

Kumar K, Prakash A (2020) How National Education Policy 2020 can be a lodestar to transform future generation in India. https://onlinelibrary.wiley.com/doi/abs/https://doi.org/10.1002/pa.2500

Kumar KS, Prabu M (2021) Open educational practices of SWAYAM programme among research scholars. Educ Inf Technol 26:1–25. https://doi.org/10.1007/s10639-021-10495-2

Mitra S, Mishra SK, Abhay RK (2023) Out-of-school girls in India: a study of socioeconomic-spatial disparities. GeoJournal 88:341–357. https://doi.org/10.1007/s10708-022-10579-7
Mohanty A (2018) Education for sustainable development: A conceptual model of sustainable education for India. Int J Dev Sci 7(9):2242–2255
Mundy K, Green A, Lingard B, Verger A (eds) (2016) The Handbook of Global Education Policy, 1st edn. Hoboken, Wiley-Blackwell
New National Education Policy 2020 of India. Introduction of NEP 2020. https://shikshan.org/nep-2020
OECD. (2019). Education at a Glance 2019: OECD Indicators. Retrieved from https://www.oecd-ilibrary.org/education/education-at-a-glance-2019_f8d7880d-en
Pandey B (2018). Achieving SDG 4 in India: Moving from Quantity to Quality Education for All. Ensure inclusive and quality education for all and promote lifelong learning. Retrieved from https://sdg-tracker.org/quality-education
Rammohan A, Vu P (2018) Gender inequality in education and kinship norms in India. Fem Econ 24(1):142–167
Saini M, Sengupta E, Singh M, Singh H, Singh J. (2023). Sustainable Development Goal for Quality Education (SDG 4): A study on SDG 4 to extract the pattern of association among the indicators of SDG 4 employing a genetic algorithm. Retrieved fromhttps://link.springer.com/article/https://doi.org/10.1007/s10639-022-11265-4#Sec2
Singh M, Adebayo SO, Saini M, Singh J (2021) Indian government E-learning initiatives in response to COVID-19 crisis: A case study on online learning in Indian higher education system. Educ Inf Technol 26:7569–7607
Tomislav K (2018) The concept of sustainable development: From its beginning to the contemporary issues. Zagreb Int Rev Econ Bus 21(1):67–94
UNESCO. (2015). Education 2030 Incheon Declaration and Framework for Action for the implementation of Sustainable Development Goal 4: Ensure inclusive and equitable quality education and promote lifelong learning opportunities for all. Retrieved from https://en.unesco.org/sites/default/files/ed-2030-framework-for-action-implementation-en.pdf
UNICEF. (2018). Progress for Every Child in the SDG Era. Retrieved from https://www.unicef.org/publications/files/Progress_for_Every_Child_in_the_SDG_Era_Summary_English.pdf
UNICEF. (2020). Education in Emergencies and Protracted Crises: Toward a Strengthened Response. Retrieved from https://www.unicef.org/media/91946/file/Education-in-emergencies-and-protracted-crises-2020.pdf
UNIFIED DISTRICT INFORMATION SYSTEM FOR EDUCATION PLUS (UDISE+) 2021–22 FLASH STATISTICS. Retrieved from https://dsel.education.gov.in/sites/default/files/statistics/report_in_PDF/udise_21_22.pdf
Unified District Information System For Education Plus (UDISE+). https://udiseplus.gov.in/
United Nations. (2015). Transforming our world: the 2030 Agenda for Sustainable Development. Retrieved from https://www.un.org/ga/search/view_doc.asp?symbol=A/RES/70/1&Lang=E
World Bank. (2018). World Development Report 2018: Learning to Realize Education's Promise. Retrieved from https://openknowledge.worldbank.org/handle/10986/28340
World Bank. (2020). The COVID-19 Pandemic: Shocks to Education and Policy Responses. Retrieved from https://www.worldbank.org/en/topic/education/brief/global-education-monitoring-report-coronavirus-covid19-education-response

[illegible]

# The Global Learning in Agriculture Community: A Digital Community of Practice to Support Education for Sustainable Development

**Melanie Miller Foster and Daniel Foster**

**Abstract** Educators are essential actors in preparing the next generation to address the world's most pressing issues and facilitating the transition to a more sustainable society. Learners need to be empowered to envision, take responsibility for, and take action towards a sustainable future. Educator professional development for sustainability education must address these complex learner and societal needs.

This chapter has three primary objectives including: (1) Discuss the strategic positionality of educators as change agents of sustainability; (2) Explore how a digital community of practice as a digital technology is being leveraged to promote the Education for Sustainable Development and (3) Uncover challenges and successes of building an accessible digital community of practice through the lived experiences of two social scientists. The chapter utilizes a duoethnography method (Sawyer and Norris, Duoethnography, Oxford University Press, Incorporated, 2012; Adams et al., Handbook of Autoethnography, Routledge, 2022), in which the co-founders of the Global Teach Ag Network share about a promising strategy in educator professional development focused on collaboration through impactful communities of practice while discussing the challenges and successes of building an accessible digital global community of practice designed to support Education for Sustainable Development.

Since 2014 the Global Learning in Agriculture Community (GLAG) has provided space for thousands of hours of professional development events and multiple points of contact between participating community members in over forty-five countries. For some members, this is the only access they have to educator-oriented professional development while for others, this is just one opportunity of many. GLAG has organically grown and developed over the years with resources dedicated by multisectoral organizational partners of the Global Teach Ag Network.

---

M. M. Foster (✉)
Department of International Agriculture and Development, Penn State University, University Park, PA, USA
e-mail: melanie@psu.edu

D. Foster
Department of Agriculture & Extension Education, Penn State University, University Park, PA, USA
e-mail: foster@psu.edu

 
W. Leal Filho et al. (eds.), *Digital Technologies to Implement the UN Sustainable Development Goals*, World Sustainability Series,
https://doi.org/10.1007/978-3-031-68427-2_8

**Keywords** Educators · Communities of Practice · Agriculture · Professional Development · Duoethnography

## 1 Introduction

Educators are essential preparing the next generation to address sustainability issues outlined in the United Nations Sustainable Development Goals such as climate change, inequalities, biodiversity loss, conflict, and food insecurity. Educators are at the heart of education systems as they interact daily with students and community members. Critically, educators have first-hand familiarity with community challenges as well as community assets. Educators have the potential to work with learners to shape the communities that learners want to live in, and influence mindsets toward a more sustainable future. To achieve the UN Sustainable Development Goals (SDGs), the strategic positionality of educators as community change makers must be harnessed.

To date, Education for Sustainable Development (ESD) has been seen as the way that educational systems must realign themselves to work toward sustainable development outcomes. The term Education for Sustainable Development emerged at the 21st Agenda of the UN in Rio de Janeiro in 1992 and has evolved to focus on transforming the educational systems worldwide to foster knowledge, skills, and attitudes that will lead to a sustainable society (Leicht et al., 2018). The concept of sustainability has evolved to encompass not just environmental movements, but to recognizes the interconnectedness of the economy and society within the biosphere (United Nations, 2015). To be active and effective partners in the complexity of ESD, educators need holistic and interdisciplinary professional development (UNESCO, 2020).

Educator professional development encompasses "all types of learning undertaken by teachers beyond the point of their initial training" (Craft, 2000, p. 9). Educators must be empowered with the necessary pedagogies, tools, partnerships, and resources to transform their teaching to align with sustainable development efforts. Unfortunately, only a small percentage of educators worldwide have access to professional development (OECD, 2019; García & Weiss, 2019), and existing professional development programs are often ineffective (Cedefop, 2018). Professional development programs employing technical-rational-top-down strategies focus on the deficits of educators and provided "one-shot" professional development sessions to fill in gaps (Caena, 2011). Educator professional development trends have shifted to more cultural-individual interactive approaches, driven by adult learning, and situated cognition theories (Caena, 2011; Creemers, et al., 2013). This more progressive model considers educators as professionals, and rests on the notion that educators are active partners in their own professional development. The result is that change is more likely to be innovative and enduring (Svendsen, 2020).

With the rapid advancements in technology and the widespread access to the internet, educators now have a wealth of resources and opportunities for professional growth at their fingertips. Research has highlighted the benefits of online platforms and virtual communities in facilitating professional growth and knowledge exchange among educators (Malik, et al., 2019), although Vrasidas and Zembylas (2004) point out that access to technology alone does not promote educator professional development. Digital platforms offer flexibility and accessibility, enabling educators to engage in continuous learning and collaborative discussions at their own pace and convenience (Archambault & Crippen, 2009; Çelen & Seferoglu, 2020; Kimmons et al., 2017).

Digital platforms provide opportunities for educators to access a wide range of resources, participate in webinars, engage in virtual workshops, and connect with educators from diverse geographical locations (Kay, 2012; Vangrieken et al., 2017). The availability of digital educator professional development is particularly crucial in addressing the time constraints and geographic barriers that often limit educators' access to traditional face-to-face professional development opportunities (Archambault & Crippen, 2009; Blaschke, 2012). Such digital professional development initiatives have been found to enhance educators' pedagogical knowledge, promote reflective practice, and foster a sense of belonging within professional communities (Shih et al., 2010). Leveraging digital professional development opportunities has emerged as a promising avenue for empowering educators and enhancing their teaching practices as well as provide space for the social and collaborative learning.

This chapter highlights a promising strategy in educator professional development focused on educator collaboration through impactful communities of practice. The chapter utilizes a duoethnography method allowing three objectives to be addressed directly by the co-founders of the Global Teach Ag Network: (1) Discuss the strategic positionality of educators a change agents of sustainability; (2) Explore how a digital community of practice as a digital technology is being leveraged to promote the Education for Sustainable Development and (3) Uncover challenges and successes of building an accessible digital community of practice through the lived experiences of two social scientists.

## 2 Method

Duoethnography is a qualitative research method that involves two researchers engaging in a collaborative and reflexive exploration of shared experiences and perspectives (Adams et al., 2022; He & Phillion, 2008; Norris, Sawyer, & Lund, 2016; Sawyer & Norris, 2012). The researchers, who often have a close personal or professional relationship, engage in a process of dialogue and reflection, employing various techniques such as in-depth conversations, interviews, and written exchanges (Etherington, 2018). Through these interactions, duoethnographers work together to uncover the complexities of their intertwined experiences and identities, examining the dynamic interplay of power relations, emotions, and personal narratives (Norris, Sawyer, & Lund, 2016). The focus of duoethnography is on co-creating knowledge and meaning through mutual interpretation and fostering a deep understanding of subjective realities within a specific context (Etherington, 2018). Duoethnography contributes to the broader field of qualitative research by generating rich and contextually grounded insights while illuminating the nuances of personal and interpersonal dynamics (He & Phillion, 2008).

In this study two researchers will engage in the duoethnographic method to explore the research question. Melanie Miller Foster received her training in rural sociology, with a particular focus on agriculture in international development. She currently is a faculty member in Ag Sciences Global, the international programs unit in Penn State's College of Agricultural Sciences. Daniel Foster is an agriculture teacher educator, working to prepare secondary agriculture teachers, with a focus on educator professional development. He is a faculty member in the Department of Agricultural Economics, Sociology and Education and associated with the Center for Professional Personnel Development in Penn State's College of Agricultural Sciences. Previously, he worked as a secondary agriculture teacher in a rural setting. Working together, the two created the Global Teach Ag Network whose signature programming is the Global Learning in Agriculture Community.

To create the duoethnography, the researchers examined notes from meetings, internet correspondence such as emails, historical documents from previous events and meetings, and documents resulting from previous collaborative writing sessions. Additionally, the researchers reviewed literature and research studies that guided their practice and identified current statistics to illustrate specific points. The following narrative was written in tandem by the researchers, with each of their thoughts on each topic labeled.

## 3 The Need for Education for Sustainable Development in Agricultural Education

### *3.1 Melanie:*

I work in the field of international agriculture and development, and for the past decade there has been quite a bit of concern about the aging farmer population. For example, the age of agricultural producers in the U.S. increased from 56.3 to 57.5 years of age between 2012 and 2017 (USDA, 2017), which mirrors the global age trend of producers. Exacerbating the situation is the youth bulge in some countries in sub-Saharan Africa and South Asia, where younger generations are an increasing share of the overall population, particularly in rural areas (FAO, 2017). Jobs in rural areas can be difficult to find, and without job opportunities the young population can lead to outmigration. Meanwhile, the share of employment in agriculture, forestry and finishing continues to decline globally, dropping 13 percentage points from 2000 to 2020, even though agriculture continues to be the second largest employment sector in the world behind the service sector (FAO, 2021).

People involved in international agriculture and development organizations are concerned about the lack of youth engagement in agriculture. I've attended many conferences where a topic of enthusiastic discussion is how agriculture can be portrayed as a more appealing occupation in order to attract youth. In many places around the world, agriculture is thought of as something that poor people do, and as the occupation of last resort. Parents encourage children to do well in school and go into high paying professions, which rarely includes agriculture. Many organizations in the agriculture arena have a desire to work with and appeal to youth but scratch their heads on how to meaningfully engage youth around the world at scale.

At first, I thought it was strange that the idea of engaging educators does not emerge more often in these conversations or in the research. Now that I've worked closely with Daniel, a teacher educator, I understand that educators are at the heart of education systems because they directly interact with youth and community members daily. From a community development standpoint, they are familiar with community challenges as well as community assets. Educators can work with youth to shape the communities that learners want to live in, and influence mindsets toward new career possibilities and a more sustainable future. It took me closely working with a person from another discipline to undercover this blind spot in my thinking, but now I can't see it any other way. I find meaning in my work by working to provide evidence about the importance of educators in the field of international agriculture development, and key actors in ESD.

## *3.2 Daniel:*

As someone with a lifelong passion for agriculture, I have always believed in its importance to society, but it took me a while to grasp the potential and opportunity of the role of agriculture in shaping a sustainable future. After working with Melanie and broadening my understanding of the global agriculture system, I better understand the role that it plays in economic and social development. I firmly believe that agricultural education is well suited to engage in and advance ESD. Agriculture, at its core, embodies the principles of sustainability, as it seeks to optimize the use of natural resources while promoting environmental stewardship and social equity. By integrating ESD into formal, non-formal and informal agricultural education systems, we can empower learners to understand the interconnectedness between agriculture, the environment, and society, fostering a holistic approach to sustainable development.

For those not familiar with formal school-based agricultural education often situated in the context of career and technical education, it is known for being strongly rooted in local communities and for preparing students for the local workforce. Agricultural education programs around the world are addressing local community issues in food security, environmental protection, and social issues, but the link between local action for global good is not well understood. There is tremendous opportunity to capture the imagination of youth to see agriculture as a discipline where they can bring their passion and creativity to work towards solving the issues that they really care about. We should also note that education about agriculture also occurs in other academic disciplines in formal school systems as a context to reach a wide variety of learning outcomes. Agricultural education also occurs outside of our school systems in non-formal and informal contexts, be this from family or community activities or as structured youth development activities incorporated into the work of extension systems.

## 4 Why Professional Development for Educators

### *4.1 Melanie:*

In the realm of international development, we acknowledge that enduring change is achieved by fostering local buy-in and implementing tailored solutions that empower communities at the grassroots level. By embracing and involving local stakeholders, we can co-create sustainable initiatives that address the specific needs and aspirations of the communities, leading to transformative and lasting impact. Despite that philosophy, I've noticed that many development projects that seek to engage youth focus on the development and creation of a new curriculum. Yet there is little acknowledgement of who and how the curriculum will be implemented, rather, just a hope that a skilled educator will appear to pick it up and run with it. If professional development does occur as part of the project, sadly, it is often a one-shot workshop with no follow up after the project completion.

To empower educators, educators need to be involved in what and how they will teach. We see in the literature that teacher empowerment is strongly correlated with task motivation (Elahi & Mahanta, 2019), meaning that empowering educators through professional development can be thought of as a more sustainable way to transform education.

### *4.2 Daniel:*

I believe it is crucial to involve educator in the process of determining what they will teach and how they will teach it. Educators possess a wealth of knowledge and expertise that can contribute to the development of effective and engaging teaching strategies. By giving educators a voice in curriculum design and instructional methods, we acknowledge their professionalism as well as tap into their creativity, passion, and understanding of their students' needs. When educators are actively involved in decision-making processes, they feel a sense of ownership and agency over their teaching practices, leading to increased job satisfaction, professional growth, and ultimately, better learning outcomes for students.

Professional development is thought to be a vital aspect to achieving Education for Sustainable Development because of the complexity of the transformation required to move the needle. An interdisciplinary approach is required to tackle the complex issues of sustainable development (Annan-Diab & Molinari, 2017), and student-centered pedagogies are at the heart of Education for Sustainable Development. In essence, educators need to move from being conveyors of knowledge to facilitators of learning working in concert with colleagues across disciplinary areas. Educators need to be supported in this endeavor through high quality professional development if they are going to be able to implement new teaching practices and philosophies.

Once Melanie and I identified the "agriculture education gap" in the Education for Sustainable Development movement, we felt called to collaborate and focus our efforts on agriculture educator professional development for educators of all disciplines and contexts helping learners explore food, fiber, and natural resources areas.

## 5 The Formation of the Global Learning in Agriculture Community

### *5.1 Melanie:*

After Daniel and I identified the need for agriculture educators to be a part of the conversation in the ESD movement, we began to design and implement educator professional development programs, seek external funding, and conduct research to understand the impacts.

In 2013, we received a Fulbright-Hays Group Project Abroad from the U.S. Department of Education. After the group immersive travel to Korea in 2014, a two-day workshop on global learning in agriculture was conducted at Penn State University as an unfunded mandated culminating grant activity. The twenty participants of the Fulbright-Hays project were expected to attend; however, other educators from across the U.S. learned about the topically focused gathering and requested to attend. Without intending to grow to such a size, the 2014 meeting attendance was fifty-seven participants! It was at this point we realized that this was a dialogue and topic that agricultural educators of all contexts wanted and needed to engage in.

Since that initial grant culminating activity in 2014, the Global Learning in Agriculture Conference (referred to as GLAG) has brought together educators and international agriculture and development organizations to share and learn from each other on an annual basis. In 2016, the GLAG conference became an annual event held online so that a greater number of in-service and pre-service educators – in agriculture and beyond – could participate from across the U.S. and around the world. Thousands of educators from around the world joined the conversation through the conference, listserv, social media, and face-to-face professional development events.

After seven years of growing the digital GLAG Conference, we realized that we had outgrown the conference model. In 2023, the Global Learning in Agriculture *Conference* transitioned into the Global Learning in Agriculture *Community* to reflect the goals of a year-round community of practice to support educators in their professional development journey. This change to an active year-round community facilitates more accessible and effective peer-to-peer sharing, and engagement in events and challenges. Importantly, the GLAG Community has transformed away from an annual dose of professional development, and into a stable community of practice where educators can find support and inspiration.

## 5.2 Daniel:

As an agriculture educator, I understand the significance of Communities of Practice (CoP) in education. According to Wenger et al. (2002), CoPs are groups of individuals who share a common concern, passion, or set of problems and engage in ongoing interaction to deepen their knowledge and expertise in that area. Being part of a CoP can offer educators a valuable learning environment to reflect, collaborate, give and receive peer feedback, and explore innovative teaching methods with colleagues.

In the seminal text on CoPs, Wenger et al. (2002) defined seven principles for cultivating CoPs. For reference, I present them in Table 1. These principles revolve around nurturing the "aliveness" or vitality of the community. This aliveness refers to the interactions within the community that generate excitement and demonstrate value to its members. While aliveness cannot be forced, it can be fostered by embracing the seven principles for cultivating CoPs.

I want to describe how we have applied each of these principles to the online GLAG Community to reflect on how these best practices can be applied to the community in order to effectively promote ESD.

***Design for evolution:*** The GLAG Conference grew out overwhelming demand to attend a single workshop on global learning in agriculture. From that initial face-to-face workshop, the conversation was moved online for a run of seven years. Recognizing the need for more interpersonal connections between GLAG attendees as well as the challenges from time zones of building a truly global community with events at specific times, the online GLAG Conference then transitioned into the Global Learning in Agriculture Community to better reflect the goals of a year-round community of practice to support educators in their ESD professional development journey. This positioned GLAG to be able to provide just-in-time professional development accessible at the time when the educator was in most need. GLAG has demonstrated the design for evolution concept by evolving over the years as we have learned about cultivating a digital community of practice, and will continue evolve as we learn more and strive to meet educator needs through thoughtful and intentional community design.

***Open a dialogue for inside and outside perspectives:*** The GLAG Community is composed of educators of any context, location, and discipline, but is also supported

**Table 1** Seven Principles for Cultivating Communities of Practice (Wenger, et al., 2002)

| Principle | Definition |
|---|---|
| Design for evolution | *Design elements should facilitate the evolution of the community over time* |
| Open a dialogue for inside and outside perspectives | *Community design should bring information from outside the community into the dialogue about what the community could achieve* |
| Invite different levels of participation | *Members should be able to participate at a level that feels comfortable to them* |
| Develop both public and private community spaces | *Communities should center on the relationships between members and facilitate different spaces in which relationships can develop and thrive* |
| Focus on value | *Members need to find value in a community because participation is usually voluntary* |
| Combine familiarity and excitement | *Communities should cultivate stability, but also a sense of adventure and excitement* |
| Create a rhythm for the community | *The frequency of events and interactions should reflect the norms of the community and the needs of the members* |

by global learning partner representing academic, private, public, non-governmental, and industry organizations. The purpose of these partnerships is to bring relevant information to educators for purposes of engaging their learners in current and authentic global issues. Additionally, global learning partners can engage educators as key stakeholders in their work by developing connections and identifying educator needs.

***Invite different levels of participation:*** Some GLAG community members are extremely active and take on leadership roles, while others are more occasional contributors and even others could be classified as lurkers, that is, watching and absorbing information by not contributing to the discourse in a visible way. All are valued members of the community. Specific examples included the opportunities for educators lead programming efforts as coordinators of educator volunteers while other educators are those volunteers, contributing their insight efforts and programming and finally, some educators simply choose to be participants.

***Develop both public and private community spaces:*** Within GLAG as a digital community of practice, there are spaces that are open to all members of the community, while there are some spaces that are maintained as closed affinity space in which they can conduct their business only among a select group of members.

***Focus on value:*** Participation in the Global Learning in Agriculture Community is completely voluntary, meaning that educators access the community because they find it valuable in some way. The real value of participation in the community may not be apparent immediately, as it takes time for teachers to reflect and integrate new knowledge and ideas into their teaching.

***Combine familiarity and excitement:*** The Global Learning in Agriculture Community provides spaces for educators to bring their ideas to other group members. One member brought the idea of writing poetry about food security topics to process the emotions that arise when discussing food insecurity, and to encourage creativity around the issue. For some educators in the community this was an idea that they had never considered before, and it gave members an opportunity to practice open-mindedness.

***Create a rhythm for the community:*** The rhythm for the Global Learning in Agriculture Community followed the evolution from a conference to a community. The conference had previously taken place in February, so the community has come to expect a catalyzing event at that time of year. Other "seasons" in the community include an annual book study and a poetry contest. Members are otherwise active throughout the year with activities such as addressing questions and sharing resources in community forums.

The Global Learning in Agriculture Community has already provided space for hundreds of hours of professional development events and thousands of points of contact between members. For some members, this is the only access they have to educator-oriented professional development while for others this is just one opportunity of many. The Global Learning in Agriculture Community has organically grown and developed over the years with resources dedicated by organizational partners of the Global Teach Ag Network.

## 6 Leveraging Digital Opportunities

### *6.1 Daniel:*

Creating a community in a digital space is in some ways very freeing from the typical constraints imposed by a face-to-face conference. Community members, many public school teachers, do not have to try take time away from the classroom to travel to a conference. Members from around the world are not constrained by limited travel funds or visa requirements. When participants walk away from a conference, they are often overwhelmed with new ideas to the point where they might only implement one new idea into their practice, and there is often no follow-up if they have questions or need more information. In this way, leveraging digital technologies has allowed for greater accessibility of participants to the community of practice with space for ongoing follow up and support.

Leveraging digital technologies in a community of practice offers immense potential for accommodating people who speak different languages, fostering inclusivity, and facilitating effective communication and collaboration across language barriers. Language translation applications and tools, such as machine translation systems and real-time language interpretation software, enable instant translation of written and spoken content, making it accessible to individuals regardless of their language proficiency. We selected a digital platform with multilingual features, allowing users to access content and participate in discussions in their preferred language. By embracing digital tools and platforms, we believe that we can bridge linguistic divides, promote cross-cultural understanding, and create an inclusive global learning environment that values and accommodates linguistic diversity. It is still a work in progress, but this is an area of great opportunity as we think about leveraging digital platforms for global educator professional development.

It should be noted that we have also run into the "digital divide", in that many educators wishing to participate do not have access to reliable internet connection with appropriate bandwidth. For example, there is good participation in the GLAG Community by Ugandan educators, but just ten percent of the population in Uganda have access to the internet, according to 2021 data from the World Bank. In the United States, around 19 million people do not have access to high-speed internet, with most of those living in rural areas, according to the Federal Communications Commission. Rural areas are very active in agriculture education, meaning that many agriculture educators struggle to access digital professional development such as the GLAG Community. Additionally, educators around the world often do not have access to high quality digital devices, and our GLAG Community analytics show that the majority of participants are accessing GLAG on their mobile phone.

We have worked to overcome some of these accessibility challenges by intentional design of the GLAG Community. For example, we have situated the community on a platform that has a well-developed phone application to accommodate members accessing the GLAG Community on their smart phone. To accommodate slow or spotty internet access, sessions are recorded and offered on demand. Doing so also allows for greater equality among time zones, which also benefits community members from around the world. Overcoming the digital divide in educator professional development is a critical endeavor to ensure equitable access to quality learning opportunities for all educators, and through the GLAG Community we are exploring potential options and best practices for the future.

## *6.2 Melanie:*

As Daniel said, we've tried to design the GLAG Community with busy educators living all over the world in mind. Features of the GLAG Community include online self-paced courses offered by our partner organizations, with successful completion of a course resulting digital badges. Courses are structured so that educators can present the materials to their governing institutions to receive continuing education units that many educators around the world are required to earn on an annual basis. Courses provide a systematic approach to cross-disciplinary content areas such as food security, or methodology such as problem-based learning.

In addition, complementary GLAG programs have the opportunity to carve out their own space in the digital community to work on specialized projects together. There are currently three educator-led programs including an annual book study called GLAGreads, educator-developed modules for youth learners called GLAGjr, and an initiative exploring creative endeavors to solve global challenges called GLAGcreates. Reflecting one of our core values of inclusiveness, we have recently added a fourth area, GLAG en español, to reach Spanish-speaking educators across the world.

The community provides a space for members to engage in collective learning experiences. The digital platform allows for asynchronous discussions to take place, and might include an educator asking a question, pointing out a useful and relevant resource, or celebrating a success in their teaching practice. Members can connect with like-minded professionals, expand their professional network, and establish relationships that extend beyond the digital community. This enables them to access diverse perspectives, potential collaborations, and opportunities for professional growth.

One important aspect of digital CoPs to note is that they require active community management to ensure smooth functioning, moderation of discussions, resolution of conflicts, and facilitation of community engagement. The GLAG Community has a dedicated program manager, who plays a crucial role in facilitating interactions, organizing activities, and maintaining the overall health and vitality of the community.

Overall, the functions of the GLAG Community provide a dynamic and collaborative platform for professionals to connect, learn, and grow together, regardless of geographical constraints. Throughout the process of building community, we have learned many lessons, but the key is to learn from those lessons and adjust our practice in order to do better in the future. It can be very humbling when an exciting and innovative idea does not work in practice like you thought it might. We have learned that we must be nimble and not be beholden to the past in order to allow us space to try new things in the future.

## 7 Challenges and Successes of the GLAG Community

### *7.1 Melanie:*

One of the primary challenges of the GLAG Community is to simply describe what it is. I think this is because there are so many initiatives out there that claim that they have the answer to whatever the problem is, and there's some type of product that they offer as the solution. In the case of the GLAG Community, we come to this humbly, not professing to have the answers, but offering a space for important conversations between educators who want to connect their learners to the world's most pressing issues. We are not playing the "sage on the stage" expert role and we do not have a set curriculum we developed to offer. Instead, we work to be the connector, and provide seats at the table for professionals in international agricultural development as well as professionals in the field of agricultural education, for collaborative discussions to engage educators in the types of issues that they care about such as food insecurity, climate change, and loss of biodiversity. *What does the GLAG Community provide?* The high-level answer is that GLAG is a space for knowledge sharing, collaborative learning, professional networking, and collective problem-solving, fostering continuous growth and development among its members, all with the aim of transforming education toward sustainability.

*How does one measure the impact of creating community, or holding space for important conversations?* You might remember, or even use, the term "social capital" – this term refers to the resources embedded within social networks, including trust, reciprocity, social norms, and connections. It feels intuitive, and a way to quantify something that some people have and others don't. However, there are many measurement challenges due to its abstract nature and multifaceted dimensions, posing difficulties in quantifying and operationalizing these complex concepts into measurable indicators. Like social capital, it is difficult to measure the abstract nature of a community of practice. Our research program has worked to measure the impact of GLAG in several ways:

1. Membership and participation: tracking the number of active members, frequency of participation, and the diversity of participants within the CoP. For the GLAG Community we have measured this through surveys, registration data, and advanced digital analytics.
2. Network analysis: analyzing the structure and connectivity of the GLAG Community has provided valuable information about the relationships and interactions among members including identifying key influencers, information flow patterns, and the strength of relationships within the GLAG Community.
3. Learning and skill development: measuring the impact of the GLAG Community on members' learning and skill development. This is primarily done through the digital badging process, where members participate in a specified activity and then complete an assessment to measure their learning.
4. Qualitative methods: to help capture the nuances and richness of the community's dynamics, the research team has gathered the stories of CoP members in order to gain deeper insights into the experiences on GLAG Community members and the impact on their teaching practice.

Our team continues to research the GLAG Community to develop a better understanding of the CoP's functioning, effectiveness, and impact. Of particular interest is to develop measures of knowledge sharing and collaboration in order to assess the quality of knowledge shared within the GLAG Community.

## *7.2 Daniel:*

One of the challenges that has taken years to overcome is finding the right digital platform. There are many digital platforms in existence, which is wonderful because we did not have the funds to develop one on our own. We started by thinking that we might be able to utilize social media as a place to grow the community, but had pushback from members on several fronts including not having social media accounts, not wanting to mix personal and work content, and not feeling free to openly discuss issues where they could be widely seen and commented upon. We then transferred the community conversation to a digital conference platform, which was very popular, but did not facilitate the lasting connections and conversations that we desired once the event was over. After years of searching, we finally found the right digital platform—a dedicated and affordable solution that prioritizes long term community building – and got it approved by our university's risk management office. Educators around the world can now join the GLAG Community after answering a few simple questions, and without a registration fee by simply clicking one web link: https://global-teach-ag-network.mn.co/.

While the GLAG Community finally has the right digital home, the difficult work continues on building a community culture. Transitioning to a new digital platform was just the first step in fostering a thriving and supportive community. Now, we must focus on cultivating a culture of engagement, collaboration, and inclusivity. This entails encouraging active participation, promoting respectful dialogue, and creating opportunities for members to connect, share their expertise, and support one another. Building a community culture requires ongoing efforts such as organizing virtual events, facilitating discussions on relevant topics, and recognizing and celebrating member contributions. An important piece of the transformation from a conference to a community involves the upskilling of our team. We've been experimenting with digital conferencing since 2016, but maintaining a digital community requires a different mindset and skillset. Our team has recently attended professional development in digital community design to understand how to cultivate a digital community more effectively. We know that by nurturing a strong community culture, we can ensure that the GLAG Community becomes a vibrant and valuable resource for all its members.

## 8 Areas of Future Research and Growth

### *8.1 Melanie:*

While we have recognized the power of a digital community of practice through the GLAG Community, we also see the potential in hybrid professional development in the post-pandemic era where face-to-face interactions are so highly prized. By blending face-to-face interactions with virtual components, hybrid professional development is both flexible and inclusive. Participants can engage in hands-on activities, collaborative discussions, and networking opportunities during in-person sessions, while also benefiting from the convenience and accessibility of online resources and virtual discussions. While we have trialed professional development that utilizes both in-person and digital methods of connection through the World

Food Prize Foundation Global Guides program, this form of professional development is very expensive and time consuming. However, we've found over the years that it is through the combination of in-person connections combined with repeated interactions in the digital space is one recipe for creating strong and lasting connections, and forging future leaders. It's a reminder that while digital is a great option, leveraging them alongside in-person opportunities also has value.

### *8.2 Daniel:*

Digital professional development has so many advantages, but it can be so time consuming to build and moderate a global digital community, and to create high quality digital professional development materials. Our small team is eager to explore the potential of leveraging artificial intelligence (AI) to streamline and automate various tasks associated with the GLAG Community. AI-powered tools could help in creating high-quality digital professional development materials, such as personalized learning paths, automated assessments, and intelligent content recommendations. By integrating AI into our processes, we are excited about the possibilities to optimize efficiency, offer more learning opportunities for community members, and ensure that our limited resources are utilized effectively to foster a vibrant and thriving digital community of practice. We are hesitant to implement some of these strategies because the lack of human touch may be damaging to community-building efforts. If and when we implement AI into our work, we will conduct research in order to understand some of the impacts on the community.

## 9 Conclusion

In this chapter, we have addressed three primary objectives including: (1) Discuss the strategic positionality of educators change agents of sustainability; (2) Explore how a digital community of practice as a digital technology is being leveraged to promote the Education for Sustainable Development and (3) Uncover challenges and successes of building an accessible digital community of practice through the lived experiences of two social scientists. In conclusion, this chapter underscores the crucial role of educators in preparing learners for a sustainable future and emphasizes the need for professional development programs that address the complex needs of both learners and society. The potential of digital communities of practice are as an effective strategy for educator professional development is highlighted.

The case of the GLAG Community exemplifies the power and impact of a digital global community of practice in providing accessible professional development opportunities for educators. By creating a space for collaboration, knowledge sharing, and networking, GLAG has facilitated thousands of hours of professional development events and meaningful interactions among educators from diverse backgrounds and geographical locations. For many members, GLAG serves as a vital resource, providing their only access to educator-oriented professional development. This highlights the importance of inclusive and accessible digital platforms that enable educators to enhance their pedagogical practices and contribute to sustainable education.

As an organic and evolving community, GLAG exemplifies the potential of dedicated resources and collaborative efforts to foster the growth and development of a vibrant professional learning community. The lessons learned from GLAG's journey emphasize the importance of continuous improvement, community engagement, and the utilization of digital platforms to facilitate professional development for educators seeking to embed Education for Sustainable Development in their teaching practice.

A limitation of the GLAG digital community of practice is the need for further exploration into the quality of the community formed among its members in the digital space. Future research should delve into understanding the dynamics, interactions, and depth of relationships within the community, shedding light on the effectiveness and impact of the digital platform in fostering a robust and supportive learning environment. The contributions of digital communities of practice are not yet well understood, as research in this area is still evolving. Despite their potential in facilitating knowledge sharing and professional development, there is a little funding allocated for the support and exploration of digital communities of practice, which hinders further investigation into their effectiveness and limits their widespread implementation and potential impact on sustainable development.

By fostering collaboration, sharing knowledge, and providing accessible professional development opportunities, communities of practice have the potential to offer a powerful mechanism for empowering educators and preparing them to address the challenges of a rapidly changing world. The experiences shared in this chapter serve as an inspiration and guide for educators, policymakers, and stakeholders interested in leveraging communities of practice to support Education for Sustainable Development and create a more sustainable future.

## References

Adams TE, Jones SH, Ellis C (2022) Handbook of Autoethnography, 2nd edn. Routledge, London

Annan-Diab F, Molinari C (2017) Interdisciplinarity: Practical approach to advancing education for sustainability and for the Sustainable Development Goals. Int J Manag Edu 15(2):73–83. https://doi.org/10.1016/j.ijme.2017.03.006

Archambault LM, Crippen KJ (2009) Examining TPACK among K-12 online distance educators in the United States. CITE 9(1):71–88

Blaschke LM (2012) Heutagogy and lifelong learning: A review of heutagogical practice and self-determined learning. IRRODL 13(1):56–71

Caena F (2011) Literature review Quality in Teachers' continuing professional development. Euro Comm 2:20

Cedefop E (2018). *Skills forecast: trends and challenges to 2030*. Luxembourg: Publications Office. Cedefop reference series; No 108. http://data.europa.eu/doi/https://doi.org/10.2801/4492
Çelen FK, Seferoglu SS (2020) Features of Effective Professional Development Practices for Teachers as ICT Users. In: Durnali M, Limon İ (eds) Enriching Teaching and Learning Environments With Contemporary Technologies. IGI Global, Pennsylvania, pp 18–38
Craft A (2000) Continuing Professional Development: A Practical Guide for Teachers and Schools, 2nd edn. London, Routledge
Creemers B, Kyriakides L, Antoniou P (2013) Teacher Professional Development for Improving Quality of Teaching. Springer, Cham. https://doi.org/10.1007/978-94-007-5207-8
Elahi A, Mahanta P (2019) Empowering The Teacher As A Curriculum Developer. Deliberative Research 41(1):50–52
Etherington K (2018) Ethnography, critical reflexivity, and research practice. Qualitative research in practice. Routledge, Paris, pp 17–37
FAO. (2017). *The Future of food and agriculture: Trends and challenges*. United Nations, Rome. https://www.fao.org/3/i6583e/i6583e.pdf
FAO. (2021). *World Food and Agriculture Statistical Yearbook*. United Nations, Rome. https://www.fao.org/3/cb4477en/cb4477en.pdf
García E, Weiss E (2019) The role of early career supports, continuous professional development, and learning communities in the teacher shortage. Washington D.C, Economic Policy Institute
He MF, Phillion J (2008) Autoethnography and duoethnography: Reimagining pedagogical and research relationships in teacher education. Stud Teach Educ 4(2):103–118
Kay RH (2012) Exploring the use of video podcasts in education: A comprehensive review of the literature. Comput Hum Behav 28(3):820–831. https://doi.org/10.1016/j.chb.2012.01.011
Kimmons R, Veletsianos G, Woodward S (2017) Institutional uses of Twitter in US higher education. Innov High Educ 42(2):97–111. https://doi.org/10.1007/s10755-016-9375-6
Leicht A, Heiss J, Byun WJ (2018) Issues and trends in education for sustainable development, vol 5. UNESCO publishing, Paris
Malik A, Heyman-Schrum C, Johri A (2019) Use of Twitter across educational settings: a review of the literature. Int J Educ Technol High Educ 16(36):1–22. https://doi.org/10.1186/s41239-019-0166-x
Norris J, Sawyer RD, Lund D (eds) (2016) Duoethnography: Dialogic Methods for Social, Health, and Educational Research. Routledge, London
OECD. (2019). *Teachers' Professional Development: Europe in international comparison*: A secondary analysis based on the TALIS dataset. Paris: OECD. https://gpseducation.oecd.org/Home
Sawyer RD, Norris J (2012) Duoethnography. Oxford University Press, Incorporated
Shih J, Chuang C, Hwang G (2010) An inquiry-based mobile learning approach to enhancing social science learning effectiveness. Educ Technol Soc 13(4):50–62
Svendsen B (2020) Inquiries into teacher professional development – What matters? Education 140(3):111–131
United Nations Educational, Scientific and Cultural Organization. (2020). *Education for sustainable development: A roadmap*. https://unesdoc.unesco.org/ark:/48223/pf0000374802.locale=en
United Nations (2015). *Sustainable Development Goals*. https://sdgs.un.org/goals
USDA. (2017). *Farm Producers: 2017 Census of Agriculture Highlights*. United States Department of Agriculture National Agricultural Statistics Service. https://www.nass.usda.gov/Publications/Highlights/2019/2017Census_Farm_Producers.pdf
Vangrieken K, Meredith C, Packer T, Kyndt E (2017) Teacher communities as a context for professional development: A systematic review. Teach Teach Educ 61:47–59. https://doi.org/10.1016/j.tate.2016.10.001
Vrasidas C, Zembylas M (2004) ICT in education: The rhetoric and the reality. Educ Inf Technol 9(3):201–216
Wenger E, McDermott RA, Snyder W (2002) Cultivating communities of practice: A guide to managing knowledge. Harvard business press, MA

# Powering Sustainable Mobility: Electric Vehicles the Future of Smart Cities in India

**Divya Verma and Vaishali**

**Abstract** Growth and progress of transportation and the economy are inextricably linked. With that being said, it is crucial not to brush aside the environment due to the transportation sector, which releases massive amounts of greenhouse gases (GHGs). The escalating number of vehicles in India has exacerbated the issue with the sector's share of GHG emissions steadily increasing, underlining the importance of adopting sustainable transportation approaches. This chapter examines the link between GHG emissions and sustainable transport alternatives to highlight the importance of sustainable transport. It investigates the unique context of electric vehicles (EVs) and public transportation in India, as well as the country's potential for adopting sustainable transportation solutions to build sustainable cities. Simultaneously, the chapter explores various interventions that can reduce energy consumption and GHG emissions within the transportation sector. By emphasizing the significance of e-mobility, the chapter highlights its potential as a key driver of sustainable growth and mobility and its contribution to achieving UN Sustainable Development Goals (SDGs).

**Keywords** Sustainable Transportation · E-mobility · Public Transportation · Electric Vehicles · SDG · Sustainable Development · GHG mitigation

## 1 Introduction

The invention and adoption of the wheel revolutionized transportation, fulfilling the humanity's need for faster travel. Wheeled transport has become an essential cornerstone of modern society, playing a significant role in the functioning and growth of our economy. Furthermore, the automation of the transportation industry

D. Verma · Vaishali (✉)
University School of Management Studies, Guru Gobind Singh Indraprastha University, Delhi 110078, India
e-mail: vaishalikasana21@gmail.com



W. Leal Filho et al. (eds.), *Digital Technologies to Implement the UN Sustainable Development Goals*, World Sustainability Series,
https://doi.org/10.1007/978-3-031-68427-2_9

has driven advancements, expanding transportation modes from land and sea to air and rail, leading to increased trade and thriving economies (Helling, 1997).

Following its liberalization in 1991, and thereafter allowance of 100 per cent foreign direct investment (FDI) by way of automatic route, the Indian automobile sector has made remarkable progress (Ministry of Heavy Industries, 2023), as evident in Fig. 1. In 1951, India had a mere 306 automobiles. However, with the opening up of the economy and the implementation of liberalization measures, there was a notable increase in the number of automobiles in 1995, with a total of 30,295 vehicles sold. By the time 2020 arrived, India witnessed an impressive sales figure of 326,299 vehicles. Nevertheless, alongside the economic benefits, the development of the transportation industry led to a rise in GHG emissions (Kumar et al., 2022). Initially, motor engines relied on carbon-based fuels like coal for operation, but over time, these fuels were replaced by petrol and diesel. This surge in emissions by this transportation sector has accelerated climate change, posing serious threats to human life and the environment (Singh et al., 2019).

The transportation sector is responsible for emitting 16.2% of global GHGs. Among the various modes of transportation, road transport has the highest contribution, accounting for 11.9% of the total emissions. Aviation follows with a contribution of 1.9%, while shipping accounts for 1.7%. Rail transport and pipelines have rather smaller shares, pitching in 0.4% and 0.3%, respectively (Ritchie, 2020). Fig 2 demonstrates the growing proportion of GHG emissions from India's transport sector. Research suggests the presence of approximately 40% 20% PM2.5, and more than 80% of NOx and CO in atmospheric air pollution in Indian cities (ET, 2022).

There is a growing emphasis on creating and using sustainable transportation options in order to reduce GHG emissions and cope with climate change. Sustainable transportation confers a set of alternatives working on two major pillars the first being, the use of alternative fuels and the other option is to improve fuel economy.

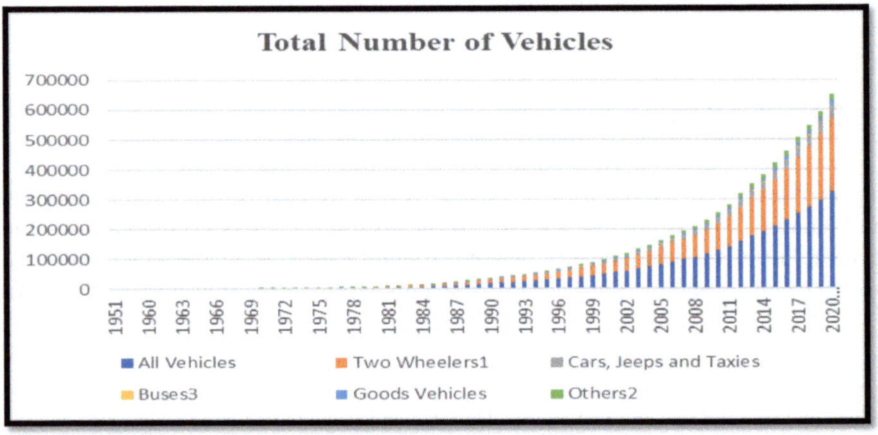


**Fig. 1** Growth in total Number of Vehicles in India. Source: Indiastat

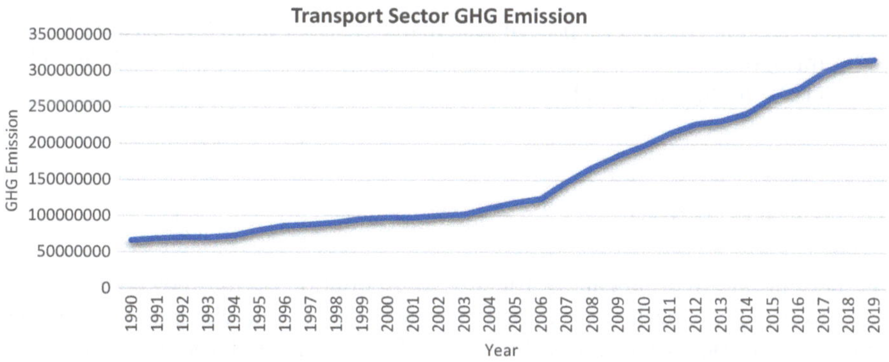


**Fig. 2** Transport Sector GHG Emission. Source: Our World in Data based on Climate Analysis Indicators Tool (CAIT)

Fostering the use of EVs as a replacement for conventional automobiles is regarded as an articulate way to the transportation sector's ecological footprint concerns. India is firmly supporting the adoption of EVs in both private and public transport aimed at promoting sustainable mobility throughout the country.

## 2 Mitigating and Adapting to Greenhouse Gas Emissions Through Sustainable Transportation

Sustainable transportation, as defined by the Asian Development Bank (ADB), "encompasses a transport system characterized by accessibility, safety, environmental friendliness, and affordability" (ADB, 2010). It has come up as an important response to the significant environmental, social, and economic challenges posed by conventional transportation systems. The need to switch to a more sustainable and resilient transportation system is becoming more and more obvious with the rise in the need for mobility and the negative impacts of fossil fuel-powered automobiles on climate change, air pollution, and urban congestion become clear (Steg & Gifford, 2008).

The notion of "sustainable transportation" refers to a variety of tactics, procedures, and regulations aimed at lowering GHG emissions, improving energy efficiency, increasing accessibility, and improving the overall quality of transportation networks. While minimising negative environmental and social effects, it aims to fulfil the mobility demands of individuals and communities (UN, 2021).

The transportation system is made up of four basic components, which are (i) control structures, (ii) terminals, (iii) vehicles, and (iv) ways. These components work together harmoniously to ensure the efficient functioning of the entire system

(Mohapatra et al., 2023). Two pivotal sustainable transportation options are EVs and public transportation, which play a crucial role in achieving sustainable mobility (Ercan et al., 2015; Wang et al., 2018).

## *2.1 Electric Vehicle*

EVs present an environmentally friendly transportation option, powered by electricity from their batteries or external charging sources like charging stations and home chargers (Rajper & Albrecht, 2020). These vehicles possess unique qualities, including zero tailpipe emissions, energy efficiency, quiet operation, regenerative braking to convert kinetic energy into electricity, low maintenance needs, potential integration with renewable energy sources, and overall superior environmental friendliness compared to traditional fossil fuel-powered cars (Carlsson & Johansson-Stenman, 2003; Falahi et al., 2013). The eco-conscious characteristics and cost-effectiveness of EVs have led to their increasing global promotion. Unlike conventional vehicles with significant tailpipe emissions and high fuel costs, EVs have reduced or zero tailpipe emissions and lower running costs. This drives their widespread adoption as a budget-friendly and environmentally responsible mode of transportation (Egbue & Long, 2012).

EVs come in four different types, each with its own distinct features, as illustrated in Fig. 3. The first type is the battery electric vehicle (BEV), highly efficient and entirely reliant on electricity for power. The second type is the hybrid electric vehicle (HEV), operating with both an internal combustion engine, usually fueled by petrol, and a battery-powered electric motor. In the HEV, the petrol engine serves the dual purpose of driving the vehicle and charging the battery. Another variant is the plug-in hybrid electric vehicle (PHEV), allowing external charging through plugs, unlike the HEV where the engine primarily charges the battery. Lastly, the fuel cell electric vehicle (FCEV) produces electric energy using a chemical substance like hydrogen. Each type of EV offers specific advantages and features, catering to different preferences and requirements (e-Amrit, n.d.).

While EVs are known for their minimal tailpipe emissions, it is important to note that the manufacturing process of their batteries contributes to GHG emissions (Kumar, 2023; Hall & Lutsey, 2018). The International Council on Clean Transportation (ICCT, 2021) conducted a lifecycle analysis of EVs and their battery production, revealing that studies suggest the production of batteries emits approximately 56–494 kg of carbon dioxide per kilowatt-hour of battery capacity (kg $CO_2$/kWh) for EVs (S, 2019).

Additionally, EVs indirectly contribute to the accumulation of GHG emissions during their operational phase. This is because the electricity required to charge these vehicles is primarily generated from fossil fuels, particularly in India, where over 50 per cent of electricity is produced in thermal power plants.

In most cases, EVs are found to be low emitting during their lifetime compared to conventional vehicles (Kumar, 2023; Koroma et al., 2022). However, in India, EVs

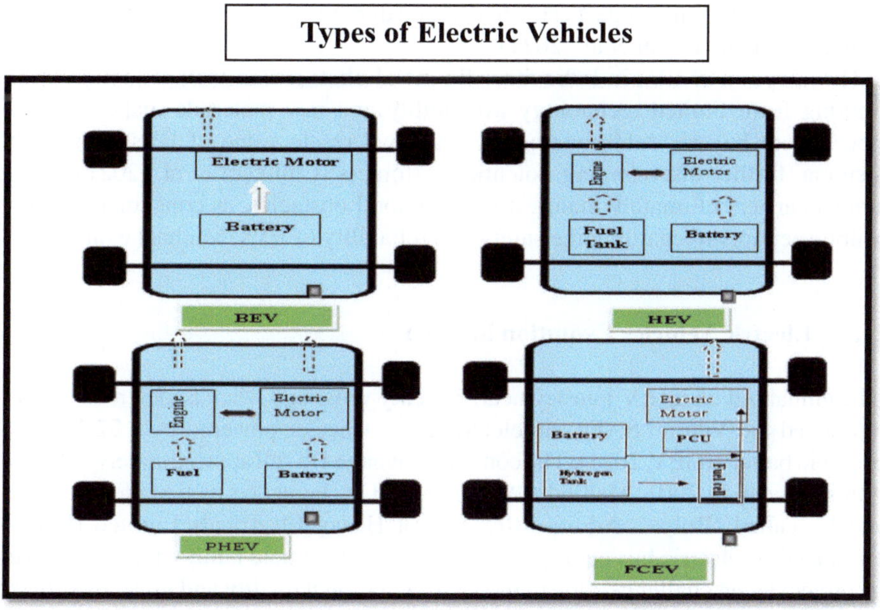


**Fig. 3** Types Of Electric Vehicles

are found to be more polluting due to the indirect effect of using electricity generated from thermal power plants (Das, 2022).

To enhance their environmental friendliness, greater emphasis should be placed on utilizing renewable energy sources for charging EVs and exploring options for reusing installed batteries (Koroma et al., 2022). In addition to the environmental benefits, EVs are found to be more cost-effective at the operational stage. The running cost of an EV, on average, is as low as Rs 1 per Km compared to Rs 9 per Km for petrol, Rs 6 per Km for cars running on diesel, and Rs 2.5 per Km for CNG cars in India (ET, 2021).

### 2.1.1 Barriers to EV Adoption

In spite of the cost-effective and environmentally friendly nature of EVs, they struggle with a low adoption rate in the Indian market, presently standing at less than 2% (Malik, 2021). There are a lot of reasons for this slow growth. A lack of **infrastructure** is the most prominent, like not having enough charging stations, not having a reliable power source, not having enough EV manufacturers, and not having enough support for maintenance and repair services (Dash, 2023).

The **technological aspect** of EVs is one of the major challenges in India. As the technology is relatively new and not yet widely known to consumers and suppliers, there is a lack of parts and components on the market. In addition to this, there are concerns related to the range, performance, and durability of EVs, creating reluctance among customers (Tarei et al., 2021).

The adoption of EVs in India faces **financial obstacles** as well. The high prices resulting from limited technology availability and raw materials make EVs less appealing to buyers. Additionally, the unknown resale value of EVs adds to the dilemma, further discouraging potential customers (Chhikara et al., 2021). These various barriers ultimately manifest as behavioral obstacles, as consumers in India exhibit skepticism regarding the safety and reliability of EVs (Michael et al., 2022).

### 2.1.2 Electric Vehicles Evolution in India

India embarked on its EV journey nearly twenty years ago when Scooters India Ltd. introduced the Vikram SAFA, an electric three-wheeler powered by a 72 Volt (V) lead-acid battery (GEV, 2021). The company managed to sell approximately 400 units of this vehicle. In 2001, Mahindra & Mahindra Ltd. launched its first electric three-wheeler called 'Bijlee'. Additionally, Bharat Heavy Electricals Limited (BHEL) introduced an electric bus equipped with a 96 Volt lead-acid battery pack. However, the project faced challenges due to inconsistent product quality and high costs. In the same year, a Bangalore-based company named 'REVA' entered the Indian EV market in the four-wheel segment and successfully sold over 1500 vehicles (Sudarshan, 2021).

The entry of companies in the two-wheeler segment and the moderate success in the three-wheeler segment led to the production of electric bikes and e-rickshaws in India. Currently, over 700,000 electric three-wheelers are operating in the country, mostly using lead-acid batteries. However, the shift towards lithium-ion batteries in electric rickshaws could open up export opportunities in South Asia and Africa (Mohanty & Kotak, 2017).

In the four-wheeler segment, Mahindra & Mahindra introduced the E2O a few years ago, offering two battery variants of 48 V and 72 V. They later expanded their EV lineup with the launch of the Mahindra Verito sedan. Tata Motors also made its entry into the electric car segment with the introduction of the Tata Tigor, featuring a 72 V battery and a maximum power of 30 KW (Cheema, 2019).

Electric buses are being integrated into the public transport fleet in various states in India, starting with Himachal Pradesh. Cities like Ahmedabad, Bengaluru, Mumbai, Hyderabad, and Kolkata have initiated the procurement process for electric buses. State governments are actively pursuing e-bus tenders, which are expected to benefit original equipment manufacturers (OEMs) and result in increased volume and lower prices. Pilot tests are also being under taken in the cargo sector (Vijaykumar et al., 2022).

## 2.2 Eco-Friendly Public Transportation

One of the most sustainable forms of transportation, public transportation has long been hailed for its many advantages that promote social, economic, and environmental well-being (Kawabata & Shen, 2007). Its beneficial effects go well beyond simply getting people from one place to another. In order for cities and regions to thrive sustainably and create prosperous and thriving communities, public transport is essential.

Public transport benefits the environment by reducing the number of vehicles on the road. It helps cut down on traffic by getting people to take public transportation instead of driving themselves, which makes for better and more efficient transportation. Eliminating private automobiles on road, reduces GHG emissions, helping tackle climate change, scoring it as a greener and more sustainable option (Greenblatt & Shaheen, 2015).

In addition, public transport networks are constantly focusing on adapting to cleaner and more efficient technology for emissions reduction, Electric and hybrid buses are becoming more and more popular in public transportation, given their lower noise levels and positive environmental impact compared to conventional buses. (Manzolli, 2022).

In societal context, public transit plays a pivotal role in providing accessibility and inclusion. Public transit acts as a lifeline to people from humble background who don't have access to personal vehicle. Convenient and affordable public transit ensures that people from all sections of society has access to essential services, jobs, and experiences. The inclusiveness affirms building a unified and equitable society in which no one is left behind (Qamhaieh & Chakravarty, 2016; Bastiaanssen et al., 2021).

In the same way, public transit networks have a distinctive ability to connect communities and bind them together (Transport for Sustainable Development, 2016). By conferring convenient and connected modes of transport, public transit is the bridge that unites communities and promotes social cohesion. This connection makes people feel more connected to each other and makes the communities stronger.

The economic contribution of public transit to economic growth cannot be underestimated. By making the most of existing infrastructure, it reduces the need for expensive new roads and parking lots. The saved resources can be used for other essential areas of development, thereby increasing overall economic efficiency.

Further, the operation and maintenance of public transit networks generate significant employment opportunities, directly and indirectly. The workforce required to operate and manage these systems injects vitality into local economies and creates a multiplier effect that supports ancillary businesses and services (Magazzino & Mele, 2020).

The urban population in India is constantly increasing, necessitating the implementation of sustainable public transportation to address traffic congestion and environmental concerns. The primary components of Indian public transport include buses, metros, and suburban rail services. The Government of India (GOI) has undertaken various policy interventions to cater to the demand for public transportation. One significant measure was the enactment of the Road Transport Corporation Act in 1956, empowering state governments to establish Transport Corporations with the aim of delivering an efficient, sufficient, cost-effective, and well-coordinated road-based public transport system across the nation (Jha, 2019). Further endeavors are underway to enhance the eco-friendliness of the Indian bus system by transitioning it to an electrified mode. Indian railways serve as another crucial means of public transportation, facilitating inter-city and interstate connectivity. Notably, almost all trains in the country have already been electrified, rendering them more environmentally friendly and cost-effective (Kuriakose & Bhattacharjee, 2021).

For enhancing intra-city connectivity, the metro rail system was introduced in India. The Metro Railways Act of 1978 was implemented to establish metro railways in major metropolitan cities, with Calcutta being the primary focus (Verma et al., 2021). Currently, all major metropolitan areas proudly offer a robust metro network, delivering a convenient, effective, and eco-conscious transportation choice for the rapidly growing urban population.

To sum up, public transportation serves as an exemplary model of sustainable mobility, embodying the values of economic effectiveness, social fairness, and ecological consciousness. Its diverse benefits play a crucial role in fortifying and empowering communities, all while preserving the environment for future generations. Looking ahead, ongoing investments and advancements in public transit systems will be pivotal in fostering the development of sustainable and inclusive societies.

## 3 Interventions for Reducing Energy Consumption and GHG Emission

The Indian government has implemented several initiatives to promote the adoption of EVs. One of these projects is the Faster Adoption and Manufacture of Hybrid and Electric Vehicles (FAME) India Scheme, which was introduced in 2011 as part of the National Electric Mobility Mission Plan. By July 2018, the scheme had supported approximately 2,18,625 EVs, with a total cost of Rs 895 crore. The National Electric Mobility Mission Plan aims to sell 6–7 million electric vehicles to reduce fossil fuel usage. The first phase of the FAME India Scheme, running from 2015 to March 2019, focused on creating demand, developing technology, launching pilot projects, and installing charging infrastructure through incentives for different vehicle types (Ministry of Heavy Industries and Public Enterprises, 2020).

**The achievements of Phase-1** of the FAME India Scheme by January 20, 2023, show support for over 2,80,000 hybrid and EVs through demand incentives, leading to fuel savings of about 116 million litres and a $CO_2$ reduction of about 289 million kg. Technological development projects worth approximately Rs. 158 crores, including testing facilities and research facilities, were approved. Incentives were also provided for 425 electric buses, estimated to save 80 million litres of gasoline and 180 million kg of $CO_2$ during their lifetime. Furthermore, the program facilitated the construction of 520 EV charging stations, with 461 already in operation. Overall, the FAME Scheme stimulated discussions on electric transportation at both the central and state levels (Ministry of Heavy Industries, 2020; Ministry of Heavy Industries and Public Enterprises, 2018; Ministry of Heavy Industries, 2019).

**FAME II**, the second phase of the program, focuses on promoting electric and hybrid vehicles through upfront incentives and charging infrastructure. The program also includes commercial and private electric two-wheelers, emphasizing accessible and environmentally friendly public transportation. Incentives for electric two-wheelers are based on demand and battery capacity. The program has been adjusted based on sector feedback to reduce upfront costs. As of December 9, 2022, Phase II has attracted the registration and revalidation of 135 electric vehicle models from 64 OEMs. A total of 6,71,644 e-2Ws, 70,912 e-3Ws, and 5,435 e-4Ws have been sold as electric vehicles. Additionally, the Ministry of Heavy Industries has approved the use of 6,315 electric buses for intra-city operations in 65 cities, state transport corporations, and government bodies. Among the 3,738 electric buses ordered, 2,435 are already in service in cities like Mumbai, Ahmedabad, Pune, and UP. Moreover, 100 e-buses will offer last-mile connections for the Delhi Metro Rail Corporation (DMRC), and 650 e-buses have been certified for intercity service. The Ministry has also consolidated the demand for e-buses in nine major cities with populations exceeding 4 million, with a total of 5,450 e-buses planned under the FAME-II scheme (Ministry of Heavy Industries, 2023).

Regarding **electric vehicle charging stations**, the Ministry has authorized 2,877 EV charging stations in 68 cities across 25 States and UTs. As of December 9, 2022, 1,822 EV charging stations have received Letters of Award, with 83 currently operational. Additionally, 1,576 charging stations are permitted along 9 motorways and 16 highways as part of FAME India Scheme Phase II (Ministry of Heavy Industries, 2021).

Furthermore, the Indian government has launched the **Production Linked Incentive scheme** 'National Programme on Advanced Chemistry Cell (ACC) Battery Storage' to encourage the establishment of sizable ACC manufacturing plants in the country. This program focuses on value-added, high-quality production, and specific capacity goals to improve India's electric mobility and battery storage ecosystem, increase manufacturing capabilities, and boost exports in the ACC industry (Ministry of Heavy Industries, 2021).

For the automobile and auto component sector, the **Production Linked Incentive (PLI)** plan aims to increase India's production capacity for advanced automotive products. The scheme, with a budget of Rs. 25,938 Crore, offers financial incentives to promote local AAT product manufacture and attract investments to the automotive value chain. It prioritizes technology like zero-emission vehicles, safety features, connected cars, emissions management, and other auto parts. Under this scheme, 85 candidates were approved, anticipating a projected investment of Rs. 67,690 crores and the creation of 1.45 lakh direct job opportunities over five years (Ministry of Heavy Industries, 2021). The Quarterly Review Report (QRR) Module and the PLI Auto Portal are two IT projects supporting the scheme's implementation.

## 4 The Potential of Sustainable Growth and Mobility and Contribution to UN SDGs

Sustainable transportation offers numerous economic, social, and environmental benefits and plays a significant role in achieving the United Nations Sustainable Development Goals (SDGs). By prioritizing sustainable mobility, cities can enhance their livability and inclusivity, making them more attractive and sustainable places to live. This is accomplished through initiatives that improve access to essential services, alleviate traffic congestion, and enhance air quality, aligning with **SDG 11**, which aims to foster sustainable cities and communities (Nikulina et al., 2017; Shokoohyar et al., 2022).

Promoting the use of low-emission alternative fuels, EVs, public transportation, walking, and cycling in place of traditional motor vehicles is crucial in reducing GHG emissions, aligning with the main objective of **SDG 13**, which emphasizes taking action to combat climate change and its impacts (Shokoohyar et al., 2022; Cheshmehzangi & Dawodu, 2018).

Adopting sustainable mobility practices offers numerous health benefits as well, as it helps mitigate the risk of harmful diseases associated with GHG emissions, aligning with the objective of **SDG 3**, which aims to ensure good health and well-being for all (Bassi et al., 2022).

Moreover, sustainable mobility promotes the use of clean energy alternatives such as EVs and renewable energy sources for public transportation, reducing reliance on fossil fuels and contributing to the objective of **SDG 7**, which focuses on affordable and clean energy (Thilakshan & Bandara, 2019).

Sustainable mobility also plays a crucial role in achieving **SDG 10**, which aims to reduce inequalities, and **SDG 8**, which focuses on decent work and economic growth. Promoting public transportation helps reduce inequality by providing accessible and affordable transportation options for everyone, regardless of socioeconomic status. Additionally, the expansion of sustainable transportation systems creates new job opportunities, fostering economic growth (Winter, 2020; Furszyfer Del Rio et al., 2023).

To achieve sustainable mobility, advanced technologies, appropriate infrastructure, and efficient transportation systems are essential. This commitment to sustainable mobility drives the development of innovative technologies, ongoing research, and the establishment of environmentally-friendly infrastructure, aligning with **SDG 9**, which emphasizes industry, innovation, and infrastructure (Furszyfer Del Rio et al., 2023; Song et al., 2021).

India's transportation sector, with its extensive and varied road network, serves the transportation needs of its large population. The country's road network is continuously expanding, facilitating the movement of goods, services, and people, contributing to sustainable urban development and economic activities (Zope et al., 2019). India's highway network density is comparable to that of the United States and higher than China or Brazil (World Bank, 2014).

## 5 Future Direction

EVs are undoubtedly the future of sustainable transportation. While the GOI is already implementing policy measures to promote the adoption of EVs in both public and private transportation, there remains ample scope for further progress in the realm of sustainable transportation.

**Intelligent Transport System (ITS)** can play a vital role in addressing traffic congestion and improving the quality of driving in India. Leveraging Artificial Intelligence (AI), ITS offers innovative solutions for urban mobility, incident detection, traffic prediction, and traffic management (Abduljabbar et al., 2019). The implementation of Artificial Neural Network (ANN) technology enables the detection of vehicle location and speed, facilitating the identification of congested areas and suggesting alternative routes to alleviate congestion (Nandal et al., 2021). Various AI algorithms such as Genetic Algorithms (GA), Simulated Annealing (SA), Artificial Immune Systems (AIS), Ant Colony Optimization (ACO), Bee Colony Optimization (BCO), and Fuzzy Logic Models (FLMs) further optimize routes, reduce travel time, fuel consumption, and enhance transportation efficiency (Ceylan & Bell, 2004; Xu et al., 2009; Timmis et al., 2000; Dochtermann, 2008; Lučić & Teodorović, 2003; Kulak et al., 2005). Encouraging the integration of AI in sustainable transportation systems is paramount (Abduljabbar et al., 2019).

**Infrastructure** plays a significant role in the adoption of EVs in India. Rapid installation of charging stations and battery swapping stations are essential steps to overcome the lack of charging infrastructure. Upgrading public transportation with the electrification of buses and taxis requires the development of charging infrastructure at relevant locations (Sachan & Singh, 2022).

To achieve a sustainable transportation system, there is a need to transition to **environmentally friendly fuels**. The Indian government is taking initiatives to promote alternative fuels such as ethanol-blended petrol and renewable energy sources for EVs (PIB, 2022; Andersen et al., 2009). It is crucial to focus on utilizing renewable energy sources for EVs. Integrating EVs with renewable energy can effectively decrease the ecological impact and contribute to the development of sustainable cities. Renewable sources like wind power and solar photovoltaic can be harnessed to charge EV batteries, further promoting environmentally friendly transportation options (Andersen et al., 2009).

**Effective end-of-life management of batteries** is essential to address the growing stock of used lithium-ion batteries. Implementing recycling technologies can extend the lifespan of components and recover valuable materials like lithium, cobalt, and nickel, thus reducing battery costs and mitigating resource strain (Labios, 2018; S, 2019). Toyota Motor Corporation has developed innovative copper recycling technology specifically for end-of-life EVs, guaranteeing a steady and sustainable source of copper for vehicle production (Copper Recycling Success for Toyota, 2014); Bhattacharjya et al., 2018). Similar recycling methods should be further explored and implemented to promote sustainable transportation and sustainable lifestyles in urban areas.

## 6 Conclusion

The chapter emphasises the necessity of putting into practice sustainable transportation practices in India, with a focus on how EVs might help to satisfy the demand for eco-friendly commuting options. The nation is, however, facing significant barriers to the adoption of EVs, such as a lack of the required infrastructure, technological constraints, and financial constraints.

The Indian government has proactively taken various measures and interventions to address these challenges and encourage the widespread adoption of EVs. Notably, the introduction of two significant initiatives, FAME I and II (Faster Adoption and Manufacturing of Electric Vehicles), aims to provide support and incentives for the transition towards electric mobility in India's transportation system.

Despite the commendable efforts of the Indian government, it is imperative to maintain a continuous focus on improving and optimizing the sustainable transportation system in the country. By addressing and resolving the existing shortcomings, a more favorable environment can be created to facilitate the smooth integration of EVs into the mainstream transportation network in India.

Through ongoing investments in charging infrastructure, advancements in EV technologies, and the implementation of suitable financial incentives, India can lay the foundation for a more sustainable and eco-friendly transportation system. Collaborative endeavors from both the government and private sectors will play a pivotal role in facilitating a seamless and prosperous transition towards a greener and more sustainable transportation future in the country.

## References

Abduljabbar R, Dia H, Liyanage S, Bagloee SA (2019) Applications of artificial intelligence in transport: An overview. Sustainability (switzerland) 11(1):189. https://doi.org/10.3390/su11010189

ADB. (2010). *SUSTAINABLE TRANSPORT INITIATIVE*. https://www.adb.org/sites/default/files/institutional-document/31315/sustainable-transport-initiative.pdf

Andersen PH, Mathews JA, Rask M (2009) Integrating private transport into renewable energy policy: The strategy of creating intelligent recharging grids for electric vehicles. Energy Policy 37(7):2481–2486. https://doi.org/10.1016/j.enpol.2009.03.032

Annual Report 2017–18. (2018). In *Ministry of Heavy Industries*. Ministry of Heavy Industries and Public Enterprises Government of India. https://heavyindustries.gov.in/ViewData/index?mid=1117

Annual Report 2018–19. (2019). In *Ministry of Heavy Industries*. Ministry of Heavy Industries and Public Enterprises Government of India. https://heavyindustries.gov.in/ViewData/index?mid=1117

Annual Report 2019–20. (2020). In *Ministry of Heavy Industries*. Ministry of Heavy Industries and Public Enterprises Government of India. https://heavyindustries.gov.in/ViewData/index?mid=1117

Annual Report 2020–21. (2021). In *Ministry of Heavy Industries*. Ministry of Heavy Industries and Public Enterprises Government of India. https://heavyindustries.gov.in/ViewData/index?mid=1117

Annual Report 2021–22. (2022). In *Ministry of Heavy Industries*. Ministry of Heavy Industries Government of India. https://heavyindustries.gov.in/ViewData/index?mid=1117

Annual Report 2022–23. (2023). In *Ministry of Heavy Industries*. Ministry of Heavy Industries Government of India. https://heavyindustries.gov.in/ViewData/index?mid=1117

Bassi AM, Pallaske G, Niño N, Casier L (2022) Does Sustainable Transport Deliver Societal Value? Exploring Concepts, Methods, and Impacts with Case Studies. Future Transportation 2(1):115–134. https://doi.org/10.3390/futuretransp2010007

Bastiaanssen J, Johnson D, Lucas K (2021) Does better job accessibility help people gain employment? The role of public transport in Great Britain. Urban Studies 59:301–322. https://doi.org/10.1177/00420980211012635

Bhattacharjya, S., Arora, N., Bakshi, S. K., & Jasuja, G. (2018). Towards resource efficient electric vehicle sector in india. In European Union's Resource Efficiency Initiative (EU-REI) Project.

Carlsson F, Johansson-Stenman O (2003) Costs and Benefts of Electric Vehicles A 2010 Perspective. JTEP 37(1):1–28

Ceylan H, Bell MGH (2004) Traffic signal timing optimisation based on genetic algorithm approach, including drivers' routing. Transportation Research Part b: Methodological 38(4):329–342. https://doi.org/10.1016/S0191-2615(03)00015-8

Cheema K (2019). India Vs China Turns Electric: The Story Of Two EV Markets. Inc42 Media. https://inc42.com/features/india-vs-china-turns-electric-the-story-of-two-ev-markets/

Cheshmehzangi A, Dawodu A (2018) Introduction: Climate Change and Cities—Perspectives Planning and People. Sustainable Urban Development in the Age of Climate Change. Palgrave Macmillan, Singapore, pp 1–22

Chhikara R, Garg R, Chhabra S, Karnatak U, Agrawal G (2021) Factors affecting adoption of electric vehicles in India: An exploratory study. Transp Res Part d: Transp Environ 100(October):103084. https://doi.org/10.1016/j.trd.2021.103084

*Copper recycling success for Toyota*. (2014). Recycling International; Recycling International. https://recyclinginternational.com/non-ferrous-metals/copper-recycling-success-for-toyota/5649/

Das J (2022) Comparative life cycle GHG emission analysis of conventional and electric vehicles in India. Environ Dev Sustain 24(11):13294–13333. https://doi.org/10.1007/s10668-021-01990-0

Dash A (2023) Adapting to electric vehicles value chain in India: The MSME perspective. Case Studies on Transport Policy 12(February):100996. https://doi.org/10.1016/j.cstp.2023.100996

Dochtermann N (2008) Ant Colony Optimization. Int J Comput Sci Net Sec 8(6):351–357

e-Amrit (n.d.). TYPES OF ELECTRIC VEHICLES. *NITI AAYOG*. Retrieved July 1, 2023, from https://e-amrit.niti.gov.in/types-of-electric-vehicles

Egbue O, Long S (2012) Barriers to widespread adoption of electric vehicles: An analysis of consumer attitudes and perceptions. Energy Policy 48(2012):717–729. https://doi.org/10.1016/j.enpol.2012.06.009

Ercan T, Zhao Y, Tatari O, Pazour JA (2015) Optimization of transit bus fleet's life cycle assessment impacts with alternative fuel options. Energy 93:323–334. https://doi.org/10.1016/j.energy.2015.09.018

ET. (2021). *As petrol nears Rs 100/litre, EVs running at one-tenth cost*. ET Auto. https://auto.economictimes.indiatimes.com/news/oil-and-lubes/as-petrol-nears-rs-100/litre-evs-running-at-one-tenth-cost/81029317

ET. (2022). Electric mobility is key enabler to decarbonise transport sector, say experts. *ET*.

Falahi M, Chou HM, Ehsani M, Xie L, Butler-Purry KL (2013) Potential power quality benefits of electric vehicles. IEEE Trans Sustain Energy 4(4):1016–1023. https://doi.org/10.1109/TSTE.2013.2263848

Furszyfer Del Rio DD, Sovacool BK, Griffiths S, Foley AM, Furszyfer Del Rio J (2023) A cross-country analysis of sustainability, transport and energy poverty. Npj Urban Sustainability 3(1):1–18. https://doi.org/10.1038/s42949-023-00121-0

GEV. (2021). *History of Electric Vehicles in India*. GET ELECTRIC VEHICLE. https://getelectricvehicle.com/history-of-electric-vehicles-in-india/

Greenblatt JB, Shaheen S (2015) Automated Vehicles, On-Demand Mobility, and Environmental Impacts. Current Sustainable/renewable Energy Reports 2(3):74–81. https://doi.org/10.1007/s40518-015-0038-5

Hall D, Lutsey N (2018). *Effects of battery manufacturing on electric vehicle life-cycle greenhouse gas emissions* (Issue February).

Helling A (1997) Transportatuon and Economic Development A Review. Public Work Management & Policy 2(1):79–93. https://doi.org/10.1177/1087724X9700200108

ICCT. (2021). A new life-cycle assessment of the greenhouse gas emissions of combustion engine and electric passenger cars in major markets. In *Communications* (Issue July). https://www.team-bhp.com/forum/attachments/electric-cars/2310712d1653112667-things-consider-when-buying-electric-car-global-lca-passenger-cars-jul2021-0.pdf

Jha P (2019). *The role of public transportation in India – Motorindia*. Motorindia. https://www.motorindiaonline.in/the-role-of-public-transportation-in-india/

Kawabata M, Shen Q (2007) Commuting Inequality between Cars and Public Transit: The Case of the San Francisco Bay Area, 1990–2000. Urban Studies 44:1759–1780. https://doi.org/10.1080/00420980701426616

Koroma MS, Costa D, Philippot M, Cardellini G, Hosen MS, Coosemans T, Messagie M (2022) Life cycle assessment of battery electric vehicles: Implications of future electricity mix and different battery end-of-life management. Sci Total Environ 831:154859. https://doi.org/10.1016/j.scitotenv.2022.154859

Kulak O, Durmuşoğlu MB, Kahraman C (2005) Fuzzy multi-attribute equipment selection based on information axiom. J Mater Process Technol 169(3):337–345. https://doi.org/10.1016/j.jmatprotec.2005.03.030

Kumar M, Shao Z, Braun C, Bandivadekar A (2022). Decarbonizing India ' S Road Transport : a Meta-Analysis of Road Transport. *The International Council of Clean Transportation, May*.

Kumar S. (2023). *Are EVs more polluting than Conventional Vehicles?* E-Vehicleinfo. https://e-vehicleinfo.com/are-evs-more-polluting-than-conventional-vehicles/

Kuriakose PN, Bhattacharjee J (2021) India's Public Transport Systems: The Role of Metro Rail. In: Mitra S, Bandyopadhyay S, Roy S, Ponce Dentinho T (eds) Railway Transportation in South Asia. Contemporary South Asian Studies. Springer, Cham, pp 131–152

Labios, L. (2018). Recycling and Reusing Worn Cathodes to Make New Lithium Ion Batteries. Today.ucsd.edu. https://today.ucsd.edu/story/recycling_and_reusing_worn_cathodes_to_make_new_lithium_ion_batteries

Lučić P, Teodorović D (2003) Computing with Bees: Attacking Complex Transportation Engineering Problems. Int J Artif Intell Tools 12(03):375–394. https://doi.org/10.1142/s0218213003001289

Magazzino C, Mele M (2020) Research in Transportation Economics On the relationship between transportation infrastructure and economic development in China. Res Transp Econ 88:100947. https://doi.org/10.1016/j.retrec.2020.100947

Malik B (2021). India's EV market to grow by 90% to touch $150 billion by 2030: Report. *Business Today*. https://www.businesstoday.in/latest/story/indias-ev-market-to-grow-by-90-to-touch-150-billion-by-2030-report-314116-2021-12-01

Manzolli A (2022). *A review of electric bus vehicles research topics – Methods and trends EU. 159*(February 2021). https://doi.org/10.1016/j.rser.2022.112211

Michael LK, Sriram KV, Hungund SS, Fernandes M (2022) Factors influencing adoption of electric vehicles–A case in India. Cogent Engineering 9(1):2085375. https://doi.org/10.1080/23311916.2022.2085375

Mohanty P, Kotak Y (2017) Electric vehicles: Status and roadmap for India. Electric Vehicles: Prospects and Challenges. Elsevier Inc., Amsterdam, pp 387–414

Mohapatra S, Mohanachandran D, Dwivedi G, Kesharvani S, Harish VSKV, Verma S, Verma P (2023) A Comprehensive Study on the Sustainable Transportation System in India and Lessons to Be Learned from Other Developing Nations. Energies 16(4):1986. https://doi.org/10.3390/en16041986

Nandal M, Mor N, Sood H (2021) An overview of use of artificial neural network in sustainable transport system. Computational Methods and Data Engineering, vol 1227. Springer Nature, Singapore, pp 83–91

Nikulina V, Baumann H, Simon D, Sprei F (2017) Sustainable Transport Futures: Analysis of the Selected Methodologies Supporting the Planning Process Towards Achieving Goal 11 Sustainable Cities and Communities. World Sustainability Series. Springer, Cham, pp 473–488

PIB. (2022). *ENCOURAGING ETHANOL BLENDING.*

Qamhaieh A, Chakravarty S (2016) Global Cities, Public Transportation, and Social Exclusion: A Study of the Bus System in Abu Dhabi. Mobilities 12:462–478. https://doi.org/10.1080/17450101.2016.1139805

Rajper SZ, Albrecht J (2020) Prospects of electric vehicles in the developing countries: A literature review. Sustainability (switzerland) 12(5):1906. https://doi.org/10.3390/su12051906

Ritchie, H. (2020, September 18). *Sector by Sector: Where do Global Greenhouse Gas Emissions come from?* Our World in Data. https://ourworldindata.org/ghgemissions-by-sector

S, S. (2019, December 17). *Are Electric Vehicles Really Green If They Don't Use Clean Energy?* Inc42 Media. https://inc42.com/features/are-electric-vehicles-really-green-if-they-dont-use-clean-energy/

Sachan S, Singh PP (2022) Charging infrastructure planning for electric vehicle in India: Present status and future challenges. Regional Sustainability 3(4):335–345. https://doi.org/10.1016/j.regsus.2022.11.008

Shokoohyar S, Gorizi AJ, Ghomi V, Liang W, Kim HJ (2022) Sustainable Transportation in Practice: A Systematic Quantitative Review of Case Studies. Sustainability (switzerland) 14(5):2617. https://doi.org/10.3390/su14052617

Singh N, Mishra T, Banerjee R (2019) Greenhouse Gas Emissions in India's Road Transport Sector. Climate Change Signals and Response. Springer, Singapore, pp 197–209

Song Y, Wu P, Hampson K, Anumba C (2021) Assessing block-level sustainable transport infrastructure development using a spatial trade-off relation model. Int J Appl Earth Obs Geoinf 105:102585. https://doi.org/10.1016/j.jag.2021.102585

Steg L, Gifford R (2008). Sustainable Transport and Quality of Life. In *Building Blocks for Sustainable Transport* (Issue January 2008). https://doi.org/10.1108/9780857245168-011

Sudarshan. (2021, September 4). *History of Electric Vehicles (EVs) in India -*. Ewd. https://electricworldofdark.in/history-of-electricvehicles-evs-in-india/#:~:text=In

Tarei PK, Chand P, Gupta H (2021) Barriers to the adoption of electric vehicles: Evidence from India. J Clean Prod 291:125847. https://doi.org/10.1016/j.jclepro.2021.125847

Thilakshan T, Bandara SJ (2019). Identification of relevant sustainable transportation links to sustainable development goals ( SDGs ) in the national context. *Annual Sessions of IESL, October*, 341–348.
Timmis J, Neal M, Hunt J (2000) An artificial immune system for data analysis. BioSystems 55(1–3):143–150. https://doi.org/10.1016/S0303-2647(99)00092-1
Transport for Sustainable Development (2016) Transport for Sustainable Development. United Nations, NY
UN. (2021). Sustainable Transport, Sustainable Development. In *Sustainable Transport, Sustainable Development*. https://doi.org/10.18356/9789210010788
Verma A, Harsha V, Subramanian GH (2021) Evolution of Urban Transportation Policies in India: A Review and Analysis. Transp Dev Econ 7(2):25. https://doi.org/10.1007/s40890-021-00136-1
Vijaykumar A, Das S, Ray A. (2022). Here's why India needs a Comprehensive 2030 Electric Bus Rollout Plan. WRI INDIA. https://wri-india.org/blog/heres-why-india-needs-comprehensive-2030-electric-bus-rollout-plan
Wang R, Wang P, Xiao G (2018) Hybrid Charging Control of Electric Vehicles. Springer Nature, Singapore. https://doi.org/10.1007/978-981-10-4250-8_7
Winter S (2020) Wayfinding and navigation research for sustainable transport. J Spat Inf Sci 20(20):103–107. https://doi.org/10.5311/JOSIS.2020.20.615
World Bank. (2014). Highways. World Bank. https://www.worldbank.org/en/country/india/brief/highways#:~:text=In%20terms%20of%20density%2C%20India%E2%80%99s%20roads%20and%20highways
Xu T, Wei H, Hu G (2009) Study on continuous network design problem using simulated annealing and genetic algorithm. Exp Sys Appl 36(2 PART 1):1322–1328
Zope, R., Vasudevan, N., Arkatkar, S., & Joshi, G. (2019). Benchmarking: A tool for evaluation and monitoring sustainability of urban transport system in metropolitan cities of India. *Sustainable Cities and Society*. https://doi.org/10.1016/J.SCS.2018.11.011.

# Artificial Intelligence of Things for Sustainable Development of Smart City Infrastructures

**Aman Kataria, Sita Rani, and Sandeep Kautish**

**Abstract** The discipline of Artificial Intelligence of Things (AIoT) is experiencing significant growth and holds the capacity to fundamentally transform urban planning, construction, and operation. AIoT is a novel paradigm that merges the capabilities of the Internet of Things (IoT) and Artificial Intelligence (AI) to generate intelligent systems capable of real-time data collection, analysis, and action. In addition to these, a vast array of smart city applications that contribute to sustainable development are being developed with the assistance of AIoT. As an illustration, intelligent parking systems, intelligent water management systems, and intelligent public illumination systems are all being developed using AIoT. Although it is a relatively new technology, AIoT has the capacity to revolutionize the way in which cities are planned, constructed, and managed. Future-generation-capable smart cities can be developed more sustainably through the application of AIoT to enhance energy efficiency, mitigate pollution, and bolster resilience. Notwithstanding the potential for AIoT to substantially enhance the sustainability of smart city infrastructures, there exist certain obstacles that necessitate resolution. This chapter deals with the requirement to develop secure and exhaustive data management systems is one obstacle. It is critical to ensure that the enormous quantities of data collected by AIoT systems are protected against unauthorized access and use. An additional obstacle is the requirement to create AI algorithms that are impartial and unbiased. It is critical to ensure that the training data utilized to train AI algorithms accurately reflects the characteristics of the population that the AI system intends to assist. Notwithstanding these obstacles, the prospects presented by AIoT in the realm of sustainable

A. Kataria
University Centre for Research and Development, Chandigarh University, Gharuan, Mohali, Punjab 140413, India
e-mail: ammankataria@gmail.com

S. Rani (✉)
Department of Computer Science and Engineering, Guru Nanak Dev Engineering College, Ludhiana-141006, Punjab, India
e-mail: cse_sita@gndec.ac.in

S. Kautish
Director-APEX CSE, Chandigarh University, Mohali, Punjab, India

 
W. Leal Filho et al. (eds.), *Digital Technologies to Implement the UN Sustainable Development Goals*, World Sustainability Series,
https://doi.org/10.1007/978-3-031-68427-2_10

smart city development are extensive. Anticipated as AIoT technology advances are further inventive and efficacious implementations that leverage AIoT to enhance the sustainability of urban areas.

**Keywords** AIoT · IoT · Sustainable development · Smart City · AI

# 1 Introduction

Smart cities are a game-changing way of thinking about urban life because they use data and technology to make cities better places to live for everyone. High-speed internet, Wi-Fi hotspots, and the IoT — a network of interconnected gadgets and sensors — are the backbone of any "smart city." The ability to collect data in real time, analyze it, and make decisions based on that data allows cities to function more efficiently and effectively (Rani et al., 2021).

Data-driven decision making is a cornerstone of smart cities. These cities use AI and ML to sift through massive amounts of data collected from sensors, monitors, and other sources related to traffic, the environment, and public services. Better urban services and resource management are the end result of using this information to optimize traffic flow, energy consumption, and resource allocation.

Smart cities place a premium on efficient and environmentally friendly transportation options. Congestion can be alleviated, pollution can be reduced, and the quality of life in cities can be improved by implementing measures like integrated public transportation systems, real-time tracking, electric vehicles, and bike-sharing programs. Energy efficiency is also a top concern, prompting the use of LED lights, smart grids, and alternative energy sources. Waste reduction, recycling programs, green building methods, and the development of parks and bike lanes are just some of the ways in which sustainable living is encouraged in smart cities (Dhanalakshmi, Anand, Sivaraman, & Rani, 2022).

In smart cities, citizens can access government services online and emergency response times are reduced. Digital advances like telemedicine and e-learning platforms also improve healthcare and education, giving more people access to it. Security and safety are increased by advanced surveillance systems, further supporting law enforcement in incident response and crime prevention.

Smart cities rely heavily on resident participation. Internet sites, social media, and mobile apps provide citizens a voice in municipal government by facilitating problem reporting, news dissemination, and decision making. However, these urban centers face difficulties in data privacy, security, digital equity, and the cost of deploying and maintaining cutting-edge technologies. It is of paramount importance that all city citizens, regardless of their economic standing, have equal opportunity to use and profit from smart city services. Ultimately, smart cities aspire to build more efficient, sustainable, and inclusive urban environments by leveraging technology and data-driven methods to address the difficulties of urban living.

The term "Artificial Intelligence of Things" (AIoT) refers to the developing paradigm that integrates AI with the IoT. Using machine learning and deep learning, it enables IoT devices to collect, analyze, and interpret data on their own. The use of AI into IoT devices improves its intelligence and flexibility. It enables gadgets to make real-time judgments, anticipate user needs, and respond intelligently to changing conditions. The convergence of AI and IoT will usher in a new era of automation, efficiency, and enhanced user experiences across many industries, including healthcare, manufacturing, transportation, and smart cities.

AIoT's involvement in smart cities is revolutionary because it provides novel approaches to old problems. The IoT and AI technologies work together to improve many facets of smart city management. To reduce gridlock, optimize traffic signals, and direct drivers to less congested routes, AIoT in transportation can leverage data from IoT sensors in real time. This makes transportation more productive and environmentally benign by decreasing both commuting times and emissions. AIoT also plays a crucial role in enhancing energy efficiency. When IoT sensors are combined with AI, smart cities can control their energy use in real time, based on factors like population density, weather, and building occupancy. The resulting energy savings and decreased carbon footprint are substantial. AIoT also adds to public safety. The use of AI-enhanced, state-of-the-art surveillance cameras and sensors allows for rapid responses to potential security breaches, accidents, or other crises. Predictive maintenance of essential infrastructure is made possible, lowering the likelihood of unanticipated breakdowns. Further, by optimizing collection routes according to the trash bins' levels of fullness, AIoT improves waste management while lowering operational costs and minimizing environmental damage caused by wasteful pickups (Silva, Khan, & Han, 2018).

The advancement of "smart cities" relies heavily on AIoT. It enables cities to make decisions based on data, increasing efficiency, boosting sustainability, and enhancing services for citizens to make cities more hospitable, responsive, and resilient.

The fundamental objectives of this book chapter are:

- To discuss the importance of smart city infrastructure in sustainable development
- To explore the essential components of smart city infrastructure.
- To present the importance of AIoT in the development of sustainable smart city ecosystem
- To discuss the various challenges and possible future research directions for the integration of AIoT in smart city infrastructure.

## 2 Smart City Infrastructures: An Overview

The primary goals of smart city development are to improve the standard of living for city dwellers, make city infrastructure more efficient and sustainable, and provide solutions to critical urban problems. The goal of a "smart city" is to leverage data, connectivity, and technological advancements to improve public transit, lessen the burden on emergency services, cut down on energy usage, and increase the use of renewable energy while also improving trash management. They place a premium on public welfare and the availability of healthcare and education. To further improve the quality of life in the city, smart cities place an emphasis on citizen participation, diversity, and equal access to services.

## 2.1 *Defining Smart City Infrastructure Components*

The term "smart city infrastructure" refers to an all-encompassing method of urban development that makes use of data and technology to better the lives of citizens, streamline municipal operations, and solve complex problems (Shah, Kothari, & Doshi, 2019). Connectivity, data analytics, sensors and the Internet of Things, transportation, electricity, waste management, and urban planning are all discussed here as essential parts of a smart city's infrastructure. Some of the most important smart city components are shown in Fig. 1.

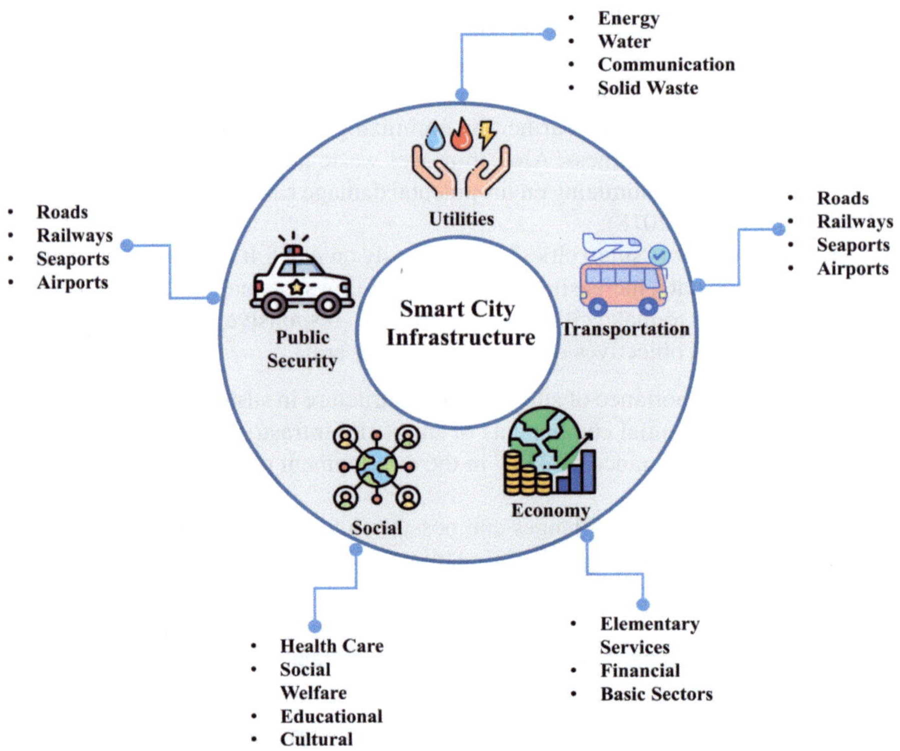


**Fig. 1** Essential components of Smart City Infrastructure

- **Connectivity.** A smart city's ability to communicate quickly and efficiently is essential to its success. 5G and other next-generation wireless technologies lay the groundwork for interconnecting smart city devices and sensors. This connectivity is critical for gathering and sending data in real-time, which is required for the proper operation of smart applications. It enables the connection and communication of objects, people, and infrastructure across the IoT.
- **Data Analytics.** Smart city efforts are propelled by data analytics. The city employs sophisticated analytics tools to sift through massive volumes of data collected from sensors, cameras, and social media in order to draw conclusions and make judgments. The data is analyzed with machine learning and AI to spot patterns, forecast results, and determine where to best direct available resources. Cities can be more proactive in their responses to problems and more efficient in their use of resources when they make decisions based on data.
- **Sensors and IoT.** Sensors and the IoT are vital to the functioning of modern "smart" cities. These tiny, networked sensors are placed all around the city to monitor traffic, air quality, and other aspects of urban life in real time. Environmental sensors can track things like temperature, humidity, and noise levels. Transportation efficiency can be improved with the help of traffic sensors that offer real-time data. Waste collection companies can be notified by smart trash cans when they are full. Sensors are widespread in smart cities, giving important data for better decision-making.
- **Transportation.** The transportation system of a smart city is one of its most crucial components. It's the use of technology to the transportation system with the goal of making it more accessible, less time-consuming, and more pleasurable for the user. This involves the introduction of autonomous and electric vehicles, real-time public transportation data, and intelligent traffic control technologies. Smart transportation systems also support environmentally friendly modes of transportation like bicycling and walking as a means of getting about.
- **Energy.** Smart cities place a premium on sustainable energy management. Reduce carbon emissions and guarantee a steady supply of energy by switching to renewable energy, energy-efficient buildings, and smart grids. Smart grid technology and energy-efficient street lighting also contribute to this goal. Also helping to lessen the city's carbon impact is the adoption of renewable energy sources like solar and wind power.
- **Waste Management.** A smart city's efforts to reduce its environmental impact are greatly aided by effective trash management. Sensors in intelligent waste management systems track when trash cans are full, allowing for more efficient garbage collection and fewer visits to the dump. Cities can also create recycling programs and promote eco-friendly garbage disposal habits by inhabitants. Smart cities help keep the environment cleaner by decreasing trash sent to landfills and increasing recycling rates.
- **Urban Planning.** Infrastructure in a smart city is planned and built with sustainability, efficiency, and quality of life in mind. Mixed land use, pedestrian-friendly surroundings, green areas, and accessibility to public transportation are just few of the features that should be taken into account while building a smart city. Cities

can create and test numerous scenarios with the help of modern urban planning tools and simulations to guarantee that future construction is in line with their long-term objectives.

- **Safety and Security.** Smart city infrastructure also plays an important role in strengthening safety and security. Surveillance cameras, facial recognition software, and automated alarm systems are just a few examples of the cutting-edge security technology that may aid in preventing and responding to criminal activity. Safety concerns, such as natural disasters and pandemics, can be anticipated and countered with the help of data analytics. Public safety and emergency services can benefit from the investments made in infrastructure by smart cities.
- **Health and Education.** Access to high-caliber medical care and educational opportunities is another need for smart city planners. Residents can gain easier access to medical care because to innovations like telehealth, healthcare monitoring devices, and data-driven healthcare systems. In particular, underserved communities benefit from increased access to educational opportunities and materials made possible by smart classrooms and digital education platforms.
- **Sustainability and Environment.** Sustainable practices and protecting the environment are at the forefront of smart city planning. They create regulations and build infrastructure to lessen pollution and safeguard the environment. Sustainable water management strategies, for example, are a part of this, as are green construction standards, urban farming programs, and other similar endeavors. The goal of developing a "smart city" is to make metropolitan areas more sustainable and secure.
- **Public Engagements and Services.** Citizen participation and service provision are important components of smart city infrastructure. With the help of online portals and mobile apps, citizens may engage with local services, obtain needed data, and offer suggestions. City departments will be better able to address problems reported by citizens, such as potholes and malfunctioning streetlights. Technology is also being integrated into public services such as libraries and community centers.
- **Accessibility and Inclusivity.** The infrastructure of a smart city should be accessible to all residents. Every effort is taken to guarantee that all citizens, especially those with disabilities and those living in underprivileged areas, have equal access to technology and services. A fairer and more welcoming urban environment can be achieved through the smart city's investment in accessible transit, public spaces, and digital accessibility.

Thus, the term "smart city infrastructure" refers to a wide range of systems and technologies that collaborate to improve the quality of life in metropolitan areas. To create a more sustainable and equitable future for their citizens and to enhance the quality of life in the present, "smart cities" leverage technology and data to solve urban problems. Smart city infrastructure are the key components in designing a typical smart city framework, shown in Fig. 2.

In smart city environment various services are provided to make the life of the residents convenient. These services are:

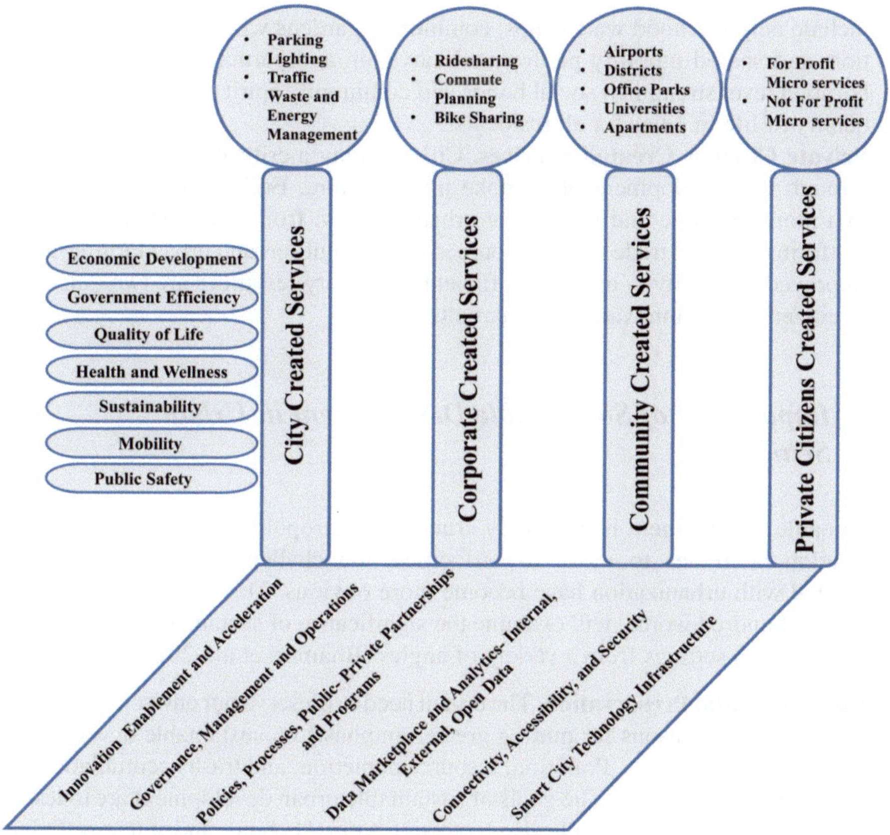


**Fig. 2** A typical smart city framework

- **City Created Services.** Modern technologies are used in "smart cities" to improve efficiency. The Internet of Things (IoT) makes garbage collection and disposal easier, and sensors improve public safety in cities. Citizen participation is simplified by digital platforms, leading to more efficient use of available resources. In a high-tech city, incorporating data-driven solutions boosts productivity, sustainability, and the quality of life for everyone who lives there.
- **Corporate Created Services.** Business contributions to smart cities include cutting-edge features like AI-powered analytics for city planning, eco-friendly waste management systems, and renewable power generation options. Partnerships between the private and public sectors improve public infrastructure, resulting in a more connected and digitally optimized urban environment. Economic development, technological advancement, and an improved standard of living all benefit from corporate involvement.
- **Community Created Services.** In smart cities, residents take an active role in providing essential services through grassroots efforts. Citizen-driven projects

include neighborhood watch apps, community gardens with sensor-based irrigation, and shared mobility platforms. Innovation and participation at the neighborhood level strengthen social bonds and community spirit while improving the quality of life in cities for all residents.

- **Private Citizens Created Services.** Citizens play a critical part in smart cities through the development of bespoke infrastructure. Each initiative contributes to a more dynamic and responsive urban ecology, from community-led digital platforms for local trade to crowdsourced traffic solutions and urban farming apps. Improved connectivity and more efficient use of city resources are two outcomes of citizen-driven innovation in smart cities.

## *2.2 Importance of Sustainable Development in Urban Settings*

Sustainable development is especially crucial in metropolitan areas. As the global population continues to move toward cities, the challenges and opportunities connected with urbanization have become more obvious. This conversation, which spans five hundred words, will examine the significance of sustainable development in metropolitan settings from a variety of angles (Bhambri et al., 2022).

- **Environmental Preservation.** The urgent need to preserve our environment is one of the primary reasons for putting greater emphasis on sustainable development in urban environments. Pollution, resource depletion, and trash accumulation are common in urban areas. The goals of sustainable urban development are to lessen cities' carbon footprints while also protecting natural habitats and making efficient use of available resources. The environmental effect of cities can be reduced by the use of sustainable building methods, the promotion of renewable energy sources, and the introduction of efficient transportation networks.
- **Mitigating Climate Change.** Greenhouse gas emissions are largely attributable to urban regions. Compact city design, public transit, and energy-efficient structures are all examples of sustainable urban development initiatives that contribute significantly to slowing global warming. Cities can help the world fight climate change and move toward a low-carbon future by cutting emissions, improving energy efficiency, and building green infrastructure (Blasi, Ganzaroli, & De Noni, 2022).
- **Quality of Life.** Improving city dwellers' standard of living is inextricably bound up with the concept of sustainable urban development. Improved air and water quality, more green areas, and less noise pollution are just a few of the ways that this plan aims to make cities more hospitable. A better quality of life is achieved when people have easy access to parks and recreational places, clean air, safe drinking water, and effective waste management systems. Cities that are environmentally responsible are ones that prioritize their residents' happiness, health, and contentment.

- **Economic Viability.** Economic benefits accompany sustainable urban development's environmental and social benefits. Cities may save money and foster economic growth by investing in green technologies, public transit, and energy-efficient buildings. Sustainability can also entice companies, investors, and visitors, all of which contribute to the city's economic growth and long-term financial security.
- **Resource Efficiency.** In order to be sustainable, urban areas must prioritize conservation. Land conservation, water management, trash reduction, and energy efficiency are all included in this strategy. Conservation of natural resources is just one benefit of sustainable development; cost savings and increased resilience are additional advantages. Smart urban design reduces sprawl, maximizes land utilization, and protects green spaces for future use.
- **Resilience to Disasters.** In a world marked by escalating climate-related calamities, metropolitan areas are particularly vulnerable. Resilience planning is an integral part of sustainable urban development because it helps communities better prepare for natural disasters like floods, storms, and wildfires. By strengthening infrastructure, adopting disaster response protocols, and implementing adaptive measures, cities may limit the impact of disasters and protect the safety of their citizens (Zeng, Yu, Yang, Lv, & Sarker, 2022).
- **Social Equity.** Those living in metropolitan areas, regardless of their socioeconomic status, should have equal access to basic amenities and opportunities, which is a goal of sustainable development. Affordable housing, easily accessible public transportation, and fully accessible infrastructure are the cornerstones of equitable urban development. The result is less inequality, a stronger sense of community, and increased opportunity for all city residents.
- **Efficient Transportation.** The integration of sustainable transportation systems in metropolitan areas dramatically decreases traffic congestion, air pollution, and energy consumption. In order to create cleaner and more efficient urban transportation networks, investing in public transit, bike infrastructure, and pedestrian-friendly architecture is essential (Mora et al., 2022).
- **Preservation of Biodiversity.** Green spaces, parks, and wildlife corridors are all ways in which cities may help sustain biodiversity. Within city bounds, the goal of sustainable urban development is to preserve and improve the natural environment. Creating green oases, planting native plants, and lowering pollution are all ways that cities may offer citizens a connection to nature while also providing habitats for local wildlife.
- **Global and Local Collaboration.** Most people on Earth now live in cities, making sustainable urban development a pressing concern for everyone. Best practices, novel solutions, and environmentally friendly technologies can only be shared effectively with others if people from all over the world work together. Cities may speed up their journey toward a more sustainable future by learning from one another's mistakes and implementing those lessons (Mahato, Kundu, Makwana, & Joshi, 2023).

Growing cities must prioritize sustainability if they are to meet the challenges of the twenty-first century and build sustainable, resilient communities that are also pleasant to live in.

## 3 Artificial Intelligence of Things (AIoT) in Smart Cities

In order to make city life better, "smart city" environments implement many technological solutions. IoT sensors collect data, big data analytics provide insights, smart grids distribute power more efficiently, and intelligent transportation systems optimize traffic flow. 5G and Wi-Fi connectivity allow high-speed internet access while also reducing environmental impact, while sustainable infrastructure and renewable energy sources lessen environmental impact. Blockchain promotes security and transparency, and AI-driven applications automate numerous civic activities. AI is essential to smart cities because it paves the way for data-driven decision making, automation, and improved services. With the use of AI, we can streamline our traffic systems and cut down on delays. Through the use of analytics and surveillance tools, it also makes communities safer. Smart grids powered by AI improve energy management by maximizing distribution, decreasing waste, and integrating renewable sources (Marine, 2023; Prabha, Mittal, Sharma, & Singh, 2022). By keeping tabs on how much of a certain resource is being used, AI helps with waste management and healthcare through remote monitoring and telemedicine. Smart cities are more efficient, sustainable, and responsive to the demands of their citizens because it improves public involvement, helps with urban planning, facilitates emergency responses, and enables predictive maintenance. In addition, civic engagement platforms encourage citizen participation, and cutting-edge urban planning instruments help city administration run more smoothly. Together, these innovations propel smart cities toward greater sustainability, efficiency, and improved quality of life. A comparative analysis of IoT and AIoT in the implementation and integration in smart city infrastructure is shown in Table 1.

### *3.1 Understanding AIoT and Its Integration in Urban Infrastructures*

AIoT presents the integration of AI and IoT. It brings together IoT device's data collection powers with AI's data analysis, learning, and decision-making prowess. The massive volumes of data produced by IoT devices like sensors, cameras, and linked items are analyzed by AI algorithms to provide insights and enable informed decisions. The fields in which AIoT can be applied are vast. Examples of where AIoT has been put to good use include smart cities, where it has been used to improve traffic flow, security, and resource management. It allows for predictive

**Table 1** Comparative analysis of IoT and AIoT in the implementation and integration in smart city infrastructure

| S.No | Parameter | IoT | AIoT |
|---|---|---|---|
| 1 | **Data Processing and Analysis** | Data collection and transfer on a fundamental level | For real-time, predictive, and prescriptive analysis, advanced analytics and machine learning |
| 2 | **Decision Making** | Capabilities for making decisions are severely restricted | Improved decision-making capabilities with the application of AI algorithms, which enables fully autonomous and adaptable systems |
| 3 | **Adaptability** | Limited capability to adjust to constantly shifting conditions | Dynamic adaptability to changes occurring in real time, self-optimization, and ongoing learning capabilities |
| 4 | **Automation** | Automation is in its most fundamental form of predetermined duties | Intelligent automation that is able to develop itself over time and learn from previous experiences and data |
| 5 | **Predictive Capabilities** | Limited predictive capabilities | Strong predictive modeling and forecasting that is based on both historical data and data collected in real time |
| 6 | **Complex Problem Solving** | A limited capacity to tackle difficult issues | Capabilities for advanced problem solving employing AI approaches, with the ability to handle difficult circumstances |
| 7 | **Energy Efficiency** | Measures fundamental energy efficiency | Optimization and predictive maintenance made possible by the use of AI to achieve greater energy efficiency |

diagnosis and remote patient monitoring in the medical field. Predictive maintenance and process optimization are two applications in manufacturing that benefit from this technology. Additionally, AIoT can aid precision farming in the agricultural sector. The promise of AIoT resides in its ability to facilitate real-time, data-driven decision-making, automation, and enhanced efficiency across industries, which will eventually improve our quality of life and fuel innovation in a wide range of settings. The continued development of AIoT, however, also brings up questions regarding data privacy, security, and ethics. An abstract view of the working of AIoT is shown in Fig. 3.

The way cities operate is changing as a result of the incorporation of AIoT into smart urban environments. This improves sustainability, efficiency, and general quality of life. The IoT and AI technologies are combined with IoT devices to gather, process, and analyze data for intelligent decision-making and automation of various municipal activities. In traffic management, real-time data from sensors and cars is used by AIoT to optimize traffic flow and lessen congestion. AI-powered smart grids, which control energy distribution, cut waste, and encourage the integration of renewable energy sources, increase energy efficiency. AIoT helps public safety by facilitating fast danger identification and real-time surveillance. AIoT also improves

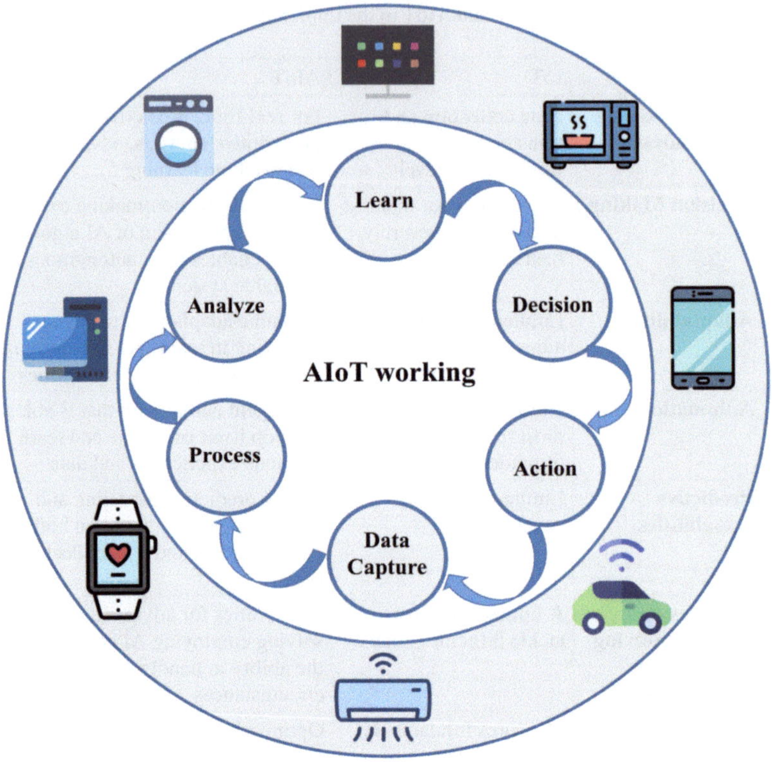


**Fig. 3** An abstract view of AIoT working

emergency response, urban planning, trash management, and environmental monitoring, offering more sustainable and effective solutions. AI-driven apps that expedite communications with local authorities enhance citizen services. The deployment of AIoT has great promise for creating smart urban environments; nevertheless, in order to secure the well-being of people and the successful development of smart cities, data security, privacy, and ethical considerations must be carefully considered.

## *3.2 Applications of AIoT in Various Aspects of Smart Cities*

Applications of AIoT in smart cities have the ability to raise productivity, raise quality of life, and solve a wide range of urban problems. Some of the most important uses of AIoT in modern smart cities include:

- **Smart Traffic Management.** Using data gathered in real time from sensors, cameras, and GPS devices, AIoT can improve traffic flow. AIoT gathers real-time information about traffic patterns, car movements, and road conditions by

placing sensors, cameras, and connected devices all over cities. Then, machine learning algorithms look at this data to find ideas that can be used. This makes it possible for adaptive control systems and predictive traffic management. AIoT makes it possible to use dynamic traffic light control, which changes the timing of signals based on how busy the roads are at the moment. In addition, it makes it easier to create smart transportation systems that can give drivers real-time traffic information so they can make smart choices and pick different paths. By using AIoT, cities can not only improve traffic flow but also have less of an effect on the climate by using less fuel and releasing fewer emissions. This will lead to more sustainable and effective urban transportation ecosystems.
- **Smart Energy Management.** By keeping tabs on usage, weather, and grid conditions, IoT systems can improve efficiency, cut down on waste, and save money on power. AIoT uses advanced analytics and machine learning techniques to make predictive maintenance possible. This lets you know how well equipment is working and when it might break down. This proactive method helps keep downtime to a minimum and makes energy infrastructure last longer. AIoT is also very important for demand-side management because it looks at real-time data to figure out how things are used and change how much energy is used on the fly. Smart devices that react to signals of demand make sure that energy resources are used wisely. When it comes to distributed energy resources (DER), AIoT makes it easier for renewable energy sources to be seamlessly connected to the grid, which constantly balances supply and demand.
- **Environmental Monitoring.** Air quality, noise levels, temperatures, and other environmental parameters can be monitored with the help of sensors dispersed across the city. AIoT uses the power of AI algorithms and the huge network of devices and sensors that are all linked to each other to give real-time information about the environment. This integration makes it possible to look at the quality of the air and water, as well as biodiversity and climate factors. Machine learning algorithms can look through huge amounts of data and find patterns, outliers, and trends that would be hard for standard monitoring methods to spot. AIoT can help find pollution events earlier, predict changes in the environment, and give useful information for long-term resource management in environmental tracking. This technology lets us make natural systems that can adapt and respond, which makes it easier to deal with problems like climate change, pollution, and cutting down trees.
- **Waste Management.** Waste collection services can save time and gas money thanks to smart bins that are fitted with IoT sensors that notify them when the bins are full. Using smart sensors, data processing, and automation, AIoT is a key part of making waste management work better. These smart systems let you keep an eye on trash cans in real time, telling you how full they are and suggesting the best ways for collection to make things run more smoothly. Predictive analytics driven by AIoT algorithms can predict how waste will be generated in the future, which helps with planning ahead and allocating resources. AIoT also helps with recycling efforts by using automatic systems to sort and separate trash, which encourages environmentally friendly habits.

- **Water Management.** The AIoT can be used to keep an eye on water quality, find plumbing problems, and conserve resources. AIoT is a key part of making water management systems more accurate and efficient by collecting, analyzing, and making decisions in real time. AI-powered smart sensors and devices let you keep an eye on water quality, usage trends, and infrastructure conditions all the time. This method is based on data, which helps find leaks, contaminated water, and weak spots in infrastructure early on. This lets people respond quickly and take preventative steps. AIoT systems use machine learning algorithms to predict patterns in water usage. This helps improve distribution networks and cut down on waste. The Internet of Things also helps create smart watering systems for farming, which make sure that water is used precisely and efficiently.
- **Public Safety.** Public safety can be improved with the help of AIoT by using it for things like video monitoring and facial recognition. AIoT applications help find threats ahead of time, respond quickly to incidents, and handle crises in general. Smart public safety systems can be set up in towns by combining AI algorithms with IoT devices and sensors. Situational awareness is made possible by real-time data from surveillance cameras, sensors, and wearable tech. This lets the police find and act quickly to incidents. Machine learning algorithms can look for trends in data to predict possible threats and take steps to stop them. AIoT also helps make smart emergency response systems possible, which improves the use of resources and cooperation during emergencies.
- **Healthcare.** Wearable IoT devices can be used in smart city applications to monitor health parameters, with AI analyzing the data to spot health patterns and epidemics. AIoT has many uses in healthcare, ranging from bettering patient care to making operations run more smoothly. In diagnostics, AIoT helps connect medical devices and sensors so that real-time data on patients can be collected for accurate tracking and early spotting of health problems. This data is also analyzed by AI systems, which gives doctors useful information for making personalized treatment plans. AIoT makes it possible to watch a patient's health from afar, so they don't have to go to the hospital as often and can get help when they need it. AIoT also helps to make smart medical equipment, which improves the effectiveness of treatments and the health of patients. In healthcare management, AIoT improves the use of resources, the control of inventory, and hospital workflows, which lowers costs and makes operations more efficient.
- **Smart Buildings.** Lighting, HVAC, and other building systems can all benefit from AIoT's ability to regulate them for maximum efficiency and occupant comfort. AIoT is a key part of reducing energy use, making surroundings smarter and more responsive, and improving security. Sensors and other connected gadgets in smart buildings collect real-time information about things like temperature, lighting, and how many people are in the building. AI algorithms look at this information and use it to make smart choices. They change heating, ventilation, and air conditioning (HVAC) systems to match user tastes and waste less energy. AIoT also helps with predictive maintenance by keeping an eye on how well building systems are working and guessing when they might break down. This cuts down on downtime and maintenance costs. AIoT is used in smart buildings

for improved security measures like facial recognition and anomaly detection, which make the space safer for people who live or work there.

- **Urban Planning.** Through the study of traffic flows, population shifts, and building footprints, AIoT can provide useful information to city planners. AIoT combines AI with the IoT to get useful information from the huge amounts of data that sensors and devices linked to the internet send and receive. AIoT is very important in a number of important areas of urban planning. For starters, it lets you make decisions based on data by giving you real-time data on things like traffic trends, energy use, air quality, and other important factors. This data-driven method helps city planners improve the overall functionality of cities, make the best use of resources, and improve the development of infrastructure. AIoT also helps build smart cities by making public services better, like transportation, waste management, and first responders in a disaster.
- **Agriculture in Urban Areas.** Monitoring soil quality, crop health, and irrigation systems with the use of AIoT can help urban farms be more efficient and environmentally friendly. AIoT technologies are very important for urban agriculture because space is limited. They help make the best use of resources, boost production, and encourage environmentally friendly practices. Soil, crops, and farming tools all have smart sensors and motors that collect real-time information about the weather, moisture levels, and nutrient content. AI programs look at this data and find useful information for precision farming. Urban areas with limited room need to make good use of land. AIoT makes vertical farming and hydroponics possible, which lets plants grow all year. AI-controlled automated systems can handle watering, check on plant health, and control the surroundings to ensure the best growth. AIoT also makes it easier to use predictive analytics to find field diseases and control pests, which lets people act quickly and cuts down on the need for chemicals.
- **Parking Management.** Finding parking can be a major source of traffic delays, but with the help of AIoT systems, cars can be directed to open spots. Smart parking solutions can give you data and insights in real time by combining AI with IoT devices. To make parking lots work better, AI programs look at data from sensors, cameras, and other IoT devices. This lets dynamic price models be made, which improves usage during rush hours and encourages people to use other forms of transportation. AIoT-based parking systems can also offer predictive analytics to make it easier for drivers to find open parking spots, which can cut down on pollution and traffic.
- **Wearable Devices.** Citizens can use their smartphones and wearable devices to engage with the city's AIoT systems, allowing them to gain access to data, make purchases, and receive notifications. In healthcare, AIoT makes personalized tracking and predictive analytics possible, which lets wearables keep an eye on vital signs, guess when health problems might happen, and send out timely alerts. In fitness, smart devices use AI to look at performance data, make personalized workout suggestions, and adjust to each person's fitness goals. AIoT also helps make smart tech more energy-efficient by reducing the amount of power used and increasing battery life. Wearables can give people more accurate and useful

information because they can connect and process data more efficiently and use AI algorithms.

These use cases show how AIoT may revolutionize smart cities by making them more efficient, sustainable, and in tune with their citizens' wants and requirements. The implementation of such technology, however, necessitates careful consideration of privacy and security issues in order to safeguard citizens' personal information and constitutional liberties.

## 4 Sustainable Development Goals in Smart City Infrastructures

Sustainable development is progress that provides for today without jeopardizing future generations' capabilities to do the same. In order to better the lives of their citizens and lessen their negative influence on the environment, smart cities implement various forms of information and communication technology (ICT). The Sustainable Development Goals (SDGs) are a collection of 17 global goals adopted by the United Nations in 2015. These objectives seek to ensure that all people are able to live in peace and prosperity and to put an end to extreme poverty. By leveraging information and communication technology to enhance urban infrastructure's effectiveness, sustainability, and resilience, smart cities can contribute significantly to the realization of the SDGs as represented in Fig. 4 (Arisdakessian, Wahab, Mourad, Otrok, & Guizani, 2022; Rani et al., 2022).

Smart cities can achieve the SDGs by focusing on the following areas:

- **Sustainable energy:** ICT can help smart cities create renewable energy while also lowering their overall energy usage, which is a step toward more environmentally friendly practices. This can be accomplished by utilizing intelligent grids, intelligent buildings, and intelligent transportation systems.
- **Sustainable transportation:** Intelligent information and communication technology (ICT) can help smart cities make their transportation systems more efficient, leading to greater sustainability. Using sophisticated traffic management systems, public transit networks, and electric automobiles are all viable options for accomplishing this goal.
- **Sustainable water management:** Smart cities can employ ICT to conserve water and minimize pollution by using these technologies. Sustainable water management. This can be accomplished through the utilization of leak detection systems, water recycling systems, and smart metering systems.
- **Waste management:** Smart cities can use ICT to reduce waste and improve recycling rates. This can be accomplished through the utilization of intelligent waste-collecting systems, waste-sorting systems, and facilities that convert waste into energy.

**Fig. 4** Sustainable Development Goals in Smart City Infrastructures

- **Affordable housing:** ICT has the potential to be used by smart cities to lower housing costs, which would benefit residents greatly. This can be accomplished through the utilization of online platforms that connect landlords and tenants, as well as through the utilization of ICT for the management of affordable housing programs.
- **Public safety:** ICT has the potential to significantly contribute to an increase in a city's overall level of public safety. This can be accomplished through the utilization of video surveillance systems, the study of real-time crime data, and predictive policing.
- **Development in Education:** Education is one area that can benefit greatly from the implementation of ICT in smart cities. Utilizing online learning platforms, adaptive learning tools, and virtual reality classrooms are all viable options for accomplishing this goal.
- **Development in Healthcare:** ICT has the potential to be utilized by smart cities in order to enhance the standard of medical care available. Telemedicine, electronic health records, and wearable health gadgets are all viable options for accomplishing this goal.

### *4.1 Exploring Sustainable Development Objectives in Urban Environments*

More than half of the world's population already resides in urban areas, and by the year 2050, that proportion is projected to increase to two-thirds. This rapid urbanization is putting a strain on the environment and resources, and it is also creating social and economic challenges (Holzinger, Langs, Denk, Zatloukal, & Müller, 2019; Puri, Kataria, & Sharma, 2021). In order for cities to effectively handle these difficulties, sustainable development techniques need to be used. A key component of sustainable development is attending to the requirements of the here and now without jeopardizing the capacity of succeeding generations to do the same for themselves (Albahri et al., 2023; Meske & Bunde, 2020). It is about striking a balance between economic progress, social fairness, and the conservation of the environment. The Sustainable Development Goals (SDGs) outlined by the United Nations offer a road map for cities to follow in order to achieve sustainable development. The Sustainable Development Goals (SDGs) are a collection of some global objectives with the goals of eradicating poverty, preserving the environment, and ensuring that all people live in peace and prosperity (Arora, Leekha, Lee, & Kataria, 2020; da Costa, Pasluosta, Eskofier, da Silva, & da Rosa Righi, 2018).

Implementing a wide variety of sustainable development policies and practices is one way for cities to work toward achieving the Sustainable Development Goals (SDGs). As shown in Fig. 5, these policies and practices have the potential to address a wide range of problems, including the following:

- **Climate change:** By investing in renewable energy, energy efficiency, and green infrastructure, cities have the ability to lessen their contribution to greenhouse gas emissions and better prepare themselves for the effects of climate change. AIoT plays a key part in moving these efforts forward. AIoT applications help fight climate change by making it easier to use less energy, handle resources smartly, and make decisions based on data. For example, AIoT can help buildings, factories, and transportation systems use less energy, which lowers their carbon footprints. AIoT also makes it possible to watch environmental parameters in real time, which helps find climate-related risks earlier and makes disaster preparedness

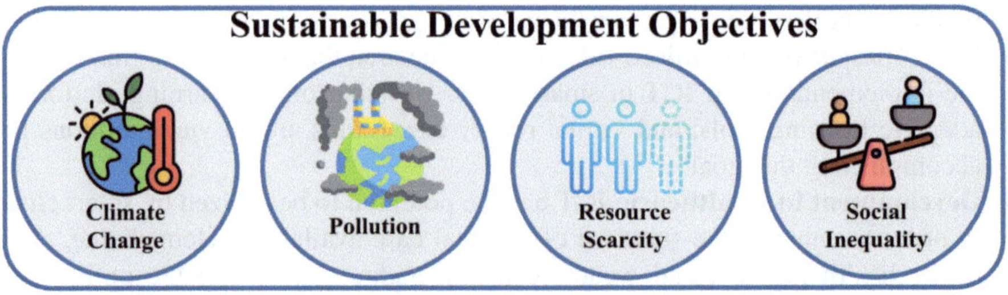


**Fig. 5** Sustainable Development Objectives

and response better. Using AIoT in climate modeling and prediction helps us learn more about complicated climate systems, which helps policymakers make better decisions (Bronner, Gebauer, Lamprecht, & Wortmann, 2021).

- **Pollution:** Urban areas have the ability to lower their levels of air, water, and soil pollution by enhancing their waste management systems, making investments in cleaner transportation, and supporting sustainable agriculture. IoT offers new ways to track, control, and reduce pollution. AIoT gives real-time information on air and water quality through a network of connected gadgets and sensors. This lets people make more accurate pollution levels assessments. A type of AI called machine learning algorithms can look at this huge amount of data to find trends, sources, and possible environmental risks in pollution. With this knowledge, policymakers can make smart choices, target their efforts, and come up with good ways to stop pollution (Yang et al., 2021).
- **Resource scarcity:** Urban areas have the ability to reduce their use of water, energy, and other resources by adopting more sustainable patterns of consumption and production. AIoT is very important in this situation because it offers new ways to use resources more efficiently and encourage environmentally friendly habits. AIoT lets resources be monitored, analyzed, and managed in real time by combining AI with the IoT. When resources like water and energy are limited, AIoT makes it possible for systems to be smart and efficient so they can adapt to new situations. For example, AIoT apps help farmers get the most out of irrigation by taking into account weather forecasts and the amount of water that is already in the ground. When used in factories, AIoT helps save energy by keeping an eye on and changing processes in real time. In addition, AIoT makes it easier to build smart grids and intelligent energy control systems, which help people use less electricity (Weber, 2023).
- **Social inequality:** Cities have the ability to alleviate poverty, inequality, and discrimination by increasing their investments in education, healthcare, and social services. AIoT is a key part of achieving these goals because it offers new ways to deal with social imbalance. AIoT helps lawmakers make better, more targeted plans by giving them data-driven insights and predictive analytics. For example, AIoT apps can make it easier for people to get medical care and make the best use of resources, which can help reduce health disparities. AIoT makes individual learning possible in education, closing gaps and encouraging everyone to be included. Additionally, AIoT helps build smart towns by improving infrastructure and services that benefit all city residents, irrespective of their socioeconomic status (Nozari, Szmelter-Jarosz, & Ghahremani-Nahr, 2022).
- **Affordable and Clean Energy:** AIoT technologies present novel approaches to improve energy efficiency, optimize the utilization of resources, and facilitate the incorporation of sustainable and environmentally friendly energy sources into urban settings. AIoT applications enable the advancement of smart infrastructures, which effectively oversee electricity distribution, reconcile supply and demand, and seamlessly incorporate renewable energy sources, all with the aim of promoting affordable and environmentally sustainable energy. By analyzing real-time data, machine learning algorithms are capable of forecasting energy

consumption patterns, which facilitates improved grid optimization and planning. Moreover, the implementation of smart meters and devices enabled by AIoT empowers consumers to monitor and regulate their energy consumption, thereby encouraging responsible behavior. By means of adaptive learning and ongoing optimization, the Internet of Things (IoT) facilitates the development of intelligent energy systems that not only increase energy accessibility but also substantially diminish environmental footprints by decreasing dependence on fossil fuels (Bano, Ud Din, & Al-Huqail, 2020).
- **Industry, Innovation, and Infrastructure:** The incorporation of AIoT technology into industrial operations improves productivity, sustainability, and efficiency. AIoT enhances infrastructure performance by means of real-time monitoring and predictive maintenance, thereby mitigating resource wastage and delay. Driven by the Internet of Things, innovations such as intelligent transportation systems and smart infrastructure contribute to improved urban mobility and sustainable energy consumption. In addition, by facilitating data-driven automation and decision-making across industries, AIoT stimulates the development of sustainable technologies, thereby promoting innovation. In urban planning, the application of AI algorithms improves infrastructure design, thereby fostering inclusive development and assuring climate change resilience (Kuo & Wu, 2023).
- **Sustainable Cities and Communities:** Innovative solutions to urban challenges and the promotion of sustainable development are provided by AIoT technologies. Amidst the domain of smart cities, applications of the Internet of Things facilitate data-informed decision-making, thereby optimizing urban infrastructure and resource utilization. By leveraging machine learning algorithms and real-time data analytics, AIoT facilitates the advancement of intelligent transportation systems, thereby reducing carbon emissions and alleviating traffic congestion. Furthermore, AIoT encourages energy efficiency and resilience by facilitating the integration of renewable energy sources and the deployment of smart grids, thereby contributing to the pursuit of affordable and sustainable energy (Kuguoglu, van der Voort, & Janssen, 2021).

**Challenges and Opportunities:** The process of establishing sustainable development in urban contexts is fraught with a great deal of difficulty. These challenges include the following:

- **Limited financial resources:** A great number of cities do not have the financial resources necessary to carry out programs relating to sustainable development.
- **Lack of political will:** Due to a lack of political will, not all municipal administrations are dedicated to the concept of sustainable development.
- **Public apathy:** Apathy on the part of the general public: Many residents are ignorant of the significance of sustainable development.
- **Rapid urbanization:** It is characterized by the rapid growth of cities, which places a strain not only on infrastructure but also on resources and services.

In spite of these obstacles, metropolitan surroundings have a wealth of opportunities for achieving sustainable development. These opportunities include:

- **Growing awareness of sustainability:** There has been a growing understanding among the general public of the significance of sustainability, and as a result, there has been a rise in demand for sustainable products and services.
- **Technological advances:** The advancement of technology is making it possible to implement efforts for sustainable development in a manner that is both more cost-effective and efficient.
- **Dominant Leadership:** Leadership that is unwavering: In some cities, the leadership is unwavering in its dedication to sustainable development.

**AIoT Contributions to Achieve Sustainable Goals**

AIoT can contribute to sustainable development in a number of ways, including:

- **Waste reduction:** Reducing trash can be accomplished with the help of AIoT systems, which can monitor the patterns of waste generation, pinpoint potential for waste reduction, and optimize waste collection routes. Smart bins that are enabled with AIoT can indicate when they need to have their contents emptied, and waste management systems that are powered by AI can optimize collection routes to reduce the amount of fuel used and emissions produced.
- **Environmental monitoring:** IoT devices may be used to track deforestation, monitor the quality of air and water, and identify potential environmental dangers. Real-time data on the circumstances of the environment can be gathered by sensors driven by AIoT that are endowed with AI, and then analyzed by models powered by AI to reveal trends and patterns.
- **Energy efficiency:** Efficient use of energy is possible with the help of AIoT systems, which can monitor and improve the efficiency of energy use in buildings, factories, and transportation networks. For instance, smart thermostats that are enabled by AIoT may modify heating and cooling settings according to occupancy and weather conditions, and AIoT-powered traffic management systems can reduce congestion as well as the amount of fuel that is consumed.
- **Water conservation:** AIoT systems can be used to detect leaks, monitor water usage patterns, and implement water conservation measures. For instance, smart water meters that are coupled with AIoT may identify water leaks in real time, and AI-powered irrigation systems can maximize the efficiency with which water is utilized in agricultural production.

## 5 AIoT Applications for Urban Infrastructure Sustainability

Cities around the world are struggling with a variety of issues in today's world, including climate change, population increase, and a lack of available resources. These challenges are putting a strain on urban infrastructure, which is often inefficient and unsustainable. The AI of things, also known as AIoT, is a relatively new technology that has the potential to solve many of the problems that cities are currently experiencing (Holzinger et al., 2019; Ullah, Al-Turjman, Mostarda, & Gagliardi, 2020). AI allows devices to collect and analyze data, while IoT allows devices to communicate with each other and with the internet. IoT can be utilized to produce a variety of applications for maintaining the sustainability of urban infrastructure. Some examples of these applications include energy efficiency, smart grid management, smart transportation systems, and traffic management. Building a sustainable urban infrastructure requires the incorporation of many essential components, including energy efficiency, smart grid management, intelligent transportation systems, and traffic management. The convergence of AIoT applications in various domains plays an essential role in improving sustainability and optimizing resource utilization (Allam & Dhunny, 2019; Clinciu & Hastie, 2019; Rožanec, Fortuna, & Mladenić, 2022).

## *5.1 AIoT in Energy Efficiency*

- **Predictive Maintenance:** AIoT applications use machine learning algorithms to forecast equipment failures and inefficiencies in energy systems, which enables proactive maintenance to be performed. This is referred to as predictive maintenance.
- **Demand-Side Management:** AIoT assists in evaluating and managing energy demand by offering real-time insights into usage patterns, hence optimizing energy consumption. This is accomplished through demand-side analytics and management.
- **Intelligent Home Appliances and gadgets:** The IoT makes it possible to integrate intelligent appliances and gadgets that are able to modify their power consumption in response to pricing and demand signals, hence improving energy efficiency.

## *5.2 AIoT in Smart Grid Management*

- **Distributed Energy Resources:** AIoT helps manage renewable energy sources like solar panels, wind turbines, etc., known as distributed energy resources (DER), by maximizing the efficiency with which they are connected to the grid and striking a dynamic balance between supply and demand.
- **Grid Optimization:** By enabling real-time monitoring, fault detection, and power flow control, AIoT applications optimize the performance and efficiency of the grid.
- **Load Forecasting:** The application of machine learning in smart grid management facilitates the precise prediction of capacity demands, thereby enhancing resource allocation and planning.

### 5.3 AIoT in Smart Transportation Systems

- **Optimization of Traffic Flow:** AIoT-based traffic management systems use real-time data from cameras, sensors, and GPS to optimize traffic flow, which in turn helps to minimize congestion and shorten travel times.
- **Optimization of Public Transportation:** The IoT makes it possible to carry out effective scheduling and monitoring of public transportation, providing passengers with information in real time and modifying routes according to demand.
- **Connected and Autonomous cars (CAVs):** AIoT is an essential component in the development of CAVs, which are able to connect with both the surrounding infrastructure and other cars to enhance both road safety and the flow of traffic.

### 5.4 AIoT in Traffic Management

- **Predictive Analytics:** Machine learning algorithms can be used to forecast traffic patterns, accidents, and road obstacles, enabling authorities to take preventative actions for better traffic management. Predictive analytics refers to the application of these algorithms.
- **Dynamic Road Pricing:** The IoT makes it possible to implement dynamic toll pricing and congestion charge systems. These systems encourage alternate travel times and modes, which in turn helps to reduce traffic congestion.
- **Intelligent Traffic Signal Control:** The IoT helps make adaptable signal control systems that change based on traffic conditions. This cuts down on idle time and traffic jams.

### 5.5 Benefits for AIoT for Urban Infrastructure Sustainability

- **Environmental Impact:** Applications of AIoT have the potential to drastically reduce carbon emissions and environmental pollution. This can be accomplished by fostering energy efficiency and cutting down on traffic congestion.
- **Resource Optimization:** Many uses of the IoT assist optimize energy usage, reducing waste, and enhancing the efficiency of transportation systems, all of which contribute to overall resource conservation.
- **Improved User Experience:** Better User Experience Efficient management of traffic and energy systems leads to a more streamlined and comfortable urban living experience for residents. This is because of the improved user experience.
- **Data-Driven Decision Making:** The IoT creates enormous volumes of data, which enables city planners and policymakers to make educated decisions about the development of infrastructure and the distribution of resources.

## 6 Major Smart City Infrastructures

Infrastructures in "smart" cities around the world show how technology and city design may come together to solve urban problems. Singapore's Smart Nation initiative stands out because it uses sensors and data analytics to improve public services like traffic management and environmental surveillance. The Internet of Things (IoT) is integrated into Barcelona's smart city model to improve garbage collection, street lighting, and parking. South Korea's Songdo was designed to be a "smart city," complete with a sophisticated smart grid, efficient waste management systems, and eco-friendly construction.

Smart parking meters and intelligent traffic management systems have been adopted in the United States, particularly in cities like San Francisco, while Smart City efforts in Dubai use blockchain for secure transactions and AI to improve public services. Green technology, digital governance, and eco-friendly city planning are prioritized in Xiong'an New Area, China.

The Smart Cities Mission in India is an initiative to improve metropolitan areas via the use of cutting-edge technology. London employs smart transportation systems and embraces data-driven government to better city living. Collectively, these global cases show how the Internet of Things (IoT), artificial intelligence (AI), and blockchain may be used to improve the quality of life in cities by stimulating innovation and reducing environmental impact.

## 7 Challenges and Future Research Directions

To successfully apply AIoT in smart city infrastructures, various hurdles must be overcome. First, smart cities' various gadgets and systems cause interoperability and standards challenges. Standardized protocols and interfaces enable disparate components to communicate and collaborate. Second, privacy and security are crucial because AIoT's increased connectivity and data interchange can make cities vulnerable to cyberattacks. Data-driven insights and data security must be balanced. AIoT devices create a lot of data, making storage, processing, and analysis difficult. Smart cities need solid big data infrastructure. Comprehensive regulatory frameworks are needed to address ethical concerns and enable responsible AI deployment in metropolitan areas. Public acceptance of AIoT may be difficult due to concerns about greater surveillance and data collection, requiring clear communication and engagement initiatives. Finally, many cities struggle to afford AIoT infrastructure rollout and upkeep, requiring strategic planning and public–private collaboration. Addressing these problems is essential to maximizing AIoT in smart city infrastructures and achieving sustainable urban development.

Significant difficulties and new research opportunities arise from the deployment of AIoTs in smart city settings.

- Protecting the large and sensitive data created by connected devices remains a top priority, necessitating innovative approaches to security. Citizens need to feel safe about how their data is being handled, thus privacy protection procedures must be put in place.
- Interoperability difficulties offer a difficulty as varied IoT devices and AI systems demand standardized communication channels for easy integration. Research into efficient resource allocation and network scalability is necessary to support the expected increase in connected devices.
- Constant efforts are needed to improve energy efficiency, which calls for research into sustainable energy sources and low-power AIOT solutions for smart city infrastructures. Comprehensive rules and frameworks are needed to lead research towards responsible AI in urban environments, which is necessary due to the ethical implications of AIOT deployments.
- Adaptive and resilient AIOT architectures that can handle changing urban environments should be the focus of future research. Successfully deploying AIOTs in smart city environments necessitates investigating edge computing for real-time processing, improving AI algorithms tailored for smart city applications, and fostering interdisciplinary collaborations between academia, industry, and policymakers. This will ensure a harmonious integration that benefits urban communities.

## 8 Conclusions

In conclusion, the importance of AIoT in achieving long-term success in smart city development cannot be overstated. The seamless integration of artificial intelligence with the Internet of Things offers a new paradigm for urban living. With the help of AIoT, cities may improve the quality of life for their residents by making better use of their resources and infrastructure. AIoT helps the environment in ways like smart energy management, trash reduction, and smart transportation systems. AIoT's real-time data insights allow urban planners and policymakers to make well-informed decisions, leading to more resilient and adaptable communities. Embracing AIoT emerges as a cornerstone for building smart cities that not only leverage technological advancements but also prioritize long-term sustainability, creating intelligent, efficient, and environmentally conscious urban environments as we navigate the challenges of urbanization and climate change.

## References

Albahri A, Duhaim AM, Fadhel MA, Alnoor A, Baqer NS, Alzubaidi L, Salhi A (2023) A systematic review of trustworthy and explainable artificial intelligence in healthcare: Assessment of quality, bias risk, and data fusion. Information Fusion 96:156–191
Allam Z, Dhunny ZA (2019) On big data, artificial intelligence and smart cities. Cities 89:80–91

Arisdakessian S, Wahab OA, Mourad A, Otrok H, Guizani M (2022) A survey on IoT intrusion detection: Federated learning, game theory, social psychology, and explainable AI as future directions. IEEE Internet Things J 10(5):4059–4092

Arora V, Leekha RS, Lee K, Kataria A (2020) Facilitating User Authorization from Imbalanced Data Logs of Credit Cards Using Artificial Intelligence. Mob Inf Syst 2020:8885269. https://doi.org/10.1155/2020/8885269

Bano A, Ud Din I, Al-Huqail AA (2020) AIoT-based smart bin for real-time monitoring and management of solid waste. Sci Program 2020:1–13

Bhambri P, Aggarwal M, Singh H, Singh AP, Rani S (2022) Uprising of EVs: Charging the future with demystified analytics and sustainable development. Decision Analytics for Sustainable Development in Smart Society 5.0: Issues, Challenges and Opportunities. Springer, Cham, pp 37–53

Blasi S, Ganzaroli A, De Noni I (2022) Smartening sustainable development in cities: Strengthening the theoretical linkage between smart cities and SDGs. Sustain Cities Soc 80:103793

Bronner W, Gebauer H, Lamprecht C, Wortmann F. (2021). Sustainable AIoT: how artificial intelligence and the internet of things affect profit, people, and planet. *Connected Business: Create Value in a Networked Economy*, 137–154.

Clinciu MA, Hastie HF (2019). *A survey of explainable AI terminology.* Paper presented at the 1st Workshop on Interactive Natural Language Technology for Explainable Artificial Intelligence 2019.

da Costa CA, Pasluosta CF, Eskofier B, da Silva DB, da Rosa Righi R (2018) Internet of Health Things: Toward intelligent vital signs monitoring in hospital wards. Artif Intell Med 89:61–69

Dhanalakshmi R, Anand J, Sivaraman AK, Rani S (2022) IoT-based water quality monitoring system using cloud for agriculture use. Cloud and Fog Computing Platforms for Internet of Things 28(3):1–14

Holzinger A, Langs G, Denk H, Zatloukal K, Müller H (2019) Causability and explainability of artificial intelligence in medicine. Wiley Interdiscip Rev: Data Min Knowl Discov 9(4):e1312

Kuguoglu BK, van der Voort H, Janssen M (2021) The giant leap for smart cities: scaling up smart city artificial intelligence of things (AIOT) initiatives. Sustainability 13(21):12295

Kuo Y-H, Wu EH-K (2023) Advanced, Innovative AIoT and Edge Computing for Unmanned Vehicle Systems in Factories. Electronics 12(8):1843

Mahato S, Kundu B, Makwana N, Joshi P (2023) Early summer temperature anomalies and potential impacts on achieving Sustainable Development Goals (SDGs) in National Capital Region (NCR) of India. Urban Climate 52:101705

Marine Y (2023) A Review: Application of AIOT in Smart Cities in Industry 4.0 and Society 5.0. Int J Smart Syst 1(1):1–4

Meske C, Bunde E (2020). *Transparency and trust in human-AI-interaction: The role of model-agnostic explanations in computer vision-based decision support.* Paper presented at the Artificial Intelligence in HCI: First International Conference, AI-HCI 2020, Held as Part of the 22nd HCI International Conference, HCII 2020, Copenhagen, Denmark, July 19–24, 2020, Proceedings 22.

Mora L, Deakin M, Zhang X, Batty M, de Jong M, Santi P, Appio FP (2022) Assembling sustainable smart city transitions: An interdisciplinary theoretical perspective. Sustainable smart city transitions. Routledge, London, pp 1–27

Nozari H, Szmelter-Jarosz A, Ghahremani-Nahr J (2022) Analysis of the challenges of artificial intelligence of things (AIoT) for the smart supply chain (case study: FMCG industries). Sensors 22(8):2931

Prabha C, Mittal P, Sharma A, Singh J (2022). *AIoT Emerging Technologies, Use Cases, and Its Challenges in Implementing Smart Cities.* Paper presented at the 2022 IEEE North Karnataka Subsection Flagship International Conference (NKCon).

Puri V, Kataria A, Sharma V (2021) Artificial intelligence-powered decentralized framework for Internet of Things in Healthcare 40. Trans Emerg Telecommun Technol 35:e4245

Rani S, Mishra RK, Usman M, Kataria A, Kumar P, Bhambri P, Mishra AK (2021) Amalgamation of Advanced Technologies for Sustainable Development of Smart City Environment: A Review. IEEE Access 9:150060–150087
Rani S, Kataria A, Chauhan M, Rattan P, Kumar R, Sivaraman AK (2022). Security and Privacy Challenges in the Deployment of Cyber-Physical Systems in Smart City Applications: State-of-Art Work. *Materials Today: Proceedings.*
Rožanec JM, Fortuna B, Mladenić D (2022) Knowledge graph-based rich and confidentiality preserving Explainable Artificial Intelligence (XAI). Information Fusion 81:91–102
Shah J, Kothari J, Doshi N (2019) A survey of smart city infrastructure via case study on New York. Procedia Computer Science 160:702–705
Silva BN, Khan M, Han K (2018) Towards sustainable smart cities: A review of trends, architectures, components, and open challenges in smart cities. Sustain Cities Soc 38:697–713
Ullah Z, Al-Turjman F, Mostarda L, Gagliardi R (2020) Applications of artificial intelligence and machine learning in smart cities. Comput Commun 154:313–323
Weber MP (2023). *AIoT Chances in Resource-Constrained Environments via Manual Pruning.* Paper presented at the International Conference on Innovations for Community Services.
Yang C-T, Chen H-W, Chang E-J, Kristiani E, Nguyen KLP, Chang J-S (2021) Current advances and future challenges of AIoT applications in particulate matters (PM) monitoring and control. J Hazard Mater 419:126442
Zeng X, Yu Y, Yang S, Lv Y, Sarker MNI (2022) Urban resilience for urban sustainability: Concepts, dimensions, and perspectives. Sustainability 14(5):2481

# Extended Reality and the Implementation of the UN Sustainable Development Goals: Opportunities, Challenges and Future Directions

**Girish Bekaroo**

**Abstract** The Sustainable Development Goals (SDGs) serve as the essential framework for tackling the complex contemporary challenges and it is imperative to take meaningful steps towards their advancement so as to foster sustainability and cultivate an improved future for global well-being. Digital technologies are considered among the key enablers for advancing the SDGs due to various reasons, including their ability to enhance data-driven decision-making, promote inclusive access to essential services, and facilitate global collaboration towards sustainable development. Among the digital technologies, extended reality (XR) technologies have rapidly evolved during recent years and are increasingly being integrated into key domains such as healthcare, education, and manufacturing. This group of innovative and immersive technologies including augmented reality, virtual reality and mixed reality can potentially play a key role in advancing the SDGs by offering immersive and interactive experiences that can raise awareness, foster empathy, and drive actions towards sustainable development. However, harnessing the full potential of XR for driving the SDGs also comes with various challenges that are related to readiness of the technologies, issues with content creation, skills requirements as well as privacy and security concerns. As such, this chapter discusses the key role that XR can potentially play in advancing the 17 SDGs. As part of this chapter, the key opportunities created by XR towards advancing the SDGs are discussed as well as the key challenges involved. In addition, the prospective solutions to these challenges are highlighted as well as the future directions of XR in advancing the UN SDGs.

**Keywords** Extended Reality · Sustainable Development Goals · Immersive Education · Future of Work · Environmental Sustainability Applications · Virtual Community Spaces

G. Bekaroo (✉)
Middlesex University Mauritius, Uniciti, Coastal Road, Flic-en-Flac, Mauritius
e-mail: g.bekaroo@mdx.ac.mu



W. Leal Filho et al. (eds.), *Digital Technologies to Implement the UN Sustainable Development Goals*, World Sustainability Series,
https://doi.org/10.1007/978-3-031-68427-2_11

## 1 Introduction

The Sustainable Development Goals (SDGs) represent the primary blueprint for addressing the multifaceted challenges being faced in the modern era and actions to advance the goals are crucial to promote sustainability and create a better future for all. As part of this transformative journey to address the global issues, digital technologies are considered as key enablers, with the potential to accelerate progress towards achievement of the 17 goals (Secundo et al., 2022). This is because digital technologies have the capacity to enhance data-driven decision-making, promote inclusive access to essential services, and facilitate global collaboration towards sustainable development (Argyroudis et al., 2022). Among the digital technologies, extended reality (XR) technologies have recently undergone rapid development and are becoming increasingly ubiquitous in various sectors including healthcare, education, manufacturing, and military, among others (Ahmad et al., 2023). Extended reality groups immersive technologies that enable the display of computer-generated objects onto the real environment or the integration of real-world elements into computer-generated virtual environments. Key technologies that are part of this umbrella term include augmented reality (AR), virtual reality (VR) and mixed reality (MR). During recent years, there has been rapid growth of the market for these technologies where for instance, according to Fortune Business Insights (2023), the global AR market, which was valued at $62.75 billion in 2023, is expected to grow to $1,109.71 billion by 2030. This also represents a compound annual growth rate (CAGR) of 50.7% during the forecast period. According to the same report, the global virtual reality market size estimated at $25.11 billion in 2023, is forecasted to grow to $165.91 billion by 2030, also representing a CAGR of 31.0% during the forecast period (Fortune Business Insights, 2023).

With such growth and with advances in complementary technologies such as artificial intelligence (AI), cloud computing and networking technologies, XR technologies has the potential to play an important role in advancing the SDGs whereby offering immersive and interactive experiences that can raise awareness and induce actions towards sustainable development. Nevertheless, effectively integrating XR technologies to advance SDG initiatives necessitates the careful consideration of various challenges and limitations. These challenges also provide avenues for further advancement of XR technologies as interdisciplinary research, policy development, and industry collaborations are crucial for exploring innovative applications of XR in addressing the SDGs. As such, this chapter discusses the opportunities, challenges, prospective solutions, and future directions in relation to the application of XR technologies in advancing the UN Sustainable Development Goals.

## 2 An Overview of Extended Reality

Extended reality is not a specific technology but encompasses technologies that change the reality by adding meaningful digital elements to the real-world or by altering the virtual world to add components of the physical world (Doolani et al., 2020). Within the acronym XR, X is a variable that represents current or future technologies that adhere to the stipulated definition or vision of extended reality, as illustrated in Fig. 1.

The key technologies that form part of XR include:

- *Augmented Reality (AR)*

Augmented reality (AR) is a technology that aims at enhancing the user's perception of reality by seamlessly integrating computer-generated elements, such as images, 3D models, and animations, onto the user's view of the real-world environment. This is typically achieved through the utilization of AR software applications running on devices like smartphones, tablets, specialized AR glasses or even contact lenses. Amongst, mobile devices commonly integrate built-in hardware components like cameras, accelerometers and Global Positioning System (GPS) that can be used for identifying and tracking objects before eventually processing and superimposing computer-generated elements onto the screen. The widespread use of mobile phones

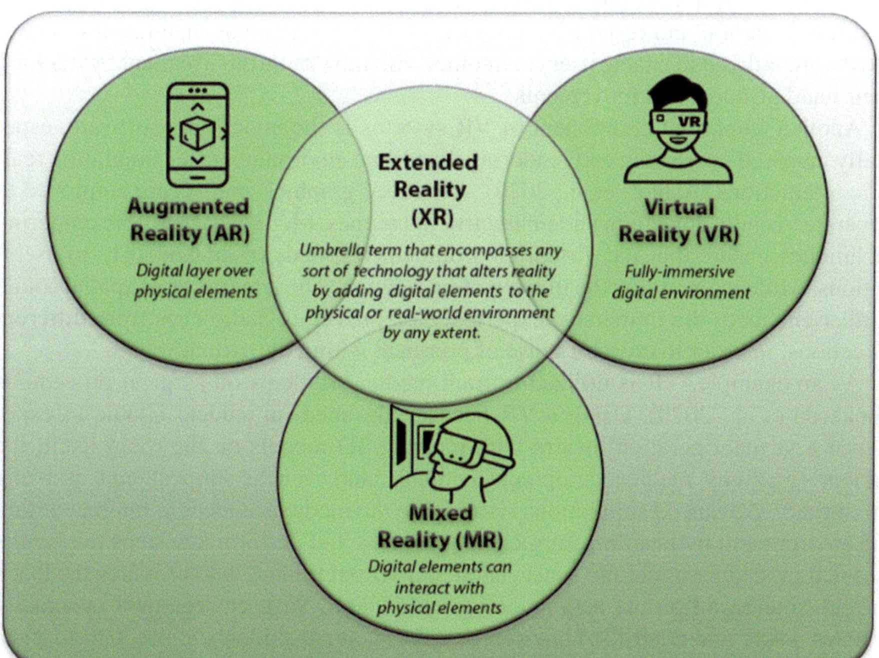


**Fig. 1** Overview of XR technologies

has the potential to make this technology accessible to everyday user (Zhang et al., 2022a, 2022b), where an increasing number of applications are available on online application stores including AR browsers and games, among others. As an illustrative example, AR is commonly used in museums whereby enabling visitors to learn about the museum and its artefacts. As users are walking through the museum and comes across artefacts, they can use their mobile phones to scan the artefact using their camera and the AR application will provide meaningful information about the target, such as descriptions, narrations and animations, among others. The user can also virtually touch and rotate the artefact examine it from various angles, thus promoting learning in an engaging, interactive and immersive way.

- *Virtual Reality (VR)*

VR enables users to experience an immersive and interactive computer-generated virtual environment within which, users can see around the virtual world, move within it, and interact with objects in the environment (Park et al., 2019). This is achieved through a combination of software and hardware components, including specialized VR headsets, equipped with high-resolution displays and lenses that are worn by users to replace their visual field with computer-generated images. As part of the immersive experience, different types of feedback are integrated within the computer-generated environment including visual, auditory, and haptic, among others. Moreover, the immersive experience is further enhanced by head-tracking systems that monitor users' head movements and adjust the displayed imagery accordingly, while also ensuring a natural and responsive interaction. In some systems, hand controllers or gloves are utilized to enable users to interact with the virtual environment by tracking their hand gestures and movements.

Another important component of VR systems is the underlying software, especially involved in creating and rendering the virtual environment and simulating realistic interactions (Ashtari et al., 2020). Advanced graphics engines are employed to generate visually appealing and interactive 3D scenes. Moreover, real-time rendering techniques ensure that the virtual environment responds instantaneously to users' actions, while maintaining the illusion of immersion. The integration of spatial sound further enhances the immersive experience by providing audio cues from different directions, in order to create a sense of presence within the virtual world.

As an example, VR is utilized to train medical students on surgical procedures (Bekaroo et al., 2020). Using a VR headset, the medical student is able to experience a virtual operation theatre with realistic 3D models on the room itself, the patient's anatomy, medical equipment available, and even the surgical team assisting the surgery. Advanced simulations also enable the medical student to interact within the environment by handling surgical instruments and perform key steps during the surgical procedure under the supervision of a virtual trainer, who provides feedback in real-time based on the actions being undertaken. Such environment eventually enables a safe and controlled environment for medical students to practice surgical procedures whereby reducing risks and at the same time developing the required skills.

- *Mixed Reality (MR)*

Mixed reality involves a combination of AR and VR technologies whereby integrating both virtual and real-world environments, within which users can interact with virtual and real-world physical objects in interactive and immersive ways (Rokhsaritalemi et al., 2020). MR systems employ advanced sensors, cameras, and tracking technology to accurately understand and map the physical environment of users whereby enabling them to seamlessly move around and interact with objects within the environments. Specialized headsets equipped with cameras and sensors provide users with a live view of their surroundings while superimposing interactive virtual elements. As an example of MR, in the same context of medical training, a surgeon wears appropriate MR equipment to overlay digital images of a patient's anatomy onto their real body during a surgical procedure, thus enhancing precision. Another example of an application of MR is within building construction, where architects and builders can use mixed reality headsets to visualize and interact with digital building models overlaid onto the real-world construction site, allowing for more accurate planning, design, and collaboration.

## 3 XR and the UN Sustainable Development Goals: Key Opportunities

With advances and convergence of different technologies such as AI and wireless communication technologies, XR has emerged as a powerful tool with the potential to progressively contribute to the achievement of the UN's SDGs. The immersive and interactive nature of XR experiences allows for new ways of addressing complex global challenges, while also creating new opportunities at various levels (Aguayo & Eames, 2023). As such, through immersive experiences, XR has the potential to drive positive change, foster sustainable development, and create a more inclusive and equitable world aligned with the SDGs. The key opportunities created by XR towards advancing the SDGs are:

- *Creating Job Opportunities and the Revolutionizing the Future of Work*

With the rapid growth of the market size of XR technologies (Fortune Business Insights, 2023), a range of job opportunities are expected to be created ranging from manufacturing of devices to development and support of XR systems (Kuleto et al., 2021), thus directly contributing to SDG 8 - Decent Work and Economic Growth and SDG Goal 9: Industry, Innovation, and Infrastructure, among others. Also, this group of technologies has the potential to revolutionize the future of work by promoting remote work within virtual workspaces, facilitating global collaboration through immersive environments, and by enabling customers to interact with products and services in virtual or augmented spaces (Peck, 2021). Only within the healthcare sector, XR can reshape existing services through telemedicine, remote medical consultations and patient education, medical training simulations and remote surgery, among others (SDG3) (Bekaroo et al., 2020). As such, the impact of these technologies on the future of work is expected to be significant such that it has been forecasted that more than 23 million jobs will be enhanced using XR technologies around the world by 2030, as compared to the 800 thousand jobs enhanced by XR technologies in 2019 (PwC, 2020).

- *Enabling Immersive Education and Training*

XR technologies are expected to reshape education and training at various levels and within different sectors by creating immersive learning environments and experiences (Cherner & Fegley, 2022). Education at different levels ranging from early childhood to higher education, can integrate AR and VR experiences, virtual classrooms and interactive simulations whereby improving engagement of learners and at the same time addressing SDG 4: Quality Education. For instance, AR has been used in early childhood to teach a range of topics including literacy, nature, mathematics and geometry, geography and the solar system and it was found that the use of such technology does not only enhance engagement during learning, but also draws learner's attention to the learning material and encourages reflection in the process (Oranç & Küntay, 2019). Similarly, XR technologies are expected to transform training across sectors, enabling trainees to engage in experiential learning and build key skills while being engaged in realistic environments. For example, XR technologies can be used to train pilots and aviation technicians on flight simulations, emergency procedures and aircraft maintenance training. Such interactive training systems can also help to raise awareness on a range of topics such as biodiversity and conservation, towards influencing behavioural change and promoting empathy. Also, such learning experience is beneficial and has shown to improve retention, enhance engagement and develop skills faster as compared to the use of traditional learning methods (Doolani et al., 2020).

- *Promoting Environmental Sustainability through XR Applications*

XR technologies have the potential to accelerate the adoption of sustainable practices (Yaqoob et al., 2023) and contribute directly to SDGs 11, 12 and 13. A range of XR applications are already available on the market that focus on environmental sustainability enhancement and key features include game elements, immersive learning components, resource management features and rewards, among others (Cosio et al., 2023). Amongst, mobile-based AR applications are available on application stores that overlay real-world environments with data on energy consumption, carbon emissions, and resource usage. These applications make environmental sustainability metrics tangible and comprehensible, while at the same time facilitating decision making and promoting reduction. Also, various XR-based simulation tools are available that guide sustainable design, reduce resource utilisation and promote environmental sustainability. For example, urban planners can use XR tools to design sustainable cities through interactive models and simulations for infrastructure development, transportation systems, and public spaces provided by such tools. In addition, remote work, collaboration within virtual environments, virtual field trips and virtual tourism have the potential to reduce travelling and carbon footprints.

- *Building Virtual Community Spaces*

The immersive environments created by XR technologies have the potential to build virtual communities and promote global collaboration, inclusivity, and engagement while at the same time addressing complex societal challenges. Such virtual community spaces can help to advance different SDGs including SDG 4 and 11. The concept of the metaverse is a key example of building virtual community spaces using XR technologies. This virtual community space can facilitate various virtual interactions ranging from social gatherings to educational conferences, while enabling users to engage in real-time discussions, knowledge sharing and problem-solving activities related to SDGs (Ning et al., 2023).

The above opportunities can be applied to advance each SDG directly or indirectly and these are summarized in Table 1.

## 4 Challenges and Prospective Solutions

While XR technologies offer various opportunities for advancing the SDGs, leveraging their full potential necessitate the careful consideration of some key obstacles. These challenges are essential to investigate and to be addressed so that stakeholders can navigate the landscape more effectively, whereby fostering the optimal utilization of XR in advancing sustainable development and promoting inclusivity in the future. This section discusses the challenges and limitations associated with unleashing the full potential of XR in driving the attainment of the SDG as well as potential solutions to these challenges.

### *4.1 Technology Readiness*

Although XR technologies are gradually advancing, they may not be fully ready in all aspects for widespread and seamless application. Consequently, this can potentially hinder its role in advancing the SDGs over the next few years. As the technology is still maturing, various software development kits (SDKs) and platforms available for developing AR, VR and MR systems are limited in terms of capabilities, requiring extensive resources and scripting in order to develop key functionalities (Zhan et al., 2020). In both AR and VR realms, much work is being done to address various issues including accurate detection and tracking of real objects, creation, registration and display of virtual objects, correction of factors such as illumination and 3D rendering, among others (Bekaroo & Dawarka, 2023), that are also key challenges to the creation of fully immersive solutions. Lack of documentation and resources to support developers also add to the challenges in addition to costs of high quality XR devices (e.g., VR headsets and AR glasses), which limit their affordability to the broader population.

**Table 1** Application of XR technologies to advance the SDGs.

| SDG | Impact of XR on the SDG | Description |
|---|---|---|
| 1 NO POVERTY | Indirect | Education is considered as a powerful medium for reducing poverty (Fenández, 2020) and XR has the potential to advance SDG 1: No Poverty, by firstly raising awareness about poverty and its causes through immersive and engaging experiences. In this endeavour, the realities faced by impoverished communities within different regions can be showcased, to foster empathy and drive action towards poverty reduction. Secondly, due to its impact on employment as discussed earlier, XR can create opportunities for disadvantaged communities, enabling them to acquire essential skills and increase employability chances. Moreover, XR can enhance remote collaboration and enable individuals in poverty-stricken regions to connect with experts and resources globally, fostering knowledge sharing and empowerment and eventually contribute towards no poverty |
| 2 ZERO HUNGER | Indirect | As a game-changing technology for promoting sustainable agriculture and food security (Anastasiou et al., 2022), XR can play an important role in advancing SDG 2: Zero Hunger. For instance, virtual training and simulations can be created for farmers, whereby supporting them to learn about modern agricultural practices, efficient resource management, climate-smart techniques, innovative measures to improve crop yields and how to better sell agricultural products and services. Moreover, XR can help to enhance public awareness and engagement in sustainable food production and consumption, whereby encouraging behavioural changes towards responsible farming and nutrition |
| 3 GOOD HEALTH AND WELL-BEING | Direct | As discussed earlier and as per the earlier examples, XR is expected to significantly impact healthcare by enhancing telemedicine, remote consultation, and medical training (Bekaroo et al., 2020). Furthermore, XR can potentially improve access to healthcare services, facilitate better diagnostics, and enhance patient outcomes through immersive therapy and rehabilitation experiences |
| 4 QUALITY EDUCATION | Direct | As discussed earlier in this section, XR has the potential to transform education and training to help simplify complex concepts, build key skills, and enable access educational resources regardless of the geographic location or socioeconomic situation of learners at different levels, to eventually advance SDG 4: Quality Education |

(continued)

**Table 1** (continued)

| SDG | Impact of XR on the SDG | Description |
|---|---|---|
| 5 GENDER EQUALITY | Indirect | XR technologies can create immersive experiences and expand collaboration in the endeavours to raise awareness on gender equality issues and empower individuals to advocate for gender equality, whereby advancing SDG 5: Gender Equality |
| 6 CLEAN WATER AND SANITATION | Indirect | XR can be used to raise awareness and educate individuals about the importance of clean water and proper sanitation. In addition, by simulating water scarcity and pollution scenarios, XR experiences can instil a sense of urgency regarding sustainable water management. Such tools can also aid in visualizing complex water systems and support data-driven decision-making for effective water resource management. As such, embracing XR can promote sustainable water practices, supporting the realization of universal access to clean water and adequate sanitation for all, towards advancing SDG 6: Clean Water and Sanitation |
| 7 AFFORDABLE AND CLEAN ENERGY | Indirect | To advance SDG 7: Affordable and Clean Energy, XR can be utilized to raise awareness and educate individuals about the benefits and challenges of clean energy solutions in immersive and engaging ways, towards promoting sustainable energy practices and encouraging behavioural changes. Also, AR, MR and VR can be applied to enhance training experiences of workers in the installation of renewable energy systems, maintenance, and operation, whereby enhancing workforce skills in the clean energy sector. Moreover, XR-based simulations tools can aid in energy infrastructure planning and optimization to improve energy management and efficiency |
| 8 DECENT WORK AND ECONOMIC GROWTH | Direct | As discussed earlier, XR technologies is expected to create numerous job opportunities and revolutionize the future of work to significantly impact SDG 8: Decent Work and Economic Growth |
| 9 INDUSTRY, INNOVATION AND INFRASTRUCTURE | Direct | Like SDG8, XR is expected to advance SDG 9: Industry, Innovation, and Infrastructure by revolutionising the future of work whereby enabling multitude of immersive and innovative services to enhance work settings in various sectors and improve productivity. Also, XR can aid in monitoring and maintenance of critical infrastructure in order to optimize operations and detect issues in a timely manner |

(continued)

**Table 1** (continued)

| SDG | Impact of XR on the SDG | Description |
|---|---|---|
| | Indirect | XR can help to advance SDG 10: Reduced Inequalities by democratizing access to information and experiences, which can consequently reduce inequalities and promote a more inclusive society. Also, XR experiences can cultivate cultural awareness, break down stereotypes, and provide accessible learning environments for individuals with disabilities or limited resources |
| | Direct | As discussed earlier, XR can be utilized to promote environmental sustainability and to directly contribute to the advancement of SDGs 11, 12 and 13. For the advancement of SDG 11: Sustainable Cities and Communities, a range of innovative solutions and simulation tools can be created for urban planning, design, and citizen engagement. Also, users can visit cities and engage with inhabitants virtually through VR, while decreasing the need to travel and reducing carbon footprint. In addition, XR can provide immersive experiences for visualizing and simulating sustainable development strategies, while involving communities in decision-making processes and fostering a sense of ownership. Moreover, XR presents the opportunity to enhance public awareness and education about sustainable practices towards promoting behaviour change and more sustainable lifestyles |
| | Direct | As for the advancement of SDG 12: Responsible Consumption and Production, XR can enable immersive experiences that promote sustainable consumption practices and educate consumers about their environmental impact. Also, XR allows businesses to simulate sustainable production processes, optimize resource utilization and reduce waste. By driving conscious decision-making and by encouraging sustainable consumption habits, XR has the potential to empower individuals and organizations to embrace more responsible lifestyles and contribute to achieving SDG 12's targets for sustainable production and consumption patterns |
| | Direct | XR provides opportunities to educate individuals at various levels on climate related courses and raise awareness about climate change and its impacts through engaging and immersive experiences. Also, by facilitating remote collaboration and communication, XR reduces the need for travel and associated carbon emissions, as discussed earlier. Hence, XR can help to inspire action and drive behavioural change to combat climate change and achieve SDG 13's targets for climate action |

(continued)

**Table 1** (continued)

| SDG | Impact of XR on the SDG | Description |
|---|---|---|
| 14 LIFE BELOW WATER | Indirect | XR-based applications can be used to raise awareness about ocean conservation and sustainable marine resource. Moreover, XR simulations can facilitate exploration of marine ecosystems, whereby enabling the population to understand key issues related to marine environments (e.g., impacts of marine litter), to eventually promote behavioural change and foster empathy for marine life. Additionally, XR aids education, research, and remote monitoring of marine environments, empowering conservation efforts and sustainable management of oceans and marine resources. As such, harnessing XR through these initiatives can inspire environmental stewardship and contribute to the preservation of our marine ecosystems, while advancing SDG 14: Life Below Water |
| 15 LIFE ON LAND | Indirect | XR technologies can create opportunities for raising awareness about biodiversity conservation and terrestrial ecosystem protection, whereby providing immersive experiences and at the same time enhancing engagement. XR simulations can provide virtual encounters with endangered species to enable the population to better appreciate the Earth's ecosystems. Additionally, XR can aid in land management and planning by offering virtual models and simulations that assist in sustainable land use practices. Hence, leveraging XR technology promotes biodiversity conservation, land restoration, and sustainable terrestrial ecosystem management, thus helping to advance SDG 15: Life on Land |
| 16 PEACE, JUSTICE AND STRONG INSTITUTIONS | Indirect | XR has the potential to contribute to SDG 16: Peace, Justice, and Strong Institutions through immersive storytelling that creates empathy and promote peaceful societies. Likewise, XR simulations can aid conflict resolution, providing platforms for dialogue and empathy-building. Also, XR supports training for law enforcement whereby promoting fair legal systems. Virtual environments can potentially monitor human rights abuses and enhance transparency and accountability |
| 17 PARTNERSHIPS FOR THE GOALS | Indirect | In addition to facilitating knowledge sharing and addressing complex challenges, the immersive virtual collaboration platforms enabled through XR technologies can promote global partnerships and cross-sector collaboration. Moreover, XR has the potential to offer innovative fundraising and advocacy opportunities, engaging virtual experiences in order to mobilize resources for sustainable development. As such, leveraging XR can enhance inclusive partnerships, technology transfer, and capacity-building, accelerating progress towards SDG 17: Partnerships for the Goals |

In addition, to achieve smooth experience, immersion, and interaction within XR applications, convergence of various technologies is required such as AI, networking technologies, computer graphics, among others (Bekaroo & Dawarka, 2023). However, many of these complementary technologies are still developing and their associated limitations can adversely impact the achievement of the goal of XR. For example, realizing XR applications that enable high quality video and 3D animations rendered in real-time necessitate high data rate and very low latency (Mumtaz et al., 2017), thus requiring wireless system with enhanced capabilities (e.g., 6G cellular networks). Moreover, XR applications often need to interface with diverse technologies whereby requiring robust data synchronization, network communication, and backend integration and in the process, integration and compatibility issues may add to the challenges (Ning et al., 2023). In other words, different XR devices and platforms may have varying requirements and capabilities, making it challenging for developers to create applications that work seamlessly across multiple devices. Such fragmentation can lead to compatibility issues and limit the reach and impact of XR applications.

To address these challenges and limitations, a multidisciplinary approach that involve collaborations among researchers, XR developers, User Experience (UX) designers and domain experts are required, to collaboratively design effective solutions in order to progressively address the highlighted challenges. In addition, establishing common development standards, best practices, and compatibility guidelines can help to promote interoperability and simplify the creation of XR applications. Furthermore, efforts to improve documentation, tutorials, and learning resources can support developers in acquiring the necessary skills and knowledge for XR development. Open-source XR development tools and platforms can also play a vital role in democratizing XR development and making it more accessible to a broader community of developers.

## *4.2 Diverse Skills Requirements*

While XR technologies are still maturing, authoring tools are progressively being improved to gradually meet application development requirements (Marques et al., 2022). Therefore, developing commercial XR applications can be time-consuming and complex whereby requiring specialized expertise in the field. This is because creating XR solutions and immersive experiences require a diverse range of skills ranging from working with various hardware and software components, to designing realistic simulations, and enabling seamless integration with existing systems. In addition, developing intuitive user interfaces and immersive experiences necessitate in-depth understanding of various concepts such as the principles of human–computer interaction, spatial computing, immersive storytelling, and gesture recognition, among others. Consequently, developers may not be able to fulfil all requirements, thus presenting challenges to companies seeking skilled professionals for the creation of innovative products. Hence, developers must stay updated and continuously enhance their competencies through continuous learning and professional development. To meet the growing demand for competencies in XR solution development, an increasing number of courses are available on online learning platforms and integrated into IT-related programmes at higher education level. These courses teach key concepts and build key skills needed to develop XR solutions, with a strong emphasis on cultivating advanced skills in 3D modelling, seamless AI integration, and the art of crafting captivating and immersive experiences.

## *4.3 Digital Divide*

Digital divide represents another important challenge that can potentially delay the realization of the full potential of XR for advancing the SDGs. Whilst XR technologies require different complementary technologies for implementing and consuming solutions, disparities in access to high-speed internet, technological infrastructures and digital literacy can limit the reach and impact of XR solutions, particularly in underserved areas (Seifert & Schlomann, 2021). For instance, various rural and remote areas within developing countries still lack adequate broadband infrastructure and have slow and unreliable internet connectivity that may disrupt the immersive experiences, real-time interactions, and data streaming required in XR solutions. Similarly, equipment such as VR headsets and AR glasses are relatively costly, thus limiting their affordability to the broader population. In addition, the varying levels of technological literacy and digital skills across different populations presents another considerable challenge. Whilst using XR technologies can be complex and requires a certain level of technical proficiency, it may be challenging for users with limited such skills to effectively use these systems, thereby hindering the intended impact on SDG-related learning and behaviour change.

To address the technological divide challenge, it is essential to consider both the social and economic factors that contribute to unequal access to XR technologies, and this requires a comprehensive and inclusive approach that take into consideration the specific needs and constraints faced by the diverse communities (Vassilakopoulou & Hustad, 2021). Some of the measures as part of such approach include infrastructural development by governmental and relevant organisations in order to improve internet access and speed in underserved regions, training programmes to empower individuals with the skills to utilize XR technologies, public–private partnerships to promote digital inclusion in underserved areas or better funding support by international agencies. All these measures can potentially drive innovation, infrastructure development, and digital inclusion in such areas to gradually reduce digital divide.

## 4.4 Slow Adoption and Acceptance of XR Solutions

Recent studies showed that acceptance of XR-based solutions by the public and enterprises is relatively slow, presenting a significant challenge to harnessing their full potential in advancing the SDGs (Bekaroo & Dawarka, 2023; Hang & Hui, 2023). The slow acceptance of XR solutions can be attributed to several factors. To start with, lack of awareness on the technology and the skills required to smoothly utilize such technologies hinder its widespread adoption (Bekaroo & Dawarka, 2023). To address this issue, it is essential to educate users about the real-world applications and advantages of XR technologies in advancing SDGs and this can be done through public awareness campaigns, workshops, and immersive demonstrations at various levels.

Another concern impeding XR adoption is the relatively short duration of use by users. For instance, in the case of AR mobile applications, it was found that many users abandon the technology due to relatively poor content quality and subpar design standards (Jha et al., 2019). Prospective solutions lie in enhancing content quality, ensuring seamless user experiences, and promoting engaging and meaningful interactions. Collaboration between XR developers, content creators, and subject-matter experts can lead to the production of compelling XR content that aligns with users' needs and interests. In addition, continuous user feedback and iterative design processes can further refine XR solutions, leading to increased adoption and sustained usage.

Furthermore, as the technology is still developing, unobtrusive fashionable appearance and the cost of XR equipment, such as gloves, helmets, or specialized devices, pose barriers to adoption. To mitigate these challenges, designers must prioritize the aesthetics of XR devices, integrating them seamlessly into everyday life. Modular and customizable XR equipment can cater to individual preferences, encouraging users to incorporate XR technologies effortlessly into their routines. Also, technological advancements and economies of scale are expected to drive down the costs of XR equipment, rendering them more accessible to a wider audience.

## 4.5 Content Development

The creation of high-quality XR contents such as images, videos, audio, and animations, among others, that align with SDGs can be challenging. Also, as discussed earlier, the duration of use of XR systems was found to be relatively shorter than anticipated and the poor quality of new and updated contents was found to be among the key reasons (Jha et al., 2019). The process of 3D content creation using current software is also regarded as slow and complex, further worsening the challenges related to content creation. As key solutions, AI and computer vision technologies can potentially improve the quality and realism of XR content, enhancing its effectiveness in addressing SDG goals. For instance, AI-driven tools such as MidJourney can be utilized to dynamically create contents for use in XR systems to provide users with better quality contents. Moreover, developments in wireless networking technologies are expected to facilitate crowdsourced content generation, enabling developers to post specific content requirements for various XR systems and content creators from around the world can contribute to fulfilling these demands. Such platforms will not only facilitate the democratization of XR content creation but also provide opportunities for new business models related to content sharing and monetization.

### 4.6 Privacy and Security Issues

Privacy and security issues are important challenges that need to be addressed so that XR-based solutions are more effectively leveraged in the endeavour towards advancing the SDGs. To create immersive experiences for end-users, XR systems need to effectively track the end-user and know his/her environment, and this is achieved by extensively collecting data through embedded sensors and cameras whereby recording details such as movement, interactions and preferences of users (Abraham et al., 2022). However, this enables XR systems to create detailed profiles of the user and the surrounding, thus presenting risks of unauthorized access, data breaches, and potential misuse of user's personal data (Wang et al., 2022). Such risks are even higher with complex XR systems that require always-on sensing and continuous camera recording. As a key solution to safeguard user privacy, it is recommended for XR developers to adopt privacy-by-design principles (ISO/IEC, 2022) and implement relevant frameworks that guide user privacy protection. Such solutions consist of the implementation of strong data protection measures including data anonymization techniques, user consent mechanisms as well as transparent data processing policies (Gumbo et al., 2022). Furthermore, ensuring compliance of relevant data protection regulations, such as the General Data Protection Regulation (GDPR), can ensure responsible data handling and accountability (Wang et al., 2022). In addition, implementation of data anonymization techniques such as blurring bystanders in real-time as well as the use of strong cryptography solutions can protect user data throughout the XR ecosystem and ensure confidentiality of data at rest or in transit.

Due the growing use of XR systems within critical domains such as healthcare and industry, cybersecurity threats such as denial of service and malware attacks are also expected to amplify (Dwivedi et al., 2022). For instance, there have been cases where malicious applications embedded within XR systems secretly read or modify digital information and provide the end-user with incorrect information while also adversely impacting the user's perception of reality (Roesner et al., 2014). As such, it is essential to implement strengthened measures and solutions to ensure that end-users of XR solutions are well-protected against such cybersecurity issues (Zhang et al., 2022a, 2022b). Some of the key solutions include adherence to secure software development practices, that recommend practices such as the implementation of robust authentication mechanisms, adoption of strong cryptography techniques and conducting regular vulnerability assessments. In addition, continuous monitoring of systems and swift response to emerging threats are crucial to maintain the integrity of XR systems and protect user data (Happa et al., 2021).

### *4.7 Physical Safety*

XR experiences can blur the boundaries between the virtual and physical worlds, potentially compromising users' physical and psychological safety (Gugenheimer et al., 2022). Disorientation and sensory overload in immersive environments may cause users to inadvertently reveal private information or engage in risky behaviours, to adversely pose safety risks. As potential solutions, XR developers and content creators must provide clear warnings and safety guidelines to users, ensuring they are aware of potential risks and the need to maintain situational awareness. Also, content creators should design XR experiences with user safety in mind, providing user-friendly interfaces and intuitive navigation. Implementing user-controlled comfort settings, such as motion sickness reduction techniques, can enhance user comfort and safety during XR interactions.

## 5 Future Directions

As discussed earlier in this chapter, XR technologies can directly or indirectly contribute to the advancement of all the 17 SDGs. However, harnessing the full potential of XR for driving the SDGs would only be possible if the challenges discussed in the previous section are effectively addressed. To address the challenges and limitations of XR, a comprehensive strategy is required whereby involving key stakeholders involved in research, development, and commercialisation of XR solutions. Such interdisciplinary cooperation among stakeholders including researchers, developers of XR platforms, tools, and applications, UX designers and end users can potentially produce impactful solutions that mitigate key challenges. This would entail addressing the challenges and limitations right at the root towards ensuring the technology is ready for the production of more effective applications and systems that would contribute to the advance of the SDGs. This also implies the need for addressing key issues with individual XR technologies such as AR, VR and MR as well as issues with complementary technologies including the application of AI and networking technologies, among others.

As XR technologies gradually mature, it becomes essential to effectively address the key issues surrounding the widespread proliferation of these technologies towards impacting the SDGs. These include training XR developers and professionals so that better polished solutions that meet key requirements are produced, while also ensuring that engaging contents are provided to end users of such technologies. Furthermore, issues related to digital divide need to be well addressed to ensure that the technology and required equipment to utilize XR solutions are more accessible to the broader population. In addition, adoption and acceptance of XR solutions need to be boosted through different measures such as public awareness campaigns, educational initiatives and platforms that promote collaboration between XR research and development communities.

With the widespread utilization of tailored solutions that align with or contribute to the advancement of the goals of sustainable development, operational issues need to be continuously addressed in a proactive manner to ascertain positive experiences when using XR solutions. For this, it is essential to ensure that XR solutions are operated and consumed in a secure manner, while also ensuring physical and psychological safety of end-users involved. Furthermore, a comprehensive framework is needed to promote sustained use of XR technologies to continuously address the SDGs whereby fostering a culture of innovation and collaboration among industry leaders, researchers, and practitioners while also ensuring the evolution and refinement of XR applications meet ever-changing needs and sustain long-term adoption.

## 6 Conclusions

Immersive technologies associated with both the real and virtual environments are expected to contribute significantly to the advancement of all the 17 sustainable development goals directly or indirectly. Firstly, XR technologies are expected to directly impact SDG 8: Decent Work and Economic Growth and SDG Goal 9: Industry, Innovation, and Infrastructure through the creation of job opportunities and by revolutionizing the future of work as the market for AR, VR and MR technologies are anticipated to grow massively over the next years. These technologies are also expected to reshape education and training within various sectors by creating immersive learning environments and experiences and directly impacting SDG 4: Quality Education. Furthermore, through a range of applications and systems that focus on environmental sustainability, XR technologies can directly contribute to SDG 11: Sustainable Cities and Communities, SDG 12: Responsible Consumption and Production and SDG 13: Climate Action. Also, the immersive environments created by XR technologies have the potential to build virtual communities and promote global collaboration, inclusivity, and engagement towards directly contributing to the advancement of SDG11: Sustainable Cities and Communities.

However, there are various challenges and limitations that adversely impact the full benefits of XR to be leveraged in the endeavour to advance the SDGs. The key challenges and prospective solutions discussed as part of this chapter include technology readiness, diverse skills requirements, digital divide, slow adoption, and acceptance of XR solutions, content development issues, privacy and security issues as well as physical safety. To effectively address the challenges and limitations of XR and to yield impactful solutions that mitigate these issues, a comprehensive strategy involving interdisciplinary collaboration among researchers, XR developers, UX designers, and end users is essential. As such, to conclude, the strategic integration of XR technologies does hold the potential to catalyse substantial advancements across diverse sectors, facilitate innovative solutions and transformative experiences that contribute to the advancement of the SDGs and foster a more sustainable and interconnected global future.

## References


Abraham M, Saeghe P, Mcgill M, Khamis M, (2022). *Implications of xr on privacy, security and behaviour: Insights from experts.* pp. 1–12.

Aguayo C, Eames C (2023) The Journal of Environmental Education. Using Mixed Reality (XR) Immersive Learning to Enhance Environmental Education 54(1):58–71

Ahmad H et al (2023) Leveraging 6G, extended reality, and IoT big data analytics for healthcare: A review. Computer Science Review 48:100558

Anastasiou E, Balafoutis A, Fountas S (2022) Applications of extended reality (XR) in agriculture, livestock farming, and aquaculture: A review. Smart Agricultural Technology 3:100105

Argyroudis S et al (2022) Digital technologies can enhance climate resilience of critical infrastructure. Clim Risk Manag 35:100387

Ashtari N et al., (2020). *Creating augmented and virtual reality applications: Current practices, challenges, and opportunities.* pp. 1–13.

Bekaroo G, Dawarka V (2023) AI-Assisted Extended Reality Toward the 6G Era: Challenges and Prospective Solutions. In: Milovanovic D, Bojkovic Z, Fowdur T (eds) Driving 5G Mobile Communications with Artificial Intelligence towards 6G. CRC Press, pp 403–423

Bekaroo G, Santokhee A, Augusto J, (2020). 5G smart and innovative healthcare services: Opportunities, challenges, and prospective solutions. In: Z. Bojkovic, D. Milovanovic & T. Tulsi Pawan Fowdur, eds. *5G Multimedia Communication: Technology, Multiservices, and Deployment.* CRC Press, pp. 279–297.

Cherner T, Fegley A, (2022). *A guided exploration of XR technologies for education: Implications for teacher education and future research.* s.l., Association for the Advancement of Computing in Education (AACE), pp. 1808–1811.

Cosio L et al., (2023). *Virtual and Augmented Reality for Environmental Sustainability: A Systematic Review.* pp. 1–23.

Doolani S et al (2020) A review of extended reality (xr) technologies for manufacturing training. Technologies 8(4):77

Dwivedi Y et al (2022) Metaverse beyond the hype: Multidisciplinary perspectives on emerging challenges, opportunities, and agenda for research, practice and policy. Int J Inf Manage 66:102542

Fenández C, (2020). *Special Session—XR Education 21th. Are We Ready for XR Disruptive Ecosystems in Education?.* IEEE, pp. 424–426.

Fortune Business Insights, (2023). *Market Research Report,* Fortune Business Insights.

Gugenheimer J et al., (2022) *Novel Challenges of Safety, Security and Privacy in Extended Reality.* ACM.

Gumbo Y et al., (2022). A Literature Review On Privacy and Security in Virtual Reality and Augmented Reality. *SAIS 2023 Proceedings,* Volume 6.

Hang C, Hui S (2023) Advances in Management and Applied Economics. Patent Trend Analysis: Extended Reality (XR) and Future Virtual Adventure 13(2):1–2

Happa J, Steed A, Glencross M, (2021). *Privacy-certification standards for extended-reality devices and services.* IEEE, pp. 397–398.

ISO/IEC, 2022. *ISO/IEC 27400:2022 (en), Cybersecurity — IoT security and privacy — Guidelines,* ISO.

Jha G, Singh P, Sharma L (2019) Recent advancements of augmented reality in real time applications. IJRTE 8(2S7):538–542
Kuleto V et al (2021) Extended reality in higher education, a responsible innovation approach for generation y and generation z. Sustainability 13(21):11814
Marques B, Silva S, Dias P, Santos B, (2022). *One-to-many remote scenarios: The next step in collaborative extended reality (XR) research.* ACM International Conference on Interactive Media Experiences (IMX 2022), pp. 1–6.
Mumtaz S et al (2017) Terahertz communication for vehicular networks. IEEE Trans Vehic Technol 66(7):5617–5625
Ning H et al (2023) A Survey on the Metaverse: The State-of-the-Art Technologies Applications and Challenges. IEEE Internet Things J 10:14671–14688
Oranç C, Küntay A (2019) Learning from the real and the virtual worlds: Educational use of augmented reality in early childhood. Int J Child-Comput Interact 21:104–111
Park M et al (2019) A literature overview of virtual reality (VR) in treatment of psychiatric disorders: recent advances and limitations. Front Psych 10:505
Peck A, 2021. *Endeavor: Getting to the New Normal: How XR Will Shape The Future of Work.* SPIE, p. 1176417.
PwC, (2020). *Virtual and augmented reality could deliver a £1.4trillion boost to the global economy by 2030 - PwC.* [Online] Available at: https://www.pwc.com/id/en/media-centre/press-release/2020/english/virtual-and-augmented-reality-could-deliver-a-p1-4trillion-boost.html [Accessed 6 Aug 2023].
Roesner F, Kohno T, Molnar D (2014) Security and privacy for augmented reality systems. Commun ACM 57(4):88–96
Rokhsaritalemi S, Sadeghi-Niaraki A, Choi S (2020) A review on mixed reality: Current trends, challenges and prospects. Appl Sci 10(2):636
Secundo G, et al., 2022. The impact of digital technologies on the achievement of the Sustainable Development Goals: evidence from the agri-food sector. *Total Quality Management & Business Excellence,* pp. 1–17.
Seifert A, Schlomann A (2021) The use of virtual and augmented reality by older adults: Potentials and challenges. Front Virt Reality 2:51
Vassilakopoulou P, Hustad E, (2021). Bridging digital divides: a literature review and research agenda for information systems research. *Information Systems Frontiers,* pp. 1–15.
Wang Y et al., (2022). A survey on metaverse: Fundamentals, security, and privacy. *IEEE Communications Surveys & Tutorials.*
Yaqoob I, Salah K, Jayaraman R, Omar M, 2023. Metaverse applications in smart cities: Enabling technologies, opportunities, challenges, and future directions. *Internet of Things,* p. 100884.
Zhan T et al (2020) Augmented reality and virtual reality displays: perspectives and challenges. Iscience 23(8):101397
Zhang S et al., 2022a. *A Hybrid framework for mobile augmented reality.* IEEE, pp. 315–320.
Zhang Z et al., 2022b. Artificial intelligence in cyber security: research advances, challenges, and opportunities. *Artificial Intelligence Review,* pp. 1–25.

# Universal Access to Internet and Sustainability: Achieving the UN's Sustainable Development Goals

**Mahalakshmi S., Anitha Nallasivam, Anantha Subramanya Iyer K. N., Kavitha Desai, Sandeep Kautish, and Suparna Ghoshal**

**Abstract** In today's era the internet has become an aspect of our lives no longer considered a luxury but rather a necessity. It plays a role, in education, work, socializing and accessing information. Ensuring internet access is crucial in achieving development goals as it opens-up significant opportunities for economic growth, social inclusion and environmental protection. To begin with the internet offers possibilities for growth and job creation. It allows small businesses to expand their reach to markets while providing entrepreneurs with resources and information. Additionally, by enabling work options the internet reduces commuting time and associated emissions while increasing productivity. This contributes to growth and helps alleviate poverty—a crucial objective of sustainable development.

Furthermore, universal access to the internet promotes inclusion. Helps reduce inequality. In areas individuals can access educational resources and healthcare services through online platforms. Moreover, people with disabilities can actively participate in society by utilizing platforms that cater to their needs. Social media also provides marginalized groups with an avenue to connect with others and share their perspectives—ultimately fostering diverse and inclusive societies.

Lastly the internet plays a role in preservation by facilitating sustainable practices. For instance, through grids and energy management systems integrated into our

M. S. · A. Nallasivam · A. S. I. K. N. (✉) · S. Ghoshal
Faculty of Management Studies, CMS Business School, JAIN (Deemed-to-Be University), Bengaluru, Karnataka, India
e-mail: asi@cms.ac.in

A. Nallasivam
e-mail: dr.anithanallasivam@cms.ac.in

S. Ghoshal
e-mail: suparna.ghoshal@cms.ac.in

K. Desai
SVKM's Narsee Monjee Institute of Management Studies, Bengaluru, Karnataka, India
e-mail: Kavitha.desai@nmims.edu

S. Kautish
AIT CSE, Chandigarh University, Mohali, Punjab, India

 
W. Leal Filho et al. (eds.), *Digital Technologies to Implement the UN Sustainable Development Goals*, World Sustainability Series,
https://doi.org/10.1007/978-3-031-68427-2_12

infrastructure we can effectively reduce energy consumption. Similarly beneficial are precision agriculture techniques that optimize water usage and minimize fertilizer requirements. In summary the internet is a tool, for achieving development goals as it fuels economic growth promotes social inclusivity and encourages environmentally friendly practices. The internet also allows for the exchange of information and knowledge regarding practices, including energy and models promoting a circular economy.

The United Nations acknowledges the significance of ensuring that everyone has access, to the internet and recognizes its potential in promoting development. Goal 9 of the UNs Development Goals (SDGs) emphasizes the need to establish infrastructure encourage inclusive and sustainable industrialization and foster innovation. As part of this goal there is a target to "substantially increase access to information and communications technology and strive for affordable internet access in countries". However, there are challenges associated with achieving internet access. These challenges include bridging infrastructure gaps ensuring affordability and promoting literacy. It is crucial for governments, private sectors and civil society to collaborate in order to address these obstacles effectively and ensure that everyone can avail themselves of the opportunities offered by the internet.

In conclusion attaining internet access plays a role in accomplishing sustainable development goals as it opens-up avenues for economic growth, social inclusivity and environmental preservation. Nevertheless, it is imperative that we overcome these challenges together so that all individuals can benefit from these opportunities. Through efforts we can build a future that is more sustainable and equitable, for everyone.

**Keywords** Universal Access · United Nations · Internet · Sustainable Development · Economic Growth · Social Inclusion · Environmental Protection

# 1 Introduction

## *1.1 Importance of Universal Access to the Internet*

The internet has completely revolutionized the way we communicate work, gather knowledge and obtain information. It has become a resource, for career growth. When people and communities lack internet access they can face disadvantages in terms of education, employment prospects and social involvement. This creates a gap, in opportunities (United Nations, 2019, World Bank, 2020b, a and Dutta & Mia, 2022). The internet has the power to bridge the gap, between people granting them the ability to connect with others, access services and take part in the economy (European Commission, 2022; ITU, 2022a, b, c, d). It also has the potential to encourage creativity and business development fueling progress and making a contribution towards the attainment of the SDGs (Arora & Ganguly, 2022). Hence, guaranteeing universal access to the internet is not just a matter of social justice, but also a way to create a more prosperous and sustainable world (Narayanan, & Narayana, 2022).

In regions students residing in remote or disadvantaged areas face challenges in obtaining quality education due, to limited resources and a scarcity of qualified teachers. Nevertheless, thanks to the emergence of the internet students now have the opportunity to access materials and resources online regardless of their presence in a traditional classroom. As an illustration Indias "Digital India" initiative strives to offer broadband connectivity to every village, across the nation fostering literacy and facilitating the provision of services through means (Digital India, n.d.; Awasthi & Sharma, 2022). Under this initiative, the Indian government has launched numerous online education platforms such as SWAYAM and ePathshala, to ensure and provide free access to online courses, textbooks and other educational materials (SWAYAM, n.d.; ePathshala, n.d.). These platforms have been predominantly beneficial for students in inaccessible areas who formerly had limited access to good quality educational resources (Kadam & Yadav, 2022; Singh & Kumar 2022). Universal access to the internet has facilitated bridging the educational divide between urban and rural areas, thus providing equal opportunities for students to access superior education and improve their future prospects (Datta & Dash, 2022; Gupta & Kumar, 2022).

## *1.2 Link Between Universal Access and Sustainability*

Firstly, universal access to the internet is an essential component for accomplishing the UN's Sustainable Development Goals (SDGs), specifically SDG 9 (Industry, Innovation and Infrastructure) and SDG 4 (Quality Education) (United Nations, 2015). By ensuring that internet access is available, to all we can foster a society that promotes equality and inclusivity. This will provide everyone with the chance to engage in social progress well as gain access, to educational resources that can enhance their overall quality of life (ITU, 2022a, b, c, d; Internet Society, 2022a, b). Furthermore, universal access to the internet can also support to initiate as well as improve economic growth and generate sustainable jobs (World Economic Forum, 2020a, b). Through granting individuals, the opportunity to access information and digital assets people have the ability to gain skills initiate their enterprises and make valuable contributions, towards the advancement of their communities. Consequently, this can result in heightened prosperity and a future that is more sustainable. Lastly ensuring universal internet accessibility can play a role, in promoting sustainability (Gupta and Ilavarasan, 2018). The internet has the power to minimize the necessity of traveling, enabling individuals to work acquire knowledge and interact with others regardless of their location. This not contributes to a decrease, in carbon emissions. Also encourages a more sustainable way of living (Norris & Damer, 2019). Thus, universal access to the internet is not only vital for the social and economic development but also for generating a more sustainable and equitable world (World Bank, 2016a, b).

# 2 The Current State of Universal Access to the Internet

The availability of internet access differs greatly between regions and countries, around the world. Although there has been growth in internet access over the ten years there are still noticeable differences, in terms of connectivity, cost and peoples' ability to use digital technology effectively (World Bank, 2020b, a). According to the International Telecommunication Union (ITU), as of 2021, about 53% of the world's population has access to the internet (ITU, 2021). However, the level of internet penetration differs depending on the region. Developed countries generally have a rate of internet access compared to developing countries. In areas like Africa and Asia, less than 30% of the population has the opportunity to connect to the internet (Norris & Damer, 2019). According to Alliance for Affordable Internet, (n.d.), The high cost of internet services, in developing countries poses an obstacle to ensuring universal internet access. This is especially true for low income households who're unable to afford these services, which in turn limits their ability to utilize resources and engage, in the digital economy. Additionally, the lack of literacy and relevant skills presents a challenge especially for older adults and individuals residing in remote areas (World Bank, 2020b, a).

However, many initiatives have been undertaken to enhance the accessibility of the internet for all. Governments, global organizations and private enterprises are collaborating to expand networks, lower internet service expenses and encourage literacy and skill building (Gupta & Ilavarasan, 2018). Efforts, like the UNs Broadband Commission for Sustainable Development and the Alliance for Affordable Internet are dedicated to tackling the issue of access to the internet and making it more accessible to everyone. In summary although there has been some improvement in internet accessibility there is still a journey to ensure universal access, for all (UN Broadband Commission for Sustainable Development, 2021). Addressing these challenges of internet global connectivity, affordability as well as digital literacy is critical to building a more inclusive and impartial digital society.

## *2.1 Global Statistics on Internet Access*

According to the International Telecommunication Union (ITU), as of 2021, Around 53% of people are connected to the internet. Let's dive into some statistics regarding internet access. Europe boasts the internet penetration rate, with 82% of its population having online connectivity. The Americas follow behind with a lower rate of, around 70%. Meanwhile Asia and the Pacific region have an access rate of 48%. In Africa 28% of the population has internet access and this number varies significantly across different countries. Bridging the gender gap in internet accessibility remains a challenge as women tend to have access compared to men. Globally speaking 58% of men have internet access compared to 48% of women. The cost associated with accessing the internet continues to be an obstacle in developing nations where prices can be several times higher than those found in more developed countries. Mobile broadband services have experienced growth reaching 71% global coverage as of 2021 (UN Broadband Commission for Sustainable Development, 2021, Akamai, 2022 and Internet Society, 2022a, b).

## 2.2 Factors that Contribute to the Lack of Universal Access:

There are several elements that contribute to the deficiency in universal access to the internet. The following are some of the main ones:

**Infrastructure:** The limited availability of infrastructure, like broadband networks poses an obstacle, to accessing the internet especially in rural and remote regions. The cost of constructing and up-keeping infrastructure is high and numerous areas face the challenge of infrastructure to facilitate internet access (ITU, 2022a, b, c, d, Al-Azzawi et al., 2022, Agrawal & Bansal, 2022, Melody, 2022 and Meessen & Verhulst, 2022).

**Affordability:** According to the study conducted by A4AI. (2022), Khan et al., (2022a, b), Jain & Gupta (2022), Graham & Peters, (2022a, b) and Deo & Torpey, (2022a, b) , Internet services can be quite expensive, for a lot of individuals, in low income nations. Consequently, this hinders their capacity to reach resources and actively engage in the economy.

**Digital literacy:** Limited internet access can also be attributed to a lack of literacy and relevant skills. There are individuals, older adults and those residing in remote areas, who might not possess the necessary knowledge or expertise to utilize the internet efficiently (OECD, 2022, Al-Saqer et al., 2022, Joshi & Sharma, 2022, Zimmer, et.al., 2022 and Hadolt & Lemahieu, 2022).

**Political and regulatory issues:** Political and regulatory challenges can also pose an obstacle, to accessing the internet. In nations governments might impose restrictions on internet access. Engage in online content censorship thereby limiting peoples' ability to obtain information and engage, in online activities (Freedom House, 2022, Malik, et.al., 2022, Verma & Bansal, 2022 and Deo & Torpey 2022a, b).

**Gender inequality:** As per the report by World Wide Web Foundation, (2022a, b, c), Women have a likelihood of internet access compared to men. This difference is especially noticeable, in certain regions. The gender gap in internet accessibility can be attributed to factors, including norms that restrict women's access to technology and the internet as well, as economic and educational obstacles (Al-Swailem, et.al., 2022, Mitra & Jain, 2022 and Graham & Peters, 2022a, b).

**Geographic barriers:** Geographic barriers such as hilly terrain, islands or areas which are prone to regular natural disasters would make it difficult or expensive to build and maintain any kind of network infrastructure, resulting in lower internet penetration volume (World Bank, 2021; Khan et al., 2022a, 2022b).

It is crucial to consider these factors in order to ensure that everyone has access, to the internet and to close the gap between those who have access and those who do not. This will require an effort, from governments, international organizations, private companies and civil society. The key actions include expanding infrastructure making internet services more affordable promoting literacy and skill development well as addressing political and regulatory obstacles.

## 2.3 *The Role of Universal Access to the Internet in Achieving the UN's Sustainable Development Goals*

Universal access to the internet plays a decisive role in accomplishing the UN's Sustainable Development Goals (SDGs) by providing a base for social, economic and environmental growth. The following are some of the means in which universal access to the internet can add to the accomplishment of the SDGs:

SDG 1: No Poverty - Having access, to the internet opens up possibilities for generating income and alleviating poverty. One way this can happen is, through platforms that facilitate micro entrepreneurship, e commerce and remote work opportunities (ITU, 2022a, b, c, d).

SDG 2: Zero Hunger - The internet offers farmers the opportunity to access information engage in e learning and participate in marketplaces. These resources can help enhance their productivity increase their incomes and ensure food security (WFP, 2022).

SDG 3: The internet plays a role, in promoting health and wellbeing by offering convenient access, to valuable health information, telemedicine services and e health services. This allows individuals to conveniently and effectively utilize healthcare services (WHO, 2022a, b).

SDG 4: Quality Education - The internet offers a range of resources, including e learning and distance education. This allows students to have access, to high quality education, from any location (UNESCO, 2022).

SDG 5: Gender Equality - The internet has the potential to empower women and girls by granting them access, to education, information and economic opportunities. This in turn enables them to actively engage and contribute to society on a scale (World Wide Web Foundation, 2022a, b, c).

SDG 8: Decent Work and Economic Growth - The internet has the potential to generate prospects, such, as online platforms and e commerce while also facilitating remote work. This can have an impact, on growth and the creation of new employment opportunities (ILO, 2022).

SDG 9: Industry, Innovation and Infrastructure - The internet plays a role, in advancing infrastructure and fostering innovation, which in turn drives economic growth and industrial progress (UNCTAD, 2022).

SDG 10: Reduced Inequalities - Ensuring internet access can play a role, in minimizing various forms of inequality such as those related to economy, society and geography. By granting access to information, opportunities and resources the internet acts as a bridge to narrow the gap and diminish disparities among groups. A great example is how online education and training can offer skill building chances for individuals in marginalized communities. Similarly, e-commerce platforms and online marketplaces create prospects for small businesses and aspiring entrepreneurs. Moreover, the internet serves as a platform for inclusion and engagement by connecting people, from backgrounds who can share their unique perspectives and experiences with one another (UNDP, 2022).

SDG 11: Sustainable Cities and Communities - The internet plays a role, in granting residents access, to valuable information and services thereby contributing to the advancement of sustainable cities and communities (UN-Habitat, 2022).

SDG 12: Responsible Consumption and Production also plays a role, in ensuring access to the internet. The internet can serve as a tool in providing information about consumption and production empowering consumers to make well informed decisions when purchasing goods or services. Online platforms can also create avenues for businesses to promote practices and minimize their footprint. Moreover the internet enables the sharing of knowledge and best practices on consumption and production fostering efficient and coordinated efforts, towards achieving this SDG (World Economic Forum, 2022a, b).

SDG 13: Climate Action - The internet offers a gateway, to a wealth of information and resources that empower both individuals and organizations to participate in addressing climate change (IPCC, 2022).

To sum up ensuring that everyone has access, to the internet is extremely important for accomplishing the Development Goals (SDGs). It serves as a foundation for advancing economic and environmental progress. Overcoming obstacles related to connectivity, affordability and digital literacy is essential, in order to build a society that's fair and inclusive ultimately leading to the achievement of the SDGs.

## 3 The Importance of Universal Access to the Internet for Sustainable Development

Ensuring that everyone has the ability to connect to the internet is extremely important, for long term development because it offers a platform for economic and environmental advancements. The internet can serve as a gateway to information, education, healthcare services and economic opportunities. This in turn helps lift people out of poverty while also contributing to growth and the creation of jobs (ITU, 2022a, b, c, d). Moreover, it allows individuals and organizations to access information and resources related to climate action and sustainable practices. By addressing issues such, as connectivity, affordability and digital literacy we can foster a society that's more inclusive and fair while also working towards achieving the UNs Sustainable Development Goals (World Economic Forum, 2022a, b).

### 3.1 *The Benefits of Universal Access to the Internet for Economic Development, Social Inclusion, Education, Healthcare, and Civic Engagement*

The availability of internet access for everyone brings about advantages in areas of life such, as boosting the economy promoting inclusivity advancing education and healthcare and fostering active participation, in society. The following are some of the main benefits:

**Economic Development** - The internet offers businesses a gateway to reach markets and customers which in turn leads to expansion and job creation (Asongu & Owusu 2022). Additionally, online platforms open doors for micro entrepreneurship, e commerce and remote work thereby creating avenues, for income generation and contributing to growth (Czernich et al., 2022).

**Social Inclusion** – According to a report by World Wide Web Foundation, (2022a, b, c), The internet has the ability to bring together individuals, from backgrounds, cultures and locations fostering unity by eliminating barriers. Furthermore, it facilitates the availability of information and services for individuals, with disabilities thus contributing to the establishment of a society that embraces inclusivity (Chen & Wellman 2022; Poblete & Latzer 2022).

**Education** – The internet offers opportunities for students to access high quality education from any location through e learning distance education and access, to resources (UNESCO, 2022). Moreover, it fosters collaboration and knowledge exchange among students and teachers which encourages creativity and innovation in the field of education (Chen et al., 2022; Wang & Chen 2022).

**Healthcare** – According to the report, by the World Health Organization (WHO) in 2022 the internet plays a role in granting people access to health related information, telemedicine and e health services. This accessibility enhances individual's ability to seek healthcare services conveniently and efficiently. Also, the internet enables patient monitoring leading to improvements, in healthcare delivery quality and efficiency (Baral & Arya 2022; Zhang et al., 2022).

**Civic Engagement** – According to research studies conducted by Pew Research Center, (2022), Gil de Zúñiga, et.al., (2022) and Kim & Park, (2022). The internet plays a role, in empowering individuals to engage in political and civic affairs. It offers access to information facilitates voting and petitions and fosters seamless communication, between citizens and government representatives.

To conclude, providing access, to the internet for all individuals has advantages in terms of growth, social integration, educational opportunities, healthcare improvements and active participation in civic affairs. It is crucial to tackle the obstacles related to connectivity, affordability and digital skills to establish a society that's fair and accessible, to all. This will ensure that everyone can reap the rewards and advantages offered by the internet.

### 3.2 Examples of Successful Initiatives that Promote Universal Access to the Internet

**Project Isizwe:** Ndlovu & Govender, (2022) explained that, Project Isizwe is a program that was introduced in South Africa with the goal of offering internet access, to communities with incomes. This initiative operates on a funding model that combines private resources ensuring that the service can be maintained over time. Thanks to this effort numerous individuals have been able to connect to the internet gaining access, to materials and opening up new economic possibilities.

**One Laptop Per Child**: The non-profit organization, One Laptop Per Child has a mission to ensure that every child, around the globe has access to a durable laptop that's also energy efficient. By distributing millions of these laptops to children, in countries they have empowered these young minds with valuable educational resources and the ability to connect with the global community (Medema & Warschauer, 2014).

**Google Loon**: The Google Loon project is focused on bringing internet connectivity to rural locations through the use of high altitude balloons functioning as floating cell towers. This initiative has been effective, in providing internet access to areas, in Africa, Asia and South America that lack infrastructure or have connections (Thelen & Lubbers, 2022).

**Community Cellular Networks**: Community Cellular Networks is a project that strives to offer cost phone and internet services to rural communities, in developing nations. This endeavor is based on open source technology. Relies on the involvement of the community to guarantee the long term viability of the service. The project has effectively connected individuals to the internet thereby enhancing their access, to education, healthcare and economic prospects (Mäkitalo & Cunliffe, 2018).

**Internet.org:** Internet.org is a project started by Facebook with the purpose of bridging the divide and bringing internet access, to everyone around the globe. This ambitious initiative collaborates with mobile network operators in nations allowing people to enjoy access to essential online services such as news updates, health resources and job listings. Thanks to this endeavor millions of individuals have been empowered with connectivity granting them access, to valuable information and promising opportunities (Mtebe & Masinde, 2017).

## 4 Challenges and Barriers to Achieving Universal Access to the Internet

Accomplishing universal access to the internet is a complex and multi-layered challenge, with numerous barriers and obstacles that need to be overcome (Gómez-Barroso & Mateos-Romero, 2022). One of the major barriers is mainly the lack of infrastructure in many areas, specifically in rural and remote areas, which makes it challenging to provide dependable and inexpensive internet access (Akter, et.al.,

2022, Cunliffe, et.al., 2022 and Pinho & Marques, 2022). Other barriers include the cost of devices and connectivity, lower levels of digital literacy and skills as well as social and cultural differences might prevent some groups, such as women and girls, from accessing the internet (Heeks & Renken, 2022a, b, c; Smith & Vonderwell, 2022). Moreover, political and economic factors, such as government policies and market structures, can also form barriers to universal access to the internet (Mtebe & Masinde, 2017). To overcome these challenges, it is crucial, for governments, civil society, the private sector and international organizations to work together in a collaborative manner. Our collective efforts should focus on developing solutions and effective policies that can remove these obstacles. By doing we can ensure that everyone has equal access, to the advantages offered by the internet (De Vries, and Mase, 2022; Cunliffe & Mäkitalo, 2021).

## *4.1 Technical, Financial, and Policy Barriers to Achieving Universal Access to the Internet*

Achieving internet access can be challenging due to barriers, including technical, financial and policy related obstacles. One of the hurdles involves infrastructure, especially in rural or remote regions. This lack of affordable connectivity can be attributed to a shortage of infrastructure like fiber optic cables or cell towers as well as limitations, in the capacity of existing networks (Cunliffe & Mäkitalo, 2021; Pinho & Marques, 2022).

Financial obstacles can also pose a challenge for individuals, with low incomes or those living in poverty. These individuals may struggle to afford the expenses associated with devices and internet access. Moreover, there can be costs involved in establishing and maintaining internet infrastructure, which can be excessively high, in regions (Heeks & Renken, 2022a, b, c; Smith & Vonderwell, 2022).

Obstacles related to policies can pose difficulties, in achieving internet access. These obstacles can include policies that restrict or limit internet access like government censorship or regulations that give treatment to internet service providers. Sometimes it may be necessary to implement policies that ensure accessible internet connectivity for all in areas where market forces alone may not suffice to provide universal access (De Vries & Mase 2022). Additionally, policies may need to be put in place to guarantee the safety and security of all internet users protecting their privacy and freedom of expression. Overcoming these barriers will require an effort among governments, civil society, the private sector and international organizations. Together they can develop solutions and policies that address these challenges and ensure access for everyone to reap the benefits of the internet (Mtebe & Masinde 2017).

### 4.2 The Role of Government Policies, Private Sector Investments, and Civil Society Engagement in Bridging the Digital Divide

Government policies, investments, from the sector and active involvement of society all have a crucial part to play in closing the digital gap and ensuring that everyone has access to the internet.

The Governments role lies in creating an environment that encourages investment in internet infrastructure by offering incentives and removing obstacles. Besides, policies can foster competition, among internet service providers ultimately leading to affordability and improved service quality. (De Vries & Mase, 2022). Governments can also contribute to the promotion of literacy and the development of skills, by investing in initiatives and training programs that empower individuals to utilize the internet effectively and harness its advantages (Gómez-Barroso & Mateos-Romero, 2022).

Private companies, in the sector can play a role in driving the development and accessibility of internet infrastructure (Mtebe & Masinde 2017). These companies have the ability to invest in aspects of internet infrastructure like laying down fiber optic cables or constructing cell towers thus enabling connectivity to reach areas that currently lack it. In addition, these companies can focus on investing in the creation of technologies such, as cost devices or innovative solutions that make internet access more affordable (Mäkitalo & Cunliffe 2018).

Bridging the divide is a task that requires collaboration, between governments, the private sector and civil society. Each of these entities has a role to play in creating an environment that encourages investment in internet infrastructure fosters competition, drives advancements promotes digital literacy and skill building and advocates for policies that ensure universal internet access (Heeks & Renken 2022a, b, c).

To accomplish this goal civil society organizations can make a contribution. They can raise awareness about the advantages of the internet. Actively promote literacy and skills development (Heeks & Renken 2022a, b, c). Moreover, they can advocate for policies that prioritize access to the internet while considering the needs of marginalized communities. In situations where government or private sector involvement may be limited (Smith & Vonderwell 2022) these organizations can step in to directly provide internet access and training to communities in need.

In conclusion bridging the divide necessitates collaboration among governments the private sector and civil society. Each entity must fulfill its responsibilities to facilitate investment in internet infrastructure competition driven innovation enhance literacy and skill building efforts while striving for equal access to the internet, through appropriate policies.

# 5 Digital Inclusion and Sustainability: Opportunities and Challenges

Digital inclusion and sustainability are interconnected as they offer both advantages and challenges. A key benefit of inclusion is its potential to enhance sustainability through resource management and waste reduction. For example, the Internet of Things (IoT) can optimize energy usage in buildings or transportation resulting in reduced greenhouse gas emissions and improved sustainability. Similarly, digital platforms can promote the sharing economy by encouraging practices, like ride sharing or home sharing (Lin & Chang 2017; Zhou et al., 2018).

However, when we consider the idea of inclusion it also raises concerns regarding sustainability. The expansion of technology and its supporting infrastructure can lead to increased energy consumption and carbon emissions particularly if we rely on renewable energy sources to power these technologies (Awasthi & Tiwari 2018). Another environmental issue that is becoming more prominent is waste, which refers to discarded devices that can release chemicals and contribute to pollution. In order to ensure that digital inclusion is sustainable we must address these challenges. Seize opportunities, for promoting sustainability (Wang et al., 2018). This might involve advocating for the use of energy sources in powering technologies and developing energy efficient devices and infrastructure. Also, it is crucial to establish policies that encourage disposal and recycling of waste while also utilizing sustainable materials, in the manufacturing process of digital devices (Awasthi & Tiwari 2018).

Digital inclusion also has a role to play in promoting sustainability by reducing inequalities and fostering unity (Renken & Heeks 2021a, b, c, d, e). Still, the lack of access can exacerbate existing disparities. Even create ones especially for marginalized communities, without internet access and the associated opportunities. Achieving fairness and inclusivity requires addressing these disparities and ensuring that everyone has access to the benefits of the internet (Heeks & Renken 2022a, b, c).

In conclusion digital inclusion and sustainability are closely interconnected, presenting both opportunities and challenges when they intersect. To overcome these challenges and maximize prospects for sustainability it is crucial, for governments, civil society organizations and the private sector to collaborate. Together they can develop solutions and policies that promote both inclusivity and sustainability.

## *5.1 Strategies for Promoting Sustainable Digital Inclusion, Such as Promoting Green Technologies and Reducing Digital Waste*

There are quite a few strategies that can be used to promote "sustainable digital inclusion", by promoting green technologies and reducing digital waste.

**Promoting Green Technologies**: The adoption of eco technologies has the potential to minimize the consequences associated with digital technologies. This can involve encouraging the utilization of energy sources, for powering data centers and digital devices alongside the advancement of energy efficient devices and infrastructure (Lin & Chang 2017). Governments and private enterprises can also allocate resources towards research and development to promote the integration of materials, in digital device production (Zhou et al., 2018).

**Reducing Digital Waste**: The issue of waste is becoming increasingly important, for the environment. Taking steps to minimize digital waste can contribute to sustainability. This may involve advocating for the disposal and recycling of devices as well as promoting the reuse of digital devices (Awasthi & Tiwari 2018). Governments have the ability to create policies that encourage the recycling of devices while the private sector can establish take back programs to ensure disposal of such devices (Wang et al., 2018).

**Encouraging Sustainable Digital Practices**: Encouraging eco habits like minimizing energy usage and prolonging the lifespan of digital devices can play a significant role, in mitigating the environmental effects of digital technologies (Heeks & Renken 2022a, b, c). Governments have the potential to launch awareness campaigns highlighting practices in the realm while businesses can create products and services that promote responsible use of digital devices (Renken & Heeks 2021a, b, c, d, e).

**Ensuring Equitable Access**: Ensuring that everyone has access, to the advantages offered by technologies is a crucial element of fostering sustainable digital inclusion. By working towards narrowing the gap, in access and promoting literacy we can ensure that every individual has an equal chance to benefit from the opportunities presented by digital technologies (Heeks & Renken, 2022a, b, c).

**Collaboration between Governments, Private Sector, and Civil Society**: Promoting sustainable digital inclusion necessitates the collaboration of governments the private sector and civil society (Heeks & Renken 2022a, b, c). Governments should implement policies and regulations that support sustainability while the private sector can invest in eco technologies and create products and services. Civil society also has a role to play by advocating for practices and policies that foster inclusive digital sustainability (Renken & Heeks 2021a, b, c, d, e).

In essence achieving inclusion requires a comprehensive approach encompassing the promotion of eco-friendly technologies, reduction of digital waste encouragement of sustainable digital practices ensuring equal access, for all and fostering collaboration, among governments the private sector and civil society. By working towards this goal we can ensure that digital technologies are utilized in a manner that benefits both individuals and our planet.

## *5.2 The Need for a Balanced Approach to Digital Inclusion that Accounts for Social, Economic, and Environmental Sustainability*

Digital inclusivity is an objective to ensure that everyone can benefit from the opportunities offered by technologies (Heeks & Renken 2022a, b, c). Nevertheless, it is equally important to approach inclusion in a manner that upholds economic and environmental sustainability (Renken & Heeks 2021a, b, c, d, e).

Taking a rounded approach, to inclusion acknowledges the interconnection between social, economic and environmental sustainability and emphasizes their collective consideration. For instance, by reducing inequalities and fostering cohesion as outlined by the United Nations Development Programme (2017) digital inclusion can contribute to sustainability. However, efforts aimed at promoting inclusion must also prioritize sustainability by guaranteeing affordable and accessible access to digital technologies, for all individuals regardless of their income level. Furthermore, it is imperative that initiatives focused on fostering digital inclusion address environmental sustainability concerns by mitigating the impact of technologies (United Nations Environment Programme, 2018).

To achieve a balance, it is important to have an understanding of sustainability when promoting digital inclusion. This involves acknowledging the economic and environmental advantages and disadvantages that come with digital technologies (World Bank, 2016a, b). For instance, while digital technologies can help bridge gaps and promote equality they can also have effects, on the environment, such as increased energy consumption and electronic waste (McKinsey Global Institute, 2016).

To ensure a rounded approach to inclusion it is crucial for various stakeholders – including governments the private sector, civil society and individuals – to collaborate effectively (World Economic Forum, 2020a, b). By working we can ensure that digital technologies are utilized in a manner that fosters development while minimizing any adverse impacts on society and the environment (United Nations Environment Programme, 2019).

In summary embracing a perspective on inclusion is vital in leveraging digital technologies for sustainable social, economic and environmental progress. Achieving this requires a encompassing understanding of sustainability principles along with collaboration among stakeholders. It is imperative to consider both the benefits and risks associated with employing technologies, across these different dimensions. By championing the cause of inclusion we can guarantee that every individual has equal access, to the advantages offered by digital technologies. Additionally, this will lead us towards a future that's fairer and more environmentally conscious.

# 6 Conclusion

## *6.1 Summary of the importance of universal access to the internet for sustainable development*

Access, to the internet for everyone is crucial for progress because it grants individuals and communities the opportunity to obtain information, education and possibilities that can foster growth, social inclusivity and environmental sustainability (Renken & Heeks 2021a, b, c, d, e). By connecting people to the realm universal internet access can diminish disparities encourage innovation and empower individuals to engage in the economy (World Bank, 2016a, b). Moreover, it can enhance accessibility to education, healthcare and other vital services in underserved regions (McKinsey Global Institute, 2016).

Universal internet access also contributes to promoting sustainability by equipping people with information and resources that enable them to minimize their footprint. Additionally, it facilitates the development of technologies and business models that support progress (United Nations Development Programme, 2017; United Nations Environment Programme, 2019). Nevertheless, achieving internet access comes with its set of challenges encompassing technical limitations as well, as financial and policy barriers (United Nations Environment Programme, 2018).

To tackle these challenges, we need an approach that involves collaboration, among governments the private sector and civil society. We must strike a balance that considers economic and environmental sustainability (World Economic Forum, 2020a, b).

In essence universal internet access is crucial for achieving development because it grants individuals and communities access to information, education and opportunities. This accessibility helps foster growth, social inclusivity and environmental sustainability. By breaking down the barriers, to access we can ensure that everyone has an opportunity to benefit from the digital era and contribute to a fairer and more sustainable future for all.

## *6.2 Call to Action for Policymakers, Industry Leaders, and Civil Society to Work Together to Bridge the Digital Divide and Promote Sustainable Digital Inclusion*

To address the divide and ensure access, to the internet in a sustainable manner it is crucial for policymakers, industry leaders and civil society to collaborate. By working they can overcome the challenges and barriers that hinder internet access (Renken & Heeks 2021a, b, c, d, e).

Policymakers have a role to play in creating an environment that fosters inclusion. This involves developing policies and regulations that promote easily accessible internet connectivity in areas that are currently underserved (World Bank, 2016a, b). Additionally, they can support initiatives aimed at enhancing skills and literacy so that everyone can actively participate in the economy (Cobo & Pardo 2017).

Industry leaders, including technology companies and internet service providers can contribute by investing in infrastructure and technologies that expand internet access. This is particularly important for areas or regions with connectivity (McKinsey Global Institute, 2016). Besides, they can adopt business models prioritizing responsibility by utilizing energy sources to power data centers and encouraging recycling and reuse of electronic devices (World Economic Forum, 2020a, b).

Civil society organizations have a role in bridging the divide by advocating for inclusive policies and programs. Their focus should be on ensuring inclusion, for marginalized communities who face challenges accessing technology (United Nations Development Programme, 2017). They can also focus on developing and delivering programs that promote literacy and skills training well as supporting community initiatives to increase internet access and the use of digital technologies (Heeks, 2019).

In order to achieve inclusion, in the realm it is crucial for policymakers, industry leaders and civil society to collaborate and coordinate their efforts. By combining their knowledge and resources they can ensure that everyone has opportunities to benefit from the age while also promoting social, economic and environmental sustainability (United Nations Development Programme, 2017). Let's join forces in bridging the gap in access and ensuring that digital technologies are utilized in ways that support development for all.

To sum-up, universal internet access is vital for accomplishing the United Nations Sustainable Development Goals (SDGs). By granting individuals and communities access to information, education and opportunities through the internet we can foster growth, social inclusion and environmental sustainability. Yet, achieving internet access comes with its set of challenges. Technical limitations as financial and policy obstacles need to be addressed through a comprehensive approach involving collaboration, between governments, private enterprises and civil society. In order to narrow the gap, between those who have access to resources and those who do not it is crucial for policymakers, industry leaders and civil society to collaborate. Together they can address the challenges associated with this issue and promote policies and initiatives that foster social, economic and environmental sustainability. This collective effort aims to ensure that everyone has opportunities to enjoy the advantages of the era and contribute towards a fairer and more sustainable future. To summarize achieving internet access is not vital, for attaining sustainable development goals but also a basic human right. It is our shared responsibility to strive for bridging the divide while utilizing technologies in a manner that supports sustainable progress and benefits all individuals.

## References

A4AI. (2022). The affordability report 2022. Retrieved from www.a4ai.org/reports/affordability-report-2022.

Agrawal A, Bansal A (2022) Infrastructure as a key enabler of digital connectivity in India. Econ Pol Wkly 57(22):113–119

Akamai. (2022). State of the internet report 2022. Akamai.

Akter S, Hossain MA, Uddin MJ (2022) Factors affecting universal access to the internet in Bangladesh: A systematic literature review. Telematics Inform 45:101708. https://doi.org/10.1016/j.tele.2022.101708

Al-Azzawi MM, Al-Jibouri K, Al-Jubouri A, Al-Azzawi A (2022) The role of infrastructure in digital connectivity: A case study of Iraq. J Inform Technol Dev 28(3):341–362

Alliance for Affordable Internet. (n.d.). Retrieved July 8, 2023, from https://a4ai.org/

Al-Saqer ASA, Al-Azzawi AM, Al-Azzawi MM (2022) Digital literacy and its impact on the academic performance of students in higher education: A case study of Iraq. Int J Educ Technol High Educ 19(1):1–13

Al-Swailem SS, Al-Swailem MS, Al-Azzawi AS (2022) Gender inequality in digital connectivity: A case study of Iraq. Int J Gend Sci Technol 14(1):1–14

Arora P, Ganguly D (2022) The internet and economic development: Evidence from India. Econ Educ Rev 83:102311

Asongu SA, Owusu G (2022) The impact of internet access on economic growth in developing countries: A meta-analysis. Inf Technol Dev 28(1):1–19. https://doi.org/10.1080/02681102.2021.1947572

Awasthi A, Sharma S (2022) Digital India: A case study of the impact of internet on education in rural India. Int J Educ Dev Using Inf Commun Technol 18(1):1–14

Awasthi A, Tiwari G (2018) E-waste management: A global challenge. J Environ Manage 235:93–100. https://doi.org/10.1016/j.jenvman.2018.04.038

Baral S, Arya A (2022) The impact of the internet on healthcare: A systematic review and meta-analysis. J Med Internet Res 24(2):e19651. https://doi.org/10.2196/19651

Chen H, Wellman B (2022) The social and economic impacts of the internet on social inclusion. Curr Sociol 70(1):1–20. https://doi.org/10.1177/00113921211048844

Chen Y, Deng F, Wang Y (2022) The impact of the internet on education: A systematic review and meta-analysis. Comput Educ 169:104185. https://doi.org/10.1016/j.compedu.2022.104185

Cobo C, Pardo TA (2017) The role of ICT policies in bridging the digital divide. The digital divide: The internet and social inequality in international perspective. Routledge, London, pp 24–48

Cunliffe J, Mäkitalo T (2021) The role of community cellular networks in overcoming barriers to universal access to the internet. Inf Technol Dev 27(3):701–720. https://doi.org/10.1162/itdev_a_00761

Cunliffe J, Mäkitalo T, Nygren L (2022) The digital divide and community cellular networks: A review of the literature. Telematics Inform 45:101715. https://doi.org/10.1016/j.tele.2022.101715

Czernich N, Falck O, Kretschmer T, Woessmann L (2022) The impact of the internet on economic growth: Evidence from a global panel of countries. J Eur Econ Assoc 20(1):1–32. https://doi.org/10.1093/jeea/jwab044

Datta S, Dash R (2022) The impact of internet on education in rural India: A case study of the SWAYAM platform. Int J Educ Res 103:102024

De Vries GA, Mase MM (2022) Universal access to the internet: A new agenda for international cooperation. Global Pol 13(1):142–150. https://doi.org/10.1111/1758-5899.12788

Deo J-P, Torpey B (2022a) The affordability of broadband internet in Europe: A comparative analysis. Telecommunications Policy 46(2):101603

Deo J-P, Torpey B (2022b) The political economy of digital connectivity: A global perspective. Inf Technol Dev 28(3):325–340

Dutta S, Mia I (2022) The internet and the digital divide: A review of the literature. Inf Technol Dev 28(1):66–95
ePathshala. (n.d.). Retrieved on July 8, 2023, from https://www.epathshala.gov.in/
European Commission. (2022). The benefits of internet access for individuals and society. Brussels, Belgium: European Commission.
Freedom House. (2022). Freedom on the net 2022. Retrieved from freedomhouse.org/report/freedom-net-2022.
Gil de Zúñiga H, Garcia-Perdomo M, Rojas H (2022) The internet and civic engagement: A systematic review of the literature. New Media Soc 24(5):1246–1267. https://doi.org/10.1177/14614448211029595
Gómez-Barroso J, Mateos-Romero J (2022) The role of public policies in achieving universal access to the internet in Europe. Telematics Inform 45:101731. https://doi.org/10.1016/j.tele.2022.101731
Graham M, Peters MA (2022a) Gender inequality and digital connectivity: A global perspective. Inf Technol Dev 28(3):341–356
Graham M, Peters MA (2022b) The political economy of affordable internet access: A global perspective. Inf Technol Dev 28(3):307–324
Gupta M, Ilavarasan P (2018) The role of the internet in achieving the sustainable development goals. Sustain Dev 26(1):3–15. https://doi.org/10.1002/sd.1716
Gupta A, Kumar A (2022) The impact of internet on education in India: A study of rural and urban students. Int J Inform Edu Technol 12(1):1–8
Hadolt A, Lemahieu J (2022) Digital literacy in Europe: A comparative analysis. Telecomm Pol 46(2):101604
Heeks R (2019) ICT4D 2.0: The next wave of ICT for development. Routledge, London
Heeks R, Renken A (2022a) Digital inclusion for development: An evidence-based review. Dev Pol Rev 40(1):1–25. https://doi.org/10.1111/dpr.12672
Heeks R, Renken A (2022b) Digital inclusion for sustainability: A systematic review of the literature. Sustainability 13(12):5698. https://doi.org/10.3390/su13125698
Heeks R, Renken A (2022c) Digital inclusion for sustainability: A research agenda. Sustainability 14(1):73. https://doi.org/10.3390/su14010073
ILO. (2022). The role of the internet in decent work and economic growth. Retrieved from www.ilo.org/global/topics/decent-work/publications/WCMS_763627/lang--en/index.htm
Digital India. (n.d.). Retrieved on July 8, 2023, from https://www.digitalindia.gov.in/
International Telecommunication Union (ITU) (2021) Measuring digital development: Facts and figures 2021 International Telecommunication Union https://doi.org/10.11884/112021
Internet Society. (2022a). The internet's value to society. Internet Society.
Internet Society. (2022b). The state of the internet 2022. Internet Society.
IPCC. (2022). The role of the internet in climate action. Retrieved from www.ipcc.ch/site/assets/uploads/sites/2/2022/02/SR16-1_FR_web_final.pdf.
ITU. (2022a). Infrastructure and its role in digital connectivity. Retrieved from www.itu.int/en/ITU-D/Statistics/Pages/publications/default.aspx.
ITU. (2022b). The role of the internet in achieving the Sustainable Development Goals. Retrieved from www.itu.int/en/ITU-D/Documents/Reports/Factsheets/Sdgs/The-role-of-the-internet-in-achieving-the-Sustainable-Development-Goals.pdf.
ITU. (2022c). The role of the internet in economic development. Retrieved from www.itu.int/en/ITU-D/Statistics/Pages/publications/default.aspx.
ITU (2022d) Measuring digital development: Facts and figures 2022 International Telecommunication Union https://doi.org/10.11884/112022
Jain A, Gupta S (2022) The affordability of broadband internet in India: A state-level analysis. Econ Pol Wkly 57(22):134–142
Joshi A, Sharma A (2022) Digital literacy in India: Challenges and opportunities. Econ Pol Wkly 57(22):120–127

Kadam S, Yadav A (2022) The impact of internet on education in rural India: A study of the ePathshala platform. Int J Edu Human Dev 8(1):1–8
Khan MA, Al-Saqer ASA, Khan MA (2022a) Geographic barriers to digital connectivity in Pakistan: A case study of Balochistan. Telematics Inform 57:101446
Khan SAA, Khan NA, Khan SU (2022b) The affordability of broadband internet in Pakistan: A case study of Lahore. Telematics Inform 57:101444
Kim S, Park H (2022) The impact of the internet on civic engagement: A meta-analysis. New Media Soc 24(5):1268–1287. https://doi.org/10.1177/14614448211030893
Lin C, Chang S (2017) The internet of things for sustainability: A systematic review and research agenda. Renew Sustain Energy Rev 79:1040–1050. https://doi.org/10.1016/j.rser.2017.04.015
Mäkitalo T, Cunliffe J (2018) Community cellular networks: A new model for rural connectivity. Telematics Inform 35(1):156–166. https://doi.org/10.1016/j.tele.2017.06.003
Malik SK, Khan MA, Khan MA (2022) Political and regulatory challenges to digital connectivity in Pakistan. Telematics Inform 57:101445
McKinsey Global Institute (2016) Digital America: A tale of two cities. McKinsey & Company, New York
Medema J, Warschauer M (2014) One Laptop Per Child: Promises and constraints. Int J Educ Dev 34(1):1–6. https://doi.org/10.1016/j.ijedudev.2013.09.003
Meessen M, Verhulst AH (2022) Digital infrastructure in Europe: Challenges and opportunities. Inf Econ Policy 59:101598
Melody WH (2022) The role of infrastructure in digital connectivity: A global perspective. Telecomm Pol 46(2):101602
Mitra T, Jain A (2022) Gender inequality in digital connectivity in India. Econ Pol Wkly 57(22):151–158
Mtebe J, Masinde T (2017) Internet.org: A critical review. Telemat Informat 34(3):523–532
Narayanan A, Narayana S (2022) The internet and economic growth: A panel data analysis of developing countries. World Dev 133:105383
Ndlovu P, Govender M (2022) Project Isizwe: Towards a socially inclusive digital future for South Africa. Inf Commun Soc 25(3):506–523. https://doi.org/10.1080/1369118X.2021.1975799
Norris P, Damer S (2019) Digital divides: Patterns and trends. Cambridge University Press, Cambridge
OECD. (2022). Digital skills for life and work. Retrieved from www.oecd.org/publications/digital-skills-for-life-and-work-9789264314007-en.htm.
Pew Research Center. (2022). The internet and civic engagement. Retrieved from www.pewresearch.org/internet/2022/01/20/the-internet-and-civic-engagement/.
Pinho V, Marques P (2022) The 'digital divide' in Portugal: Barriers to access in rural areas. Telematics Inform 45:101688. https://doi.org/10.1016/j.tele.2022.101688
Poblete M, Latzer M (2022) The internet and social inclusion: The role of digital skills, social capital, and discrimination. Inf Commun Soc 25(3):458–476. https://doi.org/10.1080/1369118X.2021.1975796
Renken A, Heeks R (2021a) Digital inclusion for sustainability: A systematic review of the literature. Sustainability 13(12):5698. https://doi.org/10.3390/su13125698
Renken A, Heeks R (2021b) Digital inclusion for sustainability: A research agenda. Sustainability 14(1):73. https://doi.org/10.3390/su14010073
Renken A, Heeks R (2021c) The role of digital technologies in achieving the Sustainable Development Goals: A systematic review of the literature. Sustainability 13(2):653. https://doi.org/10.3390/su13020653
Renken A, Heeks R (2021d) The role of digital technologies in reducing inequalities: A systematic review of the literature. Sustainability 13(4):2091. https://doi.org/10.3390/su13042091
Renken A, Heeks R (2021e) The role of multi-stakeholder partnerships in achieving the Sustainable Development Goals: A systematic review of the literature. Sustainability 13(2):654. https://doi.org/10.3390/su13020654

Singh S, Kumar R (2022) The impact of internet on education in India: A study of the Digital India initiative. J Educ Technol Soc 25(1):1–14
Smith MK, Vonderwell S (2022) Gender differences in digital literacy: A systematic review of the literature. Comput Educ 164:103799. https://doi.org/10.1016/j.compedu.2021.103799
SWAYAM. (n.d.). Retrieved on July 8, 2023, from https://swayam.gov.in/
Thelen E, Lubbers M (2022) Google Loon: The rise and fall of a connectivity solution. Inf Soc 38(1):1–15. https://doi.org/10.1080/01972243.2021.1980324
UN Broadband Commission for Sustainable Development. (2021). Retrieved July 8, 2023, from https://www.broadbandcommission.org/
UNCTAD. (2022). The role of the internet in industry, innovation and infrastructure. Retrieved from www.unctad.org/en/PublicationsLibrary/tdriinfonote2022d1_en.pdf.
UNDP. (2022). The role of the internet in reducing inequalities. Retrieved from www.undp.org/content/undp/en/home/librarypage/digital-development/the-role-of-the-internet-in-reducing-inequalities.html.
UNESCO. (2022). The role of the internet in education. Retrieved from www.unesco.org/education/en/internet-education.
UN-Habitat. (2022). The role of the internet in sustainable cities and communities. Retrieved from www.unhabitat.org/sites/default/files/publications/The-role-of-the-internet-in-sustainable-cities-and-communities.pdf.
United Nations Development Programme (2017) Human development report 2017: Beyond income, beyond averages, beyond today: Inequalities in human development in the 21st century. UNDP, New York. https://doi.org/10.18356/2a189e61-en
United Nations Environment Programme (2018) The sustainable development goals report 2018. UNEP, New York. https://doi.org/10.18356/2a189e61-en
United Nations Environment Programme. (2019). *The digital transformation of the environmental monitoring and assessment process: A global review of opportunities and challenges*. Nairobi: UNEP. https://doi.org/10.18356/2a189e61-en
United Nations. (2015). Sustainable development goals. Retrieved July 8, 2023, from https://sdgs.un.org/goals
United Nations. (2019). The digital divide: A global challenge. New York, NY: United Nations.
Verma N, Bansal A (2022) Political and regulatory challenges to digital connectivity in India. Econ Pol Wkly 57(22):143–150
Wang Y, Chen Y (2022) The internet and education: A systematic review of the literature. Education Tech Research Dev 70(1):147–170. https://doi.org/10.1007/s11423-021-09873-8
Wang S, Zhang H, Zhang W, Wang Z, Liu Z (2018) Global e-waste generation and recycling: Current status and future perspectives. J Clean Prod 183:360–371. https://doi.org/10.1016/j.jclepro.2018.02.025
WFP. (2022). The role of the internet in achieving Zero Hunger. Retrieved from www.wfp.org/publications/role-internet-achieving-zero-hunger.
WHO. (2022a). The role of the internet in achieving universal health coverage. Retrieved from www.who.int/publications/i/item/the-role-of-the-internet-in-achieving-universal-health-coverage.
WHO. (2022b). The role of the internet in healthcare. Retrieved from www.who.int/publications/i/item/the-role-of-the-internet-in-healthcare.
World Bank (2016a) Digital dividends: Leveraging the digital economy for growth and inclusion. World Bank, Washington, DC. https://doi.org/10.1596/978-1-4648-0899-3
World Bank. (2016b). The digital dividends: How the internet is transforming lives. World Bank.
World Bank (2020a) The digital divide: Understanding and addressing the economic and social consequences of unequal access to information and communication technologies World Bank https://doi.org/10.1596/978-1-4648-1440-5
World Bank (2020b) The digital dividends: How the internet is transforming lives World Bank https://doi.org/10.1596/978-1-4648-1440-5
World Bank. (2021). Connectivity for all: Powering inclusive growth. Retrieved from www.worldbank.org/en/topic/ict/publication/connectivity-for-all-powering-inclusive-growth.

World Economic Forum (2020a) The global risks report 2020. World Economic Forum, Geneva
World Economic Forum. (2020). The global impact of the internet. World Economic Forum.
World Economic Forum. (2022a). The internet economy in 2022: Thriving in the pandemic and beyond. Retrieved from www.weforum.org/reports/the-internet-economy-in-2022.
World Economic Forum. (2022b). The role of the internet in responsible consumption and production. Retrieved from www.weforum.org/reports/the-role-of-the-internet-in-responsible-consumption-and-production/.
World Wide Web Foundation. (2022a). The role of the internet in gender equality. Retrieved from www.webfoundation.org/reports/the-role-of-the-internet-in-gender-equality.
World Wide Web Foundation. (2022b). The role of the internet in social inclusion. Retrieved from www.webfoundation.org/reports/the-role-of-the-internet-in-social-inclusion.
World Wide Web Foundation. (2022c). Women's rights online. Retrieved from webfoundation.org/reports/womens-rights-online
Zhang J, Meng J, Zhang Y (2022) The internet and healthcare: A systematic review of the literature. Int J Med Informatics 146:104303. https://doi.org/10.1016/j.ijmedinf.2022.104303
Zhou J, Li J, Cheng L, Wang X (2018) The internet of things for sustainability: A review. Renew Sustain Energy Rev 82:1190–1208. https://doi.org/10.1016/j.rser.2017.09.054
Zimmer MJ, Smith DA, Warschauer M (2022) Digital literacy and the global divide: A comparative study of five countries. Inf Technol Dev 28(3):283–306

# Internet of Things and Sustainability: Opportunities and Challenges

**Mahalakshmi S., Anantha Subramanya Iyer K. N., Anitha Nallasivam, Sahana Madan, and Sandeep Kautish**

**Abstract** The Internet of Things (IoT) is rapidly transforming the way we live and work, but its impact on sustainability is not yet fully understood. This chapter aims to explore the relationship between IoT and sustainability, examining both the opportunities and challenges that arise from the intersection of these two fields. The chapter begins by discussing the potential benefits of IoT for sustainability, such as increased energy efficiency, reduced waste, and improved resource management. IoT has the potential to optimize energy consumption, reduce greenhouse gas emissions, and enhance waste management systems, leading to more sustainable practices. However, the chapter also examines the challenges and risks associated with IoT, such as privacy and security concerns, electronic waste, and the carbon footprint of IoT devices. The increased usage of IoT devices and data collection raise concerns over privacy and security. The growing amount of electronic waste, with short device lifecycles and complex systems, presents a major challenge for the sustainable disposal of electronic devices. Additionally, the energy consumption and carbon footprint of IoT devices, including the data centers that support them, add to the environmental impact of technology. The chapter looks at the policies and initiatives that have been developed to promote sustainable IoT, including eco-design guidelines, energy labeling, and recycling programs. The role of industry, government, and consumers in promoting sustainable IoT is examined, including the importance of collaboration and stakeholder engagement in achieving this goal. Finally, the chapter emphasizes the need for a holistic approach to IoT and sustainability, taking into account the entire lifecycle of IoT devices and systems. This includes sustainable product design, responsible manufacturing, and sustainable disposal practices. It calls for continued efforts to ensure that IoT contributes to a sustainable future, with

M. S. · A. S. I. K. N. (✉) · A. Nallasivam
Faculty of Management Studies, CMS Business School, JAIN (Deemed-to-Be University), Bengaluru, Karnataka, India
e-mail: asi@cms.ac.in

S. Madan
Dayanad Sagar Academy of Technology and Management, Bengaluru, Karnataka, India

S. Kautish
AIT CSE, Chandigarh University, Mohali, Punjab, India



W. Leal Filho et al. (eds.), *Digital Technologies to Implement the UN Sustainable Development Goals*, World Sustainability Series,
https://doi.org/10.1007/978-3-031-68427-2_13

stakeholders prioritizing sustainability at all stages of the product lifecycle. Overall, this chapter highlights the potential of IoT for sustainability while acknowledging the challenges and risks. It provides a comprehensive overview of the current state of IoT and sustainability and provides a call to action for stakeholders to work together towards a sustainable future.

**Keywords** Internet of Things · Sustainability · Energy efficiency · Waste reduction · Resource management · Privacy · Security · Electronic waste · Carbon footprint

## 1 Introduction

The much-touted Internet of Things (IoT) is regarded as a disruptive technology with immense potential to change several areas in our life- such as sustainability. The IoT ecosystem consists of multiple interconnected devices fitted with sensors, software algorithms and advanced technology for seamless sharing and analysis of data points (Kshetri, 2017). Therefore, this connectivity encourages the optimal coordination among them, leading increased efficiency while supporting sustainable practices by reducing waste.

In this context, IoT holds great promise for addressing some of the pressing challenges related to sustainability, such as energy management, waste reduction, and environmental monitoring (Plessis, & Clarke, 2019). According to a study by Gubbi et al. (2013), implementing IoT-based energy management systems can lead to significant energy savings and contribute to a more sustainable energy future.

For example, in the manufacturing sector, IoT-enabled sensors can monitor machine performance in real-time, detect anomalies, and enable predictive maintenance, leading to reduced downtime and increased productivity (Zhou et al., 2010). Smart bins equipped with sensors can detect when they are full and notify waste management authorities, ensuring timely collection and preventing overflow. Additionally, IoT devices can facilitate the implementation of circular economy practices by enabling the tracking and tracing of products throughout their lifecycle (Pantelopoulos & Bourbakis, 2010). This enables better resource management, minimizes waste generation, and promotes the reuse and recycling of materials.

With sensors deployed throughout various ecosystems like urban spaces or natural areas like forests or oceans equipped with IoT devices -real-time data tracking environmental parameters including water quality or air quality enabled by today's technology (Pantelopoulos & Bourbakis, 2010), has resulted in valuable insights into how we develop policy measures related to disaster management mitigating potential crises. Indeed, the Internet of Things holds immense promise for advancing sustainability efforts including waste reduction, energy management efficiency along with precise monitoring that spans various lengths whether it be over a season or years-long.

These advances will not only increase the general utilization of our resources but also encourage businesses and individuals to make better smarter choices for the environment.

## *1.1 Definition of Internet of Things (IoT)*

Advancements in technology have led to new approaches such as the Internet of Things (IoT). IoT systems comprise physical objects connected via an internet-based platform enabling communication among themselves automatically without human intervention or interaction Kaggermann, et al. (2011). These devices use specific embedded computing devices for transmitting and receiving data over a network thereby making them easily identifiable. IoT has the power to revolutionize various sectors of industry by increasing productivity and efficiency while improving data collection and analysis as described by Atzori et al. (2010).

Kevin Ashton coined the phrase "Internet of Things" back in 1999 when he envisioned an era wherein ordinary things could readily communicate using the web (Ashton, 2009). Since then, IoT's relevance has soared high making it a basic building block for our digitalized culture. Essentially aimed at making electronic devices smarter than before capable enough for independent sensing, data collection from its surroundings as well interacting within its range thereby increasing efficiency levels far higher than previously imaginable.

## *1.2 Importance of Sustainability in the Context of IoT*

IoT has tremendous potential in driving progress towards greater social responsibility and environmental stewardship if we embrace sustainability as a critical component of its development and implementation. Prioritizing ecological sustainability during IoT deployment can help us establish an equitable. Responsible future for all individuals (Kaggermann, et al., 2011). Therefore, integrating sustainable practices into our approach to deploying IoT should be our top priority.

**Resource Efficiency:** IoT has the potential to optimize the use of resources, such as energy, water, and raw materials, by enabling intelligent monitoring, control, and automation. Sustainable IoT solutions can facilitate energy-efficient operations, reduce waste, and promote the circular economy by enabling resource tracking and management (Bacciu et al., 2019). By leveraging IoT technologies, organizations and individuals can achieve significant resource savings and minimize their ecological footprint.

**Environmental Impact**: There is potential for environmental implications due to widespread use of IoT systems that include increased energy consumption, electronic waste production economy-wide resource depletion. Nonetheless, sustainable principles when developing such systems can minimize such negative effects. Ways to promote sustainability for instance would be encouraging deployment designs that promote reduced power usage in conjunction with adoption renewable power sources (Wang et al., 2017). Additionally ensuring responsible disposal habits when disposing faulty or unusable build components is crucial for mitigating any potential pollution at disposal sites The promotion of green concepts throughout creation deployment and destruction stages will enable harnessing digital solutions provided by Advancements like IOT towards solving issues linked to climate change.

**Social Equity:** IoT has the potential to address social challenges and promote equity in various domains, such as healthcare, education, and urban planning. Sustainable IoT solutions can bridge the digital divide, ensure access to information and services, and empower underserved communities (Larsson et al., 2018). By focusing on equitable IoT deployment and ensuring inclusive participation, we can harness the potential of this technology for social development and equality.

**Long-Term Viability:** IoT projects' long-term success necessitates keeping up with sustainability principles by paying attention to their life cycle impact- including from product inception through realization stages up until its disuse process for optimal effectiveness. With this holistic approach an increase in positive results would be assured without compromising socio-economic or environmental variables; by avoiding potential unintended results. Coming up with scalable adaptable resilient sustainable solutions has been an effective hack for minimizing obsolescence while exploiting its maximal potency (Lee et al., 2019).

**Regulatory and Market Demands:** The significance of embedding sustainability considerations within IoT has only amplified over time owing largely to regulatory norms and market demands. Increasing efforts by governments acknowledge the need for incorporating sustainable principles within IoT development due to pressing socio environmental challenges. Furthermore, consumer demand for ethically sound products means businesses must focus towards ensuring that their IoT solutions reflect these values (Larsson et al., 2018). Integration of sustainable policies within IoT will not only make regulatory compliance easier but will also work towards enhancing a company's' reputation and fostering competitiveness in the marketplace.

Embracing a transformative mindset towards shaping a brighter future through technologies like IoT requires an unwavering commitment towards prioritizing sustainability during its development phase - from inception till final launch stages are complete. Adopting strategies such as resource efficiency maximization while reducing environmental impact or promoting social equity helps promote an inclusive future by responding to everchanging regulatory mandates.

## 2 Opportunities for Sustainability with IoT

Multiple avenues can facilitate sustainable practices across several fields courtesy of innovative solutions provided by Internet of Things (IoT) technology. Advancements in optimal resource utilization coupled with enhanced productivity can lead to positive environmental benefits for businesses or individuals who utilize these systems effectively. We explore noteworthy opportunities that champion sustainable progress via IoT methods:

## 2.1 Increased Energy Efficiency

Internet of Things (IoT) innovations are making it possible to monitor and regulate power usage in real time leading to increased sustainability efforts by enabling better control over waste management practices (Wu et al., 2019). By incorporating IoT concepts into our building design frameworks we can optimize resource utilization through initiatives like intelligent gridding systems or connecting various electronic products such as lighting systems that automatically adjust their power output based upon surrounding light conditions (Conti et al., 2019). Ultimately through intelligent waste management practices we can reduce greenhouse emissions while generating huge cost savings benefits across multiple industries.

## 2.2 Reduced Waste

IoT can revolutionize waste management practices by enabling intelligent monitoring, collection, and recycling processes. Connected sensors and smart bins can detect fill levels, optimize waste collection routes, and minimize overflowing bins (Gomez et al., 2019). This leads to efficient resource utilization, reduced environmental impact, and cost savings in waste management operations. Additionally, IoT can facilitate waste sorting and recycling by providing real-time information on the type and quantity of waste generated (Amaral et al., 2019). Such data-driven approaches contribute to a circular economy and sustainable waste management.

## 2.3 Improved Resource Management

Resource management improvement is an area where the Internet of Things (IoT) has shown immense potential. Take into account that, smart irrigation systems integrated with moisture sensors and meteorological data could decrease water wastage while increasing crop yield in agriculture or landscaping for instance. The use of such technologies has shown tremendous efficiency gains (Zhang et al., 2020a, b).

In addition to its role in resource allocation improvements using smart transportations systems connected vehicles; as well as traffic control technologies have the punch to lessen fuel consumption by route optimization while decreasing traffic volume on congested roads (Fan et al., 2018). Similarly Smart parking solutions based on IoT are already being employed globally guiding drivers towards vacant slots which decreases polluting emissions (Milakis et al., 2016).

Furthermore, IT devices coupled with environmental parameter tracking mechanisms like air quality measures biodiversity ration among other indicators may help prevent pollution cycles by catching them early (Rajasegarar et al., 2021).

IoT presents significant opportunities for promoting sustainability across various domains. By leveraging IoT technologies for energy management, waste reduction, water conservation, environmental monitoring, and smart transportation, we can drive positive environmental and social impact. Embracing these opportunities can contribute to a more sustainable future, fostering resource efficiency, resilience, and responsible stewardship of our planet.

## 3 Challenges and Risks with IoT for Sustainability

Promoting sustainable development through the use of Internet of Things Technology (IoT) presents several challenges that need attention. While this technology provides enormous potential towards sustainable development goals, addressing its hurdles before implementation remains critical (Haller et al., 2018). Mitigating those factors ensures maximal benefit from integrating them into advancement efforts (Hu et al., 2021). Some of the challenges of implementing IoT for sustainable development are outlined below.

### *3.1 Privacy and Security Concerns*

IoT systems generate vast amounts of data, often including personal and sensitive information. Protecting this data from unauthorized access, breaches, and misuse is essential (Boyd et al., 2020). IoT devices are vulnerable to cyber-attacks, and compromised systems can have severe consequences, compromising privacy and integrity (Ray et al., 2021). Ensuring robust security measures, such as encryption, authentication, and secure data handling, is vital for maintaining user trust and protecting sensitive information.

### *3.2 Electronic Waste*

With an alarming increase in adoption rates for IoT technology comes an equally worrying trend - an abundance of electronics waste or "e-waste." As evidence shows (Kuschel et al., 2019) one key factor behind this issue is the relatively short lifespan of these devices. To prevent further environmental degradation and preserve our natural resources its essential that proper disposal and recycling procedures are implemented. Therefore. Promoting sustainable product design practices. Encouraging device upgrades and repairs. Along with responsible e waste management initiatives can go a long way in mitigating the impact that IoT has on our planet.

### *3.3 Carbon Footprint of IoT Devices*

Considering how IoT installations are made result in significant ecological implications encompassing every stage of its lifespan. From the acquisition of raw materials to produce devices to their ultimate disposition. There is carbon emission as well as overall resource depletion due to energy consumption while producing these tools (Pantelis et al., 2021). Eco-design operations could significantly mitigate concerns related to ecological perils raise consciousness around recycling initiatives, considering implementing sustainable practices like promoting renewable energy resources aimed at creating a subtler environmental footprint will benefit our planet greatly.

A holistic strategy involving collaboration across disciplines among industry incumbents alongside policymakers as well as researchers is key to tackling the challenges and risks faced by IoT effectively. By developing strong security implementations that ensure privacy breaches are minimized while uncompromising on seamless connectivity standards - along with energy efficiency strategies- one can achieve maximum impact in terms of reducing waste while promoting better e-waste management practices. Adopting ethical principles while mitigating environmental consequences should be factored into all decision-making processes pertaining to the utilization of this innovative technology.

## 4 Policies and Initiatives for Promoting Sustainable IoT

Various governmental & non-governmental establishments plus international organizations recognize that sustainable practices are vital in both developing and deploying the Internet of Things (IoT). Subsequently, measures aimed at driving sustainability agendas while addressing hindrances & encouraging responsible IoT implementation have been put forth. In a nutshell, the following are some existing IoT sustainable policies and initiatives:

### *4.1 Eco-Design Guidelines*

The EU's Eco-Design Directive oversees the development of environmentally responsible technology by setting mandatory standards for every product sold in Europe including IoT items. The directive functions as an essential tool fostering sustainability holistically considering every step involved in producing gadgets right through to their disposal management (European Commission, 2021a, b). By encouraging zero or low-energy emission IoT products, the directive aims to cultivate each consumers' environmental consciousness while promoting sustainable production practices overall.

## 4.2 Energy Labeling

Efforts towards promoting sustainable practices in relation to IoT systems can be made by investing in policy interventions such as the implementation of an effective method like 'energy labeling'. This practice works by displaying ratings or labels with information on a product's overall energy-efficiency standards allowing customers to make informed decisions when purchasing items (European Commission, 2021a, b).

Many international players are cooperating with governments across continents in introducing regulation concerning the production and promotion of labeled smart home technology ranging from intelligent televisions sets to washing machines as it lays foundation for environmental best practices (European Commission, 2021a, b). The European Commission has played its role actively introducing strict rules around how data is categorized so customers gain insight into how much electricity each device consumes helping citizens understand their carbon footprint.

The Green Electronics Council (GEC) is a non-profit organization that manages the EPEAT program. EPEAT is a global rating system for sustainable electronics, including IoT devices (Green Electronics Council, n.d.).

## 4.3 Recycling Programs

Recycling programs that emphasize sustainable practices play a fundamental role in promoting environmentally friendly processes within the Internet of Things ecosystem. Proper disposal of IoT devices at their end-of-life cycle minimizes electronic wastes' harmful effects on communities while ensuring maximum resource recovery (Kaur et al., 2021).

Such measures exhibit an admirable commitment to reducing carbon footprint while championing a circular economy approach to managing materials via recycling or reuse wherever possible (Gupta et al., 2021). Conducive to achieving these objectives requires combined efforts from government bodies, industry associations as well as eco conscious groups advocating for environmentally friendly practices (Lathuillière et al., 2020). Creating efficient infrastructures for collecting discarded IoTs as well as educating users about environment friendly disposal options are critical steps to achieving sustainable recycling programs.

Adopting models such as the EUs' Waste Electrical and Electronic Equipment Directive mandating e waste management protocols that include IoT devices has huge benefits globally (European Commission, 2012). Many similar initiatives have been introduced worldwide promoting waste management and eco-friendly practices advocating for wellbeing through environmentally mindful processes.

These policies and initiatives demonstrate the commitment of governments, organizations, and international bodies to promote sustainable IoT practices. By incorporating energy efficiency, environmental considerations, and responsible design into IoT policies, these efforts contribute to a more sustainable and resilient future.

# 5 Role of Industry, Government, and Consumers in Promoting Sustainable IoT

The promotion of sustainable initiatives relating to the Internet of Things (IoT) requires collective efforts from various stakeholders such as industry members, government representatives. And end users. Every stakeholder has a vital role to play in driving sustainability within the IoT domain. The ensuing section elaborates on how each stakeholder can contribute towards ensuring sustainable practices within IoT.

## *5.1 Industry Initiatives and Standards*

Industry players must prioritize sustainability when developing & manufacturing IoT devices due to its crucial role within such a process. This dictates that design considerations entail identifying ecofriendly materials alongside better recyclable product components, according to Li et al. (2020) study on sustainability matters within this domain. Manufacturers today must also work towards sustainable manufacturing practices, including reducing waste generation optimizing energy consumption. Implementing this lessens the use of hazardous substances in production processes; hence its importance cannot be overstated (Zhang et al., 2021).

Furthermore, establishing common standards and protocols is necessary for interoperability, energy efficiency, and environmental performance measurement of IoT devices. This comes with collaboration among industry players to ensure sustainable IoT solutions are created while achieving seamless integration across different platforms- as notes Huang et al. (2020) research.

## *5.2 Government Regulations*

Governments are pivotal actors with regard to establishing a sustainable framework for the development of IoT applications globally (EU Parliament, 2019). They drive policy formation along with regulation establishment that would govern effective usage consisting measures taking into consideration both environmental conservation as well as data integrity concerns.

Incentives such tax credits or grants may be issued to promote development in ecofriendly versions of present-day communications technology, while financial aid may be allocated to research initiatives for sustainable IoT applications (Ene et al., 2021). International cooperation can come into the picture through collaborations between governments, the corporate sector and research institutions which will allow for greater capacity building and innovation (Deakin et al., 2021).

### *5.3 Consumer Choices and Behaviors*

Consumers have the power to drive sustainable IoT practices by making conscious purchasing decisions. They can choose energy-efficient IoT devices, prioritize products with environmental certifications, and support companies that prioritize sustainability in their offerings (Zhang et al., 2020a, b).

Consumers should use IoT devices responsibly, following best practices for energy management, data privacy, and security. Additionally, they should properly dispose of end-of-life IoT devices through recycling and e-waste management programs (Zheng et al., 2021). Consumer demand for sustainable IoT solutions can influence industry practices. By expressing preferences for environmentally friendly and socially responsible IoT products and services, consumers can drive market trends and encourage industry players to prioritize sustainability (Li et al., 2021).

A sustainable future for the Internet of Things (IoT) depends on collective efforts from industry leaders, governmental bodies, and consumers alike. Collaboration and innovation are essential components needed to forge a path towards greater responsibility and resilience in the IoT ecosystem. With a shared commitment to sustainability as our driving force - success will be within reach.

## 6 Collaboration and Stakeholder Engagement for Sustainable IoT

Effective promotion of environmentally friendly practices for the development and deployment of Internet of Things (IoT) definitely requires collaboration alongside sustained engagement involving diverse groups. To achieve such an undertaking entail bringing on board all relevant parties ranging from industrialists to policymakers together with academics through civil society organisations. Our report below provides an insightful perspective into how stakeholder involvement significantly matters when it comes to achieving sustainability within IoT projects.

### 6.1 Importance of Collaboration

Collaboration between industry and academia fosters research, innovation, and knowledge transfer in sustainable IoT practices. Academic institutions contribute expertise in environmental sustainability, energy efficiency, and social impact, while industry partners provide real-world implementation scenarios and resources (Deakin et al., 2021).

Public-private partnerships bring together government entities, industry stakeholders, and NGOs to address sustainability challenges in IoT. These partnerships facilitate resource sharing, policy development, and joint initiatives for promoting sustainable IoT solutions (Ylianttila et al., 2020). Collaboration across different sectors, including technology, energy, urban planning, and transportation, allows for holistic approaches to sustainable IoT. Cross-sector collaboration enables the integration of diverse perspectives and expertise, leading to comprehensive and effective sustainability strategies (Lopez-Carmona et al., 2021).

### 6.2 Stakeholder Engagement Strategies

Working together with industry, academia and civil society is key to promoting sustainable IoT – this is why governments should actively engage them in policy-making. By doing so proponents can balance priorities from different standpoints while ensuring inclusive approaches that considers varying needs and concerns (Borges et al., 2020). To ensure sustainability from the ground up community input is vital when developing IoT solutions. This translates into appropriate initiatives that would reflect community needs while respecting social and cultural considerations (Al-Hammadi et al., 2021).

For consumers to embrace sustainable behaviors, such as for energy-efficient consumption patterns or responsibly using products requires education about conscious decision-making when purchasing IoT devices as well as raising awareness generally about sustainable practices (Soylu et al., 2020).

Effective collaboration and stakeholder engagement for sustainable IoT require open communication, trust-building, and shared goals. By working together, stakeholders can pool resources, knowledge, and perspectives to drive sustainable IoT innovation and implementation.

## 7 Holistic Approach to IoT and Sustainability

Fostering an inclusive framework for considering the many interconnected aspects of IoTs impact on sustainability is essential if we aspire towards creating long lasting and effective solutions to maintain our worlds health holistically. This means involving stakeholders at every stage of development to ensure everyone has access to relevant information as well as providing opportunities for discussion about potential challenges or unintended outcomes within the system over time. Here are some ways in which this joint effort can achieve a more meaningful and positive outcome when combined with intelligent tech implementations across various industries:

## *7.1 Consideration of Entire Lifecycle of IoT Devices and Systems*

When thinking about how anyone interacts with IoT technology while still staying mindful of our planets limited resources there must be a whole life approach should be the one that considers everything from when a device is incepted until its finally no longer used anymore. By doing so environmental footprints are minimized through the efficient use of what is available to us. Some approaches include:

**Environmentally Savvy Design Choices:** All upcoming IoT devices or systems should come equipped with an environmentally friendly mindset from the earliest stages. This might emphasize employing more ecofriendly materials placing importance on building them for recyclability as well as designing efficiently for disassembly and minimizing energy usage wherever possible (Ilic et al., 2021).

**Sustainable Manufacturing:** A significant step towards sustainability depends on efficient manufacturing methods which cut down on resource use and emissions. That can include adopting techniques like lean manufacturing, prioritizing renewable energy use, as well as implementing responsibility in supply chain management (Li et al., 2022).

**Optimizing Energy Usage:** In order to reduce carbon emissions its' vital to prioritize energy efficiency during both device usage and network implementation -- something that manufacturers themselves must keep front of mind. This can involve the use of power management techniques along with optimized algorithms to minimize power consumption plus incorporating a wider range of renewable energy resources alongside traditional sources. (Liu et al., 2021).

**End-of-Life Management:** The responsible disposal of electronic waste along with appropriately recycling devices which are no longer useful is important if environmental pollution is to be curtailed. Establishing e waste management practices and recycling programs guarantee the return of valuable resources while minimizing electronic waste impact (Kaur et al., 2021).

### 7.2 *Importance of Sustainability in Product Development and Innovation*

Prioritizing sustainability within the development of IoT products and innovation can help create environmentally-friendly solutions while ensuring a socially responsible outcome. It is therefore essential to include sustainability principles into development processes and encourage the evolution of innovative approaches to tackle sustainability-related challenges.

To achieve these sustainable goals, reducing energy consumption, optimizing resourcefulness of products along with utilizing eco-design principles have become priorities (Mörtl et al., 2022). Additionally, adopting innovation that prioritizes social responsibility such as protecting privacy rights while addressing equity requires embracing inclusive design tailored to diverse user groups (Lathuillière et al., 2020).

Adopting a circular economy model in the process of developing IoT products provides an effective means of attaining sustainable outcomes. Implementing practices aimed at repairability, product-as-a-service models with recycle-ability principals considered produces more reliable results for the environment (Gupta et al., 2021). Moreover, holistically engaging stakeholders throughout this process provides insight from various angles such as users perspectives' community needs and NGO advocacies resulting in user-centric sustainable outcomes (Rizvi et al., 2022).

By considering the entire lifecycle of IoT devices and systems and emphasizing sustainability in product development and innovation, stakeholders can contribute to a more sustainable and responsible IoT ecosystem.

## 8 Conclusion

The synergy between Sustainability & Internet of things (IoT) opens-up meaningful opportunities towards envisioning a future where better environmental responsibility meets social accountability without losing out on economic viability. Even though there are various obstructions in leveraging such possibilities, hence it becomes incumbent upon every stakeholder involved to acknowledge them earnestly & strategize how they can be dealt precisely.

This chapter has dived deep into these implications & challenges associated collectively when it comes to IoT and sustainability. Furthermore, it stresses the pivotal need to continuously pursue sustainable IoT development on a larger scale. Consequently, all industry stakeholders & governments must rally around their demand, consumers included, in prioritizing such goals towards sustainable progress.

## 8.1 Need for Continued Efforts Towards Sustainable IoT

The integration of sustainable methods into the Internet of Things (IoT) offers great potential for promoting environmental responsibility among manufacturers as well as users alike. To achieve a sustainable framework around IoT systems calls for a collaborative effort between stakeholders towards fostering an innovative approach focused on enhancing their ecological compositions through continuous research and development aimed at reducing energy consumption while optimizing resource management capacity.

Achieving this feat further involves adopting eco-friendly practices throughout their entire lifespan: from design through manufacture until end-of-life management protocols are evident- every step counts towards its longevity. Ongoing dedication is vital in dealing with emerging issues to guarantee futuristic success built on sustainable IoT practices (Nguyen et al., 2022).

## 8.2 Call to Action for Industry, Governments and Consumers to Prioritize Sustainability in IoT

**Industry:** To be responsible corporate citizens industry players need to integrate sustainability considerations into their business strategies. This entails adopting eco design principles and energy efficient technologies while embracing circular economy practices. Industry should also partner with other stakeholders to share knowledge and promote sustainable IoT initiatives (Singh et al.. 2021).

**Governments:** A collaborative effort between governments is essential in promoting environmentally conscious practicality within the Internet of Things(IoT) space supported by factors such as effective policies alongside regulations reflective of current sustainability principles. Regulations should deter environmentally harmful aspects while actively encouraging undertaking eco-friendly endeavors across sectors through financially incentivized collaborative initiatives. Public awareness campaigns targeted towards educating different communities are crucial towards changing mindsets, growing consciousness alongside undertaking responsible handling, sustainable use and disposal of IoT devices (Zhang et al., 2022).

**Consumers:** It might surprise some people just how much influence consumers hold over the creation of sustainable IoT solutions. By consciously choosing products based on their environmental impact--opting for those devices with lower power needs or higher amounts of recyclable materials--we empower ourselves as mindful shoppers who shape prevailing trends toward better outcomes. Even more impactful is ensuring responsible use when operating these connected devices, minimizing excessive energy usage, and disposing of them effectively by following appropriate waste management practices (Huang et al., 2021).

The amalgamation of Sustainability and Internet of Things (IoT) creates an opportunity to construct an environmentally responsible and socially accountable future that retains economic feasibility. Nevertheless. These opportunities face obstacles that necessitate stakeholders' earnest recognition and strategizing precise solutions.

## References

Al-Hammadi Y, Al-Dubai A, Al-Nuaimi H (2021) Internet of Things and Sustainable Development: A Case Study of the United Arab Emirates. Sustain Cities Soc 67:102737

Amaral A, Lopes R, Rodrigues JJ (2019) Internet of Things for Smart Waste Management: A Survey on Waste Bin Detection and Filling Level Monitoring techniques. J Clean Prod 214:218–230

Ashton K (2009) That 'internet of things' thing. RFiD Journal 22(7):97–114

Bacciu D, Chessa S, Gallicchio C, Micheli A, Pedrelli L (2019) Sustainable Internet of Things: A Review. ACM Comput Surv 52(6):1–35

Borges ML, Bifulco L, Brandão E, Ferreira LN (2020) Smart Cities and the Role of Stakeholders in Their Governance for Sustainable Development: The Case of Porto Maravilha. J Clean Prod 244:118739

Boyd D, Crawford K, Crawford K (2020) Critical questions for big data: Provocations for a cultural, technological, and scholarly phenomenon. Inf Commun Soc 23(9):1265–1283

European Commission. (2012). Waste Electrical and Electronic Equipment (WEEE). Retrieved from https://ec.europa.eu/environment/waste/weee/index_en.htm

European Commission. (2021). Energy Labelling for Televisions and Washing Machines. Retrieved from https://ec.europa.eu/energy/topics/energy-efficiency/energy-efficient-products/energy-labeling-televisions-and-washing-machines_en

Conti M, De Nardis L, Pietro RD (2019) IoT and Smart Living. Springer, In Internet of Things

Deakin M, Al-Sabbagh A, Griggs S, Clarke R (2021) Barriers and Enablers to Sustainable IoT: An Analysis of Business Model Perspectives. J Clean Prod 294:126221

Ene S, Vilceanu CA, Oroian IG (2021) The Role of Governments in Promoting Green Technologies: A Systematic Review. J Clean Prod 284:125354

Energy Star. (n.d.). About Energy Star. Retrieved from https://www.energystar.gov/about

EU Parliament. (2019). Guidelines on the Implementation of the EU Ecodesign Directive 2009/125/EC. Retrieved from https://www.europarl.europa.eu/RegData/etudes/IDAN/2019/603522/EPRS_IDA(2019)603522_EN.pdf

European Commission. (2021). Digital Single Market: Internet of Things. Retrieved from https://ec.europa.eu/digital-single-market/en/internet-things

Fan Y, Li H, Qian H, Deng Y (2018) Smart Transportation Systems: A Review on IoT-based Road Traffic Management Solutions. Sensors 18(11):4004

Gomez C, Paradells J, Zennaro M (2019) Smart Waste Management for Smart Cities: A Review on the Methods and Technologies for Solid Waste Bin Monitoring. Sensors 19(6):1363

Green Electronics Council. (n.d.). EPEAT. Retrieved from https://www.epeat.net/

Gubbi J, Buyya R, Marusic S, Palaniswami M (2013) Internet of Things (IoT): A vision, architectural elements, and future directions. Futur Gener Comput Syst 29(7):1645–1660

Gupta A, Mathur A, Kumar P (2021) A Systematic Review of Circular Economy in IoT Product Development. J Clean Prod 313:128600

Haller S, Karnouskos S, Schroth C (2018) The Internet of Things in an Enterprise Context. Enterprise IoT. CRC Press, Boca Roton, pp 1–28

Hu H, Zhang W, Ghosal D, Das SK, Sheng M (2021) Green Internet of Things: Principles, Standards, and Applications. IEEE Internet Things J 8(2):869–881

Huang K, Xie S, Guo B, Wang H, Chen J (2020) A Survey on Energy-Efficient Communications for Green Internet of Things. Sustain Cities Soc 61:102309

Huang Z, Ong SK, Nee AYC, Zheng X (2021) Life Cycle Assessment of the Internet of Things: A Review. J Clean Prod 278:123916

Ilic A, Mansouri SA, Gonzalez-Briones A, Silic M (2021) IoT for Sustainability: A Survey of Techniques and Enabling Technologies. Sustain Cities Soc 74:103307
Kaur H, Singh L, Gupta M (2021) Internet of Things for Circular Economy: A Review and Classification Framework. J Clean Prod 306:127442
Kshetri N (2017) The Internet of Things: A comprehensive overview. SSRN Electronic Journal.
Kuschel J, Kärkkäinen M, Virtanen S (2019) E-Waste and the Circular Economy: A Review of the Challenges and Opportunities. Sustainability 11(18):4947
Larsson TC, Gianfrate G, Prenkert F (2018) Making the Internet of Things Responsible: How to Ensure Sustainable Development in a Digitalized World. Sustainability 10(9):3019
Lathuillière RM, De Marchi M, Soriano-Marcolino L, Lamas JL (2020) Sustainability Challenges in the Internet of Things: A Review. Sustain Cities Soc 60:102247
Lee J, Bagheri B, Kao HA (2019) A Cyber-Physical Systems architecture for Industry 4.0-based manufacturing systems. Manufact Lett 19:1–5
Li Z, Li X, Lu R, Liang X (2020) Green Internet of Things: A Survey. IEEE Internet Things J 7(11):10049–10070
Li L, Feng T, Li S, Yu H, Lu R (2021) Crowdsensing-Based Sustainability: A Survey. IEEE Commun Surv Tutor 23(3):1914–1954
Li L, Feng T, Lu R, Li S (2022) Toward Sustainable IoT Manufacturing: A Review and Outlook. IEEE Trans Industr Inf 18(5):3330–3342
Liu X, Li X, Chen X, Li X, Yu H (2021) An Energy-Efficient Method for Data Transmission in Fog-Based Internet of Things. IEEE Internet Things J 8(9):7207–7216
Lopez-Carmona MA, Fernandez-Ardevol M, Herrera-Quintero LF (2021) Smart Sustainable Cities: Analysis of Success Factors and Stakeholder Engagement. J Clean Prod 280:124419
Milakis D, van Arem B, van Wee B (2016) Policy and Society. Park Policy for Automated Vehicles: A Review and Framework for Future Research 35(3):77–90
Mörtl M, Stankovic S, Haselsteiner E (2022) A Systematic Literature Review on Sustainable Internet of Things (IoT) Product Development. J Clean Prod 322:128969
Nguyen TT, Vo TTN, Nguyen TH, Dinh TQ (2022) Internet of Things (IoT) for Sustainable Development: Challenges and Opportunities. Smart Technologies for Sustainable Development. Springer, Cham, pp 171–192
Pantelis A, Bakas C, Zouridaki E, Loukas G (2021) Towards a Sustainable Internet of Things: A Review. Sustain Cities Soc 65:102595
Pantelopoulos A, Bourbakis NG (2010) A survey on wearable sensor-based systems for health monitoring and prognosis. IEEE Trans Syst Man Cyber C (applications and Reviews) 40(1):1–12
Plessis P, Clarke S (2019) The Internet of Things: A review of recent research on security and privacy. Comput Secur 85:101658
Rajasegarar S, Vu HL, Leckie C (2021) Internet of Things for Environmental Monitoring and Management: A Survey. IEEE Communications Surveys & Tutorials 23(1):340–375
Ray PP, Dash SS, Das SK (2021) Security and Privacy Issues in Internet of Things: Challenges and Solutions. Internet of Things: From Hype to Reality. Springer, Cham, pp 209–240
Rizvi SFR, Tahir M, Vukovic V, Pervaiz H, Yaqoob I (2022) Socially Sustainable Internet of Things: A Systematic Literature Review and Directions for Future Research. Sustain Cities Soc 95:103384
Singh SK, Ntalianis K, Verma P (2021) Internet of Things (IoT) for Smart and Sustainable Cities: An Analysis of Technologies, Challenges and Opportunities. Sustain Cities Soc 68:102776
Soylu A, Oztaysi B, Balcisoy S, Cagiltay NE (2020) Enhancing User Engagement in Sustainable IoT Solutions: A Gamification Approach. Sustain Cities Soc 60:102266
United Nations. (n.d.). Sustainable Development Goals. Retrieved from https://www.un.org/sustainabledevelopment/sustainable-development-goals/
Wang H, Zhang Y, Hu J, Wang X, Chou CT (2017) Green Internet of Things for Smart World. Mobile Netw Applic 22(1):70–80
Wu Z, Zhang Y, Cheng L, Chen J (2019) Smart Grids Powered by Internet of Things: Key Technologies, Architectures, and Standardization Efforts. IEEE Wirel Commun 26(1):136–143

Ylianttila M, Suorsa M, Perälä P (2020) Enhancing Sustainable IoT Ecosystems: A Survey of Public-Private Partnerships. Sustain Cities Soc 53:101942

Zhang D, Tang S, Chen H, Yang G (2020a) IoT-based Smart Irrigation Systems: An Overview and Future Directions. IEEE Internet Things J 7(11):10786–10799

Zhang Z, Yu H, Yang M, Li X, Zheng K (2020b) A Survey on Environmental Sustainability in IoT. Sustain Cities Soc 61:102335

Zhang L, Zhang Q, Zhao H, Zhang Y (2021) Sustainable Energy Management in Industrial Internet of Things: A Survey. IEEE Commun Surv Tutor 23(2):1319–1350

Zhang Q, Zeng C, Liu X, Huang L (2022) Sustainable Development of the Internet of Things: Challenges and Opportunities. Handbook of Sustainable Development. Springer, Cham, pp 1–17

Zheng P, Fu H, Tang S (2021) Socially Responsible IoT: From Sustainable Development Goals to System Design. IEEE Commun Mag 59(1):120–125

Zhou J, Leung VC, Zhang D, Song W (2010) Achieving energy-efficient and delay-aware data collection for wireless sensor networks with mobile sinks. IEEE Trans Industr Inf 6(4):690–700

# Optimizing Green Power and Green Energy Through Digital Technologies

**Anantha Subramanya Iyer K. N., Mahalakshmi S., Kavitha Desai, Hemanth Kumar S., and Sandeep Kautish**

**Abstract** The terms green power and green energy refers to the various forms of energy generated from renewable and ecofriendly resources that have a minimal impact on the environment. They represent resources and technologies that provide superior advantage to environmental sustenance. Renewable energy sources such as solar, wind, hydro, geothermal, and biomass are considered green power sources, as they generate clean energy without emitting greenhouse gases and thus reducing global warming. Cost effectiveness of green power has contributed to its popularity and competitive with other non-renewable sources.

Green power and green energy play a crucial role in achieving the United Nations Sustainable Development Goals (SDGs). Green power and energy are closely linked to several SDGs to ensure access to affordable, reliable, sustainable, and modern energy for all, climate action through protecting the planet's ecosystem, promoting sustainable economic growth and decent work for all while fostering innovation. Investments in renewable energy could save up to $4.2 trillion per year. Many countries have committed to installing 698 gigawatts (GW) of sustainable energy sources by 2030.

Digital technologies have become instrumental in energy management enabling the adoption and efficient use of green power and green energy. Digital tools such as smart grids, energy storage systems and management systems provide real-time data on energy consumption and production, enabling more efficient use of energy and greater control over costs. Smart grids also enable the integration of renewable energy sources into the existing grid.

---

A. S. I. K. N. (✉) · M. S. · H. K. S.
Faculty of Management Studies, CMS Business School, JAIN (Deemed-to-Be University), Bengaluru, Karnataka, India
e-mail: asi@cms.ac.in

K. Desai
SVKM's Narsee Monjee Institute of Management Studies, Bengaluru, Karnataka, India
e-mail: Kavitha.desai@nmims.edu

S. Kautish
AIT CSE, Chandigarh University, Mohali, Punjab, India

 
W. Leal Filho et al. (eds.), *Digital Technologies to Implement the UN Sustainable Development Goals*, World Sustainability Series,
https://doi.org/10.1007/978-3-031-68427-2_14

The other areas of intervention of digital technologies are innovative business models such as peer-to-peer energy trading platforms for trading excess energy generated from renewable energy systems. Furthermore, machine learning and artificial intelligence and blockchain technology is being explored for the creation of decentralized energy systems.

In conclusion, digital technologies are instrumental in facilitating the adoption and efficient use of green power and energy. The continued development and deployment of digital technologies will play a critical role in accelerating the awareness and transition to a sustainable energy future.

**Keywords** Optimizing Green Energy · Green Power · Sustainable Development Goals · Digital Technologies · Machine Learning · Block Chain

# 1 Introduction

As the world aspires towards building a sustainable future for humanity on this planet, implementing strategies that encompass green power and energy sources is essential. By doing so, we have the potential to considerably diminish greenhouse gases emissions overall whilst contributing positively across several sectors such as improving air quality standards globally - boosting economic growth and increasing international autonomy for countries worldwide (IRENA 2021a, b).

In this context adopting Green Power methods leaves an encouraging impact tackling environmental concerns with positive outcomes thus paving the way forward for today's institutions & societies including governments business entities alike towards securing Sustainable Growth globally through leveraging technological innovations such as clean technology options.

## *1.1 Meaning and Definition of Green Power and Green Energy*

When it comes to meeting our growing appetite for electricity in today's world, we must now consider environmentally responsible ways. This is where green power or renewable energies play significant roles around the planet. Green Power refers specifically to sustainably-generated electricity using methods that are friendly or less intrusive toward Earth's bionetworks.

By choosing this approach over traditional methods reliant on fossil fuels one could prevent increased emissions buildup while also helping alleviate pressing issues such as global warming and resource depletion. Generating heat/energy through faithful utilization, restoration and effective management towards renewable energy sources has become a means of achieving sustained growth and viable options for electricity generation.

**Green Power:** Green Power is an eco-friendly term used for electricity produced using renewable resources like solar, wind turbines hydropower plants etc. Such resources use technology with low carbon emissions footprints thus less hazardous on the environment (U.S. Department of Energy [DOE], 2020, 2021). Compared with traditional means which utilize fossils fuels this type distinguishes itself as being a clean source for electrical needs.

**Green Energy:** The concept of green energy encompasses renewable forms like electricity generation from wind & solar energies as well as fuel alternatives like biofuels (International Renewable Energy Agency [IRENA], 2018). The primary aim is to foster optimal usage practices for natural resources while decreasing our dependence on fossil fuels in the process. Green energy related innovations stand out thanks to their knack for generating power sustainably while minimizing emissions. Following are some of the important sources of green energy:

***Solar Power:*** Utilizing sunlight, solar energy is an environmentally-friendly approach towards generating electricity that functions via Photo-Voltaic (PV) panels and solar thermal technologies. PV panels convert light directly into electrical charges without creating pollution while producing no sound nor vibrations. Whereas with thermal processes, sun rays boil fluids such as oil that turn turbines which also produce electric charges without involving polluting fuels like coal-based conveyors do in conventional electric generators.

***Wind Power:*** Wind turbines extract kinetic energy from freely available wind resources and turn them mechanically via rotating blades before converting them electro-magnetically via its generators which operate silently with little maintenance costs involved but requiring much space.

***Hydroelectric Power:*** Hydroelectric dams transform the potential and kinetic energies exerted upon standing bodies of water to achieve a constant flow inside channels leading up to their generators where massive amounts of its potential energy are converted into available electrical powers resembling those of giant batteries sustaining plant growth cycles in urbanized areas today.

***Biomass Energy:*** Biomass energy is another renewable source when it comes to power generation, which results from burning organic biomass like agricultural crops and waste material heaters that make use of collected wood as fuel. In this process, materials undergo some sort of decomposition and fermentation which produces combustible gases that transfer heat into before being further transformed as biofuels like ethanol that save consumers money on gas while keeping the environment clean.

***Geothermal Energy:*** Geothermal power capitalizes on heat and pressure found deep beneath earth's crust by drilling even deeper wells to access hot underground reservoirs. Extracted water and steam are converted usable electrical energies for various purposes inspired by the limitless potential inherent in utilizing geothermal resources such as non-polluting, inexpensive electricity that never depletes.

## 1.2 Importance of Green Power and Green Energy for Sustainable Development

There are multiple reasons why there is an urgent need for transitioning towards green power and energy sources such as taking care of our environment by mitigating various climatic changes while keeping up with sustainable development goals. The following points highlight the need for adopting renewable energy sources:

**Climate Change Mitigation:** IPCC (2018) reports that human activities specifically burning of fossil fuels majorly contributes towards global warming which leads us back yet again to changing our habits for good by adopting greener solutions promptly.

Moreover, investing in renewable and cleaner technology gives us an opportunity to curtail our carbon footprint much more effectively than other traditional modes (IRENA 2021a, b).

**Environmental Preservation:** Signing about how detrimental it is for our planet when it comes to environmental harm caused by extracting & burning non-renewable fossils fuels addressed by World Wildlife Fund has severe consequences such as degraded air & water quality leading to destroyed habitats & disrupted ecosystems.

However, adopting alternative green power locations such as using wind or that of solar energy could result in less ecological impact mentioned by IEA analysing data over a course till now (2021).

**Energy Security and Independence:** Relying too much idiosyncratic material such as oil exposes any country towards global political risks together with unstable prices & resources in case of supply disruptions, IEA report (2021).

Not having to rely on finite resources and instead going for greener renewable energy usage within a diversified energy combination creates a good balance and promotes sustainable development as well as it helps preserve natural resources stated by World Bank researchers in their report (World Bank, 2020).

**Health Benefits:** The World Health Organizations (WHO) study conducted back in 2018 (WHO, 2018), indicates that using fossil fuels as an energy source comes with substantial risks such as increased air pollution levels which threaten human life; damaging respiratory systems leading up to cardiovascular issues and even premature deaths.

However, investing more resources into green power sources like hydroelectricity or solar power generation could result in fewer hazardous emissions promoting better air quality (EPA, 2021) and healthier outcomes for the public.

**Sustainable Development:** According to the Environmental Protection Agency (EPA) reports in 2021. Using renewable energy resources could lead to achieving sustainable development goals and boost an affordable energy supply that is both friendly to our environment and wallet. The United Nations' (UN) commitment to Sustainable Development Goals makes access to clean energy a priority. By encouraging global efforts towards creating a cleaner environment through more renewable options like solar panels or windmills for electricity production we can identify long term solutions that can benefit everyone's health while reducing greenhouse gas emissions that contribute to climate change (EPA, 2021).

## 2 Sustainable Development Goals and Digital Technologies

Sustainable Development Goals (SDGs) represent a core set of 17 targets aimed at fostering sustainability worldwide. The growth of technology enables significant advancements that accelerate progress towards achieving these objectives. Digital Technology has notably emerged as a driving force for transforming sustainability outcomes. This article explores digital technology's place as central to advancing sustainable development goals around the world.

**Digital Technologies and Sustainable Development Goals:** The potential gains from integration between digitalization solutions such as artificial intelligence, machine learning, and the internet of things add significant value as executing real-time monitoring towards meeting Sustainable Development Goals (SDGs). The knowledge breadth created from collaborating with diverse stakeholders could also yield innovative solutions to solve sustainably critical issues.

Digitalization powering industrial transformations ensures materials are efficiently managed and utilized maximizing production outcome reducing environmental burdens. Waste reduction initiatives extensively implemented include smart energy systems as well as circular economy principles. Precision agriculture techniques are also digitally enabled, optimizing crop yields while preserving the ecosystem and reducing dependency on hazardous substances.

Furthermore, digital technologies increase food security by efficiently supplying remote or underserved areas with relevant applicability for supply chain enhancement reducing food waste. Education, healthcare and other significant SDGs elements also can be better achieved through digital technologies such as e-Learning platforms providing equitable access to quality education where physically impossible. Telemedicine services offer remote healthcare services ensure no one is left out of basic services needed for sustainable futures.

Digital Technologies have a multitude of benefits for accelerating sustainable development goals however deployment responsibility and sustainability without discrimination must be considered. It has become a possibility enhancing accessibility towards Sustainable Development Goals (SDGs) universally.

We have the opportunity to create a future that advances fairness and sustainability through the mindful deployment of digital technologies. By acting intentionally, we can realize this vision.

## *2.1 Overview of Sustainable Development Goals*

All the Member States of United Nations agreed unanimously in 2015 on adopting the Sustainable Development Goals to offer solutions for our world's pressing social-economic-environmental concerns. These objectives serve as a comprehensive framework towards achieving an inclusive environmental future sustainably.

Aimed at providing solutions based on seventeen intertwined principles covering topics ranging from gender parity, combating socio-cultural disparities arising out of inequality, the struggle with ongoing challenges posed by global warming and conflicts while bringing about necessary changes to provide peaceful conflict resolution alongside safeguarding individual rights within societies throughout communities around the world.

Following are the 17 Sustainable Development Goals (SDGs) prosed and agreed by member countries of United Nations:

1. No Poverty
2. Zero Hunger
3. Good Health and Well-being
4. Quality Education
5. Gender Equality
6. Clean Water and Sanitation
7. Affordable and Clean Energy
8. Decent Work and Economic Growth
9. Industry, Innovation, and Infrastructure
10. Reduced Inequalities
11. Sustainable Cities and Communities
12. Responsible Consumption and Production
13. Climate Action
14. Life Below Water
15. Life on Land
16. Peace, Justice, and Strong Institutions
17. Partnerships for the Goals

In what may have been one of history's greatest feats yet, it appeared as if an unanimous message had been passed across globally in favour of sustainable development with the United Nations' Sustainable Development Goals (SDGs) of 2015, overseeing its coordination.

These authoritative regulations aim to drive towards a desirable global future made possible through stronger collaborations amongst individuals such as government representatives, social welfare groups, corporate organizations amongst others towards achieving socio-economic growth that prioritizes environmental protection while paying regards to basic human needs.

The SDGs are an important and integral part of the 2030 Agenda for Sustainable Development, a global commitment signed by 193 member states of the United Nations. With seventeen goals to pursue and achieve by 2030, the SDGs make a comprehensive framework that works in line with some previous accomplishments like the Millennium Development Goals (MDGs) that also addresses upcoming challenges and interconnected issues.

Sustainable Development Goals (SDGs) consist of comprehensive sets of targets with defined indicators, specifically aimed at achieving measurable objectives while emphasizing partnership building among governments worldwide, individual citizens, organizations and business among others to realize their mission statements and expansion demands sustainably. Collective effort and resources to be pooled together at various levels by everyone who genuinely take such issues seriously if success is required.

## *2.2 Role of Digital Technologies in Facilitating Green Power and Green Energy*

Digital technologies such as advanced data analytics, smart grids and IoT devices have played an important role in enabling the integration and optimization of renewable energy sources in the following ways:

**Renewable Energy Integration:** Digital technologies have transformed the integration of renewable sources, thus improving the reliability they play within our power supply chain. With advanced algorithms taking up cues from real-time data collection alongside weather predictions they play a crucial part in ensuring this process is seamless (IRENA, 2017).

**Smart Grids:** As part of this approach; smart grids that include integrated sensor networks work towards monitoring energy flows resulting in an overall improvement in grid stability as well as power distribution efficacy (USDE, 2021).

**Energy Management Systems:** Equally significant are digital platforms that support management systems powering effective tools suited for managing consumption levels while identifying ways through which costs can be reduced at peak moments; a tactic made possible by applications such as load shifting (Schmidt et al., 2022).

**Internet of Things (IoT) and Sensors:** Similarly, IoT devices cover a critical role both with respect to collecting reliable data pertinent to different aspects such us energy-production patterns as well as the performance abilities of equipment deployed within the system (Zanella et al., 2014). An example of how these technologies have manifested is through smart meters; allowing for real-time monitoring while supporting efficient use measures alongside flexible pricing mechanisms.

**Energy Storage Optimization:** The role that digital technology plays in enhancing our current line-up is critical. Advances have ensured batteries perform better than ever before through controlling input/output optimization via programmed algorithms; resulting in an increase in optimized system performance across battery charging/discharging cycles whilst also efficiently integrating itself into our broader power supply network (ESA, 2020).

**Distributed Energy Systems:** Digital tech's major contribution enables independent neighborhoods using new distribution methods such as peer-to-peer trading platform or smaller communities using small-scale grids for sharing energy. This empowers the end-users by giving transparency and accountability to the entire bidding process and allowing individuals to take an active part in shaping the energy market (Acemoglu et al., 2018).

## 2.3 *Affordable and Clean Energy*

The Goal-7 of the Sustainable Development Goals (SDGs) emphasizes on ensuring access to clean, affordable, reliable, sustainable and modern energy to all. It focuses on addressing the challenges of providing affordable clean energy solutions to promote economic growth and mitigating the climate change.

**Universal Access:** SDG Goal-7 emphasizes universal access to affordable, reliable, sustainable and modern renewable energies aiming at promoting economic growth while mitigating climate change effects through technological advances that provide clean energy solutions (UN, 2015).

**Renewable Energy Transition:** This will help solve one of our greatest challenges including sustaining economic development without damaging the environment Resources such as solar power, wind turbines, hydro-power and bioenergy among others are therefore very crucial for achieving clean air in various regions globally (IRENA, 2021a, b).

**Energy Efficiency:** Equally important are innovative technologies aimed at enhancing operational efficiency leading to reduction in wastage across different sectors such as transportation, industry, buildings, appliances etc., (IEA, 2021).

**Investment in Energy Infrastructure:** The Transitioning into a transition towards affordable clean energy has its challenges albeit attainable through concerted efforts between governments; private sector entities as well international organizations (UN DESA, 2019).

**Global Partnerships:** It goes without saying that strategic partnerships will play a pivotal role in actualizing this success including technology transfer capacity building while sharing knowledge in creating an overarching framework within the global community through collaborations at government, private and international levels (UNDP, 2021).

### 2.3.1 Digital Technologies for Renewable Energy

Renewable energy deployment demands constant innovation which digital technologies can provide. They facilitate the integration and optimization procedures while allowing us to monitor and manage various systems resulting in increased efficiency and reliability.

**Renewable Energy Integration**: To integrate renewable energy onto the power grid seamlessly requires digital technologies. Real time monitoring along with advanced forecasting models and control systems provide a means to integrate intermittent renewable energy generation while maintaining stable operation on the power grid (IRENA, 2019).

**Smart Grids**: Smart Grids offer highly efficient integration and management of renewable energy sources with advanced technology using sensors, communication systems and analytics. By utilizing these technologies, we improve energy distribution with demand response capabilities – ultimately leading to better grid balancing solutions. It is highlighted by the UNFCCC (2020) that smart grids play a significant role in combating climate change.

**Energy Management Systems**: Energy Management Systems facilitate real-time data collection on consumption patterns via digital platforms ensuring effective optimization of renewable resources usage while accommodating consumers' needs – such as businesses or utility services. IEA research from 2020 (IEA, 2021) states that employing these technologies can successfully promote sustainable development efforts while minimizing environmental impact.

**Internet of Things (IoT) and Sensors**: The gathering of real-time data on renewable energy systems becomes a seamless process with the use of IoT devices and sensors as stated by Zanella et al. (2014). These technologies provide remote monitoring facilities along with faulty-detection assisting in proper maintenance protocols for the efficient functioning of this infrastructure henceforth reducing any chances of unnecessary downtime so that optimal performance can be achieved.

**Data Analytics and Machine Learning**: Moreover, it was researched by Yin et al., (2020) that through digital mechanisms such as data analytics and machine learning algorithms – improving the generation processes, storage & consumption facilities – became possible along with other benefits such as forecasting for energy requirements while actively optimizing these systems.

**Energy Trading Platforms**: Individuals and even local communities can now exchange their excess renewable energy thanks to digital platforms established for such purposes. The adoption of blockchain technology assures the utmost transparency while ensuring all transactions are safe from any security threat whatsoever (Schneiders & Shipworth, 2021).

### 2.3.2 Digital Technologies for Energy Efficiency

To bolster efforts towards sustainable energy practices in different industries, digital technologies are essential tools. With features such as real-time data provision, advanced analytics and automatic processes, these technological solutions help refine energy management strategies while also facilitating informed decisions that promote conservation.

**Energy Monitoring and Management Systems**: With the advent of new age technology such as smart monitoring tools and efficient building management services it is now easier than ever before to track power usage patterns round-the-clock across multiple sectors including residential buildings or office spaces. By relying on these cyber-enabled applications for experts insights regarding current consumption trends we can zero-in on inefficiencies for potential solutions In doing so we are paving a better future that prioritizes sustainable resource use (USDOE, 2020; UNESCO, 2020).

**Internet of Things (IoT) and Sensors**: Devices interconnected through IoT alongside sensors have the ability to gather data on various surroundings such as energy usage, environmental conditions and equipment performance. Such devices empower full-fledged control in a remote environment for devices consuming energy and make a harmonious energy management system possible, eventually utilizing automation, load control and predictive maintenance (Li et al., 2019).

**Smart Grids**: One major contribution digital technologies offer towards efficient power distribution is the advancement in smart grid technology. Using advanced metering infrastructure coupled with effective communication networks allows for various benefits such as real-time feedback mechanisms to control peak demands while integrating diversified localized sources into mainstream power grid supply while achieving minimum wastage levels (USDOE, 2020). It's no doubt one technological advancement that has transformed how we manage our power needs more sustainably into the future.

**Energy Efficiency Analytics and Optimization**: Data analytics, machine learning and artificial intelligence play a central role when it comes to digital technologies. Together these innovations pave the way for identifying opportunities that save valuable energy while providing optimal levels of output usage. The possibilities involving innovative solutions like predictive modeling and energy forecasting open up further options for more advanced solutions such as dynamic control mechanisms that promote thoroughgoing efficiency improvements (Sun, et.al., 2019).

**Building Automation and Controls**: As we progress further into the digital age we are seeing advancements in several sectors that rely heavily on technology – with one such area being modern building facilities management. Thanks to sophisticated building automation methods driven by state-of-the-art digitization tools; facilities managers can reap substantial benefits via centralized automation processes aimed at developing optimal solutions for air conditioning control alongside enabling smart lighting systems etc. This new wave of integrated solutions targeted at managing eco-friendly practices through intelligent methods has fostered numerous organizations who engage in environmentally responsible practices (Pang et.al., 2018).

**Energy Performance Monitoring and Certifications**: Thanks to digital technologies' advancements today's world can monitor the upgrade on buildings' & industrial activities' capacity for improving effective ways to consume less power & resources. Consequently, allowing for further assessment through implemented standards as well as certifications called out by Energy Efficiency & Renewable Energy (EERE, 2022) promoting responsible consumption practices following transparent procedures aiding overall improvements over extended periods & instilling increased responsibility towards nature conservation alike all around us through two-fold benefits.

## 2.4 Industry, Innovation and Infrastructure

One of the key Sustainable Development Goals (SDGs) focuses on building resilience in infrastructure as well as promoting sustainable industrialization while fostering innovation. Specifically designed for green power solutions that prioritize technology advancement in clean energy sources-Goal 9 highlights the development of comprehensive measures including eco-friendly technologies; adopting greener processes; improving infrastructures as ways of governing better operational efficiency resulting in less overall emissions.

**Sustainable Industrialization**: In its bid to foster sustainability within industries, the UN highlights Goal 9 as a pathway towards adopting green industrial practices. With emphasis on resource efficiency and integration of renewable energy sources while minimizing adverse environmental impacts – industries should pursue greener options including cleaner production processes, eco-friendly technologies and sustainable supply chains (UN, 2015).

**Research and Development**: To facilitate a move towards environmentally sustainable industrial growth with green energy systems taking center stage requires innovation driven by research & technology – so says UNIDO's goal. It emphasizes the need to prioritize increased investment into R&D that promotes the creation of sustainable solutions capable of tackling future environmental challenges (UNIDO, 2016).

**Infrastructure Development**: Goal 9 accentuates the necessity for strong and durable infrastructure, specifically in regards to energy. Such infrastructure must facilitate the expansion of green power and energy. This entails developing renewable energy generation capacity, transmission and distribution networks, and storing systems for optimal reliability and efficiency in supplying energy (IRENA, 2021a, b).

**Access to Technology and Knowledge Transfer**: At the core of this goal lies a recognition of how crucial technology transfer, capacity building, and knowledge sharing are in advocating for sustainable industrial practices and eco-friendly energy solutions. To achieve this aim thoroughly requires enhanced international collaboration and partnerships that will facilitate access to ecologically sound technologies for developing countries (UNCTAD, 2021).

**Private Sector Engagement**: Goal 9 prioritizes acknowledging the critical role that the private sector plays in driving innovation, investment and sustainable industrial growth. The United Nations Global Compact (UNGC, 2019) points out its emphasis on engaging businesses such as entrepreneurs with investments that support green power initiatives promoting infrastructure development oriented for sustainability along with technology advancements.

### 2.4.1 Digital Technologies for Smart Grids

Digital technologies have a critical role in facilitating the evolution and functionality of sophisticated smart grid systems. By empowering these innovative solutions with high-tech sensing capabilities, advanced communication infrastructure as well as robust data analytics capacity it enhances not only electricity generation but also transmission distribution as well as consumption from perspectives related to efficiency reliability and environmental responsibility.

**Advanced Metering Infrastructure (AMI):** Thanks to digital technologies advanced metering infrastructure has become a reality. By utilizing smart meters that provide real time monitoring of energy usage at the consumer level demand response programs are enabled alongside precise billing and better load management in smart grids US Department of Energy (USDOE, 2020).

**Communication and Control Systems:** Digital technologies present an opportunity for smart grids to achieve optimal performance by establishing efficient communication and control networks. The real-time transfer of data between substations together with generation sources as well as consumers enables systematic management of energy flows leading to effective grid balancing (Li, et.al., 2018).

**Distributed Energy Resources (DER) Integration:** Digital technologies play a critical role in enabling the smooth management of distributed energy resources within smart grids. Through their ability to integrate multiple renewable sources of electricity such as solar or wind with storage batteries or hybrid cars seamlessly – offering a technology infrastructure that optimizes resource usage while maintaining reliable grid stability – they drive value up for all stakeholders who engage with new forms of green energy delivery via these platforms worldwide. As noted by the International Energy Agency: "Optimized approaches through end-to-end data processing will deliver a more efficient outcome" (IEA, 2021).

**Data Analytics and Grid Optimization:** Digital technologies employ an array of analytical tools which include data analytics, machine learning alongside artificial intelligence especially effective for vast quantity scrutiny from smart grid sensors, meters and various other forms. Thus, making it feasible to optimize management while handling grid operations effectively, predictive maintenance when applied enhances identification of faults & outages improving overall dependability and efficiency (Li et al., 2020).

**Demand Response and Energy Management:** Digital advancements have had profound effects on demand response programs and energy management systems within smart grids by increasing consumer engagement levels. By providing access to real-time data monitoring alongside alternate control options empowered members are more able to adjust their consumption behaviors accordingly resulting in an optimized process overall. The integration provided through such approaches enables a symbiotic relationship between users and service providers encouraging continued collaboration towards a sustainable future for all stakeholders involved (Vrettos et.al., 2018).

### 2.4.2 Digital Technologies for Energy Storage Systems

Energy storage systems' exceptional output relies heavily on digital technologies that facilitate seamless performance, effectiveness and proper integration into the system. These tools offer dynamic real-time monitoring capabilities for undeterred control mechanisms resulting in efficient optimization of operational processes. Subsequently enhancing grid stability while facilitating effective renewable power source integration while maintaining efficient energy management.

**Battery Management Systems (BMS):** In recent times there has been a significant improvement in how batteries are managed thanks to digital technology. Battery management systems (BMS) have been developed into more refined models boosting energy storage capabilities significantly. The current advanced BMS actively monitors critical areas like battery performance, state-of-charge as well as relevant health parameters in real time ensuring optimum functionability while extending the battery lifespan. The benefits include heightened safety standards making our world much safer (Dey et al., 2016).

**Control and Optimization**: Due to the digital technologies, today we have access to highly advanced control & optimization algorithms exclusively designed for optimizing our energy storage systems. This not only improves efficiency but also ensures a smart approach towards managing charging & discharging mechanism taking into account multiple variables like prevalent grid conditions- demand response signals and energy pricing techniques empowering utilities' stand over their grids & reward investors with exceptional financial benefits while maintaining essential systemic integrity throughout (Zhang et al., 2019).

**Data Analytics and Predictive Maintenance**: The use of digital technologies has facilitated the analysis of energy storage system data by employing data analytics and machine learning algorithms. These technologies offer benefits such as predictive maintenance, early fault detection and performance optimization, which enhance energy storage assets' reliability and efficiency (Marra et al., 2019).

**Integration with Smart Grids**: With their real-time communication capabilities facilitating improved coordination between energy storage assets and grid management systems, digital technologies continue to drive advancements in the integration of energy storage solutions into smart grids. This integration leads to optimal results in terms of stable grids, regulated frequencies and balanced loads that power our modern world efficiently while keeping us safe as well as on track towards our sustainability goals (Huang et al., 2018).

**Virtual Power Plants (VPPs)**: Virtual power plants are able to seamlessly combine multiple distributed energy storage systems due to the advancement in digital technologies. Through harnessing the capabilities of digital platforms, data analytics, along with control systems – VPPs provide an innovative approach towards managing and optimizing the usage of energy storage assets through effective dispatching techniques. Additionally – this lends itself to aiding grid services along with facilitating renewable energy integration (Chen et al., 2020).

## 2.5 Climate Action

The active participation of digital technologies has become indispensable to successfully addressing climate change as they contribute greatly towards monitoring. Analyzing and mitigating greenhouse gas emissions. They offer ample possibilities to explore renewable energy resources and provide necessary assistance in achieving sustainability objectives while concurrently promoting an understanding of the environments' significance.

**Emissions Monitoring and Reporting:** Cutting-edge digital technologies empower us to track greenhouse gas emissions from numerous sources such as transportation, industries and energy production more accurately than ever before. Through exploiting these advanced mechanisms, we can capture consistent data points while enabling us to conduct in-depth analysis resulting in insightful reporting; ultimately supporting efforts for implementing effectual emission reduction measures (UNFCCC, 2019).

**Renewable Energy Integration:** The successful integration of solar and wind power into the national electrical grid is made significantly easier thanks to advancements in digital technology. By enabling precise forecasting as well as real time monitoring and control over renewable energy production. These innovative tools ensure greater overall system stability while simultaneously reducing reliance on traditional non-renewable resources (IRENA, 2021a, b).

**Energy Efficiency and Resource Management:** Smart grid systems, building automation, and intelligent energy management all find vital support in the use of digital technology aimed at promoting more effective sustainability practices. With real-time data sharing capabilities that allow for detailed analytics combined with innovative automated solutions to waste reduction efforts- optimized resource management is achieved alongside significant reductions in overall power consumption (IEA, 2021).

**Climate Data Analysis and Modeling:** Thanks to state-of-the-art digital technologies that include cutting-edge big data analytics software combined with incredibly adept climate modeling applications, we can now unleash unparalleled capabilities when investigating climatic variables affecting our world today. Through this rigorous analysis process backed by powerful software resources come detailed outcomes such as fully-realized climate scenarios that provide crucial information on projected risks—and direct-action points, that policymakers can use when crafting their responses. one does not need any additional context since it won't be referenced here anyway (IPCC, 2014).

**Environmental Awareness and Engagement:** In a world where technology plays an increasingly critical role in shaping our society's narrative towards sustainability whose time is now more important than ever before. Digital technologies offer unique pathways for strengthening environmental awareness providing educational resources right at our fingertips whilst allowing us connect around shared climate change issues. Projects such as clean wave from Lemonade Foundation a UK based company is harnessing blockchain technology to mitigate the plastic waste crisis by empowering cities communities & governments working together, forming strategic alliances. Thereby demonstrating how collectively we all benefit from digitisation efforts weaving social justice into design thinking on a local & global level that has enormous potential or tackling some major planetary challenges ahead (UNEP, 2019).

### 2.5.1 Digital Technologies for Carbon Footprint Reduction

In contemporary times when reducing carbon footprints has become a pressing matter of global concern on account of the warming planet earth and its devastating consequences on life's basic needs like water security or food availability have led us to explore every avenue for solutions.

Digital technologies happen to be one promising area contributing positively to this crucial objective by ushering innovative solutions enabling wider adoptions of more ecologically responsible practices along with offering methods for monitoring and managing emissions effectively at scale across various domains from realizing higher efficiencies in energy consumption or smarter transport systems down to aiding sustainable agricultural production or raising environmental awareness levels among communities about their shared responsibility towards the earth's well-being.

**Smart Grids and Energy Management Systems:** In recent times we have seen advancement in digital technologies providing a pivotal role in revolutionizing how modern-day electricity gridworks are developed through implementing sophisticated approaches aligned with effective resource use leading to lower carbon emissions. This poses significant benefits including efficient workflows covering key phases such as generation plus distribution & consumption patterns using innovative artificial intelligence algorithms further reducing wastage providing enhanced market competitiveness. These can integrate real-time monitoring and analytics platforms, enabling fine-grained control over load balancing & demand response aiding in seamless integration of renewable energy sources optimizing the overall performance of smart grids systems (Carvalho et al., 2018).

**Internet of Things (IoT) and Building Automation:** Energy conservation remains high on our agenda even more so in recent times due to the drive towards sustainability. One way we could achieve this goal would be through active deployment of IoT technologies alongside comprehensive sensing equipment across various integrated sites within commercial or residential buildings. By leveraging these tools efficiently whilst employing modern algorithms designed for optimization of HVAC systems as well as other operational components inside the building could result in reduced energy wastage while increasing efficiency levels thereby promoting a sustainable future (Shariatzadeh et al., 2021).

**Remote Work and Video Conferencing:** The implementation of digital technologies has enabled individuals to work remotely with access to virtual meetings. As a result, this diminishes the necessity for commuting or business travel while elevating environmental sustainability efforts through lower carbon emissions from transportation. There is also an improvement in energy conservation within shared office spaces (Valente & Chieffi, 2020).

### 2.5.2 Digital Technologies for Climate Change Adaptation

Digital technologies have become increasingly crucial in supporting efficient adaptation measures for combatting climate change. By offering robust data analysis capabilities alongside valuable information dissemination platforms such as remote sensing techniques and decision-making support systems. These technological advancements play an essential role in enabling effective strategies to combat the threat posed by changing climatic conditions.

**Climate Modeling and Risk Assessment:** It is through the advancements made in digital technologies that we can now develop and leverage climate models for modelling future weather patterns. By doing so. We gain insight into how global warming may potentially impact important sectors such as agriculture, water resources management as well as our coastal ecosystems (IPCC, 2014).

**Remote Sensing and Monitoring:** With cutting edge innovations like remote sensing techniques and satellite imaging at our disposal. We now have unprecedented access to robust data detailing key climatic indicators such as temperature fluctuations, precipitation levels, and alterations to land cover. This information is essential for monitoring ongoing ecological developments with precision while pinpointing areas that may require specific support or interventions due to heightened vulnerability factors. Ultimately incorporating this type of real time feedback into adaptation planning ensures a sustainable future for everyone impacted by environmental emergencies or slowly evolving conditions alike (Huang et al., 2019).

**Decision Support Systems:** For individuals seeking reliable guidance on climate change adaptation strategies there are numerous digital tools available. Decision support systems effectively integrate myriad data sources including relevant models which allow individuals such as policymakers or organizational leaders the ability to create precise plans based on their unique needs for assessing potential risks/ vulnerability findings.

## 3 Innovative Business Models for Sustainable Development

To advance sustainable development its vital to adopt business models that prioritize environmentally friendly practices, resource efficiency and social responsibility. Innovative models like the sharing economy, circular economy and pay per use systems exemplify these priorities.

**Circular Economy**: By championing product and material reuse through repair, re-manufacturing, and recycling efforts, the circular economy business model aims to minimize waste while maximizing resource value (Geissdoerfer et al., 2017).

**Sharing Economy**: Sharing economy businesses enable individuals to rent or share underutilized assets rather than buying new ones. By promoting the reuse of existing resources. This approach supports environmentally conscious consumption patterns that can positively impact our planets future (Botsman & Rogers, 2010).

**Pay-per-Use Models**: Models that allow you to pay for access or subscribe to services instead of owning them have valuable benefits. These models promote efficient use of resources and motivate manufacturers to develop more sustainable products (Tukker et al., 2019).

Following are the newer innovative business models developed based on the concepts discussed above:

## *3.1 Peer-To-Peer Energy Trading Platforms*

Peer-to-peer (P2P) energy trading platforms enable direct energy transactions between producers and consumers, promoting decentralized and sustainable energy systems.

**Definition and Function:**

To establish a smoother movement of energy between producers and consumers P2P energy trading platforms have emerged as a solution. The use of advanced digital technologies like blockchain and smart contracts on these platforms allows for secure and transparent transactions in localized networks (Chen et al., 2017).

**Benefits and Advantages:** P2P energy trading platforms offer several benefits such as increased efficiency, renewable energy integration and empowering energy prosumers.

## *3.2 Virtual Power Plants*

Virtual power plants (VPPs) leverage digital technologies to integrate and optimize the operation of decentralized energy resources, enabling efficient energy management and grid balancing.

**Definition and Function**: With rapid advancements in new-age technology happening across various sectors around the world – the field of distributed energy resources is no exception. Virtual Power Plants (VPPs) serve as an excellent example.

VPP combines coordinating functions aimed at maximizing electricity production efficiency with smoothly managing consumption needs derived from DER sources including wind turbines, wave turbine farms or even hydrogen fuel cells (Mohsenian-Rad et al., 2010).

Thanks to their ability to use innovative monitoring techniques and smartly configured control-and-communication features. VPP environments create an efficient framework that perfectly manages these critical tasks.

**Benefits and Advantages**: Virtual power plants offer several benefits such as grid flexibility and stability, optimization of energy assets and integration of renewables and storage.

## 3.3 *Blockchain Technology*

Blockchain technology holds promise in promoting sustainable development by improving transparency and accountability through immutable records of transactions. Traceability and Sustainability by recording and verifying the origin, authenticity and movement of products and materials in supply chains supporting sustainability initiatives. Transformation and efficiency in various sectors including green energy sector by enabling peer-to-peer energy trading, efficient grid management and renewable energy certificate tracking (Tapscott & Tapscott, 2017; Yao et al., 2019).

## 3.4 *Energy-As-A-Service Models*

With consumers increasingly demanding environmentally responsible forms of power generation, Energy as a Service (EaaS) models have emerged as revolutionary business ideas for meeting these demands. By offering an array of dedicated services designed specifically to promote renewable adoption enhance efficiency and promote cleaner sources of power generation they play an important role in creating sustainable development practices within businesses.

**Definition and Function:** Customers are increasingly looking towards innovative Energy-as-a-Service (EaaS) solutions that offer great value propositions for tailored power usage requirements. The essence of an EaaS model lies in enabling clients to bypass heavy investment costs on their own infrastructural facilities and equipment required for energy production or utilization upgrades. Instead, they pay a fee solely on consumption-based delivery of desired services needed by them. Stacked with features like efficient management support systems, real-time optimsation assistance and insightful monitoring frameworks. EaaS models continue to gain momentum as alternative options within the industry (Borelli et al., 2018).

Energy-as-a-Service (EaaS) models offer benefits such as energy efficiency and optimization and renewable energy adoption (Diaz-Arriaga et al., 2019; Petersen et al., 2019).

# 4 Development of New Renewable Energy Sources for Sustainable Development

Sustainable development goals demand us to give prominence to the growth of new renewable energy sources as a priority. This will assist in augmenting our global inventory by increasing the proportion of clean and renewable forms of power usage. A wide range of accessible forms- solar panels, windmills turbines, water-based generators inspired by hydropower, biomass products generated algae, plants etc., such green technologies can significantly contribute towards ecological preservation combating climate change while safeguarding an enhanced supply chain process with enriched security levels (United Nations, 2020).

New renewable energies continue to develop rapidly thanks largely to technological advancements. Through increased efficiency in solar panels producing larger and cheaper wind turbines enhancing our capacity for energy storage as well as innovation towards exploiting power from tides, waves or currents. These drivers contribute greatly by enabling a newly viable sustainable future (IREA, 2020).

Research into ground breaking technology for creating sustainable energy is being pursued by various establishments of varying sectors – including both governmental and non-governmental institutions alike. Of particular interest in this field are emerging areas such as solar photovoltaics (PV). Advanced biofuels with high potential yield rates, innovative wave/tidal power generating tools coupled with more efficient structures to enable heightened geothermal system performance (IEA, 2021).

## *4.1 Machine Learning and Artificial Intelligence*

As we strive towards achieving sustainability worldwide. The role of Machine Learning (ML) and Artificial Intelligence (AI) cannot be overlooked. Their contributions are significant in facilitating better decision-making processes. Optimizing resource usage and improving environmental management practices.

**Decision-Making and Planning:** ML and AI algorithms play a crucial role in promoting data-driven decision-making across different sustainable development sectors. In areas such as energy utilization, transportation services access, agricultural practices optimization, and urban planning layouts evaluation, these technologies are witnessing increased adoption rates. By scrutinizing big data with precision to detect relevant patterns effectively leads to value-added insights critical for enhancing informed choice-making (Koivisto et al., 2019).

**Resource Optimization:** Efficient resource utilization is key for maintaining an eco-friendly society today – with that very objective being notably achievable through advancements in artificial intelligence or machine learning ventures. Applying refined models married with cutting-edge algorithms allows businesses to balance their resource-gathering activities aggressively while consequently minimizing the wastage involved during processing or production activities significantly (Hassani et al., 2018).

Moreover, innovations that focus on perpetuating smart infrastructure that caters adequately towards better quality control measures provide immense value-addition – inducing a decrease in environmental impact through overall high-level advancements in energy, water, and material resource management (Hassani et al., 2018).

**Environmental Monitoring and Management:** To ensure the protection and preservation of our planets natural resources requires innovative approaches that incorporate advanced technologies like AI and ML. The ability to analyze satellite imagery, sensor data. And other relevant sources has made it possible to tackle complex issues related to climate modelling. Biodiversity conservation and pollution management. Esteva et al. (2019) assert that these technological advancements are fundamental in solving ecological challenges we face today while sustaining future generations' needs as well.

## *4.2 Internet of Things (IoT)*

The utilization of the Internet of Things (IoT) is a vital driving force in achieving sustainable development since it provides channels for the efficient management of resources, environmental monitoring, and smarter infrastructure. In essence. IoT bridges the gap between technology and sustainable development by providing solutions that not only enhance our quality of life but also help to safeguard the environment.

**Efficient Resource Management:** Innovative advancements like IoT devices and sensors enable the collection of real time data on resource consumption from consuming water to electricity usage. With this information at our fingertips comes superior resource management strategies for a more optimized approach towards sustainable development. Thanks to these technologies industries such as energy efficient buildings, smart grids, precision agriculture and intelligent transportation systems continue to enjoy seamless functionalities (Atzori et al., 2017).

**Environmental Monitoring and Management:** IoT has revolutionized environmental monitoring & management through its innovative sensor deployment model that fosters continuous collection of crucial air quality parameters, water quality indexes & climatic characteristics. This critical information provides insight into potential threats while empowering us to conserve vital natural resources better & undertake effective damage control measures when disasters strike (Zanella et al., 2014).

**Smart Infrastructure:** IoT enables the development of smart infrastructure, including smart buildings, smart cities, and intelligent transportation systems. By connecting various devices and systems, IoT promotes energy efficiency, waste reduction, and optimized urban planning (Caragliu et al., 2011).

### *4.3 Digital Twin Technology*

The promotion of sustainable development calls for innovative approaches to managing resources while ensuring minimal wastage. One such approach that has proven effective is the utilization of Digital Twin Technology – a concept whereby virtual replicas are created for physical systems. By embracing this technology effectively, we can optimize our operations while enabling efficient resource management practices and predictive maintenance strategies – all necessary components that drive our goals towards sustained development as society progresses into the future.

**Efficient Resource Management:** Digital Twin Technology allows for real-time monitoring and analysis of resource consumption, such as energy and water, in various sectors, including buildings, manufacturing, and transportation. It enables better resource management, optimization, and identification of energy-saving opportunities (Fernandez et al., 2020).

**Predictive Maintenance:** Digital Twins use real-time data and predictive analytics to identify potential equipment failures and optimize maintenance activities. By detecting maintenance needs in advance, it reduces downtime, extends the lifespan of assets, and minimizes waste (Zheng et al., 2021).

**Optimized Operations:** Digital Twins simulate and optimize operations, allowing for improved energy efficiency and reduced environmental impact. It enables the testing of different scenarios and the identification of optimal strategies for achieving sustainability goals (Baines et al., 2019).

## 5 Challenges and Opportunities for Sustainable Development

Sustainable development represents an opportunity for collective efforts across the board in addressing various existing challenges affecting societies globally. One prominent concern pertains to climate change that exacerbates social inequalities by negatively affecting ecosystems' stability alongside livelihoods whilst creating pitfalls in coherent economic policies altogether. The mitigation measures against greenhouse gas emissions must be coupled with comprehensive adaptation strategies (IPCC 2018), not forgetting equally imperative goals such as eradicating poverty via equitable socio-economic policies aimed at realizing access levels for all citizens - from basic needs such as healthcare & education through inclusive economic growth initiatives alike (UNDP's Human Development Report, 2015).

Encouraging developments in technology create numerous possibilities for sustainable advancement by introducing new strategies and solutions. Take renewable energy technologies, smart infrastructure or digital tools designed for resource management as examples of such innovations (UNESCO, 2020). A transition to a circular economy that stresses the reuse of resources, minimizing wastage by recycling operations presents an ample number of opportunities for sustainable development. Additionally, to a healthy consumption pattern and productive processes that conserve resources; it promotes the management of waste materials (Ellen MacArthur Foundation, 2015).

## *5.1 Technical Challenges*

Sustainable development faces various technical challenges that require innovative solutions and advancements in technology. Technical Challenges in Sustainable Development are:

**Energy Storage:** For smooth integration of renewable power sources into the grid. Its' crucial to deploy reliable energy storing solutions. Nonetheless. Cost effective and modernized energy storing methodologies like hydrogen storing or advanced battery technologies pose significant challenges (González Díaz et al., 2020).

**Sustainable Infrastructure:** Constructing resilient infrastructure that sustains itself is no small task. Such a feat involves grappling with complex technicalities such as energy-efficient materials incorporation, clever implementation of smart technologies for monitoring and optimization purposes, and the adoption of sustainable construction methodologies (Ofori et al., 2018).

**Water Management:** Efficient and sustainable water management plays a fundamental role in achieving sustainable development. Consequently, addressing technical challenges to develop advanced technologies for purification/treatment of wastewater; implementing effective preemptive/water resource managements techniques; adopting practices for conserving freshwater resources become fundamental (Mehta et al., 2017). Indeed, it is a challenge worth tackling considering the magnitude pronounced in Mehta et al.'s work published back in 2017.

## *5.2 Policy and Regulatory Challenges*

Sustainable development encounters various policy and regulatory challenges that require effective governance and coordination among stakeholders. The following are some of the policy and regulatory challenges in sustainable development.

**Policy Alignment:** Realizing sustainable development demands policy alignment across various sectors and governance levels. Achieving this objective poses several challenges, such as ensuring coordination of energy, transportation, land use and waste management policies for coherence and integration (Bulkeley et al., 2014).

**Institutional Coordination:** Collaborative efforts between various institutions are integral to promote sustainable development. To this end, appropriate measures must be taken which includes establishing cross-sectoral coordination mechanisms and enhancing collaboration among government agencies whilst involving stakeholders in decision making process (Betsill & Bulkeley, 2006).

**Regulatory Frameworks:** In order to achieve sustainable development goals effectively, creating suitable regulatory frameworks is crucial. By addressing shortcomings in existing systems through meticulous rulemaking while prioritizing compliance with environmental and social norms will pave the way towards efficient implementation (Sustainable Development Solutions Network, 2015).

### *5.3 Opportunities for Innovation and Collaboration*

Sustainable development offers numerous opportunities for innovation and collaboration to address environmental, social, and economic challenges which are as follows:

**Technological Innovation:** The growth of technology holds promise for promoting sustainability. Among the possibilities are digital monitoring systems, renewable energy technologies, and energy efficient solutions; these advancements have paved the way towards environmentally friendly practices across various fields. UNEP (2017) highlights these technologies as encapsulating the potential that technology brings to sustainable development.

**Collaborative Partnerships:** A collaborative approach involving the participation of various stakeholders is crucial to achieve sustainable development. To this end, governments, businesses, civil societies both formally and informally along with academia must come together to form effective partnerships. These partnerships have immense potential to facilitate mutual exchange of knowledge, pooling resources and taking cohesive steps towards solving pressing sustainability issues locally regionally or globally. As indicated by United Nations (2002), this approach is indispensable, because without it attaining long term sustainability would be difficult.

**Social Innovation:** Social innovation revolves around creating novel approaches, solutions, and practices to combat societal and ecological issues. It encourages active participation from communities in decision making processes while ensuring that marginalized groups are not left out in the pursuit of sustainable development (Murray et al., 2010).

Many more challenges and opportunities to innovate and collaborate in the sustainable development processes may arise in the near future.

## 6 Conclusion

After making an exhaustive effort to understand various nuances of green power and green energy as well as role of digital technologies in optimizing the production, transmission and usage of green power and green energy more sustainably, we can draw the conclusion by summarizing the key points, peeping in to the future prospects as well as suggesting action plan to the community.

## 7 Summary of Key Points

- Now more than ever before, digital technology plays an indispensable role in optimizing green power resources by enabling better management, monitoring, and control over renewable electricity sources.
- The resulting improvements add further momentum towards wider adoption aside their undoubted environmental benefits.
- Smart grid systems work hand-in-hand with these advanced technological applications by empowering real-time scrutiny behind coalitions between generation output levels against actual consumption patterns. Findings are utilized to suggest more efficient electricity resource distribution across the grid while simultaneously reducing wasted potential.
- The seamless integration of Internet of Things (IoT) devices and sensors into renewable energy systems aid in data collection. This data assists in strategizing towards a greener and much more sustainable power generation module.
- The adoption of advanced analytics and machine learning algorithms largely eliminates dependence on other sources of energy, ensuring that resources are aptly directed with the assurance that renewable power generation is rightfully assumed. It is fast proving to be an effective route towards more efficient utilization preventing environmental damage through reduced reliance on non-sustainable energy sources.
- With digital platforms and marketplaces adoption, better-coordinated engagements entailing green energy resource producers and consumers become possible. Additionally, the widespread functionality now accorded virtual power plants - which aggregate decentralized renewable electricity sources- ensures improved access towards buying up excess electric loads from drier locations thereby ultimately supporting grid solutions when distributed more evenly as needed by end consumers.
- Lastly, energy management systems powered by digital technology have gained traction as means to provide necessary stakeholders with real-time data regarding their electricity consumption patterns catalyzed by a hunger for environmental sustenance. Optimum results have been realized over time as users can make informed decisions aimed at optimizing waste reduction practices while enhancing personal accountability.

## 7.1 *Future Outlook for Digital Technologies and Green Power & Green Energy in the Process of Sustainable Development*

The future-outlook for digital technologies and green power/energy in the context of sustainable development is promising, with significant potential for positive environmental, social, and economic impacts.

**Renewable Energy Integration:** Achieving successful integration of renewable energy sources into the grid highly depends on digital technologies that are in use today. To enable this initiative advanced monitoring and control systems alongside smart grids utilize predictive analytics as tools that synchronize with efficient renewable energy management processes including power generation optimization, secure storage as well as strategic distribution (Yan et al., 2018).

**Energy Efficiency Optimization:** Advancements in digital technology hold great promise for boosting energy efficiency across many industries. In buildings transportation systems and industrial processes, real-time monitoring can be achieved through the use of IoT devices along with machine learning algorithms and data analytics thus, allowing smarter automation control leading to better optimizations where saving energy is concerned (Zhang et al., 2018).

**Decentralized Energy Systems:** Advances in digital technology over time have unlocked new possibilities by creating decentralized power structures capable of addressing long-standing issues with current centralized models. As a result, virtual power plants are becoming a mainstream concept that enables local producers to participate in the broader market while reducing costs overall, also coupled with exciting experiments around blockchain technology which strengthen these connections even further. Ultimately this would lead towards more resilient communities built on sustainability principles through added layers of social participation (Pérez-Díaz et al., 2021).

**Data-Driven Decision Making:** A promising future awaits us as data becomes more readily available coupled with advances in artificial intelligence. It paves a path towards decision making based on reliable information whereby sustainability plays a central role in an eco-friendly world, thanks to predictive analytics. Optimization models as well as scenario analysis that allow policymakers to craft informed policies while allocating scarce resources efficiently (Fais et al., 2020).

## 7.2 *Call-To-Action for Governments, Businesses, and Individuals*

Governments, businesses, and individuals play crucial roles in promoting sustainable development through the adoption and utilization of digital technologies. The following are the actions that these entities should take up for the welfare of the people:

**Governments**: In taking strides towards sustainable developments through digital technology which respects stakeholders' interests globally; here are some essential actions worth considering by all relevant government authorities.

One key action is to work towards developing supportive social policies aimed at stimulating integration as well as deployment of digital solutions by incentivizing renewable energy investment while promoting excellent energy efficiency standards among other approaches geared towards facilitating innovations within the sector (United Nations; (UN), 2015).

The other critical measure necessary lies with partnerships alongside engagements between governments on one hand; outmatched industry players such as academia professionals/civil society organizations on the other hand -as a proactive approach- makes possible knowledge sharing with regards sustainability practices around this field; capacity building and the transfer of technology (UN, 2017).

**Businesses**: It is no exaggeration highlighting how businesses can help foster sustainable development by taking proactive steps such as investing in innovation via research & development of digital solutions worldwide. Such steps can include building energy-efficient technologies, promoting clean energy adoption along with prioritizing sustainability in product design & manufacturing processes (World Business Council for Sustainable Development, 2010).

Moreover, to maintain accountability when employing advanced technology, companies need to act transparently by sharing valuable insights including ethical practices via stakeholder communication. They should work closely with local communities while disclosing environmental impact reports (Global Reporting Initiative, 2016).

**Individuals:** Promoting sustainable development requires individual responsibility on a global scale where everyone has a role to play today so that we sustainably thrive when tomorrow comes knocking.

The first step towards this is self-education about digital technologies' potential benefits while also being aware of how their misuse contributes negatively towards planetary health through wastefulness (United Nations Development Programme, 2015).

We could further make more progress by taking concerted efforts such as adopting efficient practices that culminate into environmentally friendly lifestyles through support of renewable energy sources and responsible consumption.

Furthermore, we must actively participate in advocacy efforts and communicate our support for sustainability initiatives to policymakers. Businesses as well as communities we are part of (Sachs et al., 2019).

## References

Acemoglu D, Akcigit U, Alp H, Bloom N, Kerr W (2018) Innovation, reallocation, and Growth. American Economic Review 108(11):3450–3491. https://doi.org/10.1257/aer.20130470

Atzori L et al (2017) The Internet of Things: A Survey. Comput Netw 54:2787–2805

Baines T et al (2019) Digital Technology and Sustainability: Positive Disruption. Production Planning & Control 30(13):1071–1092
Betsill MM, Bulkeley H (2006) Cities and the Multilevel Governance of Global Climate Change. Glob Gov 12(2):141–159
Borelli F et al (2018) Energy-as-a-Service: A Survey. Appl Energy 222:545–559
Botsman R, Rogers R (2010) What's Mine is Yours: The Rise of Collaborative Consumption. HarperCollins Publishers, New York
Bulkeley H et al (2014) Urban Energy Landscapes: A Review. J Urban Des 19(1):39–60
Caragliu A et al (2011) Smart Cities in Europe. J Urban Technol 18(2):65–82
Carvalho P et al (2018) Smart Grid Technologies for Sustainable Energy Sector: A Comprehensive Review. Renew Sustain Energy Rev 81(2):2002–2022
Chen Y et al (2017) A Review of P2P Energy Trading Frameworks. Appl Energy 206:1494–1509
Chen M et al (2020) Virtual Power Plants: A Survey and Outlook in the Era of Energy Internet. Appl Energy 279:115776
Dey AK, Hussain FK, Chetty G (2016) Recent Advances in Battery Technologies for Electric Vehicles: A Review. J Power Sources 324:186–200
Diaz-Arriaga F et al (2019) Energy-as-a-Service: A Review and Comparative Analysis of Business Models for Electric Vehicle Charging Infrastructure. Energies 12(24):4761
Ellen MacArthur Foundation. (2015). Towards the Circular Economy: Opportunities for the Consumer Goods Sector.
Energy Efficiency and Renewable Energy (EERE). (2022). Energy Efficiency. Retrieved from https://www.energy.gov/eere/energy-efficiency
Energy Storage Association (ESA). (2020). Energy Storage 101: Introduction to Storage. Retrieved from https://energystorage.org/why-energy-storage/energy-storage-101/
Environmental Protection Agency (EPA). (2021). Clean Energy: Renewable Electricity Generation. Retrieved from https://www.epa.gov/cleanenergy/renewable-electricity-generation
Esteva A et al (2019) A Guide to Deep Learning in Healthcare. Nat Med 25(1):24–29
Fais B et al (2020) Data-Driven Energy Planning and Management: A Review. Renew Sustain Energy Rev 132:110090
Fernandez C et al (2020) Digital Twins in the Energy Sector: A Review of Challenges, Solutions, and Opportunities. Appl Energy 276:115402
Geissdoerfer M et al (2017) The Circular Economy – A New Sustainability Paradigm? J Clean Prod 143:757–768
Global Reporting Initiative (GRI). (2016). GRI Standards: Sustainability Reporting Guidelines. Retrieved from https://www.globalreporting.org/standards
González-Díaz B et al (2020) Technological and Economic Challenges for the Large-Scale Integration of Photovoltaic Power Generation in the European Union. Energy Policy 137:111171
Hassani A et al (2018) Artificial Intelligence for Data-driven Building Energy Management: A Review. Renew Sustain Energy Rev 81(1):1175–1187
Huang S, Chen W, Liu Z (2018) A Survey on Demand Response in Smart Grids: Mathematical Models and Approaches. Energies 11(6):1334
Huang Q et al (2019) Remote Sensing for Climate Change Research: Progress, Challenges, and Opportunities. J Geog Sci 29(7):1043–1068
IEA. (2021). World Energy Outlook 2020. Retrieved from https://www.iea.org/reports/world-energy-outlook-2020
Intergovernmental Panel on Climate Change (IPCC). (2014). Climate Change 2014: Synthesis Report. Retrieved from https://www.ipcc.ch/site/assets/uploads/2018/05/SYR_AR5_FINAL_full.pdf
International Energy Agency (IEA). (2021). Energy Efficiency 2021 - Analysis and Outlooks to 2040. Retrieved from https://www.iea.org/reports/energy-efficiency-2021
International Energy Agency. (2020). Technology Innovation Outlook 2020: Accelerating the Energy Transition Through Innovation. Retrieved from https://www.iea.org/reports/technology-innovation-outlook

International Renewable Energy Agency (IRENA). (2017). Digitalization & Renewable Energy. Retrieved from https://www.irena.org/-/media/Files/IRENA/Agency/Publication/2017/Mar/IRENA_DIGITALIZATION_RENEWABLE_ENERGY_2017.pdf

International Renewable Energy Agency (IRENA). (2018). Renewable Energy: A Definition. Retrieved from https://www.irena.org//media/Files/IRENA/Agency/Publication/ 2018/Jun/ IRENA_RE_Definition_A5_leaflet.pdf

International Renewable Energy Agency (IRENA). (2019). Innovation Landscape Brief: Digitalization and Renewable Energy. Retrieved from https://www.irena.org/-/media/Files/IRENA/Agency/Publication/2019/Nov/IRENA_Digitalization_and_Renewable_Energy_2019.pdf

International Renewable Energy Agency (IRENA). (2021a). Renewable Energy Statistics 2021. Retrieved from https://www.irena.org/publications/2021/Mar/Renewable-Energy-Statistics-2021

International Renewable Energy Agency (IRENA). (2021b). The Power to Change: Renewable Energy Policies and Incentives for a Clean Energy Future. Retrieved from https://www.irena.org/publications/2021/Jun/The-power-to-change

International Renewable Energy Agency. (2020). Renewable Power Generation Costs in 2019. Retrieved from https://www.irena.org/-/media/Files/IRENA/Agency/Publication/2020/May/IRENA_Renewable-Power-Generation-Costs-in-2019.pdf

IPCC. (2018). Global Warming of 1.5°C. Retrieved from https://www.ipcc.ch/sr15/

Koivisto MJ et al (2019) Machine Learning in Sustainable Development. Nature Sustainability 2:280–290

Li F, Li J, Zhang Q, Wang J (2018) Communication and Control System in Smart Grid: A Review. IEEE Trans Industr Inf 14(11):5057–5067

Li B, Liang Y, Zhang H, Hu J, Zeng Y (2019) Internet of Things for Energy Efficiency in Buildings: A Review. Energy and Buildings 190:89–101

Li F, Wang Z, Zhang C (2020) Data Analytics for Smart Grids: Methods, Challenges, and Opportunities. IEEE Trans Industr Inf 16(1):6–14

Marra F et al (2019) Predictive Maintenance in Energy Storage Systems: A Review. Renew Sustain Energy Rev 103:78–93

Mehta P et al (2017) Sustainable Urban Water Management: Case Study of Singapore. Water 9(5):325

Mohsenian-Rad AH et al (2010) Autonomous Demand-Side Management Based on Game-Theoretic Energy Consumption Scheduling for the Future Smart Grid. IEEE Transactions on Smart Grid 1(3):320–331

Murray R et al (2010) The Open Book of Social Innovation. Young Foundation, London

Ofori G et al (2018) Sustainable Construction and Infrastructure Development: A Case Study of Singapore. J Clean Prod 196:305–317

Pang Y, Elmenreich W, Vallée D (2018) Building Automation Systems: A Review of Concepts and Technologies. Build Environ 123:543–554

Pérez-Díaz JI et al (2021) A Review on Blockchain-Based Peer-to-Peer Energy Trading: Current Challenges and Future Opportunities. Renew Sustain Energy Rev 143:110897

Petersen NB et al (2019) Energy-as-a-Service Business Models for Energy Efficiency in the Built Environment: A Systematic Review. Energy and Buildings 204:109484

Sachs JD et al (2019) Transformations to Achieve the Sustainable Development Goals. The Lancet 393(10187):2452–2465

Schmidt M, Marrone S, Paraschakis D, Singh T (2022) Artificial Intelligence in the energy transition for solar photovoltaic small and medium-sized enterprises. Digital Humanism. Springer, Cham, pp 105–123

Schneiders A, Shipworth D (2021) Community Energy Groups: Can they shield consumers from the risks of using blockchain for peer-to-peer energy trading? Energies 14(12):3569. https://doi.org/10.3390/en14123569

Shariatzadeh N et al (2021) Internet of Things (IoT)-based Energy Management Systems for Smart Buildings: A Comprehensive Review. Sustain Cities Soc 77:102203

Sun X, Wang Z, Zhang Y, Wang Z, Lin X (2019) Application of Data Analytics and Machine Learning in Renewable Energy Systems: A Review and Outlook. Energies 12(12):2403

Sustainable Development Solutions Network. (2015). Indicators and a Monitoring Framework for Sustainable Development Goals: Launching a Data Revolution for the SDGs.

Tapscott D, Tapscott A (2017) Blockchain Revolution: How the Technology Behind Bitcoin is Changing Money, Business, and the World. Portfolio.

Tukker A et al (2019) Pay-Per-Use Business Models as a Catalyst for Sustainable Consumption? J Clean Prod 221:347–359

U.S. Department of Energy (DOE). (2020). Advanced Metering Infrastructure (AMI). Retrieved from https://www.energy.gov/oe/services/technology-development/smart-grid/advanced-metering-infrastructure-ami

U.S. Department of Energy (DOE). (2021). Green Power and Renewable Energy. Retrieved from https://www.energy.gov/eere/green-power-renewable-energy

U.S. Department of Energy. (2021). The Role of Digitalization in the Modernization of the Power Grid. Retrieved from https://www.energy.gov/sites/prod/files/2021/01/f82/BTO-Role-of-Digitalization-Modernization-of-Power-Grid-012621.pdf

United Nations (UN). (2015). Sustainable Development Goals. Retrieved from https://sdgs.un.org/goals

United Nations (2002) Report of the World Summit on Sustainable Development. United Nations, Johannesburg, South Africa

United Nations Conference on Trade and Development (UNCTAD). (2021). Technology and Innovation Report 2021: Connecting to the Value of Sustainable Development. Retrieved from https://unctad.org/system/files/official-document/tir2021_en.pdf

United Nations Department of Economic and Social Affairs (UN DESA). (2019). SDG 7: Ensure access to affordable, reliable, sustainable, and modern energy for all. Retrieved from https://sustainabledevelopment.un.org/sdg7

United Nations Development Programme (UNDP). (2015). Sustainable Development Goals: 17 Goals to Transform Our World. Retrieved from http://www.undp.org/content/undp/en/home/sustainable-development-goals.html

United Nations Development Programme (UNDP). (2021). Sustainable Development Goal 7: Affordable and Clean Energy. Retrieved from https://www.undp.org/sustainable-development-goals/goal-7-affordable-and-clean-energy

United Nations Educational, Scientific and Cultural Organization (UNESCO). (2020). Futures of Education: Learning to Become.

United Nations Environment Programme (UNEP). (2017). Towards a Pollution-Free Planet: Background Report. Nairobi, Kenya: UNEP.

United Nations Environment Programme (UNEP). (2019). Digital Transformation for Climate Action. Retrieved from https://www.unep.org/resources/report/digital-transformation-climate-action

United Nations Framework Convention on Climate Change (UNFCCC). (2019). Enhanced Reporting and Review under the Convention. Retrieved from https://unfccc.int/topics/transparency-and-reporting/the-big-picture/what-is-enhanced-reporting-and-review-under-the-convention

United Nations Framework Convention on Climate Change (UNFCCC). (2020). Digital Technologies and Renewable Energy Integration. Retrieved from https://unfccc.int/process-and-meetings/the-paris-agreement/what-is-the-paris-agreement/the-role-of-the-technology-mechanism/digital-technologies-and-renewable-energy-integration

United Nations Global Compact. (2019). SDG 9: Industry, Innovation, and Infrastructure. Retrieved from https://www.unglobalcompact.org/sdgs/9-industry-innovation-and-infrastructure

United Nations Industrial Development Organization (UNIDO). (2016). Sustainable Development Goal 9: Industry, Innovation and Infrastructure. Retrieved from https://www.unido.org/sdgs/goal-9-industry-innovation-and-infrastructure

United Nations. (2017). Partnerships for the Goals: Revitalizing the Global Partnership for Sustainable Development. Retrieved from https://sustainabledevelopment.un.org/partnerships

United Nations. (2020). Sustainable Development Goal 7: Ensure Access to Affordable, Reliable, Sustainable, and Modern Energy for All. Retrieved from https://www.un.org/sustainabledevelopment/energy/

United States Department of Energy (DOE). (2021). Smart Grid. Retrieved from https://www.energy.gov/oe/services/electricity-policy-coordination-and-implementation/smart-grid

Valente T, Chieffi S (2020) Carbon Footprint and Environmental Impact of Teleworking: A Case Study in Italy. Resour Conserv Recycl 163:105079

Vrettos E, Lymberopoulos N, Hatziargyriou N, Bousbouras C (2018) Energy Management and Demand Response in Smart Grids: Overview, Challenges, and Solutions. IEEE Trans Industr Inf 14(7):2767–2775

World Bank. (2020). Renewable Energy. Retrieved from https://www.worldbank.org/en/topic/energy/renewable-energy

World Business Council for Sustainable Development (WBCSD). (2010). Vision 2050: The New Agenda for Business. Retrieved from https://www.wbcsd.org/Programs/Sustainable-Cities/News/2020-vision

World Health Organization (WHO). (2018). Ambient Air Pollution: Health Impacts. Retrieved from https://www.who.int/airpollution/ambient/health-impacts/en/

Yan J et al (2018) Digitalization and Energy Systems Integration: A Comprehensive Review. Appl Energy 220:538–556

Yao L et al (2019) Blockchain Technology for Supply Chains: A Systematic Literature Review and Framework. J Clean Prod 240:118153

Yin L, Gao Q, Zhao L, Zhang B, Wang T, Li S, Liu H (2020) A review of machine learning for New Generation Smart Dispatch in power systems. Eng Appl Artif Intell 88:103372. https://doi.org/10.1016/j.engappai.2019.103372

Zanella A, Bui N, Castellani A, Vangelista L, Zorzi M (2014) Internet of things for smart cities. IEEE Internet Things J 1(1):22–32. https://doi.org/10.1109/jiot.2014.2306328

Zhang Y et al (2018) Digital Twin for Energy Efficiency Optimization in Buildings. Appl Energy 226:1137–1154

Zhang X, Chen CL, Hu Y (2019) Optimal Control for Energy Storage Systems in Smart Grids: A Review. IEEE Trans Industr Inf 15(1):17–28

Zheng Y et al (2021) A Review on Digital Twin Driven Predictive Maintenance. J Manuf Syst 59:383–395

# Predicting Adoption Intention and Willingness to Pay for the Electric Vehicle to Build Sustainable Cities in India: Using an Extended TPB Model

**Rishi Kant, Babeeta Mehta, and Archana Tiwari**

**Abstract** The purpose of this article is to provide a general overview of the state of EV innovation and individuals' willingness to adopt new technology. The creation of new transport technologies is necessary due to the ever-increasing pollution and depletion of natural resources. In effort to lessen the negative consequences caused by the excessive use of conventional vehicles powered by carbon, transportation authorities, and operators have recently promoted EV adoption, which has the potential to reduce constantly increasing pollution and other environmental hazards making cities more environmentally sustainable. Based on the extended theory of planned behaviour (TPB) this study aims at investigating the impact of four psychological constructs on adoption intention (AI) and willingness to pay (WTP) for it. Furthermore, the study also examines the mediating role of AI in the path of psychological constructs and WTP; gender's role as a moderator is also explored. All of the psychological constructs had a significant effect on WTP and AI, with AI playing a significant mediating role. The empirical finding was based on data collected from Indian respondents by using convenience sampling technique.

**Keywords** Electric vehicle · Adoption intention · Willingness to pay · Sustainable cities · TPB

## 1 Introduction

An enormous rise in vehicle ownership and use for urban transportation has occurred in the previous century along with the urbanization process and better quality of life in cities. Majority of private cars are using internal combustion engines (ICE) that

R. Kant · B. Mehta · A. Tiwari (✉)
Department of Commerce, University of Lucknow, Lucknow, India
e-mail: archanatiwari190619@gmail.com

R. Kant
e-mail: kant_rishi@lkouniv.ac.in



W. Leal Filho et al. (eds.), *Digital Technologies to Implement the UN Sustainable Development Goals*, World Sustainability Series,
https://doi.org/10.1007/978-3-031-68427-2_15

uses petrol or diesel as fuel (Curtale et al. 2022), which in addition to cost has many economic hazards like carbon emission, traffic jams, noise pollution and many more (Susskind et al. 2020). Considering the massive damage pollution is causing to the environment, the idea of sustainable development has become increasingly important (Cao et al. 2021; Rane et al. 2023). Adoption of electric vehicle is a crucial step in making the shift to low carbon mobility (Rogelj et al. 2018; Skyes and Axsen 2017; IEA 2021). Considers EV's to be 97% more ecological friendly then combustion vehicle. India, according to the World Health Organisation, has the poorest air quality in the entire globe (Bernard and Kazmin 2018). Integrating electric vehicles into smart cities may provide the benefits by storing energy and generating additional revenue from EV batteries. India's declining air quality is a result of vehicular pollution brought on by the growth in the number of passenger vehicles (Bernard and Kazmin 2018). To resolve these problems by 2030, the Indian transport ministry intends to completely phase out the sale of new petrol and diesel cars (PEVs) (Plötz et al. 2019). The Indian government also wants to make the country a center for the production of electric vehicles on a worldwide scale (Ghosh 2019).

Electric vehicles (EVs) include e-buses, e-bikes, e-cars, e-scooters, and e-railways, which help minimise pollution in cities and achieve deep decarbonization in the transportation sector (Kirpes et al. 2019). Indeed, EV integration in smart cities can result in synergies with smart demand-side systems, distributed generation, and renewable energies in boosting energy asset utilisation in the energy grid (Anthony 2021). As part of the "Faster Adoption and Manufacture of Electric Vehicles in India (FAME) scheme" the Indian government is offering financial incentives with the goal of accelerating the deployment of EVs in the country (Munshi et al. 2022). Despite the government and other implementing agencies' persistent efforts, the present market share of EVs is extremely limited (Dash 2023). Most experts predict that India will narrowly miss its objective of 30% PEV sales by 2030 (Dua et al. 2021). Out of its total budget for EV promotion and deployment of INR 10,000 crore 86% is reserved for customer incentives and 8% for charging infrastructure (Ministry of Heavy Industries and Public Enterprises 2019). After viewing the relevant literature and understanding the benefits of EV in creating sustainable cities this study aims at evaluating intention of Indian citizen on Adopting EV. This study takes the Theory of Planned behavior (TPB) as the theoretical framework and extends it to examine the influence of various socio-psychological factors on adoption intentions of electric vehicles. The statistical results supported that such constructs as subjective norm, personal moral norm, attitude affective and knowledge, are influencing factors on individual's intention of adopting electric vehicles (i.e., behavior intention) these constructs together with adoption intention also influence their willingness to pay for it (Kumar and Alok 2020; Noel et al. 2020; Ou et al. 2019; National Research Council 2013). The majority of the material currently available on PEV adoption focuses on North America, Europe, and China. According to a recent assessment of the literature on the adoption of EVs that included 239 articles, the USA accounted for the majority of the published articles (70), followed by China (25), Germany (13), the UK (12), and China, there were no works published on India (Kumar and Alok

2020). Consumers are more hesitant to adopt new items or technologies when they are in their early stages because they are inexperienced with them, uncertain about them, and have to pay a high price, according the technology innovation diffusion theory (Rogers 2003).

The research objectives of the study are as follows:

RO1: To evaluate the EV adoption intentions (AI) of Indian consumers.
RO2: To identify the role socio-psychological factors in forming adoption intention (AI).
RO3: To evaluate if adoption intention (AI) mediates the relation between socio-psychological construct and willingness to pay (WTP).

Table 1 mentioned below gives definition of constructs considered in this study for conceptual clarity.

The remaining paper is organised as follows, Sect. 2 deals with theoretical background of TPB and relevant literature related to TPB and EV, followed by hypotheses development. Section 3 describes the methodology adopted in this paper. Section 4 discusses the findings and their interpretation. Section 5 offers discussion of results. The final section deals with series of implication, recommendation, limitation and scope for future research.

**Table 1** Definitions of constructs (*Source*: Author's)

| Construct | Definition |
|---|---|
| Subjective norm | Subjective norm refers to "the perceived social pressure to engage in or not engage in the behaviour" (Ajzen 1991) |
| Personal moral norm | Personal norms refer to a person's belief that a particular course of action is appropriate or not. Schwartz (1977) |
| Attitude affective | The affective component of attitudes explains how strongly someone dislikes or admires a certain item (Bagozzi and Robert 1979) |
| Knowledge | Product information is a key decision factor that marketers employ to sway potential customers' opinions of the products and purchasing decisions (Tybout et al. 1981) |
| Adoption intention | Behaviour is immediate determined of intention; it offers the most precise forecast of behaviour (Ajzen and Fishbein 1980) |
| Willingness to pay | It identifies if the potential customer wants to pay additional amount for a product being introduced |

# 2 Literature Review and Hypotheses Development

## 2.1 Theoretical Background

### 2.1.1 Electric Vehicle (EVs)

The development of electric cars preceded that of engines powered by steam. The use of electric vehicles has had a long-standing slow development stage due to limited durability, charging issues, and low safety. Internal combustion engine vehicles became the de facto standard at around the same time that the petroleum industry underwent a revolutionary transformation and internal combustion engine technology advanced. However, the "oil crisis" and subsequent petrol shortage led automakers to explore for alternate energy sources, and electric vehicles once more garnered a lot of interest (Zhou et al. 2021). Electric vehicles (EVs) have gained recognition as an attractive means to cut the transportation sector's carbon emissions (Kant et al. 2024; Wang et al. 2017). The Toyota "Prius" is the very first produced in large quantities hybrid car in worldwide production and was introduced in Japan in 1997 (Zhou et al. 2021). It is propelled by an internal combustion engine with nickel-metal hydride batteries. The success of this historic event is consistent with consumer worries about growing petrol prices and carbon emissions (Yang et al. 2018; Hagman and Langbroek 2019).

### 2.1.2 Role of EVs Towards Sustainable Development

Our current use of energy must change in order to combat global climate change (Herrenkind et al. 2019), for example, by switching to electric mobility. The energy consumed by the transportation industry accounted for 28% of the total annual energy consumption in the world between 2010 and 2015. According to Tyagi et al. (2021), it used almost 60% of all oil products worldwide. EVs may assist us meet the climate change challenge of cutting greenhouse gas emissions by 50–85% by 2050 (Dalala et al. 2020) in order to avoid an increase in global temperature of 2.5 degrees Celsius (Kejun et al. 2021; Xu et al. 2021; Goel et al. 2021) (Shalender and Yadav 2017; Vidhi and Shrivastava 2018). Including two- and three-wheelers, which make up 12% of the global fleet of EVs, the fleet of EVs is predicted to reach 230 million by 2030 under the current sustainable development framework. Even if the deployment of EVs has recently been successful, it is difficult to go on a trajectory that is in line with the climate goals. Stronger motivation and action are needed from all nations. The IEA poll also reveals that the American region has a significant interest in electric vehicles (IEA 2021) (Biberman and Bajpai 2020). Delhi is a transport hub for international travel. In keeping with India's Intended Nationally Determined Contributions (NDCs), the EV policy. India's NDCs adhere to directed supply-side push policies for EVs without facing any financial difficulties (Dhar et al. 2020). EVs have attracted the attention of India's authorities as an ongoing alternative to clean technology that will enhance air quality and reduce $CO_2$ emissions (Kejun et al. 2021; Dhar et al. 2020).

### 2.1.3 Theory of Planned Behaviour (TPB)

Icek Ajzen first proposed the theory of planned behaviour (TPB) in 1985. It is an expansion of the theory of reasoned action (TRA) (Ajzen and Fishbein 1980; Ajzen and Fishbein 1980). Ajzen (1991) described TPB as a theoretically based approach that forecasts and explains human behaviour in terms of attitudes and beliefs. TPB describes behavioural intentions in terms of subjective norms, perceived behavior control, and attitude towards the behaviour (Chaudhary and Basai 2018). According to the Theory of Planned Behaviour (TPB), attitudes (positive or negative beliefs), subjective norms (perceptions that significant others would approve of use), and perceived behavioural control (PBC) (ease or difficulty of performing that behaviour) all have an impact on intentions, and these in turn have an impact on behaviour (Ajzen 1991). Current study is an extension to TPB it adds personal moral norm as an intrinsic normative factor and knowledge of EV's to better predict the behavioral intentions. It aims at examining linkage of norm attitude and knowledge in relation to adoption intention and willingness to pay for an EVs as shown in Fig. 1

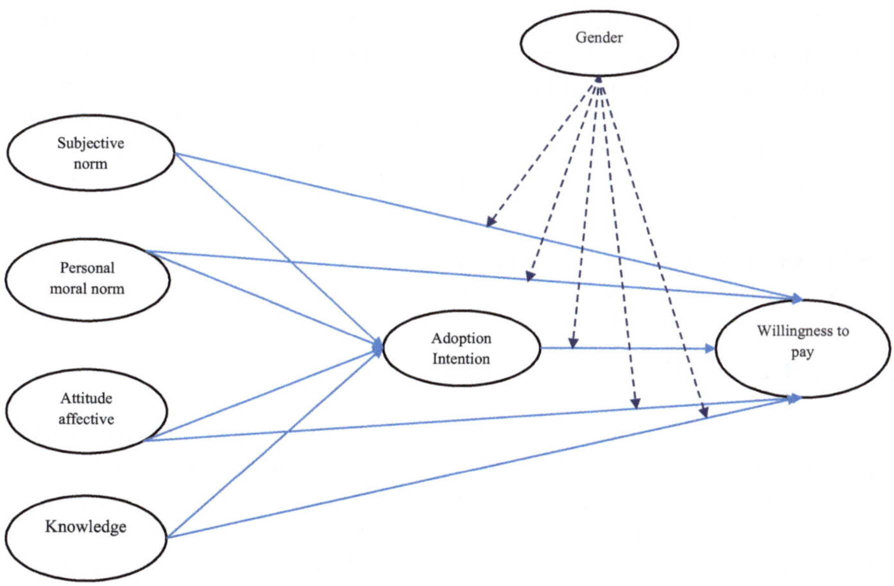


**Fig. 1** Research model

## 2.2 *Hypotheses Development*

### 2.2.1 Subjective Norm

An essential factor influencing an individual's behaviour intention is subjective norm. Individuals frequently conform to the demands or opinions of some influential people (Gao et al. 2017). In other words, a person's behaviour intention may depend on receiving the acceptance or rejection of certain others who are significant to them (Kant et al. 2024; Chen and Tung 2014). According to numerous researches, there is a strong link between subjective norms and behavioural intention, such as the desire to buy or use eco-friendly goods or electric automobiles (Wang et al. 2021). Based on the literature, this study hypothesizes that:

H1a: Subjective norm has a positive relationship with adoption intention
H1b: Subjective norm has a positive relationship with willingness to pay

### 2.2.2 Personal Moral Norm

It denotes a person's subjective judgment of whether acting in a certain way is right or wrong (Schwartz 1977). In some research, the subjective norm is used as a normative indicator, whereas the personal moral norm emphasizes internalisation and feeling as the main normative indicators n (Bamberg et al. 2007; Shi et al. 2017). Numerous academics have endorsed the value of extending TPB by including these two variables in the field of pro-environmental research. For instance, (Kaiser and Scheuthle 2003) highlighted that the inclusion of personal moral norm into TPB greatly increased the model's explanatory power and additionally that personal moral norm had a considerable impact on people's intentions to engage in ecologically conscious behaviour. Thus, it is hypothesized that:

H2a: Personal moral norm has a positive impact on adoption intention
H2b: Personal moral norm has a positive impact on willingness to pay

### 2.2.3 Affective Attitude

According to Bagozzi and Robert (1979), the emotional component of attitudes determines how much a person likes or dislikes a certain item. According to Ajzen (1991, p. 188), attitude is the extent to which a person has a favourable or unfavorable evaluation or appraisal of the behaviour in question. In other words, a person's mental state indicates whether they view a given behaviour as generally favourable or negatively. A set of more precise, salient behavioural convictions that represent the expected effects of the desired behavior give rise to attitude (Kim et al. 2013. In consumer psychology, a person's psychological inclination is frequently assessed on the basis of his or her favourable or unfavorable reactions to particular behaviour or action (kant et al. 2024; Jaiswal et al. 2021). Favourable attitude towards EV give rise to willingness to adopt (Alzahraniet al. 2019; Wang et al. 2021). The above discussions led to the following hypotheses.

H3a: Affective attitude positively influence adoption intention
H3b: Affective attitude positively influence willingness to pay

### 2.2.4 Knowledge of EV

Preferences are predicted to be influenced by knowledge and exposure to EV (Liao et al. 2017). The term "knowledge" describes a person's comprehension and familiarity with the environment and relevant topics (Chan and Lau 2000). In anticipating a person's decision-making process regarding a particular behaviour, such as the purchase of sustainable items, knowledge is a significant cognitive component (Kautish and Dash 2017; Jaiswal et al. 2022b). Scholars demonstrated that people's opinions towards buying environmentally friendly product are influenced by their knowledge, which subsequently turn influences their propensity to buy such products (Li et al. 2019; Paul et al. 2016). Additionally, there aren't many research in the literature looking at the impact of knowledge that is construed as a person's particular to the product comprehension of environmentally friendly vehicles (in this study perceived BEV knowledge). Therefore, following hypotheses are proposed:

H4a: Knowledge positively influences adoption intention
H4b: Knowledge positively influences willingness to pay

### 2.2.5 Adoption Intention and Willingness to Pay

Environmentally friendly products are generally expensive because of their high process costs (Ling 2013). Price can be one of the biggest hurdles in adoption of such eco friendly products (Gleim et al. 2013; Nasir and Karakaya 2014). Price has a significant influence in influencing buying decisions, particularly for young buyers with little purchasing power (Chaudhary and Bisai 2018). Understanding consumers' willingness to spend more is crucial for eco-friendly products because cost is one of the biggest obstacles for such purchases (Gleim et al. 2013). Researchers have previously looked at willingness to pay to understand consumer behaviour in the areas of health care (Bettiga et al. 2020), environmentally friendly purchases (Chaudhary and Bisai 2018), and hospitality (Yadav et al. 2019). Hence, we propose the following hypothesis:

H5: Adoption intention significantly influence willingness to pay

#### 2.2.6 Gender as a Moderator

Numerous studies have been conducted on the association between gender and plans to buy electric vehicles, but the results have been mixed. For green technology products like EVs in particular, gender may be a possible indicator of consumer decision-making in general (He and Hu 2022). Literature is inconsistent about the role of gender as a moderator, studies by Habich-Sobiegalla et al. (2018), Tiwari et al. (2019), She et al. (2017) revealed women are more inclined towards purchasing EV's. Whereas, Chen et al. (2020) has analyzed men in China are more participative in environmental friendly products purchasing. For the current study, adding gender as a moderating variable will help to better understand consumers' intentions to purchase electric vehicles. Therefore, following hypothesis was put up in light of the debate above.

H6a–H6e: The association between the subjective norm, personal moral norm, affective attitude, knowledge, adoption intention and willingness to pay is moderated by gender.

## 3 Research Methodology

### *3.1 Data Collection and Sample*

A well-structured questionnaire was created and circulated online to collect the sample data. The technique of purposive convenience sampling was used. The instrument was filled by 526 respondents among which 59 were found not fit for analysis and hence removed. The findings and discussion presented in the study is based on 467 usable responses.

The participants' socio demographic information is listed below in table 2. There were 269 (57.6%) men and 198 (42.4%) women among the total responders. Age-wise, 263 (56.3%) of the sample were over 26 and 204 (43.7%) of the respondents were under 26. In terms of Education 345 (73.9%) participants had at least a graduate degree, and 122 (26.1%) respondents claimed to have a master's or higher degree. Additionally, 238 (51.0%) respondents identified with the income level above INR 50,000, while 229 (49.0%) respondents claimed to be from income levels below INR 50,000. Additionally, 214 (45.7%) participants possessed two-wheelers, while 253 (54.2%) respondents owned four-wheeled personal vehicles.

**Table 2** Demographics of respondents (n = 467)

| Demographics | Frequency | Percent (%) |
|---|---|---|
| *Gender* | | |
| Male | 269 | 57.6 |
| Female | 198 | 42.4 |
| *Age* | | |
| Less than 26 | 204 | 43.7 |
| More than 26 | 263 | 56.3 |
| *Education* | | |
| Bachelor/associate degree | 345 | 73.9 |
| Masters and above degree | 122 | 26.1 |
| *Monthly household income* | | |
| < INR 50,000 | 229 | 49.0 |
| ≥ INR 50,000 | 238 | 51.0 |
| *Type of vehicle* | | |
| Two-wheeler | 214 | 45.8 |
| Four-wheeler | 253 | 54.2 |

### *3.2 Data Analysis Method*

The data was processed statistically using AMOS and SPSS, version 22. Cronbach's alpha was employed in the current study to assess each construct's reliability. Structural equation modelling (SEM) was used to analyse the gathered data. SEM is suitable for analysing multivariate models because it allows for the modelling of correlations between multiple predictor and multiple criterion variables (Chin 1998).

## 4 Findings and Results

In order to analyse the responses various statistical tools were used that includes common method variance (CMV) test, reliability and validity test, measurement model, structural model, and testing hypotheses. After estimating the mediating effect of adoption intention in independent constructs and willingness to pay, effect of gender was tested as a moderator which was carried using statistical software such as SPSS, AMOS and SEM.

## 4.1 Reliability and Validity

Table 3 provides information on the validity and reliability of various metrics. The internal consistency of all scale measures was shown by Cronbach's alpha, which ranged from 0.851 to 0.931 and was more than the cutoff point of 0.70 (Hair et al. 2015). Using item loadings, composite reliability (CR) values, and average variance extracted (AVE), convergent validity was evaluated (Hair et al. 2015). The observed value of the average variance extracted (AVE) of each measure ranged from (0.587 to 0.806), indicating that the AVE values of all measures were greater than 0.50 (Hair et al. 2015). The standardised loadings for each indicator were found to be significant and above the cutoff of 0.50. Additionally, the CR scores varied from 0.850 to 0.932, which was higher than the threshold of 0.70, indicating the convergent validity. Table 4 shows that each measure's AVE is higher than the common variance of the measures, confirming the applicability of the discriminant validity. As a result, it can be said that the reliability and validity statistical results were successfully attained.

**Table 3** Reliability and validity

| Construct | Item | Loading | Cronbach's alpha | Average variance extracted | Composite reliability |
|---|---|---|---|---|---|
| Affective attitude | AA1 | 0.891 | 0.915 | 0.782 | 0.915 |
| | AA2 | 0.871 | | | |
| | AA3 | 0.891 | | | |
| Adoption intention | AI1 | 0.859 | 0.928 | 0.773 | 0.931 |
| | AI2 | 0.941 | | | |
| | AI3 | 0.776 | | | |
| | AI4 | 0.930 | | | |
| Knowledge | KN1 | 0.921 | 0.931 | 0.775 | 0.932 |
| | KN2 | 0.904 | | | |
| | KN3 | 0.847 | | | |
| | KN4 | 0.846 | | | |
| Personal moral norms | PMN1 | 0.927 | 0.925 | 0.806 | 0.926 |
| | PMN2 | 0.860 | | | |
| | PMN3 | 0.905 | | | |
| Subjective norms | SN1 | 0.775 | 0.851 | 0.587 | 0.850 |
| | SN2 | 0.798 | | | |
| | SN3 | 0.777 | | | |
| | SN4 | 0.713 | | | |
| Willing to pay more | WTP1 | 0.869 | 0.911 | 0.773 | 0.911 |
| | WTP2 | 0.893 | | | |
| | WTP3 | 0.875 | | | |

**Table 4** Correlations and discriminant validity results.
Note: The diagonal elements (**bold**) represent the square roots of AVEs, and the off-diagonal elements indicate the correlations among constructs.

Discriminate validity results of the measurement model

| | AI | WTP | AA | PMN | KN | SN |
|---|---|---|---|---|---|---|
| AI | **0.879** | | | | | |
| WTP | 0.553 | **0.879** | | | | |
| AA | 0.353 | 0.424 | **0.884** | | | |
| PMN | 0.506 | 0.629 | 0.286 | **0.898** | | |
| KN | 0.505 | 0.646 | 0.262 | 0.590 | **0.880** | |
| SN | 0.530 | 0.511 | 0.357 | 0.423 | 0.456 | **0.766** |

## *4.2 Structural Model*

The goodness-of-fit statistics provided in Table 5 were used to test the structural equation model. Chi square/*df* = 1.947; CFI = 0.979; IFI = 0.979; NFI = 0.959; GFI = 0.938; AGFI = 0.917; RMSEA = 0.045; these are good findings for the structural model fit (Hair et al. 2015). The outcomes of straight pathways are shown in Table 6. EV adoption intention was positively and highly significantly influenced by subjective norms (H1a: = 0.295), personal moral norms (H2b: = 0.223), affective attitude (H3a: = 0.131), and knowledge (H4a: = 0.205), according to the results of this path connection. Additionally, the willingness to pay for EV was significantly and directly influenced by subjective norms (H1b: = 0.119), personal moral norms (H2b: = 0.276), affective attitude (H3b: = 0.174), and knowledge (H4b: = 0.319). The results also showed that adoption intention has a direct impact on EV purchase willingness. Accordingly, in this study, the key predictors of BEV adoption intention were socio-cognitive characteristics such subjective norms, personal moral norms, affective attitude, and knowledge. As shown in Table 6 and Fig. 2, adoption intention was an additional predictor of willingness to pay for EV.

**Table 5** Model value

| | |
|---|---|
| Chi square/*df* | 1.947098 |
| GFI | 0.938 |
| AGFI | 0.917 |
| IFI | 0.979 |
| NFI | 0.959 |
| CFI | 0.979 |
| RMSEA | 0.045 |

**Table 6** Results of hypotheses testing: direct effects

| Hypotheses | DV | | ID V | Estimate | Results |
|---|---|---|---|---|---|
| H1a | AI | ⟵ | SN | 0.295*** | S |
| H2a | AI | ⟵ | PMN | 0.223*** | S |
| H3a | AI | ⟵ | AA | 0.131*** | S |
| H4a | AI | ⟵ | KN | 0.205*** | S |
| H5a | WTP | ⟵ | AI | 0.128*** | S |
| H1b | WTP | ⟵ | SN | 0.119*** | S |
| H2b | WTP | ⟵ | PMN | 0.276*** | S |
| H3b | WTP | ⟵ | AA | 0.174*** | S |
| H4b | WTP | ⟵ | KN | 0.319*** | S |

***Significant level = 0.001

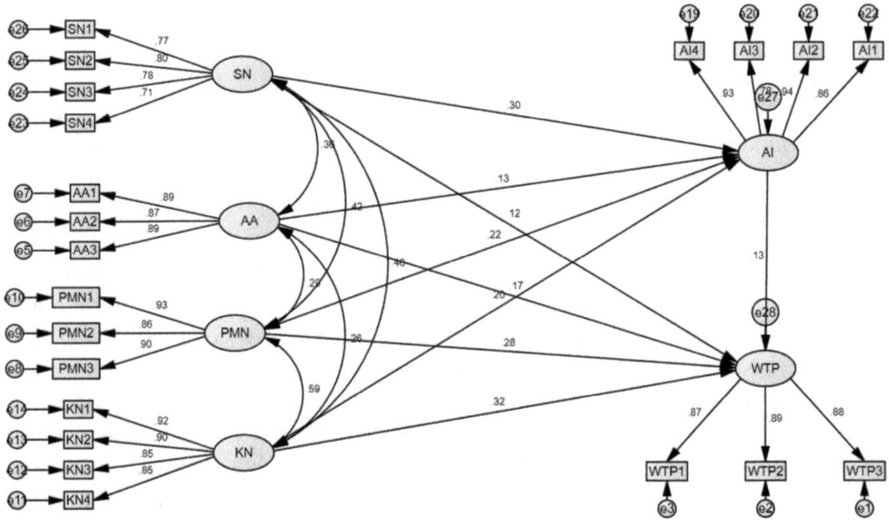


**Fig. 2** AMOS output

## 4.3 Mediation

The outcomes of the mediation effects of adoption intention are shown in Table 7. The three-step process developed by Baron and Kenny in 1986 was applied to evaluate structural mediation in a number of recent researches on electric vehicles (He and Hu 2022; Jaiswal et al. 2021). According to the suggested technique, the associations between the independent variables (IVs) and the dependent variable (DV) were first directly investigated, and the results showed that all of the factors (subjective norms, personal moral norms, affective attitude, and knowledge) significantly influenced willingness to pay. Second, the mediator (adoption intention) was examined for the

**Table 7** Results of mediation

| IV | M | DV | IV − > DV | IV − > M | IV + M − > DV | | Mediating role |
|---|---|---|---|---|---|---|---|
| | | | | | IV − > DV | M − > DV | |
| SN | AI | WTP | 0.630** | 0.618** | 0.399** | 0.322** | Partial |
| AA | AI | WTP | 0.468** | 0.396** | 0.281** | 0.448** | Partial |
| PMN | AI | WTP | 0.668** | 0.544** | 0.503** | 0.288** | Partial |
| KN | AI | WTP | 0.678** | 0.540** | 0.521** | 0.279** | Partial |

IV = Independent Variable, M = Mediator, DV = Dependent Variable, ** $p < 0.01$, *ns* = Not Significant.

effects of IV, and all results were significant. Lastly, the impact mediator (M) was examined in relation to DV (willingness to pay), which was likewise important for all mediating paths. As both IV and M significantly affect DV, they fit the criteria for partial mediation (Biswas et al. 2021; Hair et al. 2015).

## 4.4 Moderation

Table 8 indicates the results of moderation. The findings of the study revealed that the difference indicator of t value was significant for four control variables (H6a, H6b, H6c and H6d) and thus, supported the moderation of gender for the path between subjective norm, personal moral norm, affective attitude, knowledge and willingness to pay. However, the difference between adoption intention and willingness to pay (H6e) in this study did not support the moderation of gender on adoption intention.

**Table 8** Results of moderation

| Hypothesis | Effect | Particular | | | | *t*-value | Comparing results |
|---|---|---|---|---|---|---|---|
| | | Gender | | | | | |
| | | Male = 269 | | Female = 198 | | | |
| | | β | SE | β | SE | | |
| H6a | SN⟶WTP | 0.637** | 0.061 | 0.863** | 0.079 | 2.264* | Yes |
| H6b | PMN⟶WTP | 0.468** | 0.037 | 0.671** | 0.042 | 3.627** | Yes |
| H6c | AA⟶WTP | 0.280** | 0.037 | 0.410** | 0.049 | 2.117* | Yes |
| H6d | KN⟶WTP | 0.555** | 0.041 | 0.748** | 0.046 | 3.132** | Yes |
| H6e | AI⟶WTP | 0.504** | 0.043 | 0.564** | 0.054 | $0.869^{ns}$ | No |

* $p < 0.05$; ** $p < 0.01$; *** $p < 0.001$; ns = not sig

# 5 Discussion and Implications

## *5.1 Discussion*

Electric vehicles (EVs) play a major role in development of sustainable cities and make our environment cleaner and greener. A well-balanced mix of technology innovation, prudent legislation, growing environmental consciousness, lead to rising EVs penetration and, as a result, more environmentally friendly towns and cities in future. Electric vehicles (EVs) are an efficient form of transportation, and their adoption is a complex process that is influenced by a number of factors. Despite this, there are only few research examining consumers' willingness to pay and adoption intentions of electric vehicles (Irfan and Ahmad 2021). The research findings show uniformity in the context of the outcomes obtained for various types of hypothesized associations in an Indian environment. The study provided managerial recommendations for the adoption-related intention and willingness to pay of potential customers, as well as deepen understanding of theory by validating relationships among examined constructs. The current study examines the influence of subjective norms, personal moral norm, affective attitudes, and knowledge on individuals' adoption intentions and willingness to pay for electric vehicles. To strengthen the validity of the research, the study also sought to examine the mediating effects of adoption intention and the moderating effects of gender on willingness to pay. The current research findings are consistent with the prior study (Jaiswal et al. 2022a, b, c, d; Bettiga et al. 2020). All the constructs pose a direct influence on willingness to pay for electric vehicle which is in line with the previous research findings (Sang et al. 2020; Li et al. 2019).

The research findings indicate that knowledge is the strongest factor that impact willingness to pay for electric vehicle among peoples of India, and subjective norm has a least influence on willingness to pay. The findings suggest that when potential consumers have higher knowledge regarding electric vehicle, they are more likely to adopt it and there is higher readiness to pay among them. Therefore, providing people adequate information on electric vehicles will enhance their knowledge and understanding regarding them, and further ignite their adoption intention and willingness to buy electric vehicle.

The research was additionally concentrated on mediation analysis, and the findings demonstrating that adoption intention partially mediated the relationship between the independent factors' subjective norm, personal moral norm, affective attitude, and knowledge and the dependent factor, namely, willingness to pay. The study also provides a scenario for the potential of adoption intention to encourage willingness to pay as a mediator. The study provides an important finding that adoption intention significantly impacts direct and indirectly willingness to pay among potential consumers of EVs. Furthermore, the moderation results of the study suggest that, gender positively moderates the relationship between willingness to pay and subjective norm, personal moral norm, affective attitude and knowledge except for adoption intention.

### 5.2 *Theoretical and Practical Implications*

This study presents some recommendations and implications for encouraging the adoption of EVs. First, Subjective and personal moral norms significantly influence adoption intention and willingness to pay. This emphasizes the significance of friends, family, and closest one in developing favorable attitudes towards adoption intentions for EV. The premise is that a high level of social influence will be more likely to persuade consumers or prospective customers to adopt electric vehicle in a growing mobility market identical to India. Second, affective attitude and knowledge have a significant impact on adoption intention which further leads to willingness to pay for EVs. This demonstrate, that consumers who are knowledgeable about EV technology and have a positive outlook towards their adoption will be more willing to pay for EVs. Therefore, electric vehicle manufacturers should take an effort to publicize technological knowledge about EVs in order to enhance customer perceptions towards them. This implication is consistent with the prior research of Huang et al. (2021) and (Jaiswal et al. 2022a, b, c, d). Manufacturers and policymakers of electric vehicle may concentrate on younger educated generation who represent a wider segment of prospective buyers of EVs and their adoption. Additionally, the study examined the moderating effect of gender on the relationships between subjective norm, personal moral norm, affective attitude, knowledge, adoption intention and willingness to pay which provides important implication and contributions to the literature of electric vehicles.

## 6 Conclusion and Scope for Future Research

The entire globe is currently confronted with environmental challenges such as air pollution from transportation, particularly in emerging countries such as India. EVs are the most remarkable innovation in today's polluted environment, especially for individuals who live near major transportation routes, because they eliminate pollutants to a larger extent. Electric vehicles play a vital role in the development of smart cities and the reduction of $CO_2$ emissions. Renewable energy sources have a significant role in improving power quality in distribution systems. Integrating EVs with renewable energy sources, on the other hand, eliminates significant amounts of carbon emissions, are vital for the environment's future sustainability. EVs are contributing to achieve our goal to make greener and sustainable cities. Therefore, the current study focused at the urgently needed solutions for electric vehicle adoption and willingness to pay for them. This study contributes new insights into EV adoption and willingness to pay, while also confirming previous findings. The main purpose of this research is to investigate the influence of subjective norm, personal moral norm, affective attitude, and knowledge on adoption intention and willingness to pay of EVs. The results suggests that all the variables significantly influence adoption intention and willingness to pay, however knowledge of EVs has a higher influence on willingness to pay for EVs. Further, adoption intention partially mediated the relationship between subjective norm, personal moral norm, affective attitude and knowledge and willingness to pay. Additional findings of the study revealed that gender successfully moderated the relationship between all variables except for adoption intention and willingness to pay. Overall, this research provides deepen understanding of adoption intention and willingness to pay of electric vehicles as well as add to the previous findings and literature by examining the mediating role of adoption intention and moderating role of gender on willingness to pay of EVs.

Despite of providing several interesting findings and implications, the study has some significant limitations also. First, the study concentrates on consumers' adoption intentions rather than actual adoption behaviour. Second, in India, samples were obtained using a non-probability sampling method. As a result, the findings are limited to one specific country only and therefore their generalizability would be restricted. Future research could look into respondents' real adoption behaviour. Similarly, the sample data would be collected by using probability sampling technique from different countries or areas, allowing for cross-comparison of results, to confirm the more realistic accuracy of the findings.

## References


Ajzen I (1991) The theory of planned behavior. Organ Behav Hum Decis Process 50:170–211

Ajzen I, Fishbein M (1980) Understanding attitudes and predicting behavior. Prentice-Hall Inc., Englewood Cliffs

Alzahrani K, Hall-Phillips AV, Zeng AZ (2019) Applying the theory of reasoned action to understanding consumers' intention to adopt hybrid electric vehicles in Saudi Arabia. Transportation 46(1):199–215

Anthony B Jr (2021) Integrating electric vehicles to achieve sustainable energy as a service business model in smart cities. Front Sustain Cities 3:685716

Bagozzi RP, Burnkrant RE (1979) Attitude organization and the attitude–behavior relationship. J Pers Soc Psychol 37(6):913

Bamberg S, Hunecke M, Blobaum A (2007) Social context, personal norms and the use of public transportation: two field studies. J Environ Psychol 27(3):190–203

Bernard S, Kazmin A (2018) Dirty air: how India became the most polluted country on earth. Financ times 12:1

Bettiga D, Lamberti L, Lettieri E (2020) Individuals' adoption of smart technologies for preventive health care: a structural equation modeling approach. Health Care Manag Sci 23:203–214

Biberman J, Bajpai N (2020) India and the SDGs ICT India working paper# 22. CSD working paper series: towards a new Indian model of information and communications technology-led growth and development

Biswas A, Jaiswal D, Kant R (2021) Augmenting bank service quality dimensions: Moderation of perceived trust and perceived risk. Int J Product Perform Manag. https://doi.org/10.1108/IJPPM-04-2021-0196

Botetzagias I, Dima AF, Malesios C (2015) Extending the theory of planned behavior in the context of recycling: the role of moral norms and of demographic predictors. Resour Conserv Recycl 95:58–67

Cao J, Chen X, Qiu R, Hou S (2021) Electric vehicle industry sustainable development with a stakeholder engagement system. Technol Soc 67:101771
Chan RY, Lau LB (2000) Antecedents of green purchases: a survey in China. J Consum Mark 17(4):338–357
Chaudhary R, Bisai S (2018) Factors influencing green purchase behavior of millennials in India. Manag Environ Qual Int J 29(5):798–812
Chen MF, Tung PJ (2014) Developing an extended theory of planned behavior model to predict consumers' intention to visit green hotels. Int J Hosp Manag 36:221–230
Chen CF, de Rubens GZ, Noel L, Kester J, Sovacool BK (2020) Assessing the socio-demographic, technical, economic and behavioral factors of Nordic electric vehicle adoption and the influence of vehicle-to-grid preferences. Renew Sustain Energy Rev 121:109692
Chin, WW (1998) The partial least squares approach to structural equation modeling. Modern Methods for Business Research 295(2): 295–336
Curtale R, Liao F, Rebalski E (2022) Transitional behavioral intention to use autonomous electric car-sharing services: evidence from four European countries. Transp Res Part C Emerg Technol 135:103516
Dalala ZM, Alnawafa M, Saadeh O, Alnawafa E (2020) Reducing commuter $CO_2$ footprint through transit PV electrification. Sustainability 12(16):6406
Dash A (2023) Adapting to electric vehicles value chain in India: the MSME perspective. Case Stud Transp Policy 12:100996
De Leeuw A, Valois P, Ajzen I, Schmidt P (2015) Using the theory of planned behavior to identify key beliefs underlying pro-environmental behavior in high-school students: implications for educational interventions. J Environ Psychol 42:128–138
Dhar S, Pathak M, Shukla P, Gupta A (2020) Electric vehicles penetration in India for enhanced energy efficiency deployment in the transport sector. In: Energy efficiency in developing countries: policies and programmes. Routledge, London, pp 253–270
Dua R, Hardman S, Bhatt Y, Suneja D (2021) Enablers and disablers to plug-in electric vehicle adoption in India: insights from a survey of experts. Energy Rep 7:3171–3188
Gao L, Wang S, Li J, Li H (2017) Application of the extended theory of planned behavior to understand individual's energy saving behavior in workplaces. Resour Conserv Recycl 127:107–113
Ghosh M (2019) Govt steps on the gas to put India's EV mission in fast lane. https://www.livemint.com/budget/news/govt-steps-on-the-gas-to-putindia-s-ev-mission-in-fast-lane-1562350841874.html
Gleim MR, Smith JS, Andrews,D, Cronin Jr, JJ (2013) Against the green: A multi-method examination of the barriers to green consumption. J Rtail 89(1): 44–61
Goel S, Sharma R, Rathore AK (2021) A review on barrier and challenges of electric vehicle in India and vehicle to grid optimisation. Transp Eng 4:100057
Habich-Sobiegalla S, Kostka G, Anzinger N (2018) Electric vehicle purchase intentions of Chinese, Russian and Brazilian citizens: an international comparative study. J Clean Prod 205:188–200
Hagman J, Langbroek JH (2019) Conditions for electric vehicle taxi: a case study in the Greater Stockholm region. Int J Sustain Transp 13(6):450–459
Hair, NL, Hanson, JL, Wolfe, BL, Pollak, SD (2015) Association of child poverty, brain development, and academic achievement. JAMA pediatrics 169(9): 822–829
Haustein S, Jensen AF (2018) Factors of electric vehicle adoption: a comparison of conventional and electric car users based on an extended theory of planned behavior. Int J Sustain Transp 12(7):484–496
He X, Hu Y (2022) Understanding the role of emotions in consumer adoption of electric vehicles: the mediating effect of perceived value. J Environ Plan Manag 65(1):84–104
Herrenkind B, Brendel AB, Nastjuk I, Greve M, Kolbe LM (2019) Investigating end-user acceptance of autonomous electric buses to accelerate diffusion. Transp Res Part D Transp Environ 74:255–276

Huang X, Lin Y, Lim MK, Tseng ML, Zhou F (2021) The influence of knowledge management on adoption intention of electric vehicles: perspective on technological knowledge. Ind Manag Data Syst 121(7):1481–1495

IEA (2021) World energy outlook. IEA

Irfan M, Ahmad M (2021) Relating consumers' information and willingness to buy electric vehicles: Does personality matter? Transp Res Part D Transp Environ 100:103049

Jaiswal D, Kaushal V, Kant R, Singh PK (2021) Consumer adoption intention for electric vehicles: insights and evidence from Indian sustainable transportation. Technol Forecast Soc Chang 173:121089

Jaiswal D, Deshmukh AK, Thaichon P (2022a) Who will adopt electric vehicles? Segmenting and exemplifying potential buyer heterogeneity and forthcoming research. J Retail Consum Serv 67:102969

Jaiswal D, Kant R, Singh PK, Yadav R (2022b) Investigating the role of electric vehicle knowledge in consumer adoption: evidence from an emerging market. Benchmark Int J 29(3):1027–1045

Jaiswal D, Singh B, Kant R, Biswas A (2022c) Towards green product consumption: effect of green marketing stimuli and perceived environmental knowledge in Indian consumer market. Soc Bus Rev 17(1):45–65

Jaiswal D, Kaushal V, Deshmukh AK, Kant R, Kautish P (2022d) What drives electric vehicles in an emerging market? Mark Intell Plan 40(6):738–754

Kaiser FG, Scheuthle H (2003) Two challenges to a moral extension of the theory of planned behavior: moral norms and just world beliefs in conservationism. Person Ind Differ 35(5):1033–1048

Kant R, Mehta B, Jaiswal D, Kumar A (2024) Adoption intention and willingness to pay for electric vehicles: role of social-psychological attributes, fiscal incentives and socio-demographics. Manag Environ Qual Int J Instr 35(4): 945–963

Kautish P, Dash G (2017) Environmentally concerned consumer behavior: evidence from consumers in Rajasthan. J Model Manag 12(4):712–738

Kejun J, Chenmin H, Songli Z, Pianpian X, Sha C (2021) Transport scenarios for China and the role of electric vehicles under global 2 C/1.5 C targets. Energy Econ 103:105172

Kim E, Ham S, Yang IS, Choi JG (2013) The roles of attitude, subjective norm, and perceived behavioral control in the formation of consumers' behavioral intentions to read menu labels in the restaurant industry. Int J Hosp Manag 35:203–213

Kirpes B, Danner P, Basmadjian R, Meer HD, Becker C (2019) E-mobility systems architecture: a model-based framework for managing complexity and interoperability. Energy Inform 2:1–31

Kumar RR, Alok K (2020) Adoption of electric vehicle: a literature review and prospects for sustainability. J Clean Prod 253:119911

Li G, Li W, Jin Z, Wang Z (2019) Influence of environmental concern and knowledge on households' willingness to purchase energy-efficient appliances: a case study in Shanxi, China. Sustainability 11(4):1073

Liao F, Molin E, van Wee B (2017) Consumer preferences for electric vehicles: a literature review. Transp Rev 37(3):252–275

Ling, CY (2013) Consumers' purchase intention of green products: An investigation of the drivers and moderating variable. Elixir Marketing Management, 1, 14503–14509

Lizin S, Van Dael M, Van Passel S (2017) Battery pack recycling: behaviour change interventions derived from an integrative theory of planned behaviour study. Resour Conserv Recycl 122:66–82

Ministry of Heavy Industries and Public Enterprises (2018). National auto policy. https://dhi.nic.in/writereaddata/UploadFile/National%20Automotive%20Policy%20Draft%20v2.pdf

Ministry of Heavy Industries and Public Enterprises (2019) Scheme for faster adoption and manufacturing of electric vehicles in India phase II (FAME India phase II). https://www.fame-india.gov.in/WriteReadData/userfiles/file/FAMEII%20Notification.pdf

Munshi T, Dhar S, Painuly J (2022) Understanding barriers to electric vehicle adoption for personal mobility: a case study of middle income in-service residents in Hyderabad city, India. Energy Policy 167:112956

Nasir AV, Karakaya F (2014) Consumer segments in organic foods market. J Consum Mark 31(4): 263–277

National Research Council (2013) Overcoming barriers to electric-vehicle deployment. Interim Report, The National Academies Press, Washington, DC

Noel L, de Rubens GZ, Kester J, Sovacool BK (2020) Understanding the socio-technical nexus of Nordic electric vehicle (EV) barriers: a qualitative discussion of range, price, charging and knowledge. Energy Policy 138:111292

Ou S, Hao X, Lin Z, Wang H, Bouchard J, He X, Przesmitzki S, Wu Z, Zheng J, Lv R, LaClair TJ (2019) Light-duty plug-in electric vehicles in China: an overview on the market and its comparisons to the United States. Renew Sustain Energy Rev 112:747–761

Paul J, Modi A, Patel J (2016). Predicting green product consumption using theory of planned behavior and reasoned action. J Retail Consum Serv 29: 123–134

Plotz P, Axsen J, Funke SA, Gnann T (2019) Designing car bans for sustainable transportation. Nat Sustain 2(7):534–536

Rane NL, Achari A, Saha A, Poddar I, Rane J, Pande CB, Roy R (2023) An integrated GIS, MIF, and TOPSIS approach for appraising electric vehicle charging station suitability zones in Mumbai. India. Sustain Cities Soc 2023:104717

Rogelj J, Popp A, Calvin KV, Luderer G, Emmerling J, Gernaat D, ... Tavoni M (2018) Scenarios towards limiting global mean temperature increase below 1.5 C. Nat Clim Change 8(4): 325–332

Rogers EM (2003) Diffusion of innovations, 5th edn. Free Press, New York

Sang P, Yao H, Zhang L, Wang S, Wang Y, Liu J (2020) Influencing factors of consumers' willingness to purchase green housing: a survey from Shandong Province, China. Environ Dev Sustain 22(5):4267–4287

Schwartz SH (1977) Normative influence on altruism. In: Berkowitz L (ed) Advances in experimental social psychology, vol 10, pp 221–279

Shalender K, Yadav RK (2017) Promoting e-mobility in India: challenges, framework, and future roadmap. Environ Dev Sustain 20(6):2587–2607

She ZY, Sun Q, Ma JJ, Xie BC (2017) What are the barriers to widespread adoption of battery electric vehicles? A survey of public perception in Tianjin, China. Transp Policy 56:29–40

Shi H, Fan J, Zhao D (2017) Predicting household PM2.5-reduction behavior in Chinese urban areas: an integrative model of theory of planned behavior and norm activation theory. J Clean Prod 145:64–73

Susskind L, Chun J, Goldberg S, Gordon JA, Smith G, Zaerpoor Y (2020) Breaking out of carbon lock-in: Malaysia's path to decarbonization. Front Built Environ 6:21

Sykes M, Axsen J (2017) No free ride to zero-emissions: simulating a region's need to implement its own zero-emissions vehicle (ZEV) mandate to achieve 2050 GHG targets. Energy Policy 110:447–460

Tyagi B, Choudhury G, Vissa, NK, Singh J, Tesche, M (2021) Changing air pollution scenario during COVID-19: Redefining the hotspot regions over India. Environ Pollut 271: 116354

Tybout AM, Calder BJ, Sternthal B (1981) Using information processing theory to design marketing strategies. J Mark Res 18(1):73–79

Vidhi R, Shrivastava P (2018) A review of electric vehicle lifecycle emissions and policy recommendations to increase EV penetration in India. Energies 11(3):483

Wang S, Li J, Zhao D (2017) The impact of policy measures on consumer intention to adopt electric vehicles: evidence from China. Transp Res Part A Policy Pract 105:14–26

Wang XW, Cao YM, Zhang N (2021) The influences of incentive policy perceptions and consumer social attributes on battery electric vehicle purchase intentions. Energy Policy 151:112163

Xu B, Sharif A, Shahbaz M, Dong K (2021) Have electric vehicles effectively addressed $CO_2$ emissions? Analysis of eight leading countries using quantile-on-quantile regression approach. Sustain Prod Consumpt 27:1205–1214

Yadav R, Balaji MS, Jebarajakirthy C (2019). How psychological and contextual factors contribute to travelers' propensity to choose green hotels? Int J Hosp Manag 77: 385–395

Yang WH, Wong RCP, Szeto WY (2018) Modeling the acceptance of taxi owners and drivers to operate premium electric taxis: policy insights into improving taxi service quality and reducing air pollution. Transp Res Part A Policy Pract 118:581–593

Zhou M, Long P, Kong N, Zhao L, Jia F, Campy KS (2021) Characterizing the motivational mechanism behind taxi driver's adoption of electric vehicles for living: Insights from China. Transportation Research Part A

# Science Diplomacy Online: Implementing Sustainable Development in the Virtual World

**Julia M. Puaschunder**

**Abstract** Science diplomacy directly supports diplomatic processes on global issues of concern through scientific advice and evidence to inform decision-making in foreign affairs and security policies. This chapter addresses the role of Science Diplomacy Online in implementing the UN SDGs. Online science diplomacy highlights the potential of scientists trained with a sophisticated understanding of novel technological advancements. Science diplomats are versed in cutting-edge digital tools to employ for understanding global issues of concern that are too complex and large to be tackled by one nation-state. This chapter introduces the first Science Diplomacy Online Index integrating the first Science Diplomacy Index with digitalized connectivity, which is positively correlated with anti-corruption. The chapter then pays attention to strategies on how to use digitalization for the alleviation of the most pressing concerns of our times in fostering the SDGs. Most recent insights on human rights online and the rising influence of social online media dialogue for international law creation will be discussed. Flourishing science diplomacy online bestows the virtual world with sustainable development advancement potential in our common and fragile digitalized world.

**Keywords** Digitalization · International affairs · Public policy · Science diplomacy · Sustainable development goals

## 1 Introduction

In United Nations Sustainable Development Goals (UN SDGs) traditional national advocacy for reaching a global compromise is currently complemented by science diplomacy efforts. Science diplomacy directly supports diplomatic processes on

J. M. Puaschunder (✉)
Graduate School of Arts and Sciences, Columbia University, New York, NY, USA
e-mail: Julia.Puaschunder@columbia.edu
URL: http://blogs.cuit.columbia.edu/jmp2265; http://www.juliampuaschunder.com



W. Leal Filho et al. (eds.), *Digital Technologies to Implement the UN Sustainable Development Goals*, World Sustainability Series,
https://doi.org/10.1007/978-3-031-68427-2_16

global issues of concern through scientific advice and evidence to inform decision-making in foreign affairs and security policies (European Union 2022). Science diplomacy offers many advantages when being employed together with traditional global governance and policy efforts. Scientists are well-trained human capital. Science mandates advocate for independent thinking and integrity—both make scientists less susceptible to lobbying and politically-tainted, nationally-focused decision-making. Science budgets are also an additional resource in addition to conventional international development funds and governmental aid. International research communities are predestined to cross-pollinate with comparative approaches that innovate across borders. Scientific state-of-the-art standards are often based on success factors that are derived from multiple fields and various contexts. Researchers can also often inform about the newest developments to be used for the alleviation of contemporary issues of concern. In our contemporary age of digitalization, one of the most discussed scientific advancements around the globe are artificial intelligence, big data gains and robotics.

The United Nations (UN) Sustainable Development Goals (SDGs) target at finding a "shared blueprint for peace and prosperity for people and the planet, now and into the future" (United Nations Department of Economic and Social Affairs, Sustainable Development, retrieved June 19, 2023). Sprung out of the Millennium Development Goals, the United Nations General Assembly incepted the SDGs in 2015 as aspirational long-term goals (Puaschunder, forthcoming). While the initial SDGs did not specifically address the role of new technological advancements in the accomplishment of targets, in a revision, the overarching theme of digitalization was integrated. New technologies were acknowledged as an optimum source to positively change the landscape of international development. This chapter addresses the growing role of digitalization for international global governance with particular attention to the SDGs.

The chapter is structured as follows: First, science diplomacy is described to then apply the concept to the Sustainable Development Goals (SDGs) with a particular focus on digitalization. Second, science diplomacy online potentials for accomplishing and implementing the SDGs via digitalized resources are outlined. Third, an attempt to quantify science diplomacy is captured in the first Science Diplomacy Online Index. Fourth, the discussion prepares for research on the role of science diplomacy online for sustainable development. Future perspectives for science diplomacy in the digital age in light of human rights online are envisioned.

## 2 Science Diplomacy

Diplomacy has a long history. Modern diplomacy started in the eighteenth century in Europe. Since the 1930s, science diplomacy evolved as a way to use scientific discourse for accomplishing global goals aside from country politics or national interests. Scientific exchange via hard science measurable facts is coupled with diplomatic goals in science diplomacy. In a long history of science diplomacy, major accomplishments are the continued dialogue among scientific circles despite the Cold War era travel bans of the 1960s–1980s. Scientists were often allowed to travel and could maintain scientific transfers neglecting political adverse regimes' agenda. Science diplomacy networks thereby provided essential gateways for information exchange and continued dialogue on innovation transfers. Scientists have the advantage of adhering to integrity mandates and can connect over hard-science facts unbiased by national interests. Institutional support for scientists can also serve as additional funding, human capital and knowledge stream that complements traditional diplomacy. Science diplomats engage in international exchange and mutual collaboration on global challenges oftentimes using systems analysis. Systems analysis applies scientific rigor to goal-oriented approaches. Concrete tasks of systems analysis include the identification of systems and procedures that benefit the efficient accomplishment of large-scale and complex goals (Whitten and Bentle 2007). Another applied field of expertise of science diplomats is decision sciences, which employs plural viewpoints in economic sciences. In the use of systems analysis, science diplomats have always been inspired by technological advancements of the time. Most recently, science diplomats have started using advanced digitalization techniques. In the contemporary United Nations Sustainable Development Goals (SDGs), science diplomacy has been argued for providing essential leadership and support in the accomplishment of aspirational endeavors. The SDGs of the United Nations could now also benefit from attention to digitalization advantages in the use of science diplomacy.

## 3 Sustainable Development Goals (SDGs)

The United Nations (UN) Sustainable Development Goals (SDGs) comprise seventeen interrelated objectives to improve global conditions for humankind (United Nations Department of Economic and Social Affairs, Sustainable Development, retrieved June 19, 2023). Each goal identifies more specific targets and indicators as well as concrete progress measurement and tracking techniques (Puaschunder, forthcoming; United Nations Resolution, 2017). The goals address the most contemporary global concerns that are often interrelated (Puaschunder, forthcoming). The overall SDGs are further broken down into 169 individual targets (Puaschunder, forthcoming). Progress is measured by indicators estimating target accomplishments on a granular and local level (Puaschunder, forthcoming; Pierce 2018). In the rising literature on the SDGs, so far, the relation between digitalization and the SDGs is not substantially covered. Digitalization holds the clear potential to employ innovative techniques for international development. Digital tools can transport ideas and knowledge in remote areas of the world. Digital media is also the driving force for the cross-border stimulation of ideas and whole-rounded education. Social online media channels have ignited trends and international change around the world. Employing Science Diplomacy Online appears as the most innovative—yet hardly discussed or quantified—way to complement conventional goal attainment.

## 4 Applied Science Diplomacy Online

Applied science diplomacy often uses systems analysis, which studies procedures and how systems effectively help to reach common goals (Whitten and Bentle 2007). Systems analysis offers concrete problem-solving techniques used to break down complex systems to help understand how complex system parts interact. Systems analysis also informs about the interaction of complex influence factor dynamics in the process of change. For the individual decision maker, systems analysis also provides concrete estimations of risks and uncertainty based on decision science methods. Behavioral aspects of systems analysis also inform about human decision-making qualities and peculiarities in order to better understand why stylized models sometimes fail to have a concrete external validity. On a grander level, systems analysis also provides the knowledge of stakeholder analysis tools in order to predict the interaction effects of multiple parties involved in common problem solution findings. In all these features, systems analysis offers a sophisticated way to understand the success factors and pressure points for complex goals achievement strategies. Applied science diplomacy in the online context benefits from system analysis in its nature to be interdisciplinary and technology-focused. From the beginning of system analysis, this applied field studied technological advancements with a global outlook. From the beginning of science diplomacy during the Cold War era, complex interaction models were captured and measurable with the help of emerging computer systems. As digitalization evolved in the last decades, the field of systems analysis grew with it. Today, systems analysis has at its core computational sophistication and big data insights informing about complex interaction effects. The digital revolution also changed the field of behavioral sciences in the last decade to pay attention to big data insights (Puaschunder 2020). Systems analysis online is currently revolutionized in the opening up of Artificial Intelligence enhanced dynamic interaction possibilities, foremost in 5G, ChatGPT and Bard emerging on the horizon of everyday decision making. The time has come to analyze the potential of science diplomacy online for the accomplishment of the SDGs.

## 5 Science Diplomacy Online for Accomplishing the SDGs

In the initial drafting of the SDGs, digitalization and Information and Communication Technologies (ICT) for development were somewhat peripheral. The later revisions of the goals then made a clearer point about how to employ digital advancements for the SDGs. In all the literature on digitalization for sustainable development as well as in the growing body of knowledge on science diplomacy, information on the concrete use of technology for science diplomacy and science diplomacy in the online world is missing. Notable is the currently-unfolding evolution of lawmaking through digital media, foremost in the use of AI-enhanced tools and social online media. AI has become an information-gathering source in ChatGPT and Bard as well as a multitude of other search engine, communication and information-gathering plug-ins. In the transparent and open use of social online media, digitalization has become a central component of the online creation of law.

In light of these developments, classical, traditional diplomacy also changed. Transparent and fast-paced online communication on social online media has become a novel diplomacy gateway that cuts traditional red tape (Puaschunder 2023a). Digital diplomacy is characterized by being more transparent and visible to multiple stakeholders instantaneously compared to traditional diplomacy, which was often carried out behind closed doors by clearly assigned actors (Puaschunder 2023a). The open-access character of social online media diplomacy may change the dynamics of the generation of law (Puaschunder 2023a). The accessibility of the creation of law in the digital age may also bring about social and cultural dynamics in an unprecedented and rather unforeseeable way. Social online media connected large groups during the Arab Spring uprising and mobilized large-scale protests. Social online media also played a role during the January 6 U.S. Capitol riots as a January 6 committee report indicated (Zakrzewski et al. 2023).

Systems analysis helps trying to understand not only the influence factors of multiple agents on the dynamic control of systems. Systems analysis applied to measuring the social online media impetus can also offer a wealth of behavioral knowledge on animal spirits and the irrationality of groups under uncertainty (Puaschunder 2020). The shrinking role of governmental oversight and control of traditional media outlets' communication in the age of instant social online media sources as information-gathering tools appear to have accelerated a call for Science Diplomacy Online. Understanding the international dynamics of multiple actors and agents involved in creating a public discourse may allow steering crowds to engage with the help of systems analysis.

Positive advancements that the democratization of information exchange via social online media offers is the instantaneous character, the international outreach and the accessibility of tools for everyone available. Clear downsides and threats detected are the international manipulations of democratic processes (e.g., votes). Social online media control of the private sector is rather focused on industry development than governance aspects. There is also an already-noticeable negative impacts of social online media on mental health and well-being. Systems analysis could help unravel the necessary pros and cons of social online media use for diplomacy in order to make an informed decision on how to improve the quality of law and public policies in light of internet governance. Systems analysis could also address the systemic inequalities inherent in digital online media use—for instance in a stakeholder analysis and disparate impact analysis that may reveal groups with unequal access to online resources, groups with unequal capability to navigate online and groups that are discriminated by big data generated algorithmic decision making.

## 6 Quantifying Science Diplomacy Online

In the implementation of the current UN SDGs, a democratic approach is pursued. Yet given country inequalities around the world when it comes to digitalization, systems analysis could help in assigning leadership roles in Science Diplomacy Online for accomplishing the SDGs. Those countries that have established ties in science diplomacy networks have the best starting grounds to enact the SDGs (Puaschunder, forthcoming). Coupled with philosophical and economic redistribution mandates as well as responsibilities to protect schemes, country comparisons on science diplomacy offer insights into which countries have better means but also higher responsibilities to aid the world in reaching the ambitious SDG goals and targets (Nicholas 1961; Kant 1783/1993; Puaschunder, forthcoming). In a comparative systems analysis employed, this chapter addresses the science diplomacy online potential of countries around the globe.

The first index to quantify science diplomacy for 48 countries integrated (1) the academia quota per country as an indication of scientific excellence based on World Bank Educational Attainment data of at least Bachelor's or equivalent education in the population of a country from 25 years of age as cumulative percent in the population (World Bank); (2) a modified World Ranking of academic institutions based on the Web of Universities data weighted by the relevance of its academic institutions; and (3) the Lowy Global Diplomacy Index measuring diplomatic relations in embassies, consulates, or other diplomatic representations (Lowy Institute 2021; Puaschunder 2022a; forthcoming). The Science Diplomacy Index identifies which countries can lead on science diplomacy action in a comparative systems analysis approach (Puaschunder 2022a; forthcoming).

The Science Diplomacy Index for 48 countries of the world identified U.S. and Russia as key leaders in science diplomacy (Puaschunder 2022a; forthcoming). South Africa takes the lead on science diplomacy in Africa (Puaschunder 2022a; forthcoming). In Asia, Japan and India but also South Korea, Indonesia as well as the Philippines, Malaysia and Thailand, Bangladesh, Vietnam, China, Singapore, Mongolia, Myanmar, Nepal, Brunei and Bhutan are renowned to engage in science diplomacy (Puaschunder 2022a; forthcoming). Australia and New Zealand also employ science diplomats on global issues of concern (Puaschunder 2022a; forthcoming). In the Middle East, Saudi Arabia and Israel engage in science diplomacy (Puaschunder 2022; forthcoming). European science diplomacy leadership springs out of the United Kingdom, France and Germany as leaders in science diplomacy missions followed by Spain, Poland, Italy, the Netherlands, Portugal, Belgium, Greece, Hungary, Sweden, Ireland, Denmark, Norway and the Czech Republic, Austria, Finland, Lithuania, Slovakia, Latvia, Slovenia, Estonia and Iceland (Puaschunder 2022a; forthcoming). Canada complements U.S. science diplomacy leadership in North America (Puaschunder 2022a; forthcoming). In South America, Brazil, Mexico and Chile are science diplomacy advocates (Puaschunder 2022a; forthcoming).

In order to measure the disparate digitalization potential of countries around the globe, Puaschunder (2022b) created an index comprising digitalization and anti-corruption. The Anti-Corruption-AI Index Index was derived from a cross-sectional study of 79 countries' relation of Corruption Perception—measured by the Corruption Perception Index of 2019 (Transparency International 2019)—and AI—as captured by the Global Connectivity Index for 2019 (Huawei 2020; Puaschunder 2022b).

In the Anti-Corruption_AI Index, Puaschunder (2022b) outlines the relation of corruption and digitalization in 79 countries of the world. Interestingly, Puaschunder (2022b) finds Artificial Intelligence (AI)—measured by Global Connectivity—significantly positively correlated with freedom from corruption—as measured by the Corruption Perception Index of 2019 (Transparency International 2019). Digitalization, therefore, comes from parts of the world that are perceived as less corrupt (Puaschunder 2022b). AI hubs that are perceived as corruption free by their population and less digitally-connected areas of the world are viewed as more corrupt by their own inhabitants (Puaschunder 2022b).

In the comparative country approach, Asia and the Gulf region have medium levels of digitalization and corruption (Puaschunder 2022b). Singapore, Japan, South Korea and the United Arab Emirates but also Malaysia lead on AI and less corruption in Asia and the Middle East (Puaschunder 2022b). In Europe, Switzerland, Nordic countries like Sweden, Denmark, Norway and Finland are top AI and anti-corruption countries (Puaschunder 2022b). The United States of America has a top condition to lead on AI and anti-corruption (Puaschunder 2022b); while South America, especially Venezuela, Paraguay and Bolivia, rank lowest in anti-corruption coupled with digitalization (Puaschunder 2022b). In Oceania, New Zealand has a better AI and anti-corruption index performance than Australia (Puaschunder 2022b).

Digitalization therefore varies around the globe and appears as a tool to implement science diplomacy free from corruption. Digitalization-enhanced leadership offers a relatively corruption-free science diplomacy conduct, which could improve support of the SDGs in a non-corrupt systems analysis solution. Artificial global governance should therefore come from the countries with high global connectivity and low corruption that have the means to enact science diplomacy. Those countries that rank high on digitalization and anti-corruption could lead the world in science diplomacy solutions for enacting the SDGs. The parts of the world that feature better internet connectivity and high GDP levels but are also not bound by corruption but feature integrity are predestined to lead on science diplomacy online. In all these novel and emerging insights on the macroeconomic dynamics of digitalization around the world, we have no quantified information on science diplomacy online enhanced by anti-corrupt digitalization leadership.

Online science diplomacy applications are bringing together the strength of industry, government, research libraries, higher education, university consortia, and associations to discuss applied solutions to the world's most pressing contemporary problems. Science can offer quantifiable results and hard facts in order to help decision-makers from around the world to agree to develop a shared agenda and find creative strategies for a more coordinated approach to avert global issues of concern.

Actual applications comprise climate change mitigation and adaptation efforts, global pandemic prevalence, international development and inequality alleviation as well as ethical space exploration, to name a few current science diplomacy fields and foci.

Science diplomacy online is currently spearheading data analytics to accomplish shared information insights. Questions arise about how experts from different disciplines can collaborate via new media tools to gain a more holistic understanding of the many threats and vulnerabilities world populations face but also to debrief globally on the most creative cutting-edge solutions for cooperation and advancement on global scientific efforts.

Concrete areas of applied science diplomacy online comprise macroeconomic climate risk measurement of the International Monetary Fund, World Bank and the U.S. Federal Reserve (International Monetary Fund; World Bank; Federal Reserve of the United States). Data sources used by scientists about market risks include the Bloomberg Terminal and a new climate portal enabled by the United States White House (Bloomberg Terminal; The White House of the United States). Industry efforts to connect with science and provide data analytics for global concerns over global warming have been enabled via the Google Cloud Climate Engine and AT&T (AT&T; Google Cloud Climate Engine). Industry data-sharing tools with growth are Unqork and Jupyter, for example (Jupyter; Unqork). Governmental support includes the European 5$^{th}$ freedom of data codification approved by the European Parliament (European Parliament).

A key question is how experts from different disciplines can come together to gain a more holistic understanding of global issues of concern and solve them together efficiently. This chapter, therefore, introduces the first Science Diplomacy Online Leadership Index in order to detect areas of cross-disciplinary cooperation and data-informed decision-making.

## 7 Science Diplomacy Online Index

The first Science Diplomacy Online (SDO) Index was generated using a normalized version of the Anti-Corruption_AI Index for 79 countries of the world that captures the relation of corruption and digitalization (Puaschunder 2022b) factored by the Science Diplomacy Index (Puaschunder, forthcoming) for 59 countries. Corruption levels were measured by the Corruption Perception Index of 2019 (Transparency International 2019). Artificial Intelligence (AI) was operationalized by the Global Connectivity Index for 2019 (Huawei 2020; Puaschunder 2022b). Science diplomacy was integrated by using the Science Diplomacy Index comprised of (1) the academia quota per country as an indication of scientific excellence based on World Bank Educational Attainment data of at least Bachelor's or equivalent education in the population of a country from 25 years of age as cumulative percent in the population (World Bank); (2) a modified World Ranking of academic institutions based on the Web of Universities data weighted by the relevance of its academic institutions (Webometrics 2021); and (3) the Lowy Global Diplomacy Index measuring diplomatic relations in embassies, consulates, or other diplomatic representations (Lowy Institute 2021; Puaschunder forthcoming). The Science Diplomacy Online Index *SDO* is calculated based on Eq. 1, integrating the Anti-Corruption_AI Index (*AA*) of a country *c* multiplied by the Science Diplomacy Index *SDI* of a country *c*.

$$Science\,Diplomacy\,Online\,(SDO) = AA_c * SDI_c \tag{1}$$

The Science Diplomacy Online Index *SDO* results for each of the 42 countries are exhibited in Table 1.

The Science Diplomacy Online Index *SDO* results for 42 countries in comparison to each other are exhibited in Fig. 1, which ranks countries ascending from the best Science Diplomacy Online conditions.

As visible in Fig. 1, the United States has the best conditions to offer Science Diplomacy Online. In Asia, Japan is at the forefront of Science Diplomacy digital solutions followed by India, Indonesia, Malaysia, Singapore, Philippines, Thailand, China, Vietnam, and Bangladesh. Europe features a high density of Science Diplomacy Online opportunities, leading with the United Kingdom, Germany, France, Spain, Netherlands, Poland, Italy, Belgium, Sweden, Portugal, Denmark, Norway, Ireland, Finland, Austria, Greece and the Czech Republic as Science Diplomacy Online hubs. Russia, Hungary, Lithuania, Slovakia, Estonia and Slovenia are additional science diplomacy gateways in Eastern Europe. Canada and Australia, which leads before New Zealand, feature good starting grounds to help with digitalized solutions for science diplomacy. In South America, Brazil and Mexico followed by Chile are the best Science Diplomacy Online allies. In the Middle East, Saudi Arabia is the only country in the dataset to appear to offer Science Diplomacy Online credentials. Africa is only represented by South Africa in the data. Overall, it is noticeable that the dataset misses many other influential countries that are at the forefront of digitalization.

The first Science Diplomacy Online Index is the first guideline on what countries are predestined to lead the world on science diplomacy enhanced by big data science and technological advancements. The quantitatively-driven index will serve as a guideline for what countries could lead the world in finding fast and efficient solutions to the world's most pressing areas of concern. Once the leaders are established, attention needs to be paid to strategies for how to use digitalization for the alleviation of the most pressing concerns of our times in fostering the SDGs. The following chapter, therefore, outlines qualitative information on how to use science diplomacy online for implementing the SDGs.

## 8 Science Diplomacy Online for Implementing the SDGs

In a qualitative analysis of the role of Science Diplomacy Online in implementing the SDGs, the following concrete qualitative science diplomacy strengths should be considered (European Union 2022). In the pursuit of SDG Goal 1 '*No Poverty*' and

**Table 1** Science diplomacy online index values for 42 countries of the world

| | | | | | | | | | | | | | |
|---|---|---|---|---|---|---|---|---|---|---|---|---|---|
| Australia | 0.009774 | Chile | 0.001324 | France | 0.027949 | Ireland | 0.001644 | Netherlands | 0.009005 | Russia | 0.031249 | Spain | 0.010989 |
| Austria | 0.001172 | China | 0.000387 | Germany | 0.32382 | Italy | 0.005261 | New Zealand | 0.001431 | Saudi Arabia | 0.001193 | Sweden | 0.002562 |
| Bangladesh | 0.000050 | Czech Republic | 0.000982 | Greece | 0.000994 | Japan | 0.051370 | Norway | 0.001948 | Singapore | 0.000909 | Thailand | 0.000469 |
| Belgium | 0.003041 | Denmark | 0.002284 | Hungary | 0.000854 | Lithuania | 0.000430 | Philippines | 0.000565 | Slovakia | 0.000311 | United Kingdom | 0.041043 |
| Brazil | 0.009275 | Estonia | 0.000209 | India | 0.005677 | Malaysia | 0.001632 | Poland | 0.006598 | Slovenia | 0.000187 | United States of America | 0.502696 |
| Canada | 0.015435 | Finland | 0.001533 | Indonesia | 0.002692 | Mexico | 0.003343 | Portugal | 0.002394 | South Africa | 0.000163 | Vietnam | 0.000195 |

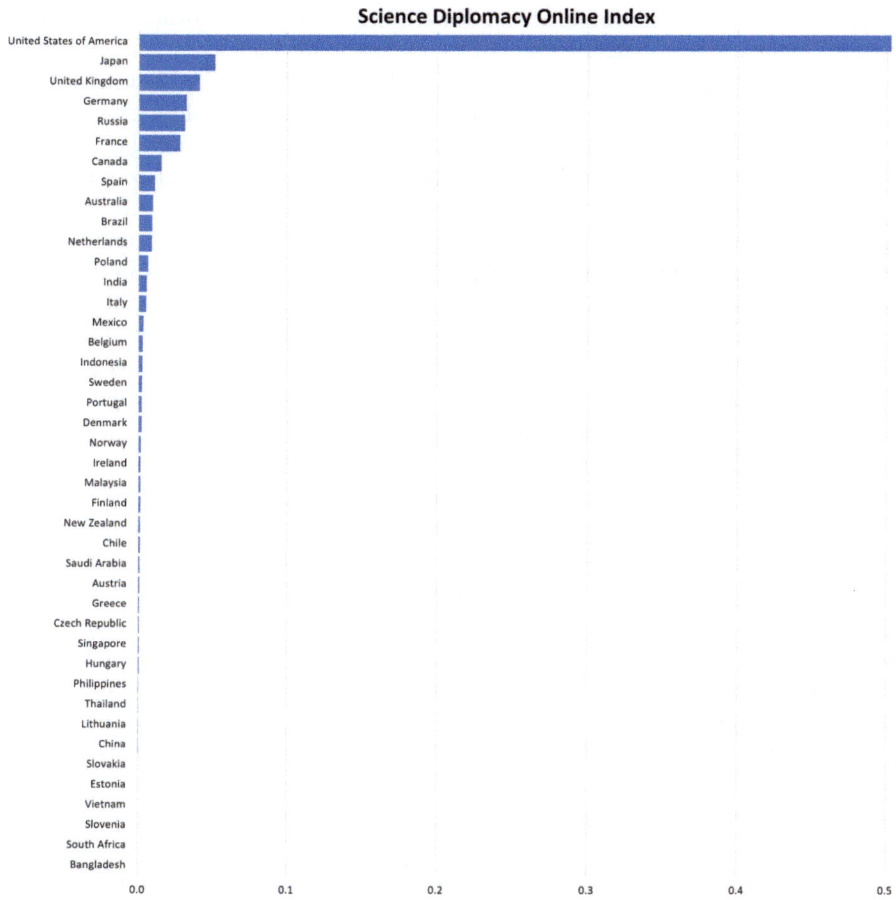


**Fig. 1** Science Diplomacy Online Index value ranking for 42 countries of the world ascending from the best Science Diplomacy Online conditions

Goal 4 '*Quality Education*' but also Goal 5 '*Gender Equality*,' digitalization and ICT for development have a long history as proven innovative drivers of change (United Nations Department of Economic and Social Affairs, Sustainable Development, retrieved June 19, 2023). Especially in territories where governmental support of education is scarce or females are not allowed to leave the house without male company, digital 'windows to the world' online have become game changers in eliciting hope of social transformation via online education (United Nations High-Level Political Forum on Sustainable Development 2022). Online classes—such as massive open online courses (MOOC)—have become a model for delivering online learning environments since 2008 (MOOC). Most recently, the COVID-19 pandemic has driven online teaching trends in platforms such as Zoom, Google Meet, Kumospace, Teachable, LinkedIn learning courses, etc. Digital inequality research is recommended to focus on online access disparities, not only in hardware accessibility but also in the use and understanding of novel communication tools.

Goal 2 '*Zero Hunger*' and GOAL 6 '*Clean Water and Sanitation*' as well as Goal 7 '*Affordable and Clean Energy*' can be enacted via Science Diplomacy Online in trying to find scarce resources and commodities with novel technologies, such as satellite imagery, like Geographic Information System (GIS), but also in the real-time tracking of changes in light of scarce resources preservation needs (National Geographic). Bluetooth digital device allows real-time tracking of supply chain movements, which enables to curb harmful bottleneck problems in underserved communities (Kremer 2019). The advantage of AI as a non-corrupt distribution tool appears relevant for international development against hunger as digitalization provides an objective, non-corrupt tool to reach those who are in need the most (Puaschunder 2022b). Real-time tracking of water quality has become a vital source of information during the COVID-19 pandemic when COVID debris in wastewater became a source of knowledge on the real spread of the virus and its multiple variants (Centers for Disease Control and Prevention CDC). Currently, tracking of medication debris in wastewater has become another concern of environmentalists who warn about the impact of the COVID-19 pandemic on clean water quality and water preservation as medication is hard to be filtered out from freshwater (Ashworth 2022; Scientific America 2011). Decentralized energy grids are additional innovative digitalization solutions to help share renewable energy efficiently between adjacent households and save energy more efficiently than in central energy conservation plants (O'Neill 2022).

Goal 3 '*Good Health and Well-being*' appears one of the major areas that can benefit from a digitalization revolution. Real-time tracking of the health status of the individual with digital devices (e.g., via implanted microchips), preventive care with big data analytics (e.g., genetic prevalence blood marker analysis) but also telemedicine offered via mobile phones in territories where access to quality medicine is scarce, are just some of the areas where Science Diplomacy Online comes in for the establishment of good health and well-being (Combi et al. 2016; Li and You 2022; Razzak et al. 2020). The COVID pandemic also opened gates, during times when hospitals were overloaded, to form online self-help groups, which currently also drive a trend towards holistic medical care (The Economist 2023). Prevention focus with the help of innovative real-time tracking tools (e.g., cancer markers real-time monitoring, air quality tracking with digital devices, clothing that measures body functions real-time) are digital innovations that prove the power of connecting science to stakeholder engagement in Science Diplomacy Online in the medical field (Ecowellness Group Salzburg Declaration 2020, 2021).

Goals 8 '*Decent Work and Economic Growth,*' Goal 9 '*Industry, Innovation and Infrastructure*' in harmony with Goal 12 '*Responsible Consumption and Production*' appear connected to science diplomacy in the innovation potential of digitalization. The AI, big data, 5G and ChatGPT revolutions are examples of digitalization changing the way people work, industries develop and consumption patterns are changing around the globe. The digital transformation is considered as the most significant innovation of our times. Social online media transparency has also changed the game for corporations to produce and serve communities via transparency and instant monitoring and large-scale whistleblower potentials in the digital age. Calls have formed to address human rights and sustainability online in the digital space (Puaschunder, 2023b).

Goal 10 '*Reduced Inequality*' and Goal 11 '*Sustainable Cities and Communities*' can both benefit from the big data-enhanced disparate impact-focused systems analysis that enables scientists to embark on. As never before in the history of research, was big data and computational power available to the extent as is today. Digitalization has opened unprecedented ways to understand the different impacts policies and city planning have on individual lives. The impact of transformative policies can be measured by longitudinal and large-scale designs today thanks to computational power and statistical advancements. For instance, in setting people on different trajectories for health, educational and economic prospects, big data analysis offers the most sophisticated evidence and reveals unknown risk factors within the individual but also in the environment (National Geographic). The COVID-19 pandemic and the different socio-psychological fallout propensities of the finance world and the real economy account for another example that was measured by the help of online information resources that allowed for deriving valuable information on how to counterbalance vulnerabilities within certain societal groups (Puaschunder 2021). The time has come to use big data insights for the alleviation of unnoticed inequality.

The accomplishment of Goal 12 '*Climate Action*' as well as Goals 13–14 '*Life below Water*' and '*Life on Land*' are all connected to digital tracking of hard science facts. The status of climate around the world, biodiversity and the variegated risks are measured by long-standing and cutting-edge technologies. Satellite imagery (e.g., enabled via GIS), Bluetooth-enabled resource tracking but also monitoring in real-time sensory are just a few examples of how digitalization can help understand ecological cycles and life stocks on earth and in water-covered areas (Kremer 2019; National Geographic). As for climate action, social online media as a crowd communication means and medium of information exchange of the young upcoming generation, predestine the internet as a major source for transparency about environmental conditions to track the status quo of the world and driving force of change. The most recent attempts of the Joseph Biden-Kamala Harris administration to measure natural resources to capture the role of natural resources for economic development and the World Bank's endeavor to link the environment to a wealth of nations, is technology data-driven (Reamer 2023; The United States Congress 2019; The White House 2023a, b). The environment-economy relationship is also part of the Green New Deal and the European Green Deal as well as the European Sustainable Finance Taxonomy, which involves quantifiable technology-enabled big data computation (European Commission; Puaschunder 2023c; The United States Congress 2019).

Goals 16 '*Peace, Justice, and Strong Institutions*' as well as Goal 17 '*Partnerships to achieve the Goal*' speak about collaboration and cooperation of international bodies and entities around the globe in the accomplishment of the ambitious SDG plan. ICT for public–private partnerships has a long tradition to employ digitalization for the harmonization and integration of multiple international actors for common endeavors. ChatGPT and 5G but also crowdsourcing of information and funding are additional resources of the future that are technology-driven success factors of tomorrow's institutions and partnerships.

## 9 Ongoing Issues or Challenges

This chapter addressed the role of Science Diplomacy Online as a novel and complementary supporting approach to making the SDGs work around the world. Technological advancements offer the most innovative attempts to foster common goals. Digitalization accounts for the most transforming common trend of our lifetimes. At the same time, as for being a novel, international and fast-changing industry development, digitalization and technological advancements may bear certain risks and call for cautious considerations that may also have to be addressed. When it comes to the efficiency question of technology use for sustainable development, little clear calculus and information exist on the carbon footprint of technology. Digitalization can often speed processes and make them more efficient—just think of all the letters that were saved in the wake of the email technology introduction. Zoom meetings during the COVID-19 pandemic saved millions of aircraft flights and train rides. Technology-enhanced preventive healthcare innovations and targeted treatments are the future hopes of breakthrough medicine to save millions of lives and countless resources. At the same time, one has to acknowledge that many technologies are massive carbon emitters and environmental polluters. 'Cloud' storage actually requires massive physical storage resources in cool locations to keep up with the massive energy consumption required to store electronic data. Space travel often viewed as one of the most futuristic endeavors actually is also one of the biggest emitters on earth when considering the massive power required to go into space. The biodiversity contagion risk of space travels—what happens if an invasive species is accidentally brought back from outer space that starts dominating life on land?—are hardly discussed in academic circles. So, the efficiency of digitalization for sustainable development is not a clear and straightforward one.

Areas of concern are privacy, for instance particularly problematic during COVID pandemic tracking. For example, healthcare data would be available for Google mapping but Google Health stopped, likely also due to industry pressures as transparency usually has a price-cutting effect (Advisory Board 2021). While national security protection can overrule privacy, also behavioral economics finds privacy concern diminishes under stress (e.g., danger, pain…) and when the information is valued as important (e.g., health). The European concept of the General Data Protection Regulation (GDPR) targets at regulating industries for improving consumer protection online with a focus on privacy among other features. It is to note that international resistance exists in some parts of the world about privacy concerns overregulating online innovations, not only in the United States. Transitioning countries and developing world appear to care less about privacy e.g., in Africa problems with risks of large-scale banking data fraud persist without clarification if the data leak is in Africa with mobile banking or the data cloud abroad. This problem is so large by now that banks do not want to refund missing funds and cloud providers abroad do not think the problem is with their cloud. China bans access to many internet platforms and privacy or intellectual property rights protection are not as important as in Western-oriented countries. Latin America is more concerned with access to the internet issues and infrastructure development. Eastern Europe has concern about internet corruption with certain regimes slowing the internet access to communities that do not vote for leadership. Remote communities in New Zealand noticed that Chat GPT is more accurate in their language than any other search engine (standard Google, Bing), which raised questions why? Is it that private social media communication data was shared? This could be a future area of privacy regulation if it is the case. Data storage in clouds is actually somewhat carbon-intensive, so if data science is increasing, this may become a problem more under scrutiny in the future.

When it comes to concrete implications of online digital worlds, fears over privacy in the European Union General Data Protection Regulation (GDPR), democracy infringements through crowd control and voting manipulation as well as mental health of teenagers on social media but also searchplace discrimination in the frustration over biased algorithms determining online searchplace results have been discussed. What this all leads to and appears to call for is a general evaluation of the potentials as well as the downfalls of digitalization for humankind. In the celebration of 75 years of the Universal Declaration of Human Rights, the need for a new dawn of a generation of human rights online is rising. It is on humankind to seize the opportunity to reap the benefits of the digital age while vigilantly guarding new emerging digital tools and choosing the use of digitalization wisely to avert the potential downfalls and shortcomings of innovation in flourishing sustainable development.

## References


Advisory Board. Google Health is dead (But this isn't the end of 'Google health') https://www.advisory.com/daily-briefing/2021/08/24/google-health

Ashworth J (2022). National History Museum. July 6, 2022. https://www.nhm.ac.uk/discover/news/2022/july/drug-pollution-threatening-water-quality-worlds-rivers.html

AT&T https://about.att.com/newsroom/2021/climate_resiliency_community_challenge_data.html

Bisanz M, Forrester JP, Tsay J. Federal Reserve of the United States. https://www.mayerbrown.com/en/perspectives-events/publications/2023/04/federal-reserve-continues-research-on-climate-risk

Bloomberg Professional Services. Bloomberg Terminal. https://data.bloomberglp.com/professional/sites/10/Climate-related-Analysis-Brochure.pdf

Centers for Disease Control and Prevention. National Wastewater Surveillance System (NWSS). https://www.cdc.gov/nwss/wastewater-surveillance.html#:~:text=By%20measuring%20SARS%2DCoV%2D2,community%20is%20increasing%20or%20decreasing

Combi C, Pozzani G, Pozzi G (2016) Telemedicine for developing countries: a survey and some design issues. Appl Clin Inform 7(4):1025–1050

EcoWellness Group (2020) Salzburg declaration: interdisciplinary conference on 'system change?! The chance of transformation of the healthcare system: Analysis and chances of the coronavirus crisis, July 14–15, 2020. Retrieved at https://www.oekowellness.de/laenderuebergreifende-konzerenz-zum-thema-system-change-die-chance-der-transformation-des-gesundheitswesens-14-07-2020/

EcoWellness Group (2021) Salzburg European Declaration from the Gasteinertal: Interdisciplinary Conference on 'System change?! The chance of transformation of the healthcare system, July 14–15, 2021. Retrieved at https://www.oekowellness.de/wp-content/uploads/2021/07/Final-Stand-5.7.-2021.07.04_Programm-14.7.-und-15.07.2021-2.pdf

European Commission. The European Green Deal. https://commission.europa.eu/strategy-and-policy/priorities-2019-2024/european-green-deal_en

European Parliament. https://www.europarl.europa.eu/news/en/press-room/20180926IPR14403/free-flow-of-non-personal-data-parliament-approves-eu-s-fifth-freedom

European Union (2022) What is science diplomacy? March 16, 2022. https://www.eeas.europa.eu/eeas/what-science-diplomacy_en

Huawei (2020) Global connectivity index. https://www.huawei.com/minisite/gci/en/index.html

Jupyter https://jupyter.org/

Kant I (1783/1993) Grounding for the metaphysics of morals. Cambridge: Hackett

Kremer M (2019) How Bluetooth is changing water pollution management. Jenco. https://blog.jencoi.com/how-bluetooth-is-changing-water-pollution-measurement

Leal Filho W, Lange Salvia A, Beynaghi A, Fritzen B, Ulisses A, Veiga Avila L, Shulla K, Vasconcelos CRP, Moggi S, Mifsud M, Anholon R, Rampasso IS, Kozlova V, Iliško D, Skouloudis A, Nikolaou I (2023) Digital transformation and sustainable development in higher education in a post-pandemic world. Int J Sust Dev World. https://doi.org/10.1080/13504509.2023.2237933

Li X, You K (2022) Real-time tracking and detection of patient conditions in the intelligent m-Health monitoring system. Front Public Health 10:922718. https://www.ncbi.nlm.nih.gov/pmc/articles/PMC9589418/

Lowy Institute (2021) Global diplomacy index. https://globaldiplomacyindex.lowyinstitute.org/country_rank.html

Mejia SA, Mrkaic M, Novta N, Pugacheva E, Topalova P. International monetary fund. https://www.imf.org/en/Publications/WP/Issues/2018/06/22/The-Effects-of-Weather-Shocks-on-Economic-Activity-What-are-the-Channels-of-Impact-45970

MOOC. https://www.mooc.org/

National Geographic. GIS (Geographic Information System). https://education.nationalgeographic.org/resource/geographic-information-system-gis/

Nicholas K (1961) Capital accumulation and economic growth. In: Lutz FA, Hague DC (eds) The theory of capital. St. Martin's Press, New York, pp 177–222

O'Neill C (2022) National renewable energy laboratory. From the bottom up: designing a decentralized power system cross-discipline team envisions the grid of the future. https://www.nrel.gov/news/features/2019/from-the-bottom-up-designing-a-decentralized-power-system.html

Pearl D, White C. Google cloud climate engine. https://cloud.google.com/blog/topics/sustainability/climate-engine-unlocks-climate-change-insights-with-google-cloud

Pierce A (2018) Why SDG 17 is the most important UN SDG? https://www.sopact.com/perspectives/sdg17-most-important-sdg

Puaschunder JM (2020) Behavioral economics and finance leadership: nudging and winking to make better choices. Springer, Cham

Puaschunder JM (2021) Focusing COVID-19 bailout and recovery. Ohio State Bus Law J 16(1):91–148

Puaschunder JM (2022a) Science diplomacy index: leadership and responsibility to act on climate change. Sci Moral Int J Multidisip Res 7(2):11–30

Puaschunder JM (2022b) The future of artificial intelligence in international healthcare: an index. In: Nagl-Docekal H, Zacharasiewicz W (eds) Artificial intelligence and human enhancement:

affirmative and critical approaches in the humanities. Austrian Academy of Sciences, Austria, European Union, De Gruyter, Berlin, pp 181–208

Puaschunder JM (2023a) Digital diplomacy. In: Proceedings of the 32nd international RAIS conference on social sciences and humanities organized by the research association for interdisciplinary studies (RAIS), June 9, 2023

Puaschunder JM (2023b) Human rights online: towards a new generation of human rights in the virtual world. 75 years of the Universal Declaration of Human Rights (UDHR) historic edition. United Nations, New York

Puaschunder JM (2023c) The future of resilient finance: finance politics in the age of sustainable development. Palgrave, New York

Puaschunder JM (forthcoming) A science diplomacy agenda for sustainable development research. In: Salvia A, Leal W (eds) An agenda for sustainable development research. Springer, London

Razzak MI, Imran M, Xu G (2020) Big data analytics for preventive medicine. Neural Comput Appl 32(9):4417–4451

Reamer A (2023) National strategy to develop statistics for environmental-economic decision: OSTP/Commerce/OMB, January 19, 2023. Retrieved at https://www.aeaweb.org/forum/3447/national-strategy-statistics-environmental-economic-decisions

Scientific America (2011) External medicine: discarded drugs may contaminate 40 million Americans' drinking water. https://www.scientificamerican.com/article/pharmaceuticals-in-the-water/#:~:text=Pharmaceutical%20drug%20contamination%20in%20our,to%20deal%20with%20such%20contaminants

Sustainable Development Goals (2022) Decade of action. https://www.un.org/sustainabledevelopment/decade-of-action/

The Economist (2023) Too many people take too many pills: Medical systems are set up to put people on drugs, not take them, April 26, 2023. Retrieved at https://www.economist.com/science-and-technology/2023/04/26/too-many-people-take-too-many-pills

The United States Congress (2019) Recognizing the duty of the Federal Government to create a Green New Deal. Retrieved at https://www.congress.gov/bill/116th-congress/house-resolution/109/text

The White House (2022) Fact sheet: Biden–Harris administration launches new climate portal to help communities navigate climate change impacts. https://www.whitehouse.gov/briefing-room/statements-releases/2022/09/08/fact-sheet-biden-harris-administration-launches-new-climate-portal-to-help-communities-navigate-climate-change-impacts/

The White House (2023a) Fact Sheet: Biden–Harris Administration releases national strategy to put nature on the nation's balance sheet, January 19, 2023. Retrieved at https://www.whitehouse.gov/ostp/news-updates/2023/01/19/fact-sheet-biden-harris-administration-releases-national-strategy-to-put-nature-on-the-nations-balance-sheet/

The White House (2023b) National strategy to develop statistics for environmental-economic decisions: A U.S. system of natural capital accounting and associated environmental-economic statistics. Office of Science and Technology Policy. Office of Management and Budget. Department of Commerce. Retrieved at https://www.whitehouse.gov/wp-content/uploads/2023/01/Natural-Capital-Accounting-Strategy-final.pdf

Transparency International (2019) CPI global highlights. https://www.transparency.org/en/news/cpi-2019-global-highlights

United Nations (2017) Resolution adopted by the General Assembly on 6 July 2017. https://en.wikipedia.org/wiki/File:A_RES_71_313_E.pdf

United Nations Department of Economic and Social Affairs, Sustainable Development. https://sdgs.un.org/goals

United Nations High-Level Political Forum on Sustainable Development (2022). https://hlpf.un.org/2022/programme/preserving-an-open-internet-to-achieve-the-sdgs-the-internet-as-an-enabler-for

Unqork. https://www.unqork.com/

Webometrics (2021). Ranking web of universities. https://www.webometrics.info/en/distribution_by_country
Whitten JL, Bentle LD (2007) Systems analysis and design for the global enterprise. Graw-Hill, London
World Bank. https://www.worldbank.org/en/topic/social-dimensions-of-climate-change
World Bank. Educational attainment, at least Bachelor's or equivalent, population 25+, total (%) (cumulative). https://data.worldbank.org/indicator/SE.TER.CUAT.BA.ZS
Zakrzewski, C, Lima C, Harwell D (2023) What the Jan. 6 probe found out about social media, but didn't report. The Washington Post, January 17, 2023. https://www.washingtonpost.com/technology/2023/01/17/jan6-committee-report-social-media/

# Pedagogical Paradigm Shift: Rethinking Education for Sustainable Development

**Suman Dahiya, Anju Kumari, and Ridhi Kochar**

**Abstract** In the era of rapid technological advancement and growing awareness of our planet's fragility, the convergence of pedagogy, sustainable development, and information technology stands as a pivotal crossroad in shaping the future of education. The title "Pedagogical Paradigm Shift: Rethinking Education for Sustainable Development" encapsulates the profound transformation that our educational systems must undergo to prepare the upcoming generations for the complexities of a dynamically changing world.

## 1 Introduction

In the era of rapid technological advancement and growing awareness of our planet's fragility, the convergence of pedagogy, sustainable development, and information technology stands as a pivotal crossroads in shaping the future of education. The title "Pedagogical Paradigm Shift: Rethinking Education for Sustainable Development" encapsulates the profound transformation that our educational systems must undergo to prepare the upcoming generations for the complexities of a dynamically changing world.

Sustainable development Jahnke and Hukkinen (2017), with its multifaceted emphasis on environmental, social, and economic equilibrium, has emerged as a paramount goal for global progress. As we confront pressing challenges such as climate change, resource scarcity, and social inequalities, the urgency to cultivate environmentally-conscious, socially-aware, and forward-thinking individuals has never been greater. This paradigm shift in pedagogy entails an evolution beyond traditional rote learning, encouraging critical thinking, problem-solving,

A. Kumari · R. Kochar
Panipat Institute of Engineering and Technology, Samalkha, Haryana, India

S. Dahiya (✉)
Gitarattan International Business School, New Delhi, India
e-mail: suman.dahiya@gitarattan.edu.in

 
W. Leal Filho et al. (eds.), *Digital Technologies to Implement the UN Sustainable Development Goals*, World Sustainability Series,
https://doi.org/10.1007/978-3-031-68427-2_17

and collaboration—skills essential for navigating the intricacies of sustainable development Lambright and McKeown (2016), Leal Filho and Pace (2017).

In this landscape of transformation, information technology emerges as a catalyst, redefining the way we acquire, disseminate, and apply knowledge. Information technology not only offers innovative tools for learning but also creates unprecedented connectivity, allowing global perspectives to enrich local discussions. As we delve deeper into this discourse, several key features come to the fore:

1. *Digital Literacy and Inclusivity*: Information technology bridges gaps in access to education by reaching remote or marginalized communities. It fosters digital literacy, empowering individuals with the skills to navigate a technology-driven world.
2. *Personalized Learning*: Technology enables tailored learning experiences, accommodating diverse learning styles and paces. Adaptive platforms and data analytics optimize learning pathways for individual students, enhancing engagement and comprehension.
3. *Interdisciplinary Exploration*: Information technology dissolves traditional subject boundaries, encouraging cross-disciplinary exploration. This approach mirrors the interconnected nature of sustainable development Leicht and Heiss (2016), promoting holistic problem-solving.
4. *Global Collaboration*: Virtual connectivity facilitates collaboration among students, educators, and experts worldwide. This interconnectedness exposes learners to varied perspectives, encouraging a deeper understanding of global challenges and solutions.
5. *Interactive and Immersive Content*: Multimedia resources, virtual simulations, and augmented reality bring abstract concepts to life, enhancing understanding and retention. These tools create dynamic learning environments that mirror real-world complexities.
6. *Data-Driven Insights*: Information technology allows educators to track individual and group progress, identify learning gaps, and refine teaching strategies. This data-driven approach enhances the efficacy of educational interventions.
7. *Environmental Impact*: Leveraging technology for education can reduce the environmental footprint associated with traditional methods, such as printed materials and commuting.

As we embark on this transformative journey, it becomes evident that the union of pedagogy, sustainable development UNESCO (2017a, b), UNESCO (2014, 2015), and information technology holds immense potential to equip learners with the knowledge, skills, and mindset needed to address the challenges and opportunities of the twenty-first century. This exploration seeks to unravel the intricate threads of this convergence, illuminating the path toward a more inclusive, responsive, and effective educational landscape.

## 2 Objectives

1. To explore the diverse array of information technology tools and platforms available for education and assess how they can be seamlessly integrated into the curriculum to enhance the learning experience.
2. To develop strategies to cultivate digital literacy among educators and learners, enabling them to navigate and leverage technology effectively for educational purposes.
3. To investigate ways to harness information technology to foster interdisciplinary learning, breaking down traditional subject silos and encouraging a more holistic understanding of complex sustainable development challenges.
4. To create interactive and immersive learning modules that utilizes multimedia, simulations, and augmented reality to engage learners in an experiential exploration of sustainable development concepts.
5. To devise methods to utilize information technology to tailor learning experiences based on individual learning styles, strengths, and needs, ensuring a more personalized and effective education for every student.
6. To explore digital platforms and tools that facilitate global collaboration among educators and students, allowing them to collaborate on projects, share perspectives, and engage in cross-cultural discussions related to sustainable development.
7. To develop mechanisms to gather and analyze data on learning outcomes facilitated by information technology, providing insights into the effectiveness of these tools in promoting sustainable development education.
8. To offer training and professional development programs for educators to build their competence in using information technology effectively in the classroom, enhancing their ability to teach sustainable development concepts.
9. To identify and address potential barriers to accessing technology, ensuring that the benefits of information technology in sustainable development education are accessible to diverse socio-economic and geographic populations.
10. To highlight the importance of ethical and responsible use of technology, emphasizing values such as digital citizenship, data privacy, and online integrity within the context of sustainable development education.
11. To evaluate the environmental impact of integrating information technology into education, considering factors such as reduced paper usage, energy consumption, and carbon footprint.
12. To advocate for policy initiatives that recognize the value of information technology in promoting sustainable development education and encourage its integration into educational systems at various levels.

These objectives collectively aim to drive a comprehensive exploration of the intersection between pedagogy, sustainable development, and information technology. By achieving these objectives, educators, policymakers, and learners can contribute to the cultivation of a more environmentally-conscious, socially-aware, and technologically-literate generation equipped to address the challenges of the present and future.

## 3 Literature Review

The literature reflects a growing recognition of the transformative potential that emerges when pedagogy, sustainable development, and information technology converge. This synthesis encapsulates innovative approaches to fostering a more responsive, inclusive, and effective educational landscape.

Scholars such as Sterling (2001) have emphasized the need for pedagogical transformation that nurtures sustainable values and critical thinking skills. The shift from traditional didactic methods to constructivist and experiential learning models aligns with the holistic ethos of sustainable development United Nations (2015). These scholars underline the importance of pedagogies that enable learners to engage actively with real-world challenges, fostering a deep understanding of sustainability concepts.

The integration of information technology in education is a focal point in the works of Siemens (2004). They explore how technology facilitates personalized learning experiences, enables real-time interaction, and empowers learners to explore concepts through diverse mediums. The rise of online learning platforms, virtual reality, and AI-powered tools offers educators and students dynamic ways to engage with sustainable development content.

Researchers Sanzgiri (2023) have underscored the importance of digital literacy in a technology-driven world. They emphasize how technological integration can bridge educational gaps, providing marginalized populations with access to quality education. The inclusion of technology also requires fostering critical digital literacy skills to navigate the vast online landscape responsibly and ethically.

Technology serves as a conduit for disseminating sustainable practices and raising global consciousness about environmental issues.

The works of research scholars explore how virtual classrooms, video conferencing, and collaborative online projects can facilitate cross-cultural learning experiences, enabling learners to grasp diverse perspectives on sustainable development challenges.

The integration of information technology also brings forth the potential for data-driven insights into education. Research discusses how data analytics can be leveraged to understand student progress, identify learning gaps, and tailor interventions in sustainable development education for greater effectiveness.

While technology offers numerous benefits, scholars highlight potential challenges. These include issues related to data privacy, digital divide, and the unintended consequences of over-reliance on technology. They call for a balanced approach that considers ethical dimensions alongside the integration of information technology in education.

In summary, the literature review showcases a rich tapestry of research that converges around the pivotal nexus of pedagogy, sustainable development, and information technology. The discourse underscores the imperative to adapt pedagogical practices to align with the ethos of sustainability and harness technology as a catalyst for transformative learning experiences. As educators, policymakers, and researchers embark on this interdisciplinary journey, the literature provides valuable insights that can guide the endeavor to shape a more enlightened and empowered generation capable of addressing the complex challenges of the modern world.

## 4 Methods

In pursuit of exploring information technology tools for enhanced learning, we embark on a comprehensive exploration of the vast landscape of information technology tools and platforms available for education. We delve into the realms of online learning management systems (LMS), educational apps, interactive platforms, and multimedia resources. Through rigorous assessment, we seek to identify the tools that align with the principles of sustainable development education. Our aim is to seamlessly integrate these technologies into the curriculum, transforming traditional learning experiences into dynamic and engaging encounters that enhance students' comprehension, participation, and critical thinking Wals and Jickling (2002).

Cultivating digital literacy for educators and learners revolves around the cultivation of digital literacy among educators and learners alike. We recognize that the modern educational landscape demands proficiency in navigating digital platforms, discerning credible online resources, and effectively utilizing technology for learning. By developing strategies that empower educators and students with these skills, we bridge the digital divide, ensuring that individuals can harness technology's potential for educational enrichment. Through workshops, tutorials, and resources, we enable both educators and learners to confidently navigate the digital landscape and embrace technology as an ally in their educational journey.

Interdisciplinary learning is paramount to understanding the multifaceted nature of sustainable development challenges. By harnessing the capabilities of information technology, we aim to dissolve traditional subject boundaries. Through collaborative projects, thematic modules, and integrated learning experiences, we encourage learners to synthesize knowledge from diverse disciplines. This approach nurtures a holistic understanding of complex sustainability issues Wiek et al. (2011), empowering students to approach challenges from multiple angles and cultivate innovative solutions that span traditional domains.

Our pursuit of creating immersive learning modules using multimedia and AR centers on creating interactive and immersive learning modules that leverage multimedia, simulations, and augmented reality (AR). By blending textual information with visual, auditory, and interactive elements, we craft an experiential journey for learners. Through AR, we enable students to immerse themselves in virtual environments, fostering a deeper connection with sustainable development concepts. These modules go beyond passive learning, inviting students to explore, interact, and engage with complex topics in an engaging and memorable manner.

With a commitment to personalized education, tailoring learning experiences through technology seeks to devise methods that utilize information technology to cater to individual learning styles, strengths, and needs. By employing adaptive learning technologies, we ensure that each student's educational journey is optimized. This approach not only enhances comprehension but also fosters self-directed learning, enabling students to take ownership of their education and embark on a path tailored to their unique abilities and interests.

Enabling global collaboration through digital platforms emphasizes the power of digital platforms in facilitating global collaboration among educators and students. By exploring virtual classrooms, online collaboration tools, and cross-cultural communication platforms, we create avenues for diverse perspectives to converge. Through collaborative projects and discussions, learners gain exposure to varied viewpoints, enhancing their cross-cultural understanding and instilling a sense of global citizenship. The digital realm becomes a bridge connecting individuals across geographical and cultural boundaries.

In pursuit of analyzing learning outcomes using information technology, we delve into the world of data analysis to evaluate the efficacy of information technology in promoting sustainable development education. By designing mechanisms to gather and analyze learning outcomes facilitated by technology, we uncover insights into the impact of these tools on students' understanding, engagement, and retention of sustainable development concepts. These data-driven insights inform future strategies, ensuring continuous improvement in educational practices.

Empowering educators is central to this objective. Through tailored training programs and professional development initiatives, we equip educators with the competence to effectively integrate information technology into the classroom. By nurturing educators' proficiency in using technology to teach sustainable development concepts, we foster a generation of instructors who can navigate the ever-evolving digital landscape with confidence, thereby enriching the learning experience for their students.

Addressing disparities in technology access is the cornerstone of overcoming barriers to technology access. By identifying and mitigating potential barriers, whether socio-economic or geographic, we ensure that the benefits of information technology are accessible to all. Through innovative solutions, partnerships, and inclusive initiatives, we strive to level the playing field, granting all learners the opportunity to engage with technology-enhanced sustainable development education.

Promoting ethical technology use and digital citizenship underscores the ethical dimension of technology integration. We emphasize the importance of digital citizenship, data privacy, and responsible online behavior within the context of sustainable development education. By fostering an awareness of digital ethics, we equip students with the tools to navigate the digital landscape ethically, ensuring that their use of technology aligns with the values of sustainable and responsible global citizenship.

Through meticulous evaluation, we assess the environmental impact of integrating technology into education. By considering factors such as reduced paper usage, energy consumption, and carbon footprint, we gauge the ecological footprint of digital education. This assessment guides us toward sustainable practices that align with the ethos of environmental stewardship.

By recognizing the value of information technology in promoting sustainable development education, we advocate for its integration into educational systems at various levels. Through collaboration with policymakers, educational institutions, and stakeholders, we aim to shape policies that acknowledge technology's transformative potential in nurturing informed, empowered, and globally conscious citizens.

In closing, these objectives exemplify our commitment to ushering in a pedagogical revolution that converges technology, education, and sustainability Wiek et al. (2011). By pursuing these goals, we embark on a journey that empowers educators and learners to transcend traditional boundaries, embrace innovation, and embrace the boundless possibilities of a technology-infused educational landscape. In doing so, we pave the way for a generation poised to address the complexities of our rapidly evolving world and steer it toward a sustainable and prosperous future.

Detailed explorations of methods that can be adopted to facilitate the pedagogical paradigm shift and rethink education for sustainable development while leveraging the potential of information technology are:

1. *Flipped Classrooms and Blended Learning*: Implementing a flipped classroom approach involves shifting the traditional lecture-based learning to self-paced online modules that students can access before class. This allows in-person sessions to be focused on collaborative discussions, problem-solving, and application of sustainable development concepts. Blended learning combines face-to-face instruction with online resources, enabling students to engage with content at their own pace while fostering deeper understanding and critical thinking.
2. *Online Learning Platforms*: Utilize online learning management systems (LMS) to host course materials, assignments, and discussions related to sustainable development. Incorporate multimedia resources, such as videos, interactive simulations, and podcasts, to make learning engaging and accessible. These platforms also enable educators to track student progress and tailor interventions based on individual needs.
3. *Virtual Reality (VR) and Augmented Reality (AR)*: Integrate VR and AR technologies to create immersive learning experiences. Students can virtually explore ecosystems, historical events, or real-world sustainability projects. For instance, a virtual field trip to a rainforest can help students understand the intricacies of biodiversity and conservation efforts.
4. *Collaborative Online Projects*: Facilitate cross-cultural and interdisciplinary collaboration through online platforms. Students from different geographical locations can collaborate on sustainability-focused projects, sharing diverse perspectives and co-creating solutions. This approach fosters a sense of global citizenship and enhances problem-solving skills.

5. *Gamification for Engagement*: Infuse gamification elements into lessons by designing educational games, quizzes, and interactive challenges related to sustainable development. Gamified content can motivate students to actively participate and explore concepts in a fun and engaging manner.
6. *Data Analysis and Visualization*: Leverage data analytics tools to analyze trends and patterns related to sustainability issues. Students can work with real-world data sets, identifying environmental trends, social inequalities, or economic impacts. Visualizing data through charts and graphs enhances understanding and empowers students to draw informed conclusions.
7. *Online Discussions and Webinars*: Organize online discussions, forums, and webinars where students can interact with experts, community leaders, and practitioners in the field of sustainable development. This exposes students to diverse viewpoints, real-world challenges, and potential solutions, enriching their understanding beyond textbooks.
8. *Personalized Learning Pathways*: Utilize adaptive learning technologies that tailor educational content based on individual learning styles and progress. As students engage with the material, the system adapts, offering additional resources or targeted interventions to address areas of difficulty.
9. *Digital Storytelling and Content Creation*: Encourage students to create digital stories Stibbe (2016), podcasts, videos, or blogs that showcase their understanding of sustainable development concepts. This fosters creativity, communication skills, and the ability to convey complex ideas effectively.
10. *Environmental Monitoring and Citizen Science*: Utilize technology for environmental monitoring and citizen science initiatives. Students can engage in data collection and analysis related to local environmental issues Tilbury et al. (2002), contributing to real-world research efforts and fostering a sense of environmental stewardship.
11. *Ethical Technology Use and Digital Citizenship*: Integrate discussions on digital ethics, responsible online behavior, and digital citizenship into the curriculum. Help students navigate the digital landscape with awareness of issues like online privacy, misinformation, and ethical use of technology.
12. *Continuous Professional Development for Educators*: Offer training and workshops for educators to enhance their technological skills, pedagogical strategies, and understanding of sustainable development concepts. Continuous professional development ensures that educators can effectively integrate technology and sustainability into their teaching practices.

By adopting these methods, educators can facilitate the envisioned pedagogical paradigm shift, fostering a holistic and innovative learning environment that empowers students to become active contributors to sustainable development Barth and Rieckmann (2012). Through the strategic incorporation of information technology, education can transcend traditional boundaries, preparing students to address the complex challenges of the modern world with creativity, critical thinking, and a global perspective.

## 5 Results and Discussion

The exploration of methods to facilitate the pedagogical paradigm shift and reshape education for sustainable development while leveraging the potential of information technology has yielded a diverse range of strategies with profound implications for transformative learning experiences. The following section delves into the outcomes and implications of each method:

1. The implementation of flipped classrooms and blended learning offers several benefits. By shifting traditional lectures online, students engage with foundational knowledge at their own pace, freeing up in-person class time for interactive discussions and collaborative problem-solving. This approach encourages active participation, critical thinking, and the application of sustainable development concepts to real-world scenarios.
2. Utilizing online learning platforms empowers both educators and students to access a wealth of resources beyond textbooks. Multimedia elements such as videos and simulations create dynamic learning experiences, catering to different learning styles. The ability to track progress and customize interventions based on individual needs enhances the efficacy of education while providing a comprehensive repository of sustainable development materials.
3. Integrating VR and AR technologies immerses students in realistic scenarios, enhancing their understanding of complex sustainable development concepts. Virtual field trips and interactive simulations foster a deeper connection with the subject matter, enabling students to explore ecosystems, historical events, and sustainable practices in an engaging and memorable way.
4. Facilitating cross-cultural and interdisciplinary collaboration through online platforms cultivates a global perspective and hones teamwork skills. Students gain exposure to diverse viewpoints and develop solutions that integrate multiple disciplines, mirroring the multifaceted nature of sustainable development challenges.
5. Gamification injects an element of playfulness into learning, enhancing engagement and motivation. Educational games and interactive challenges encourage students to actively explore sustainable development concepts, transforming learning into an enjoyable and rewarding experience.
6. Leveraging data analytics tools empowers students to derive insights from real-world data sets, enhancing their ability to understand and address sustainability issues. Visualizing data through charts and graphs aids comprehension and equips students with valuable skills for evidence-based decision-making.
7. Online discussions and webinars facilitate direct interactions with experts and practitioners in the field of sustainable development. Exposure to diverse perspectives enriches students' understanding and encourages them to think critically about the complexities of real-world challenges.

8. Adaptive learning technologies address individual learning styles and paces, ensuring that each student receives a tailored education. This personalized approach optimizes learning outcomes and caters to a diverse range of abilities and backgrounds.
9. Encouraging students to create digital content fosters creativity, communication skills, and the ability to distill complex ideas into accessible formats. This method empowers students to become effective communicators and advocates for sustainable development.
10. Engaging in environmental monitoring and citizen science initiatives fosters a sense of responsibility and connection to local and global ecosystems. Students actively contribute to real-world research efforts, reinforcing the importance of environmental stewardship.
11. Incorporating discussions on digital ethics and responsible online behavior cultivates digital citizenship, ensuring that students navigate the digital landscape ethically and safely while leveraging technology for positive impact.
12. Equipping educators with technological skills, innovative pedagogies, and a deep understanding of sustainable development concepts enhances the quality of education. Continuous professional development ensures that educators remain at the forefront of educational advancements and are capable of effectively guiding students through the dynamic landscape of sustainable development.

In conclusion, the exploration of these methods underscores the potential of information technology to drive a profound pedagogical shift that aligns education with the imperatives of sustainable development. By strategically integrating these methods, educators can create an enriched learning environment that empowers students to become agents of positive change, equipped with the skills, knowledge, and mindset to tackle the intricate challenges of our rapidly evolving world.

# 6 Conclusion

The journey through the diverse array of methods designed to catalyze a pedagogical paradigm shift in education, while harnessing the transformative power of information technology, reveals a landscape rich with opportunities for reshaping learning experiences. Each method explored in this discourse emerges as a dynamic tool poised to cultivate a generation of learners well-equipped to navigate the complexities of sustainable development in a rapidly evolving world.

From the implementation of flipped classrooms and blended learning, where the traditional teacher-centered model gives way to student-driven engagement, to the utilization of online learning platforms that transcend physical boundaries, the landscape of education takes on new dimensions. Virtual Reality (VR) and Augmented Reality (AR) create immersive experiences, transporting learners to ecosystems and historical events, fostering visceral connections to sustainable practices.

The notion of collaborative online projects transcends geographical limitations, forging a global network of minds that collaborate, communicate, and co-create solutions to intricate problems. Through gamification, learners are motivated to explore sustainable development concepts with a spirit of playfulness, unleashing their inherent curiosity. Data analysis and visualization empower students to derive insights from real-world data sets, cultivating a data-driven mindset for addressing complex sustainability challenges.

Online discussions and webinars facilitate direct engagement with experts and practitioners, embedding real-world wisdom into the educational experience. The integration of adaptive learning technologies personalizes education, ensuring that each student's unique learning journey is optimized. Digital storytelling nurtures creative expression, empowering students to communicate complex ideas effectively and advocate for sustainable development causes.

Environmental monitoring and citizen science initiatives imbue learners with a sense of environmental stewardship, catalyzing their commitment to real-world research and action. Discussions on ethical technology use and digital citizenship cultivate a responsible online presence, while equipping educators with continuous professional development ensures a cohort of knowledgeable mentors at the forefront of educational advancements.

In essence, these methods collectively illuminate the transformative potential of information technology, bridging pedagogy and sustainable development. By strategically weaving these methods into educational frameworks, educators can forge an enriched learning environment that nurtures critical thinking, ethical consciousness, and global engagement. This synthesis empowers learners not merely with information, but with the skills, mindset, and vision needed to navigate the intricate challenges of an evolving world. As the path to a sustainable future unfurls, this fusion of technology and pedagogy emerges as a guiding light, illuminating the way for a generation poised to be architects of positive change.

## References


Barth M, Rieckmann M (2012) Academic staff development as a catalyst for curriculum change towards education for sustainable development: an output perspective. J Clean Prod 26:28–36

Jahnke HN, Hukkinen JI (2017) Education for sustainable development: theoretical and conceptual frameworks. In: Encyclopedia of sustainability in higher education. Springer, London, pp 1–6

Lambright KT, McKeown R (eds) (2016) Routledge handbook of sustainable development in higher education. Routledge, London

Leal Filho W, Pace P (eds) (2017) Climate change and the sustainable development goals: linking targets and indicators. Springer, London

Leicht A, Heiss J (2016) Implementing sustainable development in higher education: challenges and strategies. Springer, London

Sanzgiri R (2023, May 31) Digital transformations in classrooms. The Times of India, Education Times, 5

Sterling S (2001) Sustainable education: re-visioning learning and change. Schumacher Briefings, Berlin

Stibbe A (2016) Ecolinguistics: language, ecology, and the stories we live by. Routledge, London
Stiggins RJ, Arter JA, Chappuis J, Chappuis S (2004) Classroom assessment for student learning: doing it right—using it well. Pearson Education, London
Tilbury D, Stevenson RB, Fien J, Schreuder D (eds) (2002) Education and sustainability: responding to the global challenge. IUCN Commission on Education and Communication, New York
UNESCO (2014) Shaping the education of tomorrow: 2014 EFA global monitoring report. UNESCO Publishing, London
UNESCO (2015) Education for sustainable development goals: learning objectives. UNESCO, London
UNESCO (2017a) Education for sustainable development goals: learning objectives. UNESCO, London
UNESCO (2017b) Education for sustainable development: experiences from the UNESCO-Japan prize on education for sustainable development. UNESCO, London
United Nations (2015) Transforming our world: the 2030 agenda for sustainable development. United Nations, London
Wals AE, Jickling B (2002) "Sustainability" in higher education: from doublethink and newspeak to critical thinking and meaningful learning. Int J Sustain High Educ 3(3):221–232
Wiek A, Withycombe L, Redman CL (2011) Key competencies in sustainability: a reference framework for academic program development. Sustain Sci 6(2):203–218

# Digital Technologies in Smart Sustainable Cities: Focal Cases in the UAE

**Anushka Lydia Issac**

**Abstract** The chapter aims to provide a conceptual review of the role of digital technologies in building smart sustainable cities and smart infrastructure in the United Arab Emirates (UAE). The chapter explores how digital technologies can be leveraged to address sustainability challenges, including energy consumption, transportation, and waste management. The main objective of this chapter is to examine how digital technologies can be integrated into urban planning and development strategies to promote sustainable development in the UAE. To achieve this objective, the chapter will employ a case study methodology. The case studies will focus on specific examples of smart infrastructure and sustainable development initiatives in the UAE. These case studies will provide insights into how digital technologies have been used to improve sustainability outcomes, and the challenges and opportunities associated with their implementation. Specifically, the chapter will examine the role of digital technologies in building smart transportation systems, smart energy grids, and smart waste management systems. Overall, this chapter contributes to the ongoing discussion on the potential of digital technologies in promoting sustainable development in urban areas. It provides insights and recommendations for policymakers, urban planners, and other stakeholders involved in the development of smart sustainable cities and smart infrastructure in the UAE and beyond.

**Keywords** Smart sustainable cities · Digital technologies · Smart infrastructure · Urban planning and development · Environmental sustainability

A. L. Issac (✉)
Lecturer and Program Manager—L3 Pearson Foundation Program, Westford University College, Sharjah, United Arab Emirates
e-mail: anushka.i@westford.org.uk



W. Leal Filho et al. (eds.), *Digital Technologies to Implement the UN Sustainable Development Goals*, World Sustainability Series,
https://doi.org/10.1007/978-3-031-68427-2_18

## 1 Introduction

In an age where the urgent call for sustainability reverberates louder than ever, the world finds itself at a pivotal juncture, grappling with a web of intricate environmental, societal, and economic quandaries. The essence of sustainability resides in the delicate balance between our present actions and their potential ramifications for the generations that will follow. As this global pursuit gains momentum, the United Nations Sustainable Development Goals (SDGs) stand as a guiding constellation, offering a roadmap towards a more equitable and sustainable world by 2030. Central to this constellation is SDG Goal 11, a beacon aimed at transforming cities and human settlements into bastions of inclusivity, safety, resilience, and sustainability.

This narrative gains potency against the backdrop of unprecedented urbanization, with more than half of humanity now congregating in urban landscapes. Within the mosaic of SDGs, Goal 11 assumes paramount importance, resonating deeply with the escalating urban challenges. The goal's essence lies in orchestrating urban development that is not merely sustainable but also equitable, ensuring access to affordable housing while erecting defenses against natural calamities and unforeseen disruptions (Pedersen 2018). The far-reaching rewards of achieving Goal 11 paint a picture of cleaner skies, unpolluted waters, and optimal resource management (Cordova and Celone 2019). Aspiring for affordable housing and essential services translates into tangible strides in poverty alleviation and equality, while building urban resilience can serve as a shield against the onslaught of nature's fury, safeguarding both lives and livelihoods. Beyond these contours, the canvas extends to economic realms, promising enhanced investments, job creation, and heightened competitive vigor (Yu et al. 2020).

Yet, in this complex and intricate tapestry of sustainability, digital technologies emerge as the unassuming protagonists. Their pervasive influence has unfurled across dimensions, redefining human interaction, economic transactions, and intellectual pursuits. Significantly, their potential to synergize with sustainability has lately kindled a burgeoning discourse. As burgeoning urban landscapes strain finite resources, the clarion call for innovative solutions reverberates with greater urgency (Del Río Castro et al. 2021).

Amid these global dynamics, the United Arab Emirates (UAE) takes center stage, a microcosm of rapid urbanization and the attendant conundrums. As the UAE's skylines morph and populations burgeon, the challenges of energy conservation, efficient transportation, waste management, and water scarcity loom as formidable foes. Responding to these calls, the UAE administration has unfurled an ambitious tapestry, entwining smart infrastructure and sustainable development.

A pivotal cornerstone in this edifice is the omnipresence of digital technologies. These enigmatic forces orchestrate the intricate ballet of data capture, analysis, and dissemination, culminating in the optimization of resource allocation and the nurturing of sustainability. The heartbeat of smart transportation resonates with real-time data, harmonizing traffic patterns, curbing emissions, and elevating public transportation efficiency. Parallelly, the realm of smart energy grids embraces digital transformation, propelling energy distribution, consumption management, and the proliferation of renewable sources. Waste management, too, succumbs to the digital realm, as intelligent data dictates waste collection, recycling, and disposal protocols (Ufua et al. 2021).

This digitally infused transformation is a global phenomenon, with success stories echoing from cities like Barcelona, Singapore, and Amsterdam, where smart transportation systems have ushered in reduced congestion, lowered emissions, and an elevated quality of life. The dance of smart energy grids has enlivened cities such as San Diego and Chattanooga, optimizing energy distribution, deflating consumption, and fanning the flames of renewable energy adoption. The UAE, resolute in its sustainability quest, has harnessed these technologies with zeal. Dubai's visionary Future Accelerators initiative forges partnerships between governmental entities and startups to forge innovative solutions—such as the ingenious smart irrigation system slashing water consumption by a staggering 30%. Meanwhile, Abu Dhabi's Estidama program interweaves green building mandates into urban planning, underpinning energy monitoring, waste reduction, and holistic urban sustainability.

The narrative that unfurls attests to digital technologies emerging as the architects of a new era, forging smart sustainable cities with an indelible imprint. Their import extends beyond technological marvels, embracing the intricate dance of resource optimization, quality of life enhancement, and mitigation of environmental impact (Evans et al. 2019).

This chapter stands as an intellectual expedition, delving deep into the conceptual realm. Its compass points unwaveringly at digital technologies and their role in sculpting smart sustainable cities and resilient infrastructures within the UAE. Rooted in the soil of SDG Goal 11, it navigates through a labyrinth of digital pathways, unraveling their potential in confronting urban challenges head-on. The journey stretches beyond theory, unfolding practical landscapes within the UAE that offer lessons and inspiration to policymakers, urban planners, and stakeholders alike. Thus, as theory and practice converge, this chapter forges a beacon of enlightenment, illuminating pathways toward a sustainable tomorrow within the UAE and beyond.

The structure of this chapter follows a deliberate trajectory, guiding readers from the panoramic view of sustainability's global context to the intricacies of digital transformation's role within the UAE. After this introduction, the subsequent sections delve into the overarching themes of digital technologies in smart sustainable cities, examining their role in transportation optimization, energy management, and waste reduction. Real-world case studies from both local and international contexts spotlight the practical impact of these technologies. The chapter culminates with a synthesis of insights, offering actionable recommendations for policymakers, urban planners, and stakeholders. Through this progression, the chapter endeavors to paint a comprehensive portrait of the potential digital technologies hold in weaving the fabric of smart sustainable cities and resilient infrastructures.

## 1.1 Methodology

The exploration of digital technologies' role in shaping smart sustainable cities and infrastructure within the UAE hinges upon a meticulous selection of case studies that embody the core objectives of this chapter. This methodology meticulously curates case studies based on their relevance, scope, and impact on sustainable development (Meyer 2001). The selection process entails a judicious evaluation of initiatives that exemplify the fusion of digital technologies with sustainability paradigms, drawing insights from their triumphs and navigating the challenges they encountered (Tellis 1997).

A pivotal criterion guiding the selection is the extent to which digital technologies were interwoven into each initiative's fabric. This entails a comprehensive examination of the technological underpinnings, ranging from data capture and analytics to the seamless integration of digital systems within existing urban frameworks (Johansson 2007). Furthermore, the scope of the initiative comes under scrutiny, encompassing both the breadth of impact and the depth of transformation. These criteria coalesce to ensure that the chosen case studies resonate harmoniously with the overarching objectives of this chapter (Meyer 2001; Tellis 1997).

As we delve into the fabric of each case study, the research question permeating our analysis is two-fold: How have digital technologies been instrumental in enhancing sustainability outcomes within the UAE's smart sustainable cities and infrastructure initiatives? Additionally, what are the challenges encountered and opportunities unlocked in the pursuit of this integration? By addressing these pivotal questions, our analysis seeks to unravel the intricate interplay between digital technologies and sustainable urban development (Al-Dabbagh 2022). The focus on the UAE as the backdrop for our case studies is a deliberate choice, underpinned by its status as a microcosm of rapid urbanization and technological innovation. The nation's proactive stance in embracing digital solutions to surmount sustainability challenges amplifies its relevance as a robust and instructive context for this study. By homing in on the UAE, we can distill invaluable insights that resonate not only within its borders but reverberate as lessons and inspirations across diverse global landscapes (Geray 2019). Drawing upon a diverse array of data sources, including governmental reports, industry publications, and scholarly articles, we endeavor to create a comprehensive tapestry that paints the landscape of digital technology integration in smart sustainable cities and infrastructure. These sources furnish us with a panoramic view of the initiatives, their objectives, methodologies, and outcomes (Ibrahim et al. 2016). The inclusion/exclusion criteria for case studies are meticulously established to ensure alignment with the chapter's objectives, thereby safeguarding the cohesion of our analysis.

In summation, the choice of case study methodology emanates from its profound advantages. By dissecting specific examples, we uncover the nuanced effectiveness and limitations of diverse approaches, enabling us to draw holistic insights (Johansson 2007). The rich tapestry of data procured from case studies empowers this analysis to serve as a lodestar for policymakers and decision-makers, offering guidance rooted in empirical realities (Tellis 1997). Furthermore, the lessons distilled from these case studies emerge as blueprints for best practices, informing future initiatives and projects. Ultimately, the case study methodology in this chapter not only promises a comprehensive review of digital technologies' role in shaping smart sustainable cities and infrastructure within the UAE but also cements itself as a reservoir of knowledge capable of steering future endeavors toward a sustainable and technologically enriched urban future (Johansson 2007; Meyer 2001; Tellis 1997).

## 2 Conceptual Framework: Digital Technologies and Smart Sustainable Cities

Smart sustainable cities represent a new paradigm in urban development, integrating digital technologies, environmental consciousness, and social well-being. Defining and understanding the characteristics of smart sustainable cities is crucial for policymakers, urban planners, and researchers aiming to create and evaluate such urban environments.

At its core, a smart sustainable city is an urban area that leverages innovative digital technologies to enhance its sustainability and livability. It goes beyond traditional urban development by adopting a holistic approach that encompasses economic, social, and environmental dimensions. These cities strive to achieve a balance between economic growth, resource efficiency, and the well-being of their residents (Martin et al. 2018).

One defining characteristic of smart sustainable cities is the use of digital technologies to enable efficient resource management. Through the integration of Internet of Things (IoT) devices, data analytics, and advanced sensors, these cities can monitor and optimize resource consumption. This includes managing energy use, optimizing transportation systems, and implementing smart waste management practices. By leveraging real-time data and predictive analytics, smart sustainable cities can make informed decisions to minimize waste, reduce energy consumption, and lower carbon emissions.

Additionally, smart sustainable cities prioritize sustainable mobility and transportation systems. They promote alternative modes of transportation such as walking, cycling, and public transit while minimizing reliance on private vehicles. Smart transportation infrastructure, including intelligent traffic management systems and electric vehicle charging stations, contributes to reduced congestion, improved air quality, and enhanced mobility options for residents.

Moreover, social inclusivity and citizen engagement are integral aspects of smart sustainable cities. They strive to create inclusive communities that provide equal access to services, amenities, and opportunities for all residents. Digital technologies are harnessed to foster citizen participation, enabling residents to actively engage in decision-making processes and contribute to the development and management of their cities (Evans et al. 2019) (Fig. 1).

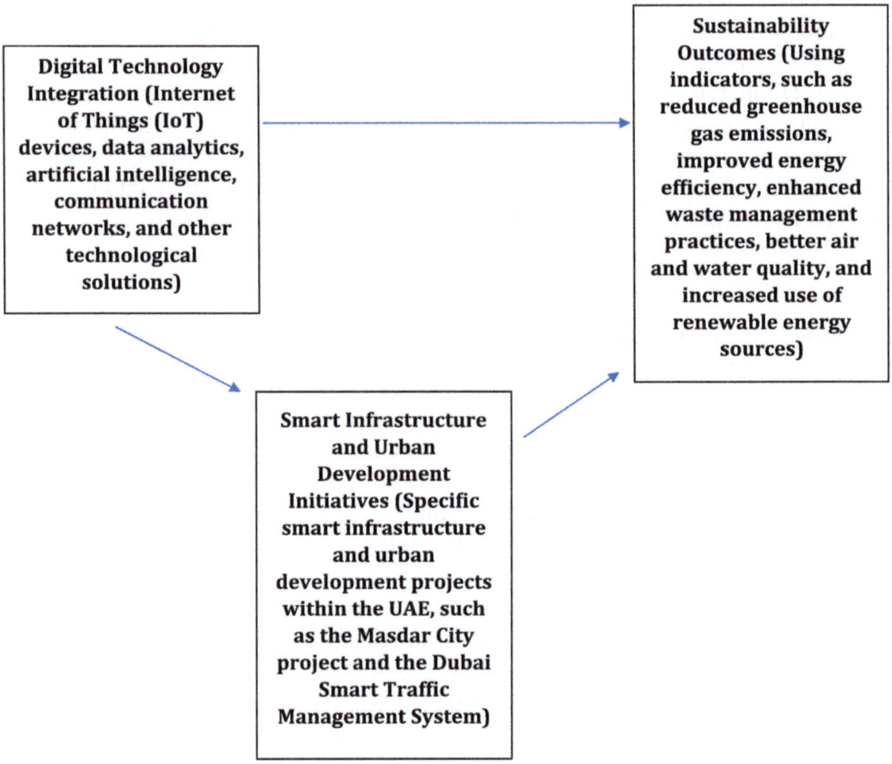


**Fig. 1** Conceptual framework organizing and analyzing the various elements of the research

Lastly, smart sustainable cities are characterized by their resilience and adaptability. They embrace climate change mitigation and adaptation strategies, integrating renewable energy sources, green infrastructure, and climate-resilient designs. By implementing strategies that enhance resilience, these cities can better withstand environmental shocks and promote long-term sustainability.

Smart sustainable cities represent a transformative approach to urban development. They combine digital technologies, environmental sustainability, and social well-being to create livable, efficient, and inclusive urban environments. By leveraging the power of data and technology, these cities strive to achieve resource efficiency, promote sustainable mobility, foster citizen engagement, and build resilience. Defining and understanding the characteristics of smart sustainable cities provides a foundation for the planning, design, and evaluation of urban environments that aim to address the complex challenges of the twenty-first century (Fig. 2).

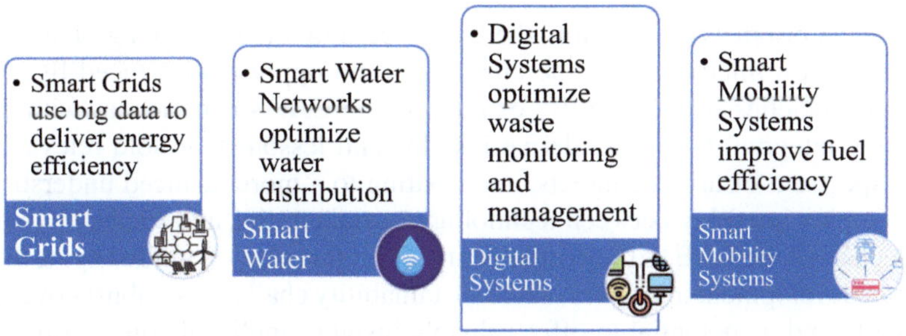


**Fig. 2** Digital technologies and smart sustainable cities framework. *Source*: Author's Interpretation, adapted from (Belli et al. 2020; Bibri and Krogstie 2020)

## 3 Rationale Behind UAE Being the Focal Point: Exploring Cultural, Social, and Economic Barriers

The development of smart sustainable cities in the UAE, while undoubtedly promising, is not devoid of complex challenges that span a spectrum of dimensions. A comprehensive understanding of the barriers rooted in cultural, social, economic, and technological aspects is imperative to chart a meaningful course toward a technologically advanced urban landscape. This section critically delves into these barriers, shedding light on the multifaceted challenges they pose, while also illuminating the potential opportunities that lie beneath the surface.

The choice to focus on the United Arab Emirates (UAE) as a pivotal context for this study is rooted in its distinctive urban development landscape, characterized by ambitious initiatives, pressing sustainability challenges, and a dynamic blend of modernization and tradition. First and foremost, the UAE's rapid urbanization and ambitious development projects provide a rich and diverse canvas to explore the multifaceted applications of digital technologies. The country has embarked on groundbreaking ventures, such as the iconic Masdar City and the Dubai Smart Traffic Management System, showcasing a commitment to innovative urban solutions. These initiatives reflect the UAE's aspiration to not only embrace technological advancements but also to address its unique challenges, including energy consumption, water scarcity, and urban congestion. Furthermore, the UAE's strategic geographical location and climatic conditions underscore the urgency of sustainable urbanization. With a desert climate and high energy demands for cooling and desalination, the UAE faces distinct sustainability challenges that necessitate creative and effective solutions. The integration of digital technologies offers a pathway to enhance resource efficiency, reduce environmental impact, and ensure the well-being of its citizens in the face of these challenges. The UAE's emphasis on visionary governance and strong public–private collaborations also positions it as a compelling case study. The collaboration between public and private sectors has led to the successful implementation of transformative projects, making the UAE an exemplar of how partnerships can drive sustainable technological progress. In addition, the UAE's cultural diversity and global connectivity render it a microcosm of the challenges and opportunities faced by rapidly urbanizing societies worldwide. By examining the UAE's experiences with digital technology integration, we can derive insights and lessons that hold relevance for a broad spectrum of urban contexts, contributing to a more nuanced understanding of the complex interplay between technology, sustainability, and societal dynamics. The selection of the UAE as the primary context for this study is guided by its distinctive urban development initiatives, acute sustainability challenges, robust governance framework, and its potential to offer valuable insights applicable on a global scale. By focusing on the UAE, this paper aims to delve deeply into the intricate relationship between digital technologies and smart sustainable cities while considering the country's unique socioeconomic and environmental context.

## 4 Case Studies of Infrastructure Projects Related to Digital Technology Implementation in UAE's Smart Cities

### *4.1 Case Study 1: Smart Transportation Systems*

In the heart of Dubai's bustling urban landscape, the implementation of the Smart Traffic Management System stands as a beacon of innovation and efficiency, addressing the pervasive challenge of traffic congestion through the strategic fusion of real-time data and cutting-edge artificial intelligence. The Dubai Smart Traffic Management System represents a paradigm shift in urban mobility (Jimenez 2018). By seamlessly integrating a vast network of cameras, sensors, and GPS devices, the city orchestrates a symphony of data, capturing real-time insights into traffic flow, road conditions, and congestion points (Saarika et al. 2017). This dynamic dataset is then channeled into sophisticated artificial intelligence algorithms, capable of predicting traffic patterns, optimizing traffic signal timings, and offering commuters alternative routes in real-time (Nassiri et al. 2018).

The system extends its impact through user-centric mobile applications, empowering individuals with personalized recommendations and live updates to navigate the city's intricate web of roads and thoroughfares. The outcomes of this pioneering endeavor have resonated across Dubai's urban landscape. Through the orchestration of dynamic traffic signal optimization and real-time route guidance, the Smart Traffic Management System has ushered in a palpable reduction in congestion, thereby elevating the overall quality of urban life (Xiong et al. 2012). Commuters now traverse the city with newfound ease, translating to heightened productivity and well-being (Ziyad et al. 2021). Beyond individual convenience, the reduction in traffic congestion has facilitated a decrease in fuel consumption and emissions, aligning with Dubai's broader sustainability objectives. Furthermore, the data-driven insights generated by the system offer an invaluable toolkit for urban planners and policymakers, fostering informed decisions on road infrastructure enhancements and public transportation advancements (Nassiri et al. 2018).

The Dubai Smart Traffic Management System imparts a series of invaluable lessons for urban centers grappling with the complex issue of traffic congestion. The first step involves identifying and integrating relevant data sources (Baker et al. 2022). These sources may include sensors, IoT devices, satellite imagery, social media feeds, and municipal databases. Each source contributes a unique perspective, enriching the overall dataset. The seamless synchronization of real-time information from diverse sources is essential for the efficacy of the system. Second, fostering public engagement and awareness is integral to unlocking the system's full potential (Debnath et al. 2014). Diverse sources contribute unique datasets, offering a comprehensive understanding of the environment or situation being monitored. By seamlessly synchronizing this data, the system can generate a holistic and accurate overview, enabling informed decision-making. Real-time synchronization ensures that decision-makers receive up-to-the-minute information. This is crucial for responding promptly to changing conditions, whether it's traffic congestion, energy consumption spikes, or waste management issues (Jimenez 2018; Saarika et al. 2017). Effective communication of its benefits and active encouragement of public utilization underpins its transformative impact. Clear and accessible communication about the system's benefits—such as reduced traffic congestion, improved air quality, cost savings, and enhanced quality of life—helps the public understand the tangible advantages of its implementation. Lastly, the continuous adaptability of the system is paramount (Nassiri et al. 2018; Saarika et al. 2017). Rapid advancements in technology necessitate the system's ability to evolve and integrate new tools and capabilities. Regular updates and enhancements ensure that the system remains cutting-edge and can harness the latest innovations. Cities are dynamic entities, and their needs, challenges, and priorities evolve over time (Baker et al. 2022; Nassiri et al. 2018). An adaptable system can flexibly accommodate shifts in urban dynamics, ensuring its solutions remain aligned with current demands. Regular updates, integration of user feedback, and an unwavering commitment to improvement are pivotal to maintaining its effectiveness in the face of evolving traffic dynamics and urban morphology (Mathew 2020a).

## *4.2 Case Study 2: Implementation of Intelligent Traffic Management Systems*

The implementation of Intelligent Traffic Management Systems (ITMS) in urban centers has emerged as a transformative approach to addressing the perennial challenge of traffic congestion (Manikonda et al. 2011). Through the strategic integration of real-time data, advanced analytics, and smart technologies, ITMS exemplifies a paradigm shift in urban mobility, offering a blueprint for efficient, sustainable, and seamlessly interconnected transportation networks (Hernández et al. 2002). The adoption of Intelligent Traffic Management Systems represents a multi-faceted endeavor that orchestrates data-driven insights with the agility of modern technology (Khan et al. 2017). A vast network of cameras, sensors, and data collection points blanket the urban landscape, capturing a wealth of real-time traffic information. This data is then channeled into sophisticated analytics platforms and artificial intelligence algorithms, capable of discerning patterns, predicting traffic congestion, and optimizing traffic flow (Khanna et al. 2019). Collaborative platforms and mobile applications extend the benefits to commuters, providing personalized recommendations, real-time updates, and alternative routes, all seamlessly interconnected through the Internet of Things (IoT) infrastructure (Hamidi and Kamankesh 2018; Jacob et al. 2018; Viswanathan and Santhanam 2013). The outcomes of implementing Intelligent Traffic Management Systems are substantial and wide-ranging.

By dynamically optimizing traffic signal timings and offering real-time route guidance, ITMS has unlocked a tangible reduction in congestion levels (Singh and Gupta 2015). Commuters experience shorter travel times, elevating productivity and enhancing quality of life. Moreover, the reduction in traffic congestion translates to reduced fuel consumption and emissions, aligning with sustainability imperatives (Farid et al. 2021). The wealth of data generated by ITMS empowers urban planners and policymakers with actionable insights, facilitating informed decisions on infrastructure investments, public transportation enhancements, and urban development strategies (Ziyad et al. 2021). The case of Intelligent Traffic Management Systems imparts pivotal lessons for urban planners navigating the complexities of traffic congestion. First, the success of ITMS hinges on data accuracy and integration. Real-time data from diverse sources must seamlessly converge for accurate insights and predictions (Mathew 2020b). Second, fostering public engagement is essential. Effective communication of the system's benefits and encouraging public participation unlock its full potential (Bilal and Jacob 2007). Lastly, the adaptability of ITMS is paramount. Regular updates, integration of user feedback, and a culture of continuous improvement ensure its resilience in an ever-evolving urban landscape (Lin et al. 2022; Majeed et al. 2019; Ning et al. 2021).

### *4.3 Case Study Example 3: Integration of Smart Mobility Solutions*

The integration of Smart Mobility Solutions (SMS) within urban landscapes has ushered in a transformative era of transportation, redefining the way people move within cities and offering a vision of seamless, efficient, and sustainable mobility. The adoption of Smart Mobility Solutions represents a multifaceted integration of advanced technologies to enhance urban transportation (Bıyık et al. 2021). A network of sensors, connected vehicles, and data collection points form the foundation of SMS, capturing real-time insights into traffic patterns, road conditions, and user preferences. This data is harnessed by sophisticated algorithms and artificial intelligence, capable of optimizing routes, predicting demand, and coordinating multimodal transportation options. User-friendly mobile applications serve as the interface, offering personalized journey planning, real-time updates, and integrated payment solutions (Nielsen 2013). The outcomes of integrating Smart Mobility Solutions resonate across urban environments, redefining the urban commuting experience. By seamlessly coordinating diverse transportation modes, SMS reduces congestion and travel times, fostering efficient mobility. Commuters benefit from real-time guidance, ensuring smoother journeys and enhanced productivity (Faria et al. 2017). The reduction in vehicular congestion leads to lower emissions and fuel consumption, aligning with sustainability goals. Additionally, SMS empowers urban planners with data-driven insights, enabling informed decisions on transportation infrastructure, traffic management, and future urban development (Riva Sanseverino et al. 2017). The case of integrating Smart Mobility Solutions imparts valuable lessons for cities navigating the evolving landscape of urban transportation. Firstly, data accuracy and real-time integration are pivotal for success. Reliable data feeds underpin the effectiveness of SMS (Prajeesh and Pillai 2022). Secondly, user engagement and education are instrumental. Effective communication of benefits and seamless user experiences encourage adoption. Lastly, the adaptability and scalability of SMS are essential. Regular updates, integration of emerging technologies, and flexibility in accommodating evolving urban dynamics ensure long-term relevance (Maldonado Silveira Alonso Munhoz et al. 2020).

## *4.4 Case Study 4: Masdar City—The Eco-Innovation Oasis*

Designed as a pioneering smart sustainable city, Masdar City integrates cutting-edge digital technologies to create an urban environment that is not only environmentally conscious but also technologically advanced. At the heart of Masdar City's implementation lies the seamless integration of digital innovation. Internet of Things (IoT) sensors are strategically placed throughout the city, collecting data on energy consumption, waste management, and transportation patterns (Cugurullo 2016). This wealth of data is then channeled into sophisticated data analytics systems, which allow for real-time monitoring and optimization of various city systems. Notably, the city's energy management is orchestrated through smart grids, dynamically adjusting energy distribution based on real-time demand (Bishr 2019). Moreover, Masdar City's urban planning fosters walkability and accessibility, with smart mobility solutions, such as autonomous electric vehicles, enhancing both convenience and environmental consciousness. The outcomes of Masdar City's digital technology integration have been both impressive and far-reaching (Reiche 2010). Foremost, the city has managed to significantly reduce its carbon footprint, thanks to the integration of renewable energy sources and energy-efficient technologies. This feat has not only garnered international recognition but has also contributed tangibly to the UAE's commitment to combatting climate change (Cugurullo 2013).

Additionally, the utilization of advanced energy management systems has translated into considerable energy savings, leading to reduced utility costs and long-term financial sustainability. Beyond the environmental and economic spheres, Masdar City has emerged as an innovation hub, attracting researchers, entrepreneurs, and investors eager to be part of its sustainable development journey (Cugurullo 2016). The city's focus on pedestrian-friendly design and abundant green spaces has also enhanced the quality of life for its residents, fostering a strong sense of community and well-being. The experience of Masdar City underscores several critical lessons for the broader endeavor of building smart sustainable cities. First, the power of public–private partnerships cannot be underestimated (Cugurullo 2016; Reiche 2010). The successful realization of Masdar City relied on the collaboration between government entities, private sector stakeholders, and academia, demonstrating that shared expertise and resources are instrumental in overcoming complex challenges. Second, a steadfast and unwavering commitment to sustainability and innovation is imperative. Masdar City's long-term vision, unyielding despite the complexities of implementation, serves as a beacon for other cities aspiring to embark on similar transformative journeys (Madakam and Ramaswamy 2016). Finally, the lessons learned from Masdar City extend beyond its borders. As the UAE continues to share its experiences and knowledge, other regions and cities can draw inspiration from its successes and challenges, propelling the global movement toward smart, sustainable urbanization.

## *4.5 Case Study 5: Dubai Smart Traffic Management System-Navigating Congestion*

In the bustling metropolis of Dubai, addressing traffic congestion and optimizing transportation efficiency has been a top priority. To tackle this challenge head-on, the Dubai Smart Traffic Management System was introduced, leveraging digital technologies to transform the urban mobility landscape (Rashid 2016). The Dubai Smart Traffic Management System exemplifies the fusion of real-time data and artificial intelligence. Equipped with an extensive network of cameras, sensors, and GPS devices, the system captures and processes real-time data on traffic flow, road conditions, and congestion points (Dana et al. 2022). This data is then fed into advanced AI algorithms capable of predicting traffic patterns and suggesting optimal routes. The system interfaces with mobile applications, empowering commuters with up-to-the-minute updates and personalized recommendations for smoother journeys (Poon 2021). The outcomes of this innovative traffic management system have been nothing short of transformative. By dynamically optimizing traffic signal timings and providing real-time route guidance, the system has succeeded in significantly reducing congestion levels across the city (Al Mamlook et al. 2023). Commute times have been notably shortened, leading to increased productivity and improved quality of life for residents. Beyond the convenience for individuals, the reduction in traffic congestion has resulted in lower fuel consumption and reduced emissions, contributing to Dubai's sustainability goals (Alsarhan et al. 2018).

Moreover, the data-driven insights generated by the system have empowered urban planners and policymakers to make informed decisions about road infrastructure improvements and public transportation enhancements (Chandana et al. 2017). The Dubai Smart Traffic Management System offers valuable insights for urban centers grappling with traffic congestion and mobility challenges. First and foremost, the accuracy and reliability of data collection are paramount. The success of the system hinged on the real-time accuracy of information from various sources, highlighting the importance of investing in robust data collection mechanisms (Noori et al. 2020). Second, public engagement and awareness are crucial. Citizens must be educated about the benefits of the system and encouraged to actively participate in it to realize its full potential. Lastly, the system's ongoing success depends on its adaptability (Mandhare et al. 2018). Regular updates, feedback integration, and continuous improvement are vital to ensure that the system remains effective in addressing evolving traffic patterns and urban dynamics (Mavropoulos et al. 2015).

## 4.6 Key Sustainability Challenges in the UAE

The United Arab Emirates (UAE) faces several key sustainability challenges that demand urgent attention in the pursuit of constructing smart sustainable cities and resilient infrastructure. These challenges stem from a confluence of factors, including rapid urbanization, population growth, finite natural resources, and a heavy reliance on fossil fuels. An exemplar among these challenges is the formidable energy consumption and resulting carbon emissions prevalent in the UAE. With a notably high per capita energy consumption rate, largely attributed to the demand for cooling and desalination due to the arid climate, the UAE's energy reliance on fossil fuels has led to substantial carbon emissions. This issue is underscored by the ambitious energy-efficient initiatives undertaken, such as transitioning to cleaner and renewable energy sources, enhancing energy efficiency in buildings, and championing sustainable transportation alternatives. For instance, the Masdar City project serves as a beacon of sustainable urban development, leveraging cutting-edge renewable technologies to create a carbon–neutral environment (Madakam and Ramaswamy 2016).

Water scarcity serves as another poignant illustration of a pressing challenge in the UAE's sustainability landscape. Confronted with meager freshwater resources, the nation predominantly depends on energy-intensive desalination processes to satiate its water requirements. However, this approach comes at an environmental cost (Chandana et al. 2017). Pioneering solutions are evident through the incorporation of innovative water management strategies. The case of the Dubai Smart Traffic Management System showcases an intricate mesh of data-driven technologies to optimize transportation, minimizing energy consumption and concurrently alleviating pressure on water resources through reduced traffic congestion (Mandhare et al. 2018; Ziyad et al. 2021). The complex interplay between economic advancement and environmental conservation is emblematic of the challenges in urban planning and development. The balancing act between these factors becomes evident in endeavors like Masdar City, hailed as an "Eco-Innovation Oasis". This project not only exemplifies sustainable urban planning but also demonstrates the potential for harmonizing economic growth and environmental stewardship (Madakam and Ramaswamy 2016). Through fostering biodiversity, promoting green spaces, and embracing sustainable architectural practices, Masdar City exemplifies a paradigm of sustainable urban development amidst rapid urbanization.

Waste management represents yet another compelling facet of sustainability challenges in the UAE. The nation grapples with mounting waste generation, accentuated by diminishing landfill capacity. Against this backdrop, initiatives such as waste-to-energy plants emerge as promising solutions, leveraging technology to address waste concerns while simultaneously contributing to energy generation (Hrabec et al. 2020). For instance, the implementation of waste-to-energy facilities in the UAE underscores the transformative potential of integrating waste management with energy production. Moreover, the imperative of social inclusivity and addressing social disparities is a central theme in the UAE's sustainability discourse. As cities burgeon, ensuring access to affordable housing, quality healthcare, and education for all residents becomes paramount. The holistic approach to social inclusivity is epitomized by the multifaceted Masdar City project, which extends beyond environmental sustainability to embrace economic and social dimensions, manifesting in provisions such as educational institutions and healthcare services (Hrabec et al. 2020). In summation, the UAE's journey toward building smart sustainable cities and resilient infrastructure is distinctly characterized by challenges and proactive solutions that provide meaningful insights. By navigating the complexities of energy consumption, water scarcity, urban development, waste management, and social inclusivity through innovative initiatives like Masdar City and data-driven systems like the Dubai Smart Traffic Management System, the UAE has demonstrated a commitment to shaping a sustainable future that is not only environmentally responsible but also socially and economically inclusive.

## 5 Future Prospects and Recommendations

### *5.1 Exploration of Future Trajectory and Potential Advancements in Smart City Development in the UAE: Navigating Opportunities, Challenges, and Concrete Projections*

The United Arab Emirates (UAE) has established itself as a trailblazer in the realm of smart city development, harnessing the potential of digital technologies to redefine urban living. As the nation propels forward, it finds itself at a pivotal juncture, where the path of digital technologies in smart city development offers both promising opportunities and intricate challenges that demand careful navigation.

**Opportunities and Concrete Projections:** Looking ahead, the future trajectory of digital technologies in the UAE's smart city development reveals several concrete advancements that are set to reshape the urban landscape. The integration of Artificial Intelligence (AI) holds immense potential, particularly in predictive analytics that optimize energy consumption, waste management, and traffic flow. A noteworthy example is the Dubai Electricity and Water Authority's (DEWA) use of AI to enhance energy efficiency by predicting peak demand and optimizing power distribution (DEWA 2021). Furthermore, the Internet of Things (IoT) is poised to transform into a dynamic network of interconnected devices, enabling real-time monitoring of urban ecosystems. Dubai's 'Smart Dubai' initiative has already seen the deployment of smart sensors for environmental monitoring, waste management, and public safety (Al-Dabbagh 2022).

**Challenges and Ethical Considerations:** Alongside these opportunities, navigating the path of digital transformation presents complex challenges. As technology becomes more deeply ingrained in urban life, concerns over data privacy, security, and ethical usage come to the forefront. While data-driven innovation holds great promise, the UAE is actively addressing these concerns through initiatives such as the Dubai Data Law, which safeguards data privacy and regulates its use (Government of Dubai, 2015). The digital divide is another challenge that demands attention, as ensuring equitable access to technology is crucial. Initiatives like the 'Mohammed bin Rashid Smart Learning Program' demonstrate the UAE's commitment to providing digital education tools to all students (MBRSLP, n.d.).

**A Vision for the Future:** To navigate this multifaceted landscape, the UAE's vision for smart city development must remain centred on inclusivity and accessibility. While embracing AI, IoT, and other emerging technologies, the nation's commitment to fostering digital literacy becomes paramount. Collaborative efforts between the government, industry, and academia are essential for co-creating solutions that align with the UAE's unique cultural fabric. As the UAE advances further, it will be guided by the principle of using technology not only to reshape its cities but also to enhance the well-being and quality of life for all its residents.

# 6 Conclusion

The combination of digital technologies and sustainable urban planning has created a harmonious blend of systems that save energy, improve transportation, and make life better for citizens. As we delved into data-driven decisions, delving into the nuances of data-driven decision-making, intelligent traffic management, smart mobility solutions, and eco-innovations, a remarkable narrative unfolded—one in which the UAE has emerged as a trailblazer, forging a novel urban paradigm that seamlessly integrates technological progress with environmental stewardship. Nevertheless, this narrative is not devoid of its intricacies and challenges. While the UAE takes bold strides in advancing digital urbanization, it must proceed with a vigilant awareness of potential hurdles and ethical quandaries. The integration of digital technologies should go hand in hand with robust data governance protocols, ensuring utmost privacy and security for its citizens. The pursuit of technological innovation must remain firmly rooted in an unwavering commitment to inclusivity, aimed at bridging the digital divide and ensuring that every segment of society partakes in and benefits from this transformative journey. As we reflect on this conceptual review, it becomes apparent that it marks not the culmination, but rather the catalyst for an ongoing discourse, exploration, and concrete action. It extends an invitation for collective imagination and collaborative effort, urging us to envision and contribute to a world where smart sustainable cities, nurtured by the ingenious fusion of digital technologies and environmental consciousness, serve as the hallmark of urbanization in the dynamic landscape of the twenty-first century.

## References

Al Mamlook RE, Zahrawi M, Gharaibeh H, Nasayreh A, Shresth S (2023) Smart traffic control system for Dubai: a simulation study using YOLO algorithms. In: 2023 IEEE international conference on electro information technology (EIT), pp 254–264. https://doi.org/10.1109/eIT57321.2023.10187271

Al-Dabbagh R (2022) Dubai, the sustainable, smart city. Renew Energy Environ Sustain 7:3. https://doi.org/10.1051/rees/2021049

Alsarhan A, Al-Dubai AY, Min G, Zomaya AY, Bsoul M (2018) A new spectrum management scheme for road safety in smart cities. IEEE Trans Intell Transp Syst 19(11):3496–3506. https://doi.org/10.1109/TITS.2017.2784548

Baker T, Asim M, Samwini H, Shamim N, Alani MM, Buyya R (2022) A blockchain-based Fog-oriented lightweight framework for smart public vehicular transportation systems. Comput Netw 203:108676. https://doi.org/10.1016/j.comnet.2021.108676

Belli L, Cilfone A, Davoli L, Ferrari G, Adorni P, Di Nocera F, Dall'Olio A, Pellegrini C, Mordacci M, Bertolotti E (2020) IoT-enabled smart sustainable cities: challenges and approaches. Smart Cities 3(3):1039–1071. https://doi.org/10.3390/smartcities3030052

Bibri SE, Krogstie J (2020) Environmentally data-driven smart sustainable cities: applied innovative solutions for energy efficiency, pollution reduction, and urban metabolism. Energy Inform 3(1):29. https://doi.org/10.1186/s42162-020-00130-8

Bilal JM, Jacob D (2007) Intelligent traffic control system. In: 2007 IEEE international conference on signal processing and communications, pp 496–499. https://doi.org/10.1109/ICSPC.2007.4728364

Bishr AB (2019) Dubai: a city powered by blockchain. Innov Technol Gover Global 12(3–4):4–8. https://doi.org/10.1162/inov_a_00271

Bıyık C, Abareshi A, Paz A, Ruiz RA, Battarra R, Rogers CDF, Lizarraga C (2021) Smart mobility adoption: a review of the literature. J Open Innov Technol Mark Complex 7(2):146. https://doi.org/10.3390/joitmc7020146

Chandana KK, Meenakshi SS, Swamy MN (2017) A smart traffic management system for congestion control and warnings using internet of things (IoT). Rev Article Saudi J Eng Technol. https://doi.org/10.21276/sjeat

Cordova MF, Celone A (2019) SDGs and innovation in the business context literature review. Sustainability 11(24):7043. https://doi.org/10.3390/su11247043

Cugurullo F (2013) How to build a sandcastle: an analysis of the genesis and development of Masdar City. J Urban Technol 20(1):23–37. https://doi.org/10.1080/10630732.2012.735105

Cugurullo F (2016) Urban eco-modernisation and the policy context of new eco-city projects: where Masdar City fails and why. Urban Stud 53(11):2417–2433. https://doi.org/10.1177/0042098015588727

Dana L-P, Salamzadeh A, Hadizadeh M, Heydari G, Shamsoddin S (2022) Urban entrepreneurship and sustainable businesses in smart cities: exploring the role of digital technologies. Sustain Technol Entrepreneurship 1(2):100016. https://doi.org/10.1016/j.stae.2022.100016

Debnath AK, Chin HC, Haque MdM, Yuen B (2014) A methodological framework for benchmarking smart transport cities. Cities 37:47–56. https://doi.org/10.1016/j.cities.2013.11.004

Del Río Castro G, González Fernández MC, Uruburu Colsa Á (2021) Unleashing the convergence amid digitalization and sustainability towards pursuing the sustainable development goals (SDGs): a holistic review. J Clean Prod 280:122204. https://doi.org/10.1016/j.jclepro.2020.122204

Evans J, Karvonen A, Luque-Ayala A, Martin C, McCormick K, Raven R, Palgan YV (2019) Smart and sustainable cities? Pipedreams, practicalities and possibilities. Local Environ 24(7):557–564. https://doi.org/10.1080/13549839.2019.1624701

Faria R, Brito L, Baras K, Silva J (2017) Smart mobility: a survey. In: 2017 international conference on internet of things for the global community (IoTGC), pp 1–8. https://doi.org/10.1109/IoTGC.2017.8008972

Farid AM, Viswanath A, Al-Junaibi R, Allan D, Van der Wardt TJT (2021) Electric vehicle integration into road transportation, intelligent transportation, and electric power systems: an Abu Dhabi case study. Smart Cities 4(3):1039–1057. https://doi.org/10.3390/smartcities4030055

Geray O (2019) An impact-driven smart sustainable city framework to address urban challenges: smart Dubai experience. In: Smart cities in the Gulf. Springer, Singapore, pp 13–39. https://doi.org/10.1007/978-981-13-2011-8_3

Hamidi H, Kamankesh A (2018) An approach to intelligent traffic management system using a multi-agent system. Int J Intell Transp Syst Res 16(2):112–124. https://doi.org/10.1007/s13177-017-0142-6

Hernández JZ, Ossowski S, García-Serrano A (2002) Multiagent architectures for intelligent traffic management systems. Transp Res Part C Emerg Technol 10(5–6):473–506. https://doi.org/10.1016/S0968-090X(02)00032-3

Hrabec D, Kůdela J, Šomplák R, Nevrlý V, Popela P (2020) Circular economy implementation in waste management network design problem: a case study. CEJOR 28(4):1441–1458. https://doi.org/10.1007/s10100-019-00626-z

Ibrahim M, El-Zaart A, Adams C (2016) Paving the way to smart sustainable cities: transformation models and challenges. J Inform Syst Technol Manag. https://doi.org/10.4301/S1807-17752015000300004

Jacob S, Rekh A, Manoj G, Paul J (2018) Smart traffic management system with real time analysis. Int J Eng Technol (UAE) 7:348–351

Jimenez JA (2018) Smart transportation systems. In: Smart cities. Springer, London, pp 123–133. https://doi.org/10.1007/978-3-319-59381-4_8

Johansson R (2007) On case study methodology. Open House Int 32(3):48–54. https://doi.org/10.1108/OHI-03-2007-B0006

Khan M, Woo M, Nam K, Chathoth P (2017) Smart city and smart tourism: a case of Dubai. Sustainability 9(12):2279. https://doi.org/10.3390/su9122279

Khanna A, Goyal R, Verma M, Joshi D (2019) Intelligent traffic management system for smart cities, pp 152–164. https://doi.org/10.1007/978-981-13-3804-5_12

Lin H-H, Hsu I-C, Lin T-Y, Tung L-M, Ling Y (2022) After the epidemic, is the smart traffic management system a key factor in creating a green leisure and tourism environment in the move towards sustainable urban development? Sustainability 14(7):3762. https://doi.org/10.3390/su14073762

Madakam S, Ramaswamy R (2016) Sustainable smart city: Masdar (UAE) (a city: ecologically balanced). Indian J Sci Technol. https://doi.org/10.17485/ijst/2016/v9i6/87669

Majeed FA, Murad B, Khalil K, Mohamed A, Ali F, Obaid S, Ahmed S (2019) Smart traffic management system for foggy weather conditions. In: 2019 advances in science and engineering technology international conferences (ASET), pp 1–3. https://doi.org/10.1109/ICASET.2019.8714327

Maldonado Silveira Alonso Munhoz PA, da Costa Dias F, Kowal Chinelli C, Azevedo Guedes AL, Neves dos Santos JA, da Silveira e Silva W, Pereira Soares CA (2020) Smart mobility: the main drivers for increasing the intelligence of urban mobility. Sustainability 12(24):10675. https://doi.org/10.3390/su122410675

Mandhare PA, Kharat V, Patil CY (2018) Intelligent road traffic control system for traffic congestion a perspective. Int J Comput Sci Eng 6(7):908–915. https://doi.org/10.26438/ijcse/v6i7.908915

Manikonda P, Yerrapragada AK, Annasamudram SS (2011) Intelligent traffic management system. In: 2011 IEEE conference on sustainable utilization and development in engineering and technology (STUDENT), pp 119–122. https://doi.org/10.1109/STUDENT.2011.6089337

Martin CJ, Evans J, Karvonen A (2018) Smart and sustainable? Five tensions in the visions and practices of the smart-sustainable city in Europe and North America. Technol Forecast Soc Chang 133:269–278. https://doi.org/10.1016/j.techfore.2018.01.005

Mathew E (2020a) Intelligent transport systems and its challenges, pp 663–672. https://doi.org/10.1007/978-3-030-31129-2_61

Mathew E (2020b) Intelligent transport systems and its challenges, pp. 663–672. https://doi.org/10.1007/978-3-030-31129-2_61

Mavropoulos A, Björk H, Gratton R, Kumar Sharma R, Kamuk B, Agrawal VS, Gupta N, Kumar Ojha A (2015) Individualization of the public space: the othering of the poor in our cities. J Develop Manag Commun 2(3):1

Meyer CB (2001) A case in case study methodology. Field Methods 13(4):329–352. https://doi.org/10.1177/1525822X0101300402

Nassiri N, Rajan A, Ahamed J, Akre V, Ali A (2018a) Sustainable smart transport system in pursuit of conserving energy. In: 2018 5th HCT information technology trends (ITT), pp 342–347. https://doi.org/10.1109/CTIT.2018.8649516

Nielsen Y (2013) Building sustainability into the UAE: ESTIDAMA. AEI 2013:946–955. https://doi.org/10.1061/9780784412909.092

Ning Z, Zhang K, Wang X, Obaidat MS, Guo L, Hu X, Hu B, Guo Y, Sadoun B, Kwok RYK (2021) Joint computing and caching in 5G-envisioned internet of vehicles: a deep reinforcement learning-based traffic control system. IEEE Trans Intell Transp Syst 22(8):5201–5212. https://doi.org/10.1109/TITS.2020.2970276

Noori N, Hoppe T, de Jong M (2020) Classifying pathways for smart city development: comparing design, governance and implementation in Amsterdam, Barcelona, Dubai, and Abu Dhabi. Sustainability 12(10):4030. https://doi.org/10.3390/su12104030

Pedersen CS (2018) The UN sustainable development goals (SDGs) are a great gift to business! Proc CIRP 69:21–24. https://doi.org/10.1016/j.procir.2018.01.003

Poon STF (2021) Designing for urban mobility: the role of digital media applications in increasing efficiency of intelligent transportation management system, pp 181–195. https://doi.org/10.1007/978-3-030-60922-1_9

Prajeesh CB, Pillai AS (2022) Indian smart mobility ecosystem—key visions and missions. 050005. https://doi.org/10.1063/5.0124596

Rashid MB (2016) A smart city for public value digital transformation through agile governance—the case of smart Dubai.https://papers.ssrn.com/sol3/papers.cfm?abstract_id=2733632

Reiche D (2010) Renewable energy policies in the gulf countries: a case study of the carbon-neutral "Masdar City" in Abu Dhabi. Energy Policy 38(1):378–382. https://doi.org/10.1016/j.enpol.2009.09.028

Riva Sanseverino E, Riva Sanseverino R, Vaccaro V, Macaione I, Anello E (2017) Smart cities: case studies, pp 47–140. https://doi.org/10.1007/978-3-319-47361-1_3

Saarika PS, Sandhya K, Sudha T (2017) Smart transportation system using IoT. In: 2017 international conference on smart technologies for smart nation (SmartTechCon), pp 1104–1107. https://doi.org/10.1109/SmartTechCon.2017.8358540
Singh B, Gupta A (2015) Recent trends in intelligent transportation systems: a review. J Transp Lit 9(2):30–34. https://doi.org/10.1590/2238-1031.jtl.v9n2a6
Tellis WM (1997) Application of a case study methodology. In: The qualitative report, vol 3, Issue 3. http://nsuworks.nova.edu/tqr/vol3/iss3/1
Ufua DE, Emielu ET, Olujobi OJ, Lakhani F, Borishade TT, Ibidunni AS, Osabuohien ES (2021) Digital transformation: a conceptual framing for attaining sustainable development goals 4 and 9 in Nigeria. J Manag Organ 27(5):836–849. https://doi.org/10.1017/jmo.2021.45
Viswanathan V, Santhanam V (2013, March 17) Intelligent traffic signal control using wireless sensor networks. In: Advances in electrical and electronics engineering (ICAEE'2013)
Xiong Z, Sheng H, Rong W, Cooper DE (2012) Intelligent transportation systems for smart cities: a progress review. Sci China Inf Sci 55(12):2908–2914. https://doi.org/10.1007/s11432-012-4725-1
Yu S, Sial MS, Tran DK, Badulescu A, Thu PA, Sehleanu M (2020) Adoption and implementation of sustainable development goals (SDGs) in China—agenda 2030. Sustainability 12(15):6288. https://doi.org/10.3390/su12156288
Ziyad M, Naranje V, Suresh S, Mishra VP, Salunkhe S (2021a) Intelligent traffic system for sustainable mobility. In: 2021 international conference on computational intelligence and knowledge economy (ICCIKE), pp 504–507. https://doi.org/10.1109/ICCIKE51210.2021.9410725

# Investigating the Integration of Industry 4.0 and Circular Economy Practices for Sustainable Manufacturing

**Anshita Bhojwani and Anshu Gupta**

**Abstract** With increasing industrialisation and changing consumption patterns, environmental damage is increasing. In order to attain competitive edge, increase the effectiveness of the supply chains and minimize the environmental impact organizations are increasingly encouraged to adopt sustainable practices. Scholars have explored how technology can be leveraged for accelerating the shift towards Circular Economy (CE), however the extant research lacks a clear cognizance of the association, challenges and impact on sustainability. While techno solutions for the circular economy may offer good opportunities for sustainable manufacturing, certain barriers such as limited infrastructure, resources, lack of trust and collaboration are likely to impede the effective rollout of the circular economy model. This study aims to present the I4.0 and CE paradigms and discuss the associations and related challenges to accomplish the sustainable development goals (SDGs) in the realm of sustainable manufacturing. The study further delves into the directions for overcoming these challenges while underlining the future research avenues. This research finds implications for industry practitioners and decision makers for developing insights for realizing SDGs through technology enabled circular economy and identifying the ways to deal with related roadblocks.

**Keywords** Industry 4.0 · Circular Economy · Sustainability · Sustainable Development Goals · Sustainable Manufacturing

A. Bhojwani · A. Gupta (✉)
School of Management, Dr. B. R. Ambedkar University Delhi, Delhi, India
e-mail: guptaanshu.or@gmail.com

A. Bhojwani
e-mail: abhojwani.20@stu.aud.ac.in



W. Leal Filho et al. (eds.), *Digital Technologies to Implement the UN Sustainable Development Goals*, World Sustainability Series,
https://doi.org/10.1007/978-3-031-68427-2_19

# 1 Introduction

The increasing population, industrialization, and consumption is crossing the planetary boundaries and leading to environmental degradation. To ensure sustainable development, concepts such as circular economy and sustainability are being advocated by international organizations, governments, and environmental leaders. Technological solutions fostering virtual and intelligent systems are appreciated as pivotal drivers for shift towards circularity and realization of sustainability objectives (Ahmed et al. 2022). The concept of sustainability endeavours to protect ecosystems, enhance the standard of living, and conserve natural resources for the welfare of future generations. Sustainable development strives for an inclusive society and environmental preservation while advancing towards economic performance. Organizations in the present times should increasingly adopt holistic approaches and incorporate sustainability throughout the value chain. Sustainable Development Goals (SDGs) were set forth by the United Nations in 2015 to overcome the environmental, economic and social challenges facing the world (Modgil et al. 2020). The circular economy in the manufacturing context is a framework that minimizes resource usage and environmental impact simultaneously accelerating the efficiency of the economic resources and, reducing emissions and wastes (Patyal et al. 2022). Given the rate of environmental degradation the linear system should be replaced with a circular model. Technological advancement can be pivotal for enabling the shift to circularity from the traditional linear systems. The benefits realized by adopting Industry 4.0 technologies include profitable business, reduced costs, increased efficiency, and improved workplace conditions. Industry 4.0 should be capitalized as an opportunity to accelerate sustainable growth. Technology enabled value chain can be instrumental for advancing sustainability goals (Stock and Seliger 2016). On this background, it is pivotal to examine how the paradigms technology driven industrial revolution and circularity principles can be integrated to accelerate the path of sustainable development. Supply chain management exhibits strong relations with the circular economy concept and triple bottom line (Birkel and Müller 2021). The incorporation of circular economy practices is vital for ensuring innovation within the supply chain (Govindan and Hasanagic 2018).

India is the second most attractive manufacturing hub in the world (Madhukalya 2021). The 'Make in India' movement launched in 2014 in India endeavours to further boost the domestic manufacturing sector (Srivastava 2019). The manufacturing sector generates significant positive multiplier effects in the economy. On the other hand, the manufacturing processes lead to high levels of emissions in the environment. Given the environmental crisis, reducing emissions and optimal resource usage is vital. Sustainable manufacturing is increasingly getting attention as a holistic approach that not only mitigates the harmful effect of production of goods and services on the ecology and society but also fosters long-term profitability and resilience. Government and industry regulations, need for resource conservation and recycling, rise in customer awareness and demand for eco-friendly products, are among some of the major driving forces towards sustainable manufacturing

(Bag and Pretorius 2022). Organizations are trying to use their resources efficiently by employing sustainable manufacturing and supply chain practices. Technology is recognised as a powerful enabler in this direction (Enyoghasi and Badurdeen 2021). In order to incorporate sustainable manufacturing processes in an organization using an information technology-enabled system, a skilled workforce and motivation among employees to adopt sustainable practices is crucial. Sustainable operations and digitalization in an organization require linking Industry 4.0 technologies with circular economy dimensions (Kumar et al. 2021a). Digital transition offers potential to minimize environmental impact through cleaner production processes, supported through technologies such as Additive Manufacturing, Artificial Intelligence (AI), ecosystem of connected devices and Big Data for waste management, emission reduction, decision-making, and sustainable production (Awan et al. 2021; Rosa et al. 2020).

Existing research has explored how Industry 4.0 can embrace circular practices (Nascimento et al. 2019). Scholars have reviewed the existing literature on circular economy, I4.0, and sustainable manufacturing (Bag and Pretorius 2022; Patyal et al. 2022). Research has delved into the avenues and obstacles associated with the synergy between Industry 4.0, characterized by its digital and technological advancements, and the circular economy, guided by principles of efficient resource use (Hennemann and Sehnem 2022). However, the existing research finds limitations in providing a proficient understanding in the context of the connection, obstacles, and outcomes of combining the dimensions of Industry 4.0 with circular practices within the realm of sustainable manufacturing. The research elucidates the I4.0 and CE paradigms, explores the relationships for realizing sustainable development goals and identifies the associated challenges in the scope of sustainable manufacturing. Further the study provides suggestions for overcoming these challenges and highlights future research areas.

The subsequent sections of this chapter are structured as follows. Section 2 presents the concepts and literature review of Sustainability, Circular Economy, and Industry 4.0. Section 3 provides insight into the research gap and the rationale. The interconnection between Industry 4.0 and Circular Economy in the context of achieving sustainable development goals is elaborated in Sect. 4. In Sect. 5, the challenges that hinder the implementation of the digital-enabled circular economy are discussed. Section 6 provides direction to overcome the obstacles faced in the implementation of technology enabled circular economy. Section 7 elucidates the transition towards technology enabled Circular and Sustainable Manufacturing and discusses future research direction. The paper ends with the concluding remarks in Sect. 8, followed by the list of references.

## 2 Theoretical Background and Review of the Literature

### *2.1 Sustainability*

Ensuring resource efficiency is imperative to stay within the limits of planetary boundaries. The concern for climate change and biodiversity loss has compelled the world to embrace sustainable practices in business operations. To maintain the earth's ecosystem, it is necessary for society to make sustainable choices. The aim is to ensure that resources remain available for both current and future generations.

Sustainability is defined as "development that meets the needs of the present without compromising the ability of future generations to meet their own needs" by Brundtland Commission (Keeble 1988). The 3 pillars of sustainability are "people, planet and profit" (Elkington 1998). The harmonious incorporation of social, environmental, and economic performance is advocated by the triple bottom line concept. The United Nations framed and adopted the Sustainable Development Goals (SDGs) in 2015 to overcome the global challenges including "poverty, inequality, gender equality and conservation as well as the necessity for cross-sectorial and intergovernmental cooperation towards systemic change" (Dantas et al. 2021). It includes 17 sustainable development goals, each comprising a set of 169 targets to be achieved by the year 2030 (Dantas et al. 2021).

Sustainable manufacturing involves a holistic approach to producing goods and services such that the adverse impact of manufacturing on society, environment and economy is minimal. Sustainable Production/Manufacturing is defined as "the creation of goods and services using processes and systems that are non-polluting; conserving of energy and natural resources; economically viable; safe and healthful for employees, communities and consumers; and socially and creatively rewarding for all working people" by Lowell Center for Sustainable Production (LCSP) (Viles et al. 2022). Generic characteristics that sustainable manufacturing should incorporate are "environmental consciousness throughout the culture of the whole organization, both product and process design addressing sustainable issues, maximum use and reuse of recycled components and materials and product life-cycle concepts applied to the whole manufacturing system" (O'Brien 1999). The literature suggests several approaches the manufacturing organisation can adopt to improve their performance on sustainability including lean and clean manufacturing, process re-engineering and new technology adoption with focus on sustainability issues, continuous improvement through Kaizen, and extending manufacturing support to life cycles (O'Brien 1999).

Traditional manufacturing processes should be substituted with cleaner technologies to minimize harmful impacts on the environment. Successful implementation of sustainable manufacturing processes is dependent on technological factors and human-centric factors such as "strategic alignment, communication, green motivation, management leadership and customer relationship management" (Ahuja et al. 2019). A sustainable production system improves employee safety, increases profitability, and reduces operational costs (Rosário and Dias 2022).

The existing literature on sustainable operations management has categorized research papers under "closed-loop supply chains, low-carbon economy, environmental management and performance, innovation, and social responsibility" (Atasu et al. 2019). Economic and managerial obstacles pose significant challenges for the sustainable manufacturing settings (Malek and Desai 2019). The implementation of novel governance mechanisms can play a pivotal role in addressing sustainability challenges (Alexander 2020). The literature demonstrates that sustainable manufacturing practices requires the optimization of processes with the assistance of decision support systems (DSS) (Bag and Pretorius 2022). The extant literature has identified the drivers, challenges, the use of cleaner technology and performance measurement system for sustainable manufacturing (Bag and Pretorius 2022).

## 2.2 Circular Economy

The linear economy advocates the throwaway culture and follows a take-make-dispose model. However, given the rate of environmental degradation the linear system should be replaced with a circular economy (CE) model. CE is poised to serve as a tool for addressing environmental issues such as pollution, resource depletion, and loss of biodiversity. The objective of CE is "decoupling of economic growth from natural resource depletion and environmental degradation" (Murray et al. 2017).

By integrating circular practices, the economy can transition towards a regenerative and restorative system that will safeguard the environment. The circular economy will solve global environmental issues and build a healthy ecosystem.

The introduction and evolution of the concept cannot be attributed to a specified date or single author. The circular economy concept has gained recognition since 1970 (EMF 2013). In 1966 Kenneth Boulding suggested a shift from a cowboy economy to a spaceship economy (Blomsma and Brennan 2017). In a cowboy economy, endless frontiers suggested no boundaries on resource consumption or on disposing of waste whereas in a spaceship economy, resources are considered limited and humans have no option but to reconnect with a cyclical ecological system and reuse materials constantly. In 1976, Walter Stahel and Genevieve Reday introduced the concept of a loop economy, which has an impact on job creation, resource savings, waste prevention, and economic competitiveness (EMF 2013). In 1990, CE concept was introduced by David Pearce (Andersen 2007). Pearce and Turner posit that the term 'circular economy' was initially used in Western literature during the 1980s to portray a closed system of interactions between the economy and the environment (Murray et al. 2017). They described that natural resources are used as input for both production and consumption and as outputs in the form of waste (Pearce and Turner 1989).

The circular economy is defined as "a regenerative system in which resource input and waste emission and energy leakage are minimized by slowing, closing, and narrowing material and energy loops" (Geissdoerfer et al. 2017). Ellen MacArthur Foundation (EMF) defined circular economy as "an industrial economy that is restorative or regenerative by intention and design" (EMF 2013). The 3 basic principles of the circular economy are regeneration, circulation, and elimination of waste and pollution (EMF 2015). EMF proposed the ReSOLVE framework involving 6 business actions for the circular economy to help organizations apply circular economy principles (EMF 2015). The actions included—"regenerate, share, optimize, loop, virtualize, and exchange". Regenerate involves usage of biological cycles and a shift to renewable energy sources and materials. Share involves the sharing of assets and goods among individuals, with a focus on designing products to have longer lifespans and providing maintenance services to enable reuse and extension of the product's lifecycle. Optimise involves the utilization of digital manufacturing technologies. Loop involves the usage of biological and technical cycles. Virtualise replaces physical products with virtual and dematerialized products. Exchange entails using renewable goods instead of old and non-renewable goods (Patyal et al. 2022).

The circular economy rests on the foundation of both biological and technical cycles (Farooque et al. 2019). The biological cycle minimizes excessive extraction of natural resources, reuse energy, promotes the usage of renewable resources, and reusing organic waste, and thereby regenerates ecosystems. The technical cycle promotes product lifestyle extension through reuse, remanufacturing, recycling, refurbishment, and repair. CE can be implemented at three levels including the micro level (individual firm level), meso level (eco-industrial parks), and macro (development of the eco-city, eco-municipality, or eco-province) levels (Yang et al. 2023).

Circular economy has positive effects on the supply chain (Govindan and Hasanagic 2018). The managers can gain social, economic and environmental benefits in the supply chain by adopting circular practices (Sharma et al. 2019). Circular supply chain approach utilizes sophisticated technology to trace environmental footprints (Abideen et al. 2021). Leveraging Industry 4.0 technology for incorporating circular practices throughout supply chain activities and services, has drawn increased attention from researchers (Kumar et al. 2021b).

Germany pioneered introducing the circular economy into national legislation by passing the "Closed Substance Cycle and Waste Management Act" in 1996 (Su et al. 2013). Subsequently other countries also introduced related laws, such as in 2002, Japan introduced the "Basic Law for Establishing a Recycling-Based Society" (Geissdoerfer et al. 2017), while in 2009 China enacted the "Circular Economy Promotion Law of the People's Republic of China" (Su et al. 2013).

## *2.3 Industry 4.0*

The landscape of manufacturing processes has changed due to advancements in technology and innovation. The fourth industrial revolution has brought about a profound and transformative impact on all sectors across the globe. Industry 4.0 technologies increase efficiency and improve productivity. Organizations across industry sectors including manufacturing firms should invest in Industry 4.0 technologies and train the workforce needed to smoothen the transition to Industry 4.0. Successful adoption of Industry 4.0 will lead to a positive impact on markets, industry, and the economy. It will demand the development of novel business models, production techniques, changes in business strategy and the working environment (Pereira and Romero 2017). Industry 4.0 has affected competitiveness by customizing the offerings to the customers.

Industry 4.0 is an initiative introduced in Germany in 2011 at Hannover Fair and the term I4.0 was introduced by Klaus Schwab (Prisecaru 2017; Vogel-Heuser and Hess 2016). It advances the concept of the Cyber-Physical System into Cyber-Physical Production System. The connected smart factories facilitate communication between production equipment and product. Industry 4.0 technology combines digital manufacturing technologies with information and communication technologies (Kang et al. 2016). The main features of I4.0 mentioned by (Shrouf et al. 2014) includes creating a link between customers, machines, suppliers, and orders using IoT and electronic devices which assists firms in manufacturing products using autonomous systems and decentralizes decision-making process (Lopes de Sousa Jabbour et al. 2018; Lasi et al. 2014).

The principles of I4.0 include virtualization (virtual replica of the physical world into digital data), decentralization (the ability of interconnected systems to make independent decisions), interoperability (the capability of components and equipment to interconnect with each other using the internet), real-time capability (real-time exchange of information between systems), service orientation (system's ability to offer stakeholders service and other functionalities using internet), and modularity (ability of system to adapt to change or dysfunction) (Fuertes et al. 2022).

I4.0 technologies can be categorized as digital and physical technologies (Bai et al. 2020). Digital technologies comprise information and communication technologies while physical technologies include manufacturing technologies. Technologies such as Additive Manufacturing, the Internet of Things (IoT), Big Data, Artificial Intelligence (AI), Robotics, Blockchain, Cloud Computing, Virtual, and Augmented Reality are considered major I4.0 technologies. Industry 4.0 technology application runs through the entire value and supply chains. I4.0 technologies facilitate the production process by providing real-time information, helping in decision-making, tracing, and tracking the supply chain. Industry 4.0 technologies enable improved customer responsiveness, efficient use of resources, and monitoring material flow (Lopes de Sousa Jabbour et al. 2018). With the use of Industry 4.0 technologies data about the product can be attained throughout the product's lifecycle. Industry 4.0 offers several advantages, including profitable business, reduced costs, increased efficiency, and improved workplace conditions.

Countries have launched I4.0 initiatives aimed at promoting the development and adoption of Industry 4.0 technologies that includes Germany's "High-Tech Strategy 2025", the United States' "Advanced Manufacturing Partnership (AMP 2.0)", China's "Made in China 2025" and India's "SAMARTH (Smart Advanced Manufacturing and Rapid Transformation Hub) Udyog Bharat 4.0" (Haleem et al. 2022; Federal Ministry of Education and Research 2021).

Industry 4.0 technologies present both challenges and opportunities for supply chain management and operations. The literature on I4.0 technologies reflects its potential for achieving circularity (Farooque et al. 2019). Digital technologies possess the capacity of managing the marketing processes and supply chain management interface. The use of Industry 4.0 technologies facilitates information sharing and data collection throughout the supply chain, increases transparency in product's lifecycle and aids in decision making (Esmaeilian et al. 2020). Industry 4.0 should be capitalized as an opportunity to accelerate social, economic, and environmental growth. Incorporating industry 4.0 technologies in the value chain helps achieve sustainability (Stock and Seliger 2016).

### *2.4 Essential Aspects of Circular Economy and Industry 4.0*

Successful implementation of circular economy and Industry 4.0 technologies requires understanding of the goals, principles, characteristics and business models. The Table 1 presents a summary of their key characteristics, principles, goals and business models.

Organizations should be built on the foundation of these principles and features of CE and I4.0 and adopt the relevant business models to develop a framework of sustainable development and innovation in their manufacturing operations.

## 3 Research Gap and Contribution of the Study

Previous research has explored distinct principles for sustainable production for transitioning towards smart industry model and circular economy (Viles et al. 2022). Scholars have conducted review of the literature regarding the prospects of Industry 4.0 within the context of the triple bottom line (Birkel and Müller 2021). The application of advanced technologies in traditional supply chains and how these technologies can be applied in circular supply chains have been studied by researchers (Abideen et al. 2021). Literature review to explore the application of Industry 4.0 technologies for circular procurement (Rejeb et al. 2022), and on topics related to Industry 4.0 and environmental aspects of sustainability have been conducted (Oláh et al. 2020). Researchers have discussed business models to integrate I4.0 with circular economy practices (Nascimento et al. 2019).

Research has been conducted to explore the pathways and roadblocks linked with CE and I4.0 (Hennemann and Sehnem 2022), Industry 4.0 implications on manufacturing landscape (Enyoghasi and Badurdeen 2021). Existing literature also discusses the merits and obstacles of deploying specific Industry 4.0 technologies for circular practices (Tavares et al. 2023; Rejeb et al. 2022). Scholars have reviewed the extant literature on Industry 4.0, circular economy and sustainable manufacturing and its relation with SDGs (Bag and Pretorius 2022; Patyal et al. 2022). However, the existing

**Table 1** Key aspects of CE and I4.0—characteristics, principles, goals, and business models

| Basis | Circular Economy | Industry 4.0 |
|---|---|---|
| Key Characteristics | The fundamental characteristic of CE is its deliberate focus on being inherently restorative and regenerative product design (EMF 2015) | Industry 4.0 features include "interoperable, integrated, adapted, service-oriented, and optimized manufacturing processes" (Lu 2017) |
| Principle | The principles of CE include value preservation, resource efficiency, diversification, design for longevity, circular supply chains, and waste prevention (EMF 2015) | The principles of I4.0 include "virtualization, interoperability, decentralization, real-time capability, service orientation, and modularity" (Fuertes et al. 2022) |
| Business Model | Circular business models aim to maximize the utilization and extending the longevity of products and materials. Circular business model should have elements that slows (encourages prolonged use and reuse), intensifies (promotes value intensive use), narrows (focuses on resource efficiency), dematerialises (shifting from product centric to service centric models) and closes loop (advocates reusing material) (Hazen et al. 2021; Geissdoerfer et al. 2018) | The business models include expertization, integration, and servitization (Weking et al. 2020). Expertization "is a hybrid of product- and process-focused business models, which includes consulting services and multi-sided platforms" (Weking et al. 2020). Integration introduces novel processes and integrates parts of the value chain in the business model. Servitization offers a new product service system instead of just offering tangible products |
| Goals | The primary goal is to form a closed loop economic system that cut back on resource input and outflow of waste, reduces energy usage and emissions from the system (Geissdoerfer et al. 2017) | It aims to revolutionize the industrial manufacturing system through digitalization and the use of new technologies (Rojko 2017) |

research lacks a precise comprehension of the relationship, challenges and impact of association of Industry 4.0 technologies and circular economy practices within the context of sustainable manufacturing. This study discusses the association between I4.0 and CE for sustainable manufacturing to accomplish SDGs and roadblocks that hinder their implementation. The study also provides suggestions for overcoming these challenges and highlights future research areas.

## 4 The Convergence of CE and I4.0 for Achieving SDGs

The convergence of Circular Economy and Industry 4.0 is increasingly recognised as a persuasive impetus for accelerating sustainable development. In today's dynamic and globalized society, the CE and I4.0 are acknowledged as the transformative paradigms for improving resource efficiency and harnessing digital technologies to revolutionize industrial processes, respectively (Agrawal et al. 2022; Lopes de Sousa Jabbour et al. 2018). Exploring the linkages between these paradigms as a path towards achieving the SDGs – a nexus that holds the potential to stimulate economic growth and promote environmental stewardship while addressing the global challenges, is crucial.

## *4.1 Mapping Relationship Between Circular Economy Principles and I4.0 Technologies*

The principles of CE include regeneration, circulation, and elimination of waste and pollution (EMF 2015). Each principle can be abided by applying Industry 4.0 technology as explained below.

**CE Principle 1: Eliminate waste and pollution:** The linear economy involves using the products and then disposing them off after end of use. The product ends up in landfill or incineration. Given the fact that resources are finite, judicious usage of resources should be promoted and waste should be eliminated (EMF 2015).

### 4.1.1 How I4.0 Technology Helps in Eliminating Waste and Pollution?

The path towards circularity begins from the design of the product. In order to eliminate waste and pollution the focus should be directed towards the circular product design. The circular design targets developing products such that resource consumption and waste is minimal throughout their life cycle and ensures the longevity of the product (Farooque et al. 2019). I4.0 tools and techniques can be used for real-time monitoring, predictive maintenance, optimal resource consumption, mass customization, enabling remote updates and life cycle assessment thus ensuring extended product durability and reduced environmental impact (Ghobakhloo 2020). Leveraging technologies enhances supply chain efficiency and transparency by eliminating waste and pollution across all supply chain stages.

AI contributes to waste and pollution reduction through process optimization, optimal production, and data driven decisions (Rejeb and Appolloni 2022). Sensor and satellite data can be analysed using AI to inspect the environmental conditions which can help to identify pollution and waste generative activities ultimately encouraging manufacturers to switch to sustainable practices. With the use of AI, the need for maintenance and possibility of equipment failures can be determined which can help in repairing the equipment and reducing the need for disposal of equipment. Deploying AI powered algorithms for identification of inventory needs reduces understocking or excess stocking thereby reducing the stockouts or unsold goods. AI powered robotics systems can sort recyclable and nonrecyclable products which leads to reduction in the amount of waste directed towards landfills.

Additive manufacturing products should be designed to promote reuse, recycling, and remanufacturing which in turn would lead to waste reduction (Tavares et al. 2023). Additive manufacturing involves production of goods using only the material needed in contrast to traditional manufacturing where materials are drawn from large blocks or sheets which helps in minimising the material waste. Decentralized additive manufacturing where local or on-site production can be done reduces the need for transportation thereby eliminating transport emission and need for product packaging. Additive manufacturing can be used for product customisation or producing one of the parts instead of mass production which reduces the production of excess stock. Additive manufacturing facilitates repair services for intentionally redesigned products, making repair and component replacement more accessible and efficient (Rejeb et al. 2022).

Big data analytics helps in efficient allocation of resources by analysing historical data, using predictive models and real time information (Bag and Pretorius 2022; Awan et al. 2021). The product can be monitored throughout its lifecycle which can enable decision making about repair, reuse and recycling of the product. Big data analytics can be employed to analyse the supply chain to identify inefficient transportation routes that the organization can eliminate to reduce unnecessary transport emissions and pollution.

IoT sensors can be used to monitor the product conditions for scheduling repair. The products, components, and materials can be traced throughout their lifecycle using IoT which enables waste collection, waste recovery, and reuse (Fatimah et al. 2020). Manufacturers can address anomalies by monitoring real-time data supplied by the sensors, enhancing transparency and traceability within the supply chain.

**CE Principle 2: Circulate products and materials:** This principle advocates that material should be used as long as possible in the form of either raw material or product. Products and materials can be kept in circulation through biological or technical cycles. In the technical cycle, products are recycled, remanufactured, repaired, or reused. Biodegradable substances are naturally broken down and reintegrated into the environment in biological cycles (EMF 2015).

### 4.1.2 How Does I4.0 Technology Help in Circulating Products and Materials?

The design of the product is the key and therefore should be taken into consideration to ensure that products can be circulated in the technical or biological cycle (EMF 2013). Products should be designed with the intent of circulating them longer by sharing or maintaining, reusing or redistributing, refurbishing or remanufacturing, and recycling as last resort (EMF 2015).

AI helps in ensuring optimal resource consumption and maximizing product efficiency through application of green information technology (Rejeb et al. 2022). AI facilitates keeping the products in circulation for as long as possible by managing the products from design, manufacturing to end of life stage by creating the products which can be repaired easily, disassembled for remanufacturing and refurbishing and as a last resort can be recycled. Online platforms for selling and buying used products can be powered by AI. To encourage customers to return used products and purchase refurbished products AI tools can be used. AI facilitates the planning, scheduling, and optimization of remanufacturing processes (Rejeb et al. 2022). To aggregate and visualize information for remanufacturing, virtual reality and augmented reality can be employed (Kerin and Pham 2020).

The design of products manufactured using additive manufacturing should be such that it is easy to repair, disassemble and reuse later (Tavares et al. 2023). Utilizing recycled materials as inputs for additive manufacturing processes can help reduce the consumption of virgin materials (Chong et al. 2015). When products break down then additive manufacturing can be deployed to produce replacement parts in order to keep the product in use instead of discarding (Tavares et al. 2023). Similarly for obsolete products, replacement parts can be manufactured to keep the product in circulation.

IoT sensors can provide information about product use and product performance which can facilitate making relevant changes in product design. The product quality can be monitored using IoT to enable proactive repair of the products and equipment thereby reducing the need for replacement, and keeping the product in circulation (Esposito 2017). IoT sensors are used to track the products by suppliers and manufacturers throughout their lifecycle to ensure that products are maximally utilized; and to make decisions about reuse, material recovery and remanufacturing (Rejeb et al. 2022; Awan et al. 2021). Reverse logistics process is eased using IoT by tracking the return process to ensure proper sorting, refurbishing and recycling.

Quality control processes can be automated using smart contracts within blockchain so that products which don't meet the quality criteria can be returned for either repair or recycling. Customers can gain detailed information about the product disposal practices using Radio Frequency Identification (RFID) technology.

**CE Principle 3: Regenerate nature:** This principle advocates that natural processes should be supported. Such practices should be adopted that help in building natural capital (EMF 2015). Resources should be used optimally and such processes and technologies should be deployed that either use resources with improved performance or use renewable resources (EMF 2015).

### 4.1.3 How I4.0 Technology Helps in Regenerating Nature?

I4.0 technologies improve energy efficiency which implies lower greenhouse gas and reduced strain on the environment. By employing Industry 4.0 technologies the environmental impacts of production processes can be determined, thus encouraging timely corrective measures.

Additive manufacturing leads to conservation of resources by producing goods using only the material needed in contrast to traditional manufacturing where materials are drawn from large blocks or sheets. Design optimization and rapid prototyping can help in manufacturing eco-friendly products using additive manufacturing. Additive manufacturing processes can involve using biodegradable materials as inputs to support the earth's regenerating capacity (Tavares et al. 2023).

Blockchain technologies enhance supply chain transparency and help in ensuring that products are sourced sustainably (Esmaeilian et al. 2020). Carbon offset projects undertaken by organizations can be tracked by employing blockchain technology. Blockchain can facilitate data sharing among researchers, practitioners and the government to take action for resource conservation.

Leveraging IoT in the production process aids in decision-making by providing real-time information that managers can utilize to minimize waste and, consequently, foster the regeneration capacity (Patyal et al. 2022). IoT sensors help to monitor the ecosystem and identify areas in need of restoration.

### *4.2 Contribution of Industry 4.0 Technologies and Circular Economy to SDGs*

The circular economy paradigm can contribute to achieving SDGs (SDG 6, 7, 8, 12 and 13) directly related to hygiene, energy consumption, economy growth, sustainable manufacturing and consumption, and environmental impact. (Schöggl et al. 2020). Circular economy practices demonstrate the most significant associations with the targets established in SDG 7, SDG 8, and SDG 12 (Schroeder et al. 2019). Dantas et al. (2021) explored how the implementation of Industry 4.0 can directly support achieving the targets of SDG 7, SDG 8, SDG 9 (focussing on innovation, industry, and infrastructure), SDG 11 (promotes inclusivity, safety, resilience, and sustainability of cities), SDG 12 and SDG 13. The discussion in the following section draws insights on the synergies of the circular economy and Industry 4.0 to accomplish SDG 7, SDG 8, SDG 9, SDG 11, SDG 12 and SDG 13.

- **SDG 7**: It calls for investment in wind, solar, and thermal power; upgrading technology, and expanding infrastructure given its focus on factors for affordable and clean energy (Dantas et al. 2021).

  Integrating circularity and technology in industrial settings results in beneficial outcomes (Oláh et al. 2020). SDG 7 promotes the use of energy efficient technologies such as the Industrial IoT, data-driven analytics, big data, and cyber-physical production systems in manufacturing processes (Modgil et al. 2020). These technologies enable the collection and transmission of data, which in turn helps enhance energy and resource efficiency, while also expanding the utilization of clean energy sources (Modgil et al. 2020). Industry 4.0 significantly promotes a clean energy system by offering tracking capabilities and providing solutions to build a sustainable society.
- **SDG 8**: SDG 8 promotes sustainable economic growth, increased productivity, and technology innovation. To accomplish this SDG, job creation and entrepreneurship are vital. The objective is to promote decent work for both men and women

and achieve full and productive employment (Dantas et al. 2021; Fatimah et al. 2020).

Circular economy and Industry 4.0 propose optimization of production processes which leads to efficient systems thereby increasing the competitiveness of companies and impacting overall economic growth positively. The nexus of circular economy and Industry 4.0 impacts the job market. There is a need to train employees about the principles and paradigms of Industry 4.0 and circular economy (Ada et al. 2021; Dantas et al. 2021). Sustainable manufacturing requires adoption of circular practices and advanced technologies by skilled and trained workers. It is apparent that a different set of skills would be required which can lead to income disparity. The training and development activities should assist in supporting the workers towards the transition to technology-enabled circular economy (Phung 2019).

- **SDG 9**: In order to ensure sustainable development investment in research, innovation and promotion of sustainable industries is crucial. Investment in infrastructure and innovation drives development and economic growth. Technology progress can help in addressing environmental and economic challenges (Dantas et al. 2021).

  The goals listed in SDG 9 are aligned with Industry 4.0 since the objective is to build the structure vital for promoting innovation, technology advancement and sustainable industries. The technology and innovation are applied to industrial settings, to optimize the manufacturing process and utilize energy and resources efficiently to achieve this SDG (Modgil et al. 2020). The application of the industrial internet of things, big data, cyber-physical production systems, additive manufacturing and data-driven analytics help in decreasing energy and material flows in industry. Use of I4.0 technologies into manufacturing processes aligns with the SDG 9 for stimulating innovation. Taking SDG 9 into account, the CE-I4.0 nexus promotes environmental sustainability.

- **SDG 11:** The pursuit of sustainable communities and cities entails creating opportunities for businesses and careers, developing resilient societies and economies, and providing safe and affordable housing facilities (Dantas et al. 2021).

  Industry 4.0 can assist in building sustainable communities and cities through the implementation of circular practices. I4.0 technologies such as automation, cloud computing, big data, and cybersecurity are instrumental in enhancing city-wide logistics, transportation, security, and lighting systems (Dantas et al. 2021). Circular economy practices can be applied using industry 4.0 technologies to manage waste and mitigate climate change thereby helping in building a sustainable and resilient urban society (Hossain et al. 2020; Lausselet et al. 2017). In accordance with SDG 11, manufacturers will be prompted to adopt sustainable practices and employ modern technology to build eco-friendly infrastructure which in turn will result in waste reduction, environmental compliance and energy efficiency. An illustrative case studied in Singapore demonstrated that circular economy practices applied using IoT and industrial symbiosis enhanced the establishment of Zero Waste Management system (Kerdlap et al. 2019).

- **SDG 12**: To ensure economic growth and sustainable development there is an urgent need to lessen the ecological impact by shifting the way goods and resources are produced and consumed. To achieve this SDG natural resources should be managed efficiently and ways of disposing of toxic waste and pollutants should be examined. Industries, businesses and consumers should be motivated to reduce and recycle waste in order to lead the pathway of developing countries towards sustainable consumption by 2030 (Dantas et al. 2021; Fatimah et al. 2020)

  Industry 4.0 aids in strengthening the technological and scientific capabilities of developing nations to lead the path toward sustainable consumption and production. Technologies are revolutionizing the manufacturing processes. Resources and energy usage is optimized and waste is minimized by using Industry 4.0 technologies to gather data and for constant monitoring which aligns with the circular economy principle of reducing resource usage and eliminating pollution and waste (Inoue et al. 2020; Andrews 2015). Circular economy promotes sustainable production and consumption. Product-Service Systems (PSS) proposed by (Vasantha et al. 2015) suggest products and services should be combined in order to fulfil the needs of customers (Tukker and Tischner 2006). PSS promotes sustainable production since it reduces the number of products to be produced consequently decreasing the raw materials requirement. Business models framed upon PSS are based on service digitalization which further reduces the resource requirement.
- **SDG 13**: Countries across the globe are experiencing adverse effects due to climate change. Greenhouse emissions are 50% higher compared to 1990. Given the focus of this SDG on climate action, the development strategies of nations should integrate human security, disaster risk measures, and sustainable natural resource management (Dantas et al. 2021; Fatimah et al. 2020).

  Utilizing technologies such as blockchains, big data, additive manufacturing, and IoT can help in decreasing the material and energy flow within production processes by monitoring the resource usage and therefore can help in reducing emissions (Andoni et al. 2019; Bonilla et al. 2018). Integration of CE and I4.0 helps to implement climate change initiatives into planning, policies, and strategies. The emphasis is placed on government incentives for encouraging the adoption of Industry 4.0 and at the macro level of the CE. The CE-I4.0 nexus will work towards accomplishing systematic shifts aimed at mitigating climate change.

## 5 Roadblocks in Transitioning to a Digital-Enabled Circular Economy

I4.0 technologies are instrumental in accelerating the shift towards circular and sustainable manufacturing. However certain barriers impede this transition, as follows.

- **Lack of motivation:** Organizations lack the motivation to adopt Industry 4.0 and CE nexus since they are unaware of the benefits. Also, the financial benefits of such a transition will be earned in the long run which leads to resistance in organizations to take actions for this transition (Kumar et al. 2021b).
- **Collaboration gaps between supply chain partners:** Digital-enabled circular supply chain requires collaboration between partners and suppliers for data sharing, processes, circular practices and technology adoption which is challenging to achieve given trust issues and differing objectives of parties in the supply chain (Agrawal et al. 2022).
- **Trade off related to energy consumption:** Adoption of technologies can lead to increased energy consumption and generation of electronic waste if managed inefficiently (Birkel and Müller 2021). Smart production systems will lead to a significant surge in energy consumption, primarily driven by the construction of large-scale data centers (Birkel and Müller 2021). To fully harness the advantages of a digital-enabled circular economy, it is imperative to assess its impact on sustainability.
- **Dispersed supply chains:** To take benefit of cost economies and better quality, the supply chain operations are dispersed in different geographical locations which makes it challenging to support circular activities like collection, reuse and remanufacturing and trace products life cycle throughout the supply chain (Lewandowski 2016).
- **Inadequate government support:** The government should actively promote the utilization of Industry 4.0 technologies and the circularity paradigm for transition towards a circular economy and foster sustainable manufacturing practices. However, currently there is a lack of proper legislation, and control, government support, a global framework which hinders this transition (Govindan and Hasanagic 2018)
- **Investment in technology:** Sizable investments are required to set up the technological infrastructure in line with the Industry 4.0 technologies, which poses a financial challenge for establishing a sustainable manufacturing setting. The organizations lack the necessary investment and are uncertain about the returns that would be generated by investing in Industry 4.0 technologies (Kumar et al. 2021a).
- **Skill gap:** Employees are required to possess the requisite knowledge and skills for adoption and implementation of Industry 4.0, circular economy and sustainable manufacturing. However, there is a lack of skilled workers (Fuertes et al. 2022) and a lack of knowledge about how to manage data (Ozkan-Ozen et al. 2020) which creates an obstacle in this direction. Adoption of digital technologies requires employees to be aware and equipped with the skills to ensure data protection and compliance with regulations, which is usually a limitation.
- **Resistance to change:** Organizations hesitate to switch from traditional manufacturing practices to digital and sustainable practices. There is resistance in organizations to share information with supply chain partners due to lack of trust

(Birkel and Müller 2021). Also, there are challenges related to the fear of technology adoption in terms of job loss and cultural shift. Employees may also show resistance to change and skill development.

- **Lack of awareness:** The absence of awareness among stakeholders regarding the advantages of integrating Industry 4.0 and Circular Economy impedes organizations from implementing Industry 4.0 technologies to facilitate circular practices (Brozzi et al. 2020; Sharma et al. 2019).
- **Absence of strategy:** The systematic adoption and implementation of I4.0 technologies to deploy circular practices and ensure sustainable manufacturing requires proper strategy. However, lack of comprehension of the concepts, and absence of a proven framework acts as a roadblock (Kumar et al. 2021a).
- **Short-term goals:** Often organizations are oriented towards short-term goals that focus on making profits and disregard the consequences of their actions on society and the environment. Due to this short-term orientation, organizations don't invest in I4.0 technologies that accelerate the circular transition and assist sustainable manufacturing (Kumar et al. 2021b).

A holistic approach towards circular transformation seeks collaborative participation of businesses, governments, and technology providers. Focus on research and innovations can provide robust solutions to overcome such obstacles and ease shifts towards a digital-enabled circular economy.

## 6 Navigating the Challenges for Digital-Enabled Circular Economy

To surmount the obstacles that hinder the successful integration of I4.0 technologies aimed at transitioning to a CE and establishing sustainable manufacturing practices following actions are recommended based on literature exploration.

- **Government action:** Government support can help in transitioning to CE and sustainable manufacturing (Awan et al. 2021). Government can promote the usage of technologies, provide financial aid, incentive and technological assistance, and spread awareness about the merits of I4.0 and circularity integration. Huge investment and long-term commitment is required for adoption of technology and circular approach (Viles et al. 2022; Kumar et al. 2021b). The government with the assistance of the other stakeholder can create an ecosystem that can provide the necessary technical and financial support in this direction, including the provision of availability of assets on lease to reduce the financial burden (EMF 2013).
- **Setting clear objectives and long-term goals:** The environmental damage caused by the linear model of production will have serious consequences. Organisations are required to define clear objectives and outcomes and adopt a long-term orientation for shift to circular and sustainable models to safeguard the planet (EMF 2015). Moreover, this transformation requires the support of the leadership and

their commitment for the initiative, along with a company wide approach for this transition.

- **Development of infrastructure:** Organizations should develop a comprehensive strategy to integrate Industry 4.0 technologies in their operational framework. Collaborations among the supply chain partners can foster the momentum and promote the development of shared infrastructure to support the technology adoption in line with the principles of circular transition. They may explore the funding sources, grants, and incentives for sustainable initiatives (Agrawal et al. 2022).
- **Financial access:** Access to finance for investing in I4.0 technologies is necessary (Ada et al. 2021). Banks should be open to granting loans to organizations to fund investment in advanced technologies. Provisions of tax subsidies and interest free loans can further accelerate the momentum.
- **Training and development:** Organizations should invest in training to upskill the employees to enhance their skills in operating and maintaining technological systems, with a specific focus on educating them about issues related to data privacy, security and regulation compliance. Acquisition of the necessary skills and knowledge will equip the employees to assist in this transition (Govindan and Hasanagic 2018).
- **Spreading awareness:** Promoting awareness regarding the benefits of embracing the synergy between technology and circular practices will inspire organizations to align their effort towards circularity and adoption of Industry 4.0 technologies (Kumar et al. 2021a). Stakeholders should embrace circularity in product and process development for ensuring durability, repairability, and recyclability.
- **Global cooperation:** The coordination and cooperation amongst countries regarding strategy development, actions, policy instrument, financial aid for circular practices, and sustainability will drive actions towards a successful transition towards a sustainable future (Yang et al. 2023; Kumar et al. 2021b).

## 7 Transitioning to Technology Enabled Circular and Sustainable Manufacturing: Future Research Agenda

The manufacturing sector fosters innovation, drives the economy and plays an essential role in advancing progress across various domains. Over the years it has undergone numerous transformations spurred by technological advancements, accelerating the progress of the sector. In the contemporary landscape, the manufacturing industry confronts distinctive challenges, characterized by heightened competition and the imperative to enhance the sustainability of its operations. As manufacturers devise novel business models capable of withstanding these challenges, the concurrent trends of digitalisation and the circularity are poised to exert significant influence. Utilizing technologies such as virtual reality, additive manufacturing, IoT, big data, cloud computing, AI, robotics, blockchain, and augmented reality has streamlined operations, enhanced efficiency, reduced human errors, boosted production output, and enabled uninterrupted manufacturing processes.

The circular economy is regarded as the preferred industrial model for achieving sustainable development. Shifting to a circular economy from the current throw-away culture is hindered due to technological, financial, regulatory, and sociocultural barriers. Industry 4.0 is perceived as an enabler to overcome these barriers (Albaladejo et al. 2023). For instance, in Spain the solid waste and recycling company *AllWaste* uses IoT for waste collection and cleaning the street (Albaladejo et al. 2023). I4.0 technologies lead pathways towards circular and sustainable transformation through real-time monitoring, predictive maintenance, optimal resource consumption, mass customization, life cycle assessment, and enabling remote updates.

Drawing from a literature review, we suggest several promising avenues for future research. Scholars can explore the role of SMEs in implementing Industry 4.0 technologies to deploy circular economy practices (Birkel and Müller 2021; Hennemann and Sehnem 2022). Further study can be conducted to understand how supplier and buyer firms influence and collaborate for the implementation of Industry 4.0 practice throughout the entire supply chain network (Ada et al. 2021). The significance of cooperation among supply chain partners in building the necessary infrastructure for the shift toward a digital-enabled circular economy should be studied in depth (Ada et al. 2021; Ozkan-Ozen et al. 2020). The challenges in involving customers to facilitate the implementation of Industry 4.0 technologies to transition to a circular economy should be researched (Awan et al. 2021). Adoption and implementation of I4.0 and CE nexus in companies working in different sectors should be studied (Nascimento et al. 2019). Future studies can be conducted to understand how the public sector can combine Industry 4.0 and circularity practices to ensure sustainable manufacturing (Kumar et al. 2021a). The need for global collaboration in shaping cohesive policies and strategies for a successful transition to a sustainable future should be examined (Kumar et al. 2021b).

## 8 Conclusion

This chapter explores the sustainability, CE, and I4.0 paradigms in detail. It delves into the interplay between I4.0 and CE, focusing on their synergy to attain sustainable development goals and the obstacles that impede their successful implementation. The chapter also outlines strategies for surmounting these challenges while underscoring the potential avenues for future research in this domain.

Utilizing digital technologies for circular practices offers substantial benefits in terms of optimizing organizational processes for heightened efficiency, waste reduction, and prolonging product lifespans. Industry 4.0 technologies can be harnessed to align with the principles of circularity. The association of I4.0 and circularity contribute to the accomplishment of targets outlined in SDGs (SDG 7, 8, 9, 11, 12, and 13).

To transition toward circular economy practices and ensure sustainable manufacturing, it is imperative to leverage Industry 4.0 technologies. However, several barriers obstruct this transition. These include collaboration gaps between supply chain partners, making trade-offs related to energy consumption, managing dispersed supply chains, insufficient government support, substantial technology investments, skill gap, limited awareness, resistance to change, the absence of a strategic framework, short-term orientation, and a lack of motivation. To successfully implement digital technologies for transitioning towards circularity and ensuring sustainable manufacturing, certain recommendations are proposed. These encompass government initiatives, infrastructure development, improved access to finance, comprehensive training and development programs, spreading awareness, setting clear objectives and long-term goals, global collaboration, incentivization, and the availability of technology assets through leasing arrangements.

The current year marks the halfway point for the SDG's target to be achieved till 2030. It is increasingly evident that digitalization plays a pivotal role (Ahmed et al. 2022). Modern technologies should be utilized for operations in organizations to gradually transition towards circular practices, recognizing their paramount significance in this endeavour.

## References

Abideen, A. Z., Pyeman, J., Sundram, V. P. K., Tseng, M. L., & Sorooshian, S. (2021). Leveraging capabilities of technology into a circular supply chain to build circular business models: A state-of-the-art systematic review. Sustainability, 13(16), 8997. https://doi.org/10.3390/su13168997

Ada, N., Kazancoglu, Y., Sezer, M. D., Ede-Senturk, C., Ozer, I., & Ram, M. (2021). Analyzing barriers of circular food supply chains and proposing industry 4.0 solutions. Sustainability, 13(12), 6812. https://doi.org/10.3390/su13126812

Agrawal R, Wankhede VA, Kumar A, Upadhyay A, Garza-Reyes JA (2022) Nexus of circular economy and sustainable business performance in the era of digitalization. Int J Product Perform Manag 71(3):748–774. https://doi.org/10.1108/IJPPM-12-2020-0676

Ahmed AA, Nazzal MA, Darras BM (2022) Cyber-physical systems as an enabler of circular economy to achieve sustainable development goals: a comprehensive review. Int J Precis Eng Manuf Green Technol 9:955–975. https://doi.org/10.1007/s40684-021-00398-5

Ahuja J, Panda TK, Luthra S, Kumar A, Choudhary S, Garza-Reyes JA (2019) Do human critical success factors matter in adoption of sustainable manufacturing practices? An influential mapping analysis of multi-company perspective. J Clean Prod 239:117981. https://doi.org/10.1016/j.jclepro.2019.117981

Albaladejo M, Mirazo P, Alvins JCP, Martín CR (2023) Industry 4.0: A cornerstone of the Circular Economy | Industrial Analytics Platform. https://iap.unido.org/articles/industry-40-cornerstone-circular-economy

Alexander R (2020) Emerging roles of lead buyer governance for sustainability across global production networks. J Bus Ethics 162:269–290. https://doi.org/10.1007/s10551-019-04199-4

Andersen MS (2007) An introductory note on the environmental economics of the circular economy. Sustain Sci 2(1):133–140. https://doi.org/10.1007/s11625-006-0013-6

Andoni M, Robu V, Flynn D, Abram S, Geach D, Jenkins D, McCallum P, Peacock A (2019) Blockchain technology in the energy sector: A systematic review of challenges and opportunities. Renew Sustain Energy Rev 100:143–174. https://doi.org/10.1016/j.rser.2018.10.014

Andrews D (2015) The circular economy, design thinking and education for sustainability. Local Econ 30(3):305–315. https://doi.org/10.1177/0269094215578226

Atasu A, Corbett CJ, Huang X, Beril Toktay L (2019) Sustainable operations management through the perspective of manufacturing and service operations management. Manuf Serv Oper Manag 22(1):146–157. https://doi.org/10.1287/msom.2019.0804

Awan U, Sroufe R, Shahbaz M (2021) Industry 4.0 and the circular economy: A literature review and recommendations for future research. Bus Strategy Environ 30(4):2038–2060. https://doi.org/10.1002/bse.2731

Bag S, Pretorius JHC (2022) Relationships between industry 4.0, sustainable manufacturing and circular economy: proposal of a research framework. Int J Organ Anal 30(4):864–898. https://doi.org/10.1108/IJOA-04-2020-2120

Bai C, Dallasega P, Orzes G, Sarkis J (2020) Industry 4.0 technologies assessment: A sustainability perspective. Int J Prod Econ 229:107776. https://doi.org/10.1016/j.ijpe.2020.107776

Birkel H, Müller JM (2021) Potentials of industry 4.0 for supply chain management within the triple bottom line of sustainability—A systematic literature review. J Cleaner Prod 289:125612. https://doi.org/10.1016/j.jclepro.2020.125612

Blomsma F, Brennan G (2017) The emergence of circular economy: A new framing around prolonging resource productivity. J Ind Ecol 21(3):603–614. https://doi.org/10.1111/jiec.12603

Bonilla, S. H., Silva, H. R. O., da Silva, M. T., Gonçalves, R. F., & Sacomano, J. B. (2018). Industry 4.0 and sustainability implications: A scenario-based analysis of the impacts and challenges. Sustainability, 10(10): 3740. https://doi.org/10.3390/su10103740

Brozzi R, Forti D, Rauch E, Matt DT (2020) The advantages of industry 4.0 applications for sustainability: Results from a sample of manufacturing companies. Sustainability 12(9):3647. https://doi.org/10.3390/su12093647

Chong S, Chiu HL, Liao YC, Hung ST, Pan GT (2015) Cradle to cradle® design for 3D printing. Chem Eng Trans 45:1669–1674. https://doi.org/10.3303/CET1545279

Dantas TET, De-Souza ED, Destro IR, Hammes G, Rodriguez CMT, Soares SR (2021) How the combination of Circular Economy and Industry 4.0 can contribute towards achieving the Sustainable Development Goals. Sustain Product Consumpt 26: 213–227. https://doi.org/10.1016/j.spc.2020.10.005

Elkington J (1998) Cannibals with forks: the triple bottom line of 21st century business. New Society Publishers, London

EMF (2013) Towards the circular economy: an economic and business rationale for an accelerated transition, vol 1. https://ellenmacarthurfoundation.org/towards-the-circular-economy-vol-1-an-economic-and-business-rationale-for-an

EMF (2015) Towards a circular economy: business rationale for an accelerated transition. https://ellenmacarthurfoundation.org/towards-a-circular-economy-business-rationale-for-an-accelerated-transition

Enyoghasi C, Badurdeen F (2021) Industry 4.0 for sustainable manufacturing: Opportunities at the product, process, and system levels. Resour Conserv Recycling 166: 105362. https://doi.org/10.1016/j.resconrec.2020.105362

Esmaeilian B, Sarkis J, Lewis K, Behdad S (2020) Blockchain for the future of sustainable supply chain management in Industry 4.0. Resour Conserv Recycling 163:105064. https://doi.org/10.1016/j.resconrec.2020.105064

Esposito M (2017) Where the circular economy meets the Internet of Things | Trellis. https://trellis.net/article/where-circular-economy-meets-internet-things/

Farooque M, Zhang A, Thürer M, Qu T Huisingh D (2019) Circular supply chain management: A definition and structured literature review. J Clean Prod, 228: 882–900. https://doi.org/10.1016/j.jclepro.2019.04.303

Fatimah YA, Govindan K, Murniningsih R, Setiawan A (2020) Industry 4.0 based sustainable circular economy approach for smart waste management system to achieve sustainable development goals: a case study of Indonesia. J Cleaner Prod 269:122263. https://doi.org/10.1016/j.jclepro.2020.122263

Federal Ministry of Education and Research (2021) Federal government report on the high-tech strategy 2025. https://www.bmbf.de/SharedDocs/Publikationen/de/bmbf/FS/657232_Bericht_zur_Hightech-Strategie_2025_en.html

Fuertes G, Zamorano J, Alfaro M, Vargas M, Sabattin J, Duran C, Ternero R, Rivera R (2022) Opportunities of the technological trends linked to industry 4.0 for achieve sustainable manufacturing objectives. Sustainability 14(18):11118. https://doi.org/10.3390/su141811118

Geissdoerfer M, Savaget P, Bocken NMP, Hultink EJ (2017) The circular economy—A new sustainability paradigm? J Clean Prod 143:757–768. https://doi.org/10.1016/j.jclepro.2016.12.048

Geissdoerfer M, Morioka SN, de Carvalho MM, Evans S (2018) Business models and supply chains for the circular economy. J Clean Prod 190:712–721. https://doi.org/10.1016/j.jclepro.2018.04.159

Ghobakhloo M (2020) Industry 4.0, digitization, and opportunities for sustainability. J Cleaner Prod 252:119869. https://doi.org/10.1016/j.jclepro.2019.119869

Govindan K, Hasanagic M (2018) A systematic review on drivers, barriers, and practices towards circular economy: a supply chain perspective. Int J Prod Res 56(1–2):278–311. https://doi.org/10.1080/00207543.2017.1402141

Haleem A, Javaid M, Singh RP, Khan S (2022) Guest editorial: Industry 4.0 special issue. Industrial Robot 49(3): 385. https://doi.org/10.1108/IR-05-2022-459

Hazen BT, Russo I, Confente, I Pellathy D (2021) Supply chain management for circular economy: Conceptual framework and research agenda. Int J Logist Manag 32(2): 510–537. https://doi.org/10.1108/IJLM-12-2019-0332

Hennemann-Hilario da Silva T, Sehnem S (2022) The circular economy and Industry 4.0: synergies and challenges. Rev Gestao 29(3): 300–313. https://doi.org/10.1108/REGE-07-2021-0121

Hossain MU, Ng ST, Antwi-Afari P, Amor B (2020) Circular economy and the construction industry: Existing trends, challenges and prospective framework for sustainable construction. Renew Sustain Energy Rev 130: 109948. https://doi.org/10.1016/j.rser.2020.109948

Inoue M, Yamada S, Miyajima S, Ishii K, Hasebe R, Aoyama K, Yamada T, Bracke S (2020) A modular design strategy considering sustainability and supplier selection. J Adv Mech Des Syst Manuf.14(2). https://doi.org/10.1299/jamdsm.2020jamdsm0023

Kang HS, Lee JY, Choi S, Kim H, Park JH, Son JY, Kim BH, Noh SD (2016) Smart manufacturing: Past research, present findings, and future directions. Int J Precis Eng Manuf Green Technol 3: 111–128. https://doi.org/10.1007/s40684-016-0015-5

Keeble BR (1988) The Brundtland report: 'Our common future'. Medicine and War 4(1): 17–25 https://doi.org/10.1080/07488008808408783

Kerdlap P, Low JSC, Ramakrishna S (2019) Zero waste manufacturing: A framework and review of technology, research, and implementation barriers for enabling a circular economy transition in Singapore. Resour Conserv Recycl 151:104438. https://doi.org/10.1016/j.resconrec.2019.104438

Kerin M, Pham DT (2020) Smart remanufacturing: a review and research framework. J Manuf Technol Manag 31(6): 1205-1235 https://doi.org/10.1108/JMTM-06-2019-0205

Kumar P, Singh RK, Kumar V (2021a) Managing supply chains for sustainable operations in the era of industry 4.0 and circular economy: Analysis of barriers. Resour Conserv Recycling 164: 105215. https://doi.org/10.1016/j.resconrec.2020.105215

Kumar S, Raut RD, Nayal K, Kraus S, Yadav VS, Narkhede BE (2021b) To identify industry 4.0 and circular economy adoption barriers in the agriculture supply chain by using ISM-ANP. J Cleaner Prod 293: 126023. https://doi.org/10.1016/j.jclepro.2021.126023

Lasi H, Fettke P, Kemper HG, Feld T, Hoffmann M (2014) Industry 4.0. Bus Inform Syst Eng 6: 239–242. https://doi.org/10.1007/s12599-014-0334-4

Lausselet C, Cherubini F, Oreggioni GD, del Alamo Serrano G, Becidan M, Hu X, Rørstad PK, Strømman AH (2017) Norwegian waste-to-energy: Climate change, circular economy and carbon capture and storage. Resour Conserv Recycl 126: 50–61. https://doi.org/10.1016/j.resconrec.2017.07.025

Lewandowski M (2016) Designing the business models for circular economy-towards the conceptual framework. Sustainability 8(1): 43. https://doi.org/10.3390/su8010043

Lopes de Sousa Jabbour AB, Jabbour CJC, Godinho Filho M, Roubaud D (2018) Industry 4.0 and the circular economy: a proposed research agenda and original roadmap for sustainable operations. Ann Oper Res 270(1–2): 273–286. https://doi.org/10.1007/s10479-018-2772-8

Lu Y (2017) Industry 4.0: A survey on technologies, applications and open research issues. J Ind Inf Integr 6: 1–10. https://doi.org/10.1016/j.jii.2017.04.005

Madhukalya A (2021) India overtakes US to become second-most sought-after manufacturing destination—BusinessToday. https://www.businesstoday.in/latest/economy/story/india-overtakes-us-to-become-second-most-sought-after-manufacturing-destination-305048-2021-08-24

Malek J, Desai TN (2019) Prioritization of sustainable manufacturing barriers using Best Worst Method. J Clean Prod 226: 589–600. https://doi.org/10.1016/j.jclepro.2019.04.056

Modgil S, Gupta S, Bhushan B (2020) Building a living economy through modern information decision support systems and UN sustainable development goals. Production Planning & Control 31(11–12): 967–987. https://doi.org/10.1080/09537287.2019.1695916

Murray A, Skene K, Haynes K (2017) The circular economy: an interdisciplinary exploration of the concept and application in a global context. J Bus Ethics 140: 369–380. https://doi.org/10.1007/s10551-015-2693-2

Nascimento DLM, Alencastro V, Quelhas OLG, Caiado RGG, Garza-Reyes JA, Rocha-Lona L, Tortorella G (2019) Exploring Industry 4.0 technologies to enable circular economy practices in a manufacturing context: A business model proposal. J Manuf Technol Manag 30(3): 607–627. https://doi.org/10.1108/JMTM-03-2018-0071

O'Brien C (1999) Sustainable production—a new paradigm for a new millennium. Int J Prod Econ 60-61: 1–7. https://doi.org/10.1016/S0925-5273(98)00126-1

Oláh J, Aburumman N, Popp J, Khan MA, Haddad H, Kitukutha N (2020) Impact of Industry 4.0 on Environmental Sustainability. Sustainability 12(11): 4764. https://doi.org/10.3390/su12114674

Ozkan-Ozen YD, Kazancoglu Y, Mangla SK (2020) Synchronized barriers for circular supply chains in industry 3.5/industry 4.0 transition for sustainable resource management. Resour Conserv Recycling 161: 14986. https://doi.org/10.1016/j.resconrec.2020.104986

Patyal VS, Sarma PRS, Modgil S, Nag T, Dennehy D (2022) Mapping the links between Industry 4.0, circular economy and sustainability: a systematic literature review. J Enterpr Inform Manag 35(1): 1–35. https://doi.org/10.1108/JEIM-05-2021-0197

Pearce DW, Turner RK (1989) Economics of natural resources and the environment. Johns Hopkins University Press, London

Pereira AC, Romero F (2017) A review of the meanings and the implications of the industry 4.0 concept. Proc Manuf 13: 1206–1214. https://doi.org/10.1016/j.promfg.2017.09.032

Phung CG (2019) Implications of the circular economy and digital transition on skills and green jobs in the plastics industry. Field Actions Sci Rep (Special Issue 19):100-107

Prisecaru P (2017) The challenges of the industry 4.0. Global Econ Obser 5(1): 66–72. http://www.globeco.ro/wp-content/uploads/vol/GEO_Vol_5_No_1.pdf

Rejeb A, Appolloni A (2022) The Nexus of Industry 4.0 and Circular Procurement: A Systematic Literature Review and Research Agenda. Sustainability 14(23): 15633. https://doi.org/10.3390/su142315633

Rejeb A, Suhaiza Z, Rejeb K, Seuring S, Treiblmaier H (2022) The Internet of Things and the circular economy: A systematic literature review and research agenda. J Cleaner Prod 350: 131439. https://doi.org/10.1016/j.jclepro.2022.131439

Rojko A (2017) Industry 4.0 Concept: Background and Overview. Int J Interact Mobile Technol 11(5): 77

Rosa P, Sassanelli C, Urbinati A, Chiaroni D, Terzi S (2020) Assessing relations between Circular Economy and Industry 4.0: a systematic literature review. Int J Product Res 58(6): 1662–1687. https://doi.org/10.1080/00207543.2019.1680896

Rosário AT, Dias JC (2022) Sustainability and the Digital Transition: A Literature Review. Sustainability 14(7): 4072. https://doi.org/10.3390/su14074072

Schöggl JP, Stumpf L, Baumgartner RJ (2020) The narrative of sustainability and circular economy—A longitudinal review of two decades of research. Resour Conserv Recycl 163: 105073. https://doi.org/10.1016/j.resconrec.2020.105073

Schroeder P, Anggraeni K, Weber U (2019) The Relevance of Circular Economy Practices to the Sustainable Development Goals. J Ind Ecol 23(1): 77–95. https://doi.org/10.1111/jiec.12732

Sharma YK, Mangla SK, Patil PP, Liu S (2019) When challenges impede the process: For circular economy-driven sustainability practices in food supply chain. Manag Decis 57(4): 995–1017. https://doi.org/10.1108/MD-09-2018-1056

Shrouf F, Ordieres J, Miragliotta G (2014) Smart factories in Industry 4.0: A review of the concept and of energy management approached in production based on the Internet of Things paradigm. In IEEE International Conference on Industrial Engineering and Engineering Management, pp 697–701. https://doi.org/10.1109/IEEM.2014.7058728

Srivastava R (2019) Impact of "Make in India" in Indian Economy. Int J Trend Sci Res Develop 3(4): 429–432

Stock T, Seliger G (2016) Opportunities of Sustainable Manufacturing in Industry 4.0. Proc CIRP 40: 536–541. https://doi.org/10.1016/j.procir.2016.01.129

Su B, Heshmati A, Geng Y, Yu X (2013) A review of the circular economy in China: moving from rhetoric to implementation. J Clean Prod 42: 215–227. https://doi.org/10.1016/j.jclepro.2012.11.020

Tavares TM, Ganga GMD, Filho MG, Rodrigues VP (2023) The benefits and barriers of additive manufacturing for circular economy: A framework proposal. Sustain Prod Consumpt 37: 369–388. https://doi.org/10.1016/j.spc.2023.03.006

Tukker A, Tischner U (2006) Product-services as a research field: past, present and future. Reflections from a decade of research. J Clean Prod 14(17): 1552–1556. https://doi.org/10.1016/j.jclepro.2006.01.022

Vasantha GVA, Roy R, Corney JR (2015) Advances in designing product-service systems. J Indian Inst Sci 95(4): 429–447

Viles E, Kalemkerian F, Garza-Reyes JA, Antony J, Santos J (2022) Theorizing the Principles of Sustainable Production in the context of Circular Economy and Industry 4.0. Sustain Prod Consumpt 33: 1043–1058

Vogel-Heuser B, Hess D (2016) Guest editorial industry 4.0-prerequisites and visions. IEEE Trans Autom Sci Eng 13(2): 411–413. https://doi.org/10.1109/TASE.2016.2523639

Weking J, Stöcker M, Kowalkiewicz M, Böhm M, Krcmar H (2020) Leveraging industry 4.0—A business model pattern framework. Int J Prod Econ. 225: 107558 https://doi.org/10.1016/j.ijpe.2019.107588

Yang M, Chen L, Wang J, Msigwa G, Osman AI, Fawzy S, Rooney DW, Yap PS (2023) Circular economy strategies for combating climate change and other environmental issues. Environ Chem Lett 21(1): 55–80. https://doi.org/10.1007/s10311-022-01499-6

# RETRACTED CHAPTER: Responsible Digital Technologies: Shaping Sustainable Practices

**Rachna Rana, Pankaj Bhambri, and Sandeep Kautish**

**Abstract** Responsible digital change may be characterized as the cognitive process of group action digital technology into a commercial enterprise in an philosophy, property, and humble of human-like belief and social group. In the dynamic field of scientificmodification, correspond to philosophy and sustainable code of conduct is crucial. The importance of responsible digital technology is highlighted in this chapter, which stresses the need for collaboration between academia, NGOs, CSOs and individuals to develop innovative business models that generate synergies while dealing with social problems. The essay emphasizes the advent of corporate digital responsibility (CDR) and the change from industry 4.0 to industry 5.0, with an emphasis on partnerships between humans and artificial intelligence (AI) and human-centric methods. The need for multidisciplinary studies and methodical strategies which take into account all aspects of sustainability is also stressed. By desegregation property ICT concepts into digital modification activeness, organization can change to a more sustainable and accountable digital forthcoming.

**Keywords** Digitization · Sustainability · Technological change · Responsibility · Sustainable practices



---

The original version of this chapter was retracted: The retraction note to this chapter is available at https://doi.org/10.1007/978-3-031-68427-2_24

R. Rana
Ludhiana Group of Colleges, Ludhiana, Punjab, India

P. Bhambri (✉)
Guru Nanak Dev Engineering College, Ludhiana, Punjab, India
e-mail: pkbhambri@gndec.ac.in

S. Kautish
Chandigarh University, Mohali, Punjab, India



W. Leal Filho et al. (eds.), *Digital Technologies to Implement the UN Sustainable Development Goals*, World Sustainability Series,
https://doi.org/10.1007/978-3-031-68427-2_20

# 1 Introduction

Academics and practitioners alike continue to be very interested in the digital shift while creating environmentally conscious cultures is a primary goal for all the numerous stakeholders involved. Help researching emerging technologies and their uses can result in the (Mikalef et al. 2020) development of new or updated theoretical tool-kits, techniques, and philosophical perspectives (Kraus et al. 2022) that can be applied to the design and development of better solutions that have social and commercial value for the community (Pappas et al. 2018).

Collaborative efforts between academics, businesses of all sizes, social organizations, and the public are vital to exploit this potential and effect commercial and societal transformation fully. Businesses need to take advantage of modern models that not only increase efficiency and accuracy, but also tackle social problems with a view to creating mutual value for all interested parties. The fact that value can be produced in a variety of ways and measured with several measurements needs to be established. For all stakeholders involved in digital transformation, the creation of shared or collective value can result in greater benefits.

Digital transitions ought to be planned responsibly, benefiting every individual involved, to accomplish more general goals beyond those that are economically connected. It will take at least a few years to assess the impact of Covid-19, which witnessed an increase in digital modificationhuman action as a bodily function to the pandemic's problems, on businesses and society. There is general agreement that responsible technology design, development, and application are necessary, but there is still much to learn about the steps needed to make this happen. Unrealadministrative unit (UAU) is now the main topic of IS literature.

This includes responsible artificial intelligence (RAI) (Shneiderman 2021; Mikalef et al. 2022; Dennehy et al. 2022), explanatory artificial intelligence (XAI) (Barredo Arrieta et al. 2020), human-centric artificial intelligence (Shneiderman 2020) and pervasive artificial intelligence (Nouri 2021). Vassilakopoulou et al. (2022) highlight and define three main streams of RAI research, including electronic computerdiscipline, human-likedata processor interaction, and (Meske et al. 2022) humanities and motivation. Although there are certain commonalities depending on the disciplines, different parts of artificial intelligence are covered.

However, the phrase "responsible digital transformation" is not well-defined and is not as widely used as it might be (Zhang and Hon 2020). However, given how digital transformation has been impacted by developing technologies, particularly AI, and vice versa, various RAI-related tomessages would be relevant for answerable digital modification. Responsible digital modification may be undefined as the cognitive process of desegregation digital technology into a business organization in an philosophy, property, and social group responsible manner. It entails weighing the possible effects of disciplineencouragement on people and world organization and taking action to lessen any unfavorable effects.

While investigation on digital modification has increased in the last few decades, the word "responsible technological change" is rarely utilized in research journals.

A ResearchScholarly personinvestigationoutputs about 100 hits, of which fewer than one-half are slender related. A comparable Database search yielded just ten hits, every one of which has been released beyond 2020, in journals, papers from conferences, or book chapters. The term "responsible digitization," which is also receiving considerable attention, has findings similar to these. Concurrently, the World Economic Forum (WEF) implemented many programs that emphasized conscientious digital modification, as a part of their endeavors to mold the mechanical phenomenon of the digital efficiency and social group (O'Halloran and Griffin 2019).

Five new digital improvement—cyber repercussion, informationisolation, the Internet of Things (IoT), block chain, and unrealintelligence information (UII)—are discussed together with a methodology for analyzing their effects. Ample acquiringbusiness firm like Accenture and Deloitte have also been pushing responsible digital transformation, either to increase net financial gain in the commercial enterpriseemploymentbusiness enterprise (O'Reilly 2021) or, more broadly, to achieve social impact (Accenture 2019).

Nowadays, companies and groups are creating their corporate digital responsibilities (CDRs), which are defined as common standards (Wirtz et al. 2022) and values of the development and use of extremityapplication and collection (Lobschat et al. 2021). These CDRs go beyond traditional CSR strategies. Simultaneously, the European Commission aims to lead in the information business with its data strategy, utilizing the growing quantity of data for both commercial and societal benefits (European Commission 2020).

Organizations and businesses are currently (Wirtz et al. 2022) working on corporate digital accountability (CDA), which refers to jointbelief and standards for the creative activity and procedure of digital practical application and data (Lobschat et al. 2021). Simultaneously, the European Commission aims to lead the information business with its data strategy, utilizing the growing quantity of data for both commercial and societal benefits (European Commission 2020). Makingdividedeconomic value that may goodnessbusiness organization, organizations, customers, and (Pappas et al. 2018) society in general is required to allow answerable digital arrangement, needed to cope with the massive amounts of consumer information as well as the intrinsic ubiquitous.

Nevertheless, considering how futurepractical application, particularlyunrealadministrative body, have influenced digital modification and contrariwise, variouscomponentsincidental to RAI would be essential for a answerable digital modification. Responsible digital modification can be characterized as the cognitive process of desegregation digital engineering into business concerntransaction in the philosophy, property and ethnic responsible personal manner.

This requires considering the possible effects of scientificimprovement on people and people and fetching measures to reduce possible negative effects. Although the study of digital transformation has increased in recent decades, the word "responsible technological change" It is rarely used in scientific journals. A Google Scholar search returns about 100 hits, less than half of which are related in any way.

A comparable database search yielded only ten hits in journals, conference papers or book chapters, each published after 2020. The term "responsible digitization" which also receives a lot of attention, has similar conclusions. At the aforementionedperiod, the WEF has enforced many programs accentingself-aware digital modification as part of its exertion to configuration the mechanical phenomenon of the digital system and social group (O'Halloran and Griffin 2019).

5 new digital evolution—artificial intelligence (AI), cyber repercussion, block chain, informationsecurity, the Internet of Things (IoT), and—are discussed, along with methods to analyze their impact. Major consulting firms such as Accenture and Deloitte have also called for responsible digital transformation, either to increase net income in the financial services sector or more generally to achieve social impact (Accenture 2019). Today companies and groups create corporate digital commitments (CDR), defined as common standards and values for the development and use of digital technologies and data. These CDRs go beyond traditional CSR strategies. At the same time, with its information strategy, the European Commission aims to be at the forefront of the information business, (European Commission 2020) using the growing amount of data for commercial and social benefits. Organizations and companies are currently working on corporate digital responsibility (CDA), which mentions to mutualbelief and regulations in the conception and utilization of digital practical application and aggregation of information (Lobschat et al. 2021; Wirtz et al. 2022).

At the aforementionedperiod, the purpose of the European Commission is to getsupervising the contentsbusiness organizationthrough with its informationscheme, using the increasing amount of information for mercantile and social group benefits (European Commission 2020). Creating shared value that benefits businesses, organizations, customers and social group is indispensable to sanction responsible digital manufacture, which are needful to deal with the massive amount of customer data and its inherent ubiquity (Pappas et al. 2018).

For example, data markets are places where companies can buy and sell data allowing to create value by enabling the conversation of information between various parties, directing to new vision and business organisation perspectives. This means that companies must ensure the responsible use of data, which includes, among other things, protecting consumer privacy, maintaining transparency and avoiding bias or discrimination. The transition from automation and digitization from Industry 4.0 to Industry 5.0 is happening as a result of the rapid development of digital technology.

Industry 5.0 creates on Industry 4.0 by choosing a more human-centric conceptualization and trying to pool the rewards of modern digital engineering with the ability and flexible-determinationaccomplishment of manlikeperson (Vassilakopoulou et al. 2023). Engineering (AI, IoT, robotics) can assistpersonjoin forces, introduce and create new methods to convoluted challenges, directing to anthropomorphic-AI business organization and invention in activity and companyhigh technology (Faraj et al. 2018).

This raises the question of how to train for responsible digital modification (Dennehy et al. 2022). The former is critical, (Mikalef et al. 2020) considering the accelerator essential of scholarly person for contentengineering in the last 10 years (Blumberg et al. 2023). The currentbrainstorm in Open-AI, Chatting and Google's Decorate are fabulous representative of how productive AI can modify (Grøder et al.

2022) such business organization, modificationskillfulness, lead to labor migration and create new and diverse jobs (Hatzius et al. 2023). Therefore, deeper research on the effects of productive AI on cognitionacquiring and digital modification (Dwivedi et al. 2023a; b) is required to healthierrealize how they can alter responsible digital modification and (Pappas et al. 2017) modify to propertysocial group.

## 2 Regarding How Sustainability and Digitization Interact

(Brenner and Hartl 2021), sustainability and digitization are fundamental economic and social phenomena that (Del Río Castro et al. 2021) require significant changes. The label and digital technology and "sustainability" appear to be unrelated concepts (George et al. 2020). According to authors such as (Gebhardt 2017), they cause a modeldisplacement in social group and environment scheme because they contendthoughts. However, (Osburg 2017) points out that these concepts are game changers and importantdesperate that can lead to bettermodifications.

Digital sustainability is the mode in which conversion, as a fundamentalcomponent part of the fourth industrial revolution, can achieve world sustainability cognitive contents. Similarly, (George et al. 2020) characterize digital sustainability as "collaboration to advance sustainability goals through innovations that produce, use, transmit or generate digital information innovative use. And quot; People often believe that sustainability is in conflict with other goals. or goals. In this situation, 'digital sustainability' can combine the two main goals of ecological protection and technological change to create good social and environmental improvements, rather than simply minimizing them.

## 3 Transformation Digitally

DX creates a new opportunity for MSMEs by requiring substantial changes in business processes, operational routines, and organizational skills. Strategic and tactical elements, such as entering new markets or leaving current ones, are all part of its methodical and ongoing progress.

DX's effect extends beyond individual organizations; it can affect the entirety of the corporate world and society as a whole through the development of new internet-based technology. As a result, DX can transform existing sectors and fundamentally alter how value is produced and distributed to consumers.

DX is an biological process that utilization digital practical application and capacity to green grocerieseconomic value by transforming business organisation models, operational operations, and consumer experiences.

By combining digital technology and novel approaches to business, organizations can undergo significant changes in how they operate and can offer clients value-added services. Businesses may use a variety of technology, such as mobile phones, social networking sites, and huge amounts of data. By making use of these technologies, MSMEs may significantly enhance their operations, boost customer satisfaction, and develop fresh business ideas.

The executive level chooses DX as a strategic move to boost organizational competitiveness, maximize resource utilization, and streamline operational procedures. Effective information management, lower operational expenses, and real-time data analysis have made DX more and more important in today's highly competitive and economically demanding world. Consequently, entities can attain more adaptability, reduced expenses, and enhanced ability to make decisions.

The introduction of digital technology at various stages in an organization requires a DX strategy. The plan should be centered on transforming the company mindset. Under this new paradigm, managers must grasp their context, including situational possibilities and limits that may affect economic development, working conditions, and production processes. We may separate the setting into internal and external components. The former requires particular attention to expenses and attempts at altering the company philosophy, while the second needs improvements to the electoral, environmental, and economic spheres.

## 4 Digital Transformation for Sustainability

Advances in digital technology such as robots, 3D printing, the Internet of Things and process automation have changed the way multinational companies operate and helped them succeed in global markets. Sustainability is a key component of DX today, with a significant impact on many different industries, including agriculture. Agricultural companies understand the value of implementing effective waste management, ecological production methods and pollution control. As a result, this revolutionary change will significantly affect the management, efficiency and strategic management of SMEs as they strive for environmental sustainability.

The concept of sustainable DX must provide both internal and external advantages for the firm. It should encourage more environmental, economic, human, and social responsibility in corporate activities. Sustainable ICT-based processes may assist businesses with creating solutions that make a meaningful impact in their reduction, reuse, and recycling strategies while safeguarding corporate margins and expenses. It may also be used by businesses to reinvent business procedures, provide sustainability advantages to the whole organization, and enhance employee well-being.



## 5 Capacity for Technology Absorption

Achieving long-term digital transformation in SMEs depends on absorption. An organization's ability to find, acquire and use external technology is called its absorptive capacity. This speaks to their ability to acquire and implement external technology.

Several studies have shown the importance of absorptive capacity in improving the performance of SMEs and fostering innovation. It creates a framework for technical learning in companies, enabling them to identify and use information from the outside world.

For small and medium-sized in particular, the impact of corporate culture on technology is significant. Organizational culture influences how employees interact with one another and carry out their tasks, which can help or impede organizational learning and transformation. Furthermore, MSMEs' relationships and contacts with outside groups are essential for enhancing their ability to absorb information and performance.



## 6 Subject Matter of the Special Issue

The main objective of this special issue is to provide theoretical discussion and empirical support for a better understanding of the digital transformation of a sustainable society in the twenty-first century and to define an agenda for further research. Despite the large number of papers, not all were suitable for the special issue due to the paper request. Fifteen works were finally accepted after two or three rounds of evaluation. Only nine of the fifteen journals considered by the guest editors were selected for this special issue of Advances in Technology due to space limitations. The regular publication Information Systems Frontiers has the rest. The articles shown here cover many perspectives on how many new technologies (AI, ML, big data and business analytics), their uses and every party involved in the process can open the doors to effective digital transformation and sustainable societies. The articles examine different aspects of the model of digital transformation and sustainable development (DTS) (data actors, capability, value creation, business and social changes) (Pappas et al. 2018), use different analytical methods (mixed methods, quantitative, qualitative).).)., fsQCA) and promote digital transformation for a sustainable society in the twenty-first century. The digital divide, or the unequal distribution of access to information and communication resources, including the Internet, computers and smartphones, is the subject of study #039 by Vassilakopoulou and Hustadand in 2023. Addressing the digital divide requires more than simply provide access to these tools, the authors argue. They argue that non-digital inequality, based on socioeconomic resources such as money and educational attainment, correlates with digital inequality. The authors emphasize the importance of closing digital gaps in building sustainable digital societies. The study reviews previous research on the digital divide in economically developed countries and states. The authors outline the reasons for the gap and propose solutions. Based on the findings, the authors propose a research agenda that includes the development of new models to better understand the digital divide, a deeper examination of the effects of interventions designed to bridge the digital divide, and strengthening the link between digital divide research and sustainable development. Research Together, the study highlights the importance of understanding and mitigating digital divides, as they can have profound implications for both socioeconomic equality and the development of sustainable digital communities.

## 7 Responsible Artificial Intelligence and Its Application in the Digital Age

Explore the social, environmental and ethical implications of AI-driven digital revolutions. Research can focus on creating and improving AI accountability frameworks, identifying best practices and evaluating their impact on different stakeholders, different industries and society as a whole.

1. How can digital change processes based on artificial intelligence be ethical, socially responsible and environmentally sustainable?
2. What best practices can be used to implement responsible AI across sectors and industries?
3. What methods are used to assess and quantify the impact of responsible AI on stakeholders and the public?
4. What are the main barriers and obstacles to using appropriate artificial intelligence in digital transformation projects?
5. How could interdisciplinary collaboration contribute to the creation and effective use of responsible AI frameworks?

Human-centered Approaches to Industry 5.0: Explore the implications of combining modern digital technology with human creativity and problem-solving skills. Research can explore how its human-centered methodology affects teamwork, creativity and the creation of new solutions to complex problems, ultimately influencing the conscious adoption of digital transformation.

1. How can companies effectively combine people's creativity and problem solving with cutting-edge digital technology in Industry 5.0?
2. What factors influence the successful collaboration of people and artificial intelligence in the Industry 5.0 environment?
3. How can human-centered approaches to Industry 5.0 affect creativity, productivity and employee satisfaction?
4. What are the possible threats and difficulties associated with the implementation of Industry 5.0 human-centered strategies and how to solve them?
5. How can education and training programs adapt to equip workers for Industry 5.0 human-centric digital transformation?

Corporate Digital Responsibility (CDR): Explore the development, implementation and impact of CDR strategies in companies with an emphasis on shared values and standards in the production and use of digital technologies and data. The study can explore how companies can incorporate social, ethical and environmental concerns into their digital transformation strategies while meeting their financial goals.

1. How can companies create and implement CDR plans that successfully combine financial goals with moral, social and environmental considerations?
2. What are the main obstacles that companies face in implementing CDR and what are the critical factors driving it?
3. How can CDR tactics affect corporate competitiveness, stakeholder satisfaction, and corporate performance?
4. How should organizations evaluate and disseminate the results of their CDR activities to stakeholders and the public?

Data Protection, Security and Responsible Data Markets: Discover Challenges and Practical Solutions to Ensure Fair Use of Data in the Age of Data Markets. Data protection guidelines, legal frameworks and technological advances that ensure privacy, transparency and avoidance of bias or discrimination in data-based decision-making can be the subject of research.

1. What are the most important problems of data protection and data storage companies in information markets and how to solve them?
2. How can legislative frameworks and industry standards promote secure communication in markets?
3. What are the best ways to promote privacy, transparency and avoidance of bias in data-driven decision-making processes?
4. How can cutting-edge technology like blockchain support secure, open and accountable data markets?

Education for Responsible Digital Transformation: Explore how institutions can incorporate responsible digital transformation ideas into their curricula and training programs. Research can look at creating new pedagogies, interdisciplinary approaches and creative teaching techniques that equip students with the skills and knowledge they need to contribute to a sustainable society in the digital age.

1. What are the best ways for educational institutions to integrate the idea of responsible digital transformation into their educational programs and curricula?
2. Which multidisciplinary methodologies, pedagogies and teaching strategies are most effective in preparing future managers and professionals for responsible digital transformation?
3. How can teachers assess whether responsible digital change programs are effective in education?
4. What are the main obstacles and problems that educational institutions face when introducing responsible education on digital transformation and how to solve them?.

## 8 Challenges Facing in Shaping Sustainable Practices

The problems we confront necessitate inventive solutions and a shared commitment to a more sustainable future. A few major concepts will influence the environmental agenda in 2024. Let's look at these disruptive ideas, which will surely have an impact on our environmental and social duties.

1. **Circular Economy 2.0:** Closing the Loop. Although the idea of a circular economy is not new, there has been substantial progress as of 2024. In addition to encouraging recycling and decreasing waste, Circular Economy 2.0 highlights a more comprehensive strategy. Businesses are embracing regenerative methods more and more, emphasizing the development of goods with end-of-life concerns and the use of creative reuse models. This change reflects a dedication to developing closed-loop systems with the least amount of negative environmental effect.
2. **Natural Approaches to Climate Resilience:** Solutions rooted in the natural world are becoming more and more popular as effective weapons in the battle against climate change. We predict an increase in afforestation, reforestation, and sustainable land management initiatives in 2024 that take advantage of nature's resources. These programs safeguard ecosystems, improve biodiversity, create climate-resilient communities, and trap carbon. A progressive approach to sustainability that respects the planet's innate resilience is to embrace nature-based solutions.
3. **Technology for Good:** Leveraging Innovation for Sustainability The convergence of technology and sustainability is becoming more and more important. We predict that technical advancements targeted at resolving environmental issues will soar in 2024. From artificial intelligence maximizing energy efficiency to blockchain assuring supply chain transparency, technology will play a critical role in promoting sustainable practices across businesses. Embracing and investing in technological solutions may open up new options and redefine our approach to global concerns.
4. **Sustainable Inclusivity:** Closing Social Divides Sustainability involves more than simply preserving the environment; it also involves promoting social justice. In 2024, there is a greater emphasis on inclusive sustainability, which acknowledges the connection of social and environmental challenges. Businesses and groups are implementing initiatives that support diversity, fair labor standards, and community involvement.
5. **Conscientious Spending:** The Emergence of Conscientious Buying With growing consumer knowledge, we predict a move to more conscientious and responsible consumption by 2024. With knowledge at their fingertips, customers are increasingly making decisions that reflect their values. Businesses that value transparency, ethical sourcing, and sustainable processes will not only exceed customer expectations, but will also help to drive a cultural change toward more sustainable living. Customers are now more equipped to take an active role in the sustainability movement thanks to this trend.

## 9 Conclusion and Future Scope

Finally, ethical digital transformation for sustainable societies is critical, since it can change our world for the better. But it's impossible to overestimate how complicated it is since it entails balancing the complex interactions of environmental concerns, ethical issues, social ramifications, and technology improvements. We can address these challenges and use digital technology to create a more fair, sustainable, and inclusive future for all by adopting an interdisciplinary and holistic approach. Creating inclusive solutions that are accessible to everyone is necessary for the latter. This begins with offering access to a varied collection of users, followed by assuring their active engagement, which will finally lead to empowered success.

The use of information systems is a crucial component in the shift towards a sustainable society. This special issue aims to present cutting-edge research that will expand our knowledge on how to build and maintain such a society. The research included will be theoretically sound and will comprehensively examine the effects of information systems. To support this research, several topics have been identified that require immediate and special attention from both practitioners and researchers. These topics will be accompanied by thought-provoking research questions to aid in the development of sustainable solutions.

**Sustainable ICT:** Understanding how people and ICT interact synergistically and harmoniously to create long-term digital ecosystems that address environmental, social and economic challenges. Research into the creation, use and evaluation of digital solutions that promote sustainability, responsible digital transformation and long-term social impact requires greater attention.

1. How to use multidisciplinary approaches to identify and address environmental, social, economic and technological aspects of sustainable ICT and responsible digital transformation?
2. What are the best practices and critical success factors for creating and implementing human-AI collaboration models that promote sustainable digital ecosystems and ethical digital transformation initiatives?
3. How can companies create a comprehensive framework that considers the life cycle of ICT products and services when assessing the viability of digital transformation initiatives?
4. How could inclusive stakeholder participation and co-creation lead to new solutions for sustainable ICT and responsible digital transformation?
5. How can capacity building and training promote responsible digital transformation, sustainable ICT, multidisciplinary collaboration, critical thinking and problem-solving skills?

## References




Abad-Segura E, González-Zamar M (2021) Sustainable economic development in higher education institutions: a global analysis within the SDGs framework. J Clean Prod 294:126133. https://doi.org/10.1016/j.jclepro.2021.126133

Accenture Names Kanioura Chief Analytics Officer (2019) News Group. https://doi.org/10.1287/lytx.2019.02.18n

Barredo Arrieta A, Díaz-Rodríguez N, Del Ser J, Bennetot A, Tabik S, Barbado A, Garcia S, Gil-Lopez S, Molina D, Benjamins R, Chatila R, Herrera F (2020) Explainable artificial intelligence (XAI): concepts, taxonomies, opportunities and challenges toward responsible AI. Inform Fus 58:82–115. https://doi.org/10.1016/j.inffus.2019.12.012

Blumberg S, Krawina M, Mäkelä E, Soller H (2023) Women in tech: The best bet to solve Europe's talent shortage. McKinsey & Company. https://www.mckinsey.com/capabilities/mckinsey-digital/our-insights/women-in-tech-the-best-bet-to-solve-europes-talent-shortage/

Brenner B, Hartl B (2021) The perceived relationship between digitalization and ecological, economic, and social sustainability. J Clean Prod 315:128128. https://doi.org/10.1016/j.jclepro.2021.128128

Carnerud D, Mårtensson A, Ahlin K, Slumpi TP (2020) On the inclusion of sustainability and digitalisation in quality management—an overview from past to present. Total Qual Manag Bus Excell. https://doi.org/10.1080/14783363.2020.1848422

Del Río Castro G, González Fernández MC, Uruburu Colsa A (2021) Unleashing the convergence amid digitalization and sustainability towards pursuing the sustainable development goals (SDGs): a holistic review. J Clean Prod 280:122204. https://doi.org/10.1016/j.jclepro.2020.122204

Dennehy D, Griva A, Pouloudi N, Dwivedi YK, Mäntymäki M, Pappas IO (2022) Artificial intelligence (AI) and information systems: perspectives to responsible AI. Inf Syst Front 25(1):1–7. https://doi.org/10.1007/s10796-022-10365

Dwivedi YK, Ismagilova E, Rana NP, Raman R (2023a) Social media adoption, usage and impact in business-to-business (B2B) context: a state-of-the-art literature review. Inform Syst Front. https://doi.org/10.1007/s10796-021-10106-y

Dwivedi YK, Kshetri N, Hughes L, Slade EL, Jeyaraj A, Kar AK, Baabdullah AM, Koohang A, Raghavan V, Ahuja M (2023b) So what if ChatGPT wrote it?" Multidisciplinary perspectives on opportunities, challenges and implications of generative conversational AI for research, practice and policy. Int J Inf Manag 71:102642

European Commission (2020) A European Strategy for data. Retrieved March 2023 from https://digital-strategy.ec.europa.eu/en/policies/strategy-data

Faraj S, Pachidi S, Sayegh K (2018) Working and organizing in the age of the learning algorithm. Inf Organ 28(1):62–70

Gebhardt C (2017) Humans in the loop: the clash of concepts in digital sustainability in smart cities. CSR Sustain Ethics Govern. https://doi.org/10.1007/978-3-319-54603-2_7

George G, Merrill RK, Schillebeeckx SJ (2020) Digital sustainability and entrepreneurship: how digital innovations are helping tackle climate change and sustainable development. Entrep Theory Pract 45(5):999–1027. https://doi.org/10.1177/1042258719899425

Grøder CH, Schmager S, Parmiggiani E, Vasilakopoulou P, Pappas I, Papavlasopoulou S (2022) Educating about responsible AI in IS. Designing a course based on Experiential Learning

Hatzius J, Briggs J, Kodnani D, Pierdomenico G (2023) The potentially large effects of Artificial intelligence on economic growth. Goldman Sachs, London

Kraus S, Durst S, Ferreira JJ, Veiga P, Kailer N, Weinmann A (2022) Digital transformation in business and management research: an overview of the current status quo. Int J Inf Manag 63:102466

Liu J, Ji L, Sun Y, Chiu Y, Zhao H (2024) Unleashing the convergence between SDG 9 and SDG 8 towards pursuing SDGs: Evidence from two urban agglomerations in China during the 13th five-year plan. J Clean Prod 434:139924. https://doi.org/10.1016/j.jclepro.2023.139924

Lobschat L, Mueller B, Eggers F, Brandimarte L, Diefenbach S, Kroschke M, Wirtz J (2021) Corporate digital responsibility. J Bus Res 122:875–888

Matsa M (2020) Climate change and lessons from world Indigenous minority communities. Sustain Develop Goals Ser. https://doi.org/10.1007/978-3-030-51346-7_2

Meske C, Bunde E, Schneider J, Gersch M (2022) Explainable artificial intelligence: objectives, stakeholders, and future research opportunities. Inf Syst Manag 39(1):53–63

Mikalef P, Pappas IO, Krogstie J, Pavlou PA (2020) Big data and business analytics: a research agenda for realizing business value. Inform Manag 57(1):1

Mikalef P, Conboy K, Lundström JE, Popovič A (2022) Thinking responsibly about responsible AI and 'the dark side' of AI, vol 31. Taylor & Francis, London, pp 257–268

Nouri S (2021) Diversity And Inclusion In AI. Forbes Technology Council. http://www.forbes.com/sites/forbestechcouncil/2021/03/16/diversity-and-inclusion-in-ai/?sh=2404e2db5823

O'Halloran D, Griffin W (2019) Our shared digital future–responsible digital transformation. World Econ. Forum

O'Reilly M (2021) Accelerating digital transformation, responsibly. Deloitte Insights. https://www2.deloitte.com/uk/en/insights/industry/financial-services/responsible-digital-transformation-in-financial-services.html

Osburg T (2017) Sustainability in a digital world needs trust. CSR Sustain Ethics Govern. https://doi.org/10.1007/978-3-319-54603-2_1

Pappas IO, Giannakos MN, Jaccheri L, Sampson DG (2017) Assessing student behavior in computer science education with an fsQCA approach: the role of gains and barriers. ACM Trans Comput Educ (TOCE) 17(2):10

Pappas IO, Mikalef P, Giannakos MN, Krogstie J, Lekakos G (2018) Big data and business analytics ecosystems: paving the way towards digital transformation and sustainable societies. IseB 16(3):479–491. https://doi.org/10.1007/s10257-018-0377-z

Pappas IO, Mikalef P, Dwivedi YK, Jaccheri L, Krogstie J (2023) Responsible digital transformation for a sustainable society. Inf Syst Front 25(3):945–953. https://doi.org/10.1007/s10796-023-10406-5

Shneiderman B (2020) Human-centered artificial intelligence: reliable, safe & trustworthy. Int J Hum Comput Interact 36(6):495–504

Shneiderman B (2021) Responsible AI: bridging from ethics to practice. Commun ACM 64(8):32–35

Vassilakopoulou P, Parmiggiani E, Shollo A, Grisot M (2022) Responsible AI: concepts, critical perspectives and an information systems research agenda. Scand J Inf Syst 34(2):3

Velasco Herrejón P, Bauwens T (2023). Are energy transitions reproducing inequalities? Power, social stigma and distributive (In)Justice in Mexico. https://doi.org/10.2139/ssrn.4407968

Velasco-Herrejón P, Bauwens T, Calisto Friant M (2022) Challenging dominant sustainability worldviews on the energy transition: lessons from Indigenous communities in Mexico and a plea for pluriversal technologies. World Dev 150:105725. https://doi.org/10.1016/j.worlddev.2021.105725

Wirtz J, Kunz WH, Hartley N, Tarbit J (2022) Corporate digital responsibility in service firms and their ecosystems. J Serv Res 2022:10946705221130468

Zhang J, Hon HW (2020) Towards responsible digital transformation. Calif Manag Rev 62(3):1

# Technological Advancements in Promoting Ecosystem Health

**Pankaj Bhambri and Sandeep Kautish**

**Abstract** This chapter explores the significant impact of digital technology in addressing and supporting the health of ecosystems, as outlined by the UN Sustainable Development Goals (SDGs). In the face of imminent environmental difficulties, technology advancements arise as potent allies in promoting sustainable behaviors and alleviating the consequences of climate change. Through case studies and examples, it highlights how these advancements enable precise data collection, analysis, and informed decision-making, facilitating the preservation and restoration of biodiversity, water quality, and overall ecological balance. By harnessing the potential of digital tools, this chapter underscores the pivotal role technology plays in achieving SDGs related to life on land and below water, offering a roadmap for a harmonious coexistence between technological progress and environmental stewardship. Furthermore, the discussion delves into the ethical considerations and challenges associated with the widespread adoption of digital technologies for ecosystem health. The chapter emphasizes the need for responsible innovation, inclusive decision-making processes, and global cooperation to ensure that technological advancements contribute positively to the long-term well-being of our planet. It explores the potential risks, such as data privacy concerns and unintended environmental consequences, and proposes strategies to navigate these challenges. By critically examining the intersection of technology and ecology, this chapter aims to guide policymakers, researchers, and practitioners in leveraging digital tools effectively to attain the SDGs, fostering a balanced and sustainable relationship between technology and the natural world.

**Keywords** Ecosystem monitoring · Digital innovation · Environmental sustainability · Biodiversity preservation

---

P. Bhambri (✉)
Guru Nanak Dev Engineering College, Ludhiana, Punjab, India
e-mail: pkbhambri@gndec.ac.in

S. Kautish
Chandigarh University, Mohali, Punjab, India



W. Leal Filho et al. (eds.), *Digital Technologies to Implement the UN Sustainable Development Goals*, World Sustainability Series,
https://doi.org/10.1007/978-3-031-68427-2_21

# 1 Introduction

The significance of technological advancements in promoting ecosystem health is paramount in addressing the pressing environmental challenges of our time (Carrington 2019). As the world grapples with issues such as climate change, biodiversity loss, and pollution, innovative technologies offer a unique and powerful arsenal to monitor, manage, and restore ecosystems (IPCC 2019). Through the integration of digital tools, we gain unprecedented insights into the intricate dynamics of ecosystems, enabling us to assess their health in real-time and make informed decisions for sustainable management.

One crucial aspect of the significance lies in the precision and scalability that technology brings to ecosystem monitoring and conservation efforts (Mora and Sale 2011). Remote sensing technologies, for instance, allow us to observe vast areas with a level of detail and accuracy that was once unimaginable. This capability is instrumental in tracking changes in land cover, identifying endangered species habitats, and detecting early signs of environmental degradation. Furthermore, the advent of the Internet of Things (IoT) facilitates the creation of interconnected networks of sensors, providing continuous data streams for comprehensive ecosystem assessments (Riggio et al. 2013). These developments further our comprehension of ecosystems and enable us to carry out focused interventions and conservation plans, which supports the goals of sustainable development and the Sustainable Development Goals (SDGs) of the United Nations (Sodhi et al. 2010). The synergy between technology and ecosystem health is not merely a convenience but a critical imperative for safeguarding the planet's ecological balance for current and future generations.

# 2 Foundations of Ecosystem Health

The foundations of ecosystem health lie in the intricate web of interactions among living organisms and their physical environment, forming a delicate balance that sustains life on Earth (Watson et al. 2014). Ecosystem health is a measure of the vitality and resilience of these interconnected systems, encompassing both biotic and abiotic components (Bonney et al. 2014). Biodiversity, a cornerstone of ecosystem health, ensures the stability and adaptability of ecosystems to environmental changes. Additionally, the efficient cycling of nutrients, energy flow, and the maintenance of ecological processes contribute to the overall well-being of ecosystems. Human activities, however, pose significant threats to ecosystem health, including habitat destruction, pollution, and climate change (Gärdenfors et al. 2001). To safeguard the foundations of ecosystem health, it is imperative to adopt sustainable practices, conserve biodiversity, and promote ecosystem resilience, fostering a harmonious coexistence between human societies and the natural world.

## 2.1 Definition and Concept of Ecosystem Health

Ecosystem health is a holistic and comprehensive concept that encapsulates the overall well-being and functioning of ecological systems. It goes beyond the traditional focus on individual species or specific environmental parameters and considers the interconnectedness of all living organisms and their environments (Lai et al. 2019). Ecosystem health emphasizes the capacity of ecosystems to maintain ecological processes, support biodiversity, and resist or recover from disturbances. It incorporates the dynamic interactions between biotic and abiotic components, recognizing that the health of an ecosystem is reflected in its ability to adapt to changes, sustain life, and provide essential services to both human societies and the broader natural world (Sayer et al. 2013).

The concept of ecosystem health also underscores the intrinsic value of biodiversity and the importance of ecological integrity. Healthy ecosystems exhibit resilience, meaning they can absorb and recover from disturbances while maintaining their essential functions (Sousa et al. 2013). Ecosystem health is closely tied to the sustainable use of natural resources and the conservation of biodiversity, recognizing that the well-being of human societies is intimately linked to the health and functioning of the ecosystems they depend on. As human activities continue to exert pressures on the environment, understanding and promoting ecosystem health becomes crucial for achieving a harmonious balance between human needs and the preservation of the Earth's diverse and interconnected ecological systems (Dandois et al. 2010).

## 2.2 Importance of Maintaining Ecosystem Health

Ecosystems provide essential services such as clean air and water, pollination of crops, climate regulation, and habitat for countless species. As technological advancements continue to shape our world, they play a crucial role in promoting and preserving ecosystem health. Here are several ways in which technology contributes to maintaining ecosystem health:

- Monitoring and Surveillance: Advanced satellite technology allows for real-time monitoring of large-scale ecosystems, enabling scientists to track changes in land use, deforestation, and environmental conditions (Akbar and Seyedin 2019). This information is crucial for early detection of potential threats to ecosystems. Remote sensing technologies help assess the health of ecosystems by capturing data on vegetation, land cover, and biodiversity. This data aids in identifying areas at risk and implementing targeted conservation efforts.
- Data Analysis and Modeling: Artificial Intelligence (AI) and Machine Learning: These technologies can analyze vast amounts of ecological data to identify patterns, predict ecosystem changes, and assess potential risks. AI models can be trained to recognize anomalies and provide early warnings for issues such

as deforestation, pollution, or climate-related impacts. Advanced modeling techniques help scientists simulate the impacts of various scenarios on ecosystems (Daily et al. 2009). This allows for better-informed decision-making in land management, conservation planning, and resource allocation.
- Conservation and Restoration: Technologies such as Geographic Information System (GIS) mapping and GPS tracking enable precise identification of areas that require conservation efforts. This targeted approach allows for efficient resource allocation and maximizes the impact of conservation initiatives. Advances in genetics and biotechnology contribute to ecosystem health by supporting efforts to restore endangered species and enhance the resilience of ecosystems. Genetic tools can help address issues such as invasive species and diseases that threaten biodiversity.
- Green Technologies: The transition to renewable energy sources, facilitated by technological advancements, reduces dependence on fossil fuels and minimizes environmental degradation. This shift contributes to a healthier ecosystem by decreasing air and water pollution and mitigating climate change. Precision farming technologies, including IoT sensors and data analytics, help optimize agricultural practices, minimizing the environmental impact of farming while promoting soil health and biodiversity.
- Public Awareness and Education: The widespread use of the internet and social media facilitates the dissemination of information about the importance of ecosystem health (Kati et al. 2004). This increased awareness can lead to public support for conservation initiatives and sustainable practices. Virtual Reality (VR) and Augmented Reality (AR) technologies can be used to create immersive educational experiences, helping people understand the interconnectedness of ecosystems and the impact of human activities on the environment.

### *2.3 Role of Technology in Ecosystem Management*

The role of technology in ecosystem management has become increasingly crucial, as technological advancements play a significant role in promoting and maintaining ecosystem health. Here are several ways in which technology contributes to ecosystem management:

- Data Collection and Monitoring: Technologies like satellite imagery and drones enable the monitoring of large areas, helping scientists and ecologists collect data on ecosystem health, biodiversity, and land cover changes (Pimm et al. 1995). Deploying sensor networks in ecosystems allows for real-time data collection on environmental parameters, such as temperature, humidity, and pollution levels.
- Big Data and Analytics: Advanced analytics and big data tools help process vast amounts of ecological data. This allows researchers to identify patterns, trends, and anomalies, aiding in more informed decision-making for ecosystem management. Technology facilitates the development of models that predict the impact of

various factors on ecosystems, helping authorities plan and implement measures proactively.

- GIS (Geographic Information System): GIS technology assists in mapping and analyzing spatial relationships within ecosystems. This helps in understanding the distribution of species, identifying critical habitats, and planning conservation strategies.
- Artificial Intelligence and Machine Learning (ML): AI-powered image recognition systems can identify species from photos, aiding in biodiversity monitoring. Machine learning algorithms can analyze historical data to predict future changes in ecosystems, helping managers take preventive measures.
- Environmental DNA (eDNA) Technology: eDNA technology allows scientists to detect traces of genetic material shed by organisms into the environment (Su and Ran 2018). This can be used to monitor the presence of specific species and assess ecosystem health.
- Blockchain Technology: In the context of ecosystem management, blockchain can enhance transparency in supply chains, helping to trace the origin of products and prevent illegal or unsustainable practices that harm ecosystems.
- Citizen Science and Crowdsourcing: Technology facilitates citizen engagement through mobile apps, allowing individuals to contribute data on local ecosystems. This crowdsourced information enhances the overall understanding of ecosystem dynamics.
- Communication and Education: VR and AR technologies can be used to create immersive experiences for education and outreach, helping people better understand the importance of ecosystems and the impact of human activities.

## 3 Current Challenges in Ecosystem Health

Current challenges in ecosystem health include the negative environmental impacts of certain technologies, such as electronic waste and pollution from manufacturing processes. The rapid pace of technological innovation can outstrip our ability to assess and mitigate potential ecological risks, leading to unforeseen consequences for biodiversity and ecosystem stability (Kissling et al. 2018). Additionally, the lack of comprehensive regulations and standards for emerging technologies may result in their indiscriminate deployment without proper consideration of environmental sustainability. To address these challenges, there is a need for interdisciplinary collaboration between technologists, environmental scientists, and policymakers to develop and implement technologies that prioritize ecological well-being, minimize environmental footprints, and actively contribute to the restoration and preservation of ecosystems.

### 3.1 Biodiversity Loss

The continuous decrease in biodiversity is predominantly caused by diverse human activities, including habitat degradation, pollution, changes in the climate, and excessive utilisation of natural resources. These variables, such as disturbances or disruptions, disturb the intricate equilibrium of ecosystems, resulting in the extinction of species and the deterioration of their ecosystems (Singh and Bhambri 2023). As species disappear, the intricate web of interactions within ecosystems is disrupted, affecting the stability and resilience of these systems. Biodiversity loss has far-reaching effects on ecosystems, directly affecting vital ecosystem services like pollination, purification of water, and disease regulation, that are crucial for the survival of humans.

In addition to its ecological implications, biodiversity loss also poses socio-economic challenges. Many communities, especially those dependent on agriculture, fisheries, and forestry, rely on diverse ecosystems for their livelihoods. The depletion of biodiversity threatens the availability of resources and can lead to the loss of income and food security for these communities (Rachna and Chhabra 2022). Furthermore, the loss of biodiversity contributes to the vulnerability of ecosystems to external pressures, making them less resilient to environmental changes and more susceptible to the spread of diseases. Addressing the current challenges in ecosystem health requires comprehensive and collaborative efforts to mitigate the drivers of biodiversity loss, promote sustainable practices, and enhance conservation measures to ensure the long-term health and resilience of ecosystems.

### 3.2 Climate Change Impacts

Climate change poses significant challenges to ecosystem health, with far-reaching impacts on biodiversity, ecosystem services, and the overall functioning of natural systems (Barbier 2012). One major consequence is the alteration of habitat conditions, as rising temperatures and shifting precipitation patterns can lead to the loss of suitable habitats for many species. This can result in changes to species distribution, migration patterns, and overall ecosystem composition. Furthermore, the escalating frequency and severity of climate change-induced extreme weather phenomena, such as hurricanes, flooding, and wildfires, can cause catastrophic impacts on ecosystems. These occurrences can result in the destruction of habitats, reductions in population, and potentially the total extinction of certain species. The disruption of ecosystems can have cascading effects, affecting the intricate web of relationships between species and compromising the stability and resilience of ecosystems.

Furthermore, climate change exacerbates existing stressors on ecosystem health, such as pollution, habitat destruction, and overexploitation of natural resources (Folke et al. 2004). For example, warmer temperatures can enhance the negative impacts of pollutants on aquatic ecosystems, leading to declines in water quality and the health of aquatic organisms. Additionally, the loss of sea ice and rising ocean temperatures can threaten marine ecosystems and the species that depend on them. In terrestrial ecosystems, changes in temperature and precipitation can affect agricultural productivity, disrupting food chains and impacting both human and wildlife populations (Lemieux et al. 2008). Overall, the complex interplay of climate change with other stressors poses a formidable challenge to ecosystem health, requiring comprehensive and coordinated efforts to mitigate and adapt to these changes.

### *3.3 Pollution and Contamination*

Industrial activities, urbanization, and agricultural practices release pollutants such as heavy metals, chemicals, and toxins into air, water, and soil. Air pollution from emissions like greenhouse gases contributes to climate change, while water pollution, often from industrial runoff and improper waste disposal, harms aquatic ecosystems and endangers the health of both marine and freshwater organisms (Secretariat of the Convention on Biological Diversity 2020). Soil contamination, resulting from the use of pesticides and fertilizers, affects not only the health of plants but also infiltrates food chains, impacting the entire ecosystem. These pollutants can have cascading effects on biodiversity, disrupting the intricate web of interactions among species and potentially leading to the decline or extinction of certain organisms.

Furthermore, pollution and contamination have profound implications for human health. The release of pollutants into the environment can contaminate food and water sources, exposing humans to harmful substances that may lead to a range of health issues, from respiratory problems to chronic diseases. The interconnectedness of ecosystems means that pollution in one area can have far-reaching consequences, affecting not only local wildlife but also communities that depend on these ecosystems for their livelihoods. Tackling pollution and contamination requires concerted global efforts, involving stricter regulations, sustainable practices, and the development of innovative technologies to mitigate and reverse the damage already inflicted on ecosystems (Turner et al. 2015). Only through holistic and collaborative approaches can we hope to address the multifaceted challenges posed by pollution and contamination and safeguard the health of our planet's ecosystems for future generations.

## 4 Technological Innovations in Ecosystem Monitoring

Advances in satellite technology, remote sensing, and data analytics have enabled real-time, high-resolution monitoring of ecosystems on a global scale. Satellite imagery provides detailed information about land cover, vegetation health, and changes in ecosystems over time (Kremen and Miles 2012). Unmanned aerial vehicles (UAVs) or drones offer the flexibility to collect precise data at a local level, allowing for targeted and on-demand monitoring. Sensor networks, equipped with various environmental sensors, offer continuous data streams that can be integrated and analyzed to assess ecosystem health. Moreover, the use of artificial intelligence and machine learning algorithms improves the effectiveness of data processing, facilitating the extraction of significant insights from extensive datasets. These technology breakthroughs enable the timely identification of environmental changes, such as deforestation or biodiversity loss, and also assist in making informed decisions based on evidence for the sustainable management of ecosystems and conservation initiatives.

### *4.1 Remote Sensing Technologies*

Remote sensing technologies have revolutionized ecosystem monitoring by providing a non-invasive and comprehensive means of collecting data over large geographical areas (Buytaert and De Bievre 2012). These innovations encompass a range of tools, including satellites, drones, and airborne sensors, enabling scientists to observe and analyze ecosystems at various spatial and temporal scales. Satellite imagery, for instance, offers a global perspective, allowing the monitoring of changes in land cover, vegetation health, and biodiversity. Drones provide high-resolution data for more localized and specific assessments. These technologies facilitate the tracking of ecological indicators, such as deforestation, habitat loss, and climate change impacts, enhancing our understanding of ecosystems and supporting informed conservation and management decisions. The real-time and historical data acquired through remote sensing contribute significantly to sustainable ecosystem management and biodiversity preservation.

### *4.2 Internet of Things Applications*

The IoT has significantly impacted ecosystem monitoring by providing real-time data collection, analysis, and communication capabilities. Here are several IoT applications in ecosystem monitoring:

- Smart Agriculture: The IoT sensors and devices are utilised to monitor the levels of moisture in the soil, the temperature, and the overall health of crops (Kapos et al. 2008). This data assists farmers in enhancing irrigation, fertilisation, and overall crop management. IoT-enabled wearables on animals track their health, location, and behavior, allowing farmers to ensure the well-being of their livestock.
- Environmental Monitoring: IoT devices provide real-time measurements of air pollutants, facilitating the monitoring and mitigation of pollution levels. This is crucial for assessing the impact of industrial activities and traffic on air quality (Bhambri and Khang 2024). Sensors in water bodies provide continuous data on

parameters like pH, temperature, and nutrient levels. This is essential for managing water resources and identifying pollution sources.

- Wildlife Tracking: IoT-enabled tags or collars on wildlife provide insights into their movements, habitats, and behaviors (Liu et al. 2007). This helps conservationists in studying and protecting endangered species.
- Forest Fire Detection: Integrated systems use sensors, drones, and satellite data to monitor forests for signs of fire. Early detection enables rapid response, minimizing the damage caused by wildfires.
- Weather Monitoring: IoT Weather Stations are installed in multiple locations to gather up-to-the-minute information on humidity, wind speed, temperature, and precipitation. This information aids in predicting and responding to weather-related events.
- Ocean and Marine Ecosystem Monitoring: IoT devices are used to monitor ocean temperatures, salinity, and marine life (Sala et al. 2000). This information is vital for understanding climate change, managing fisheries, and preserving marine ecosystems.
- Urban Green Spaces: IoT sensors in urban green spaces monitor factors such as soil moisture, plant health, and foot traffic. This information helps city planners and environmentalists enhance green infrastructure and ensure the well-being of urban flora and fauna.
- Remote Sensing and Satellite Technology: IoT technologies complement satellite data for comprehensive ecosystem monitoring (Hilty and Merenlender 2003). This combination provides a broader perspective and helps monitor large, remote areas.
- Biodiversity Conservation: Automated cameras equipped with IoT technology capture images and videos of wildlife. This helps in biodiversity monitoring and understanding population dynamics.
- Energy Efficiency in Monitoring Devices: Energy-efficient IoT devices ensure prolonged battery life, enabling continuous monitoring without frequent replacements (Secretariat of the Convention on Biological Diversity 2010). This is particularly important in remote or inaccessible areas.

### *4.3 Big Data Analytics for Ecosystem Assessment*

Big Data Analytics has emerged as a powerful technological innovation in ecosystem monitoring, providing researchers and environmentalists with new tools to assess and understand ecosystems at unprecedented scales. Here are some ways in which Big Data Analytics is being utilized for ecosystem assessment:

- Data Integration and Fusion: Big Data Analytics allows the integration of diverse data types, such as satellite imagery, climate data, sensor data, and social media feeds, to provide a comprehensive view of ecosystems. By combining data from various sources, researchers can gain a more holistic understanding of ecosystem dynamics and identify correlations that may not be apparent when analyzing individual datasets.

- Real-time Monitoring: Big Data Analytics enables the processing and analysis of large volumes of data in real-time. This is crucial for monitoring ecosystems continuously and responding promptly to changes or disturbances. Real-time monitoring allows for early detection of environmental issues, such as deforestation, pollution, or habitat loss, enabling rapid intervention and mitigation efforts.
- Predictive Analytics: Big Data techniques, including machine learning and predictive modeling, enable the development of models that can forecast ecosystem changes based on historical data (Bakshi et al. 2021). Predictive analytics can help anticipate potential threats, such as the impact of climate change on biodiversity, allowing for proactive conservation strategies.
- Spatial Analysis: GIS and spatial analysis tools, integrated with Big Data Analytics, enable the mapping and monitoring of ecosystems in three dimensions. Researchers can analyze spatial patterns, identify hotspots of biodiversity, and assess the impact of land-use changes on ecosystem health.
- Crowdsourced Data and Citizen Science: Big Data Analytics facilitates the processing of large volumes of crowdsourced data from citizen science initiatives. This data can supplement traditional research efforts (Millennium Ecosystem Assessment 2005). Citizen scientists can contribute observations and data through mobile apps or online platforms, providing valuable information for ecosystem assessment at a broader scale.
- Scalability and Efficiency: Big Data technologies are designed to handle large-scale datasets efficiently. This scalability allows researchers to analyze vast amounts of information, covering extensive geographical areas and long time periods. This efficiency is crucial for managing and processing the massive datasets generated by remote sensing technologies, environmental sensors, and other monitoring tools.
- Data-driven Decision Making: Big Data Analytics empowers decision-makers with actionable insights derived from large datasets. This data-driven approach enhances the effectiveness of conservation strategies and resource allocation. Policy decisions related to land use, conservation planning, and natural resource management can be informed by evidence-based insights derived from Big Data Analytics.

## 5 Digital Solutions for Conservation and Restoration

Digital solutions play a pivotal role in advancing conservation and restoration efforts, fostering ecosystem health through innovative technological applications. Cutting-edge tools such as satellite imaging, artificial intelligence, and data analytics enable precise monitoring of ecosystems, allowing conservationists to identify and address environmental threats in real-time. These technologies facilitate the collection and analysis of vast amounts of ecological data, empowering scientists to make informed decisions for effective conservation strategies (Tallis et al. 2008).

Additionally, mobile applications and online platforms engage the public in citizen science initiatives, promoting awareness and involvement in conservation activities. The integration of technology not only enhances the efficiency of monitoring and restoration efforts but also facilitates collaboration between stakeholders, creating a comprehensive approach towards safeguarding biodiversity and promoting the overall well-being of ecosystems in our rapidly evolving digital age.

## *5.1 AI in Conservation*

AI-driven tools and technologies are revolutionizing the way conservationists monitor and protect biodiversity. Machine learning algorithms analyze vast datasets, such as satellite imagery and sensor data, to track changes in ecosystems, identify endangered species, and predict potential threats (Ficetola and Padoa-Schioppa 2009). AI-powered predictive models help in mapping the impact of climate change on various habitats, enabling proactive conservation strategies. Additionally, automated systems are aiding in the detection of illegal activities like poaching and deforestation, allowing for swift intervention. By harnessing the power of AI, conservationists can make more informed decisions, allocate resources efficiently, and implement targeted measures to preserve and restore ecosystems, ultimately contributing to the sustainable coexistence of humanity and the natural world.

## *5.2 Blockchain for Sustainable Resource Management*

Blockchain technology has emerged as a powerful tool for sustainable resource management, particularly in the context of technological advancements aimed at promoting ecosystem health. One key application lies in the transparent and decentralized nature of blockchain, which enables the creation of immutable and auditable ledgers for tracking and verifying the provenance of natural resources (Schmitt and Holbrook 2003). By implementing blockchain in supply chains, stakeholders can ensure the responsible extraction, production, and distribution of resources. This transparency reduces the risk of illegal or unsustainable practices, fostering accountability among participants in the supply chain. In addition, smart contracts, which are contracts that execute automatically based on predetermined rules, can be incorporated into the blockchain to streamline compliance with environmental legislation and standards. This guarantees that resource management techniques align with sustainability objectives.

Furthermore, the use of blockchain in sustainable resource management contributes to the development of tokenized ecosystems and incentivizes environmentally friendly practices. Through the creation of tokenized assets representing ecological contributions, stakeholders can be rewarded for adopting sustainable measures, such as reforestation efforts, waste reduction, or renewable energy initiatives (Bose et al. 2021). This tokenization of environmental benefits enables a more direct and traceable connection between conservation actions and their impact, fostering a sense of responsibility among participants. As technological advancements continue, the integration of blockchain in sustainable resource management not only enhances the efficiency and transparency of processes but also establishes a framework that aligns economic incentives with ecological well-being, ultimately promoting a healthier and more resilient ecosystem.

## *5.3 VR and AR in Environmental Education*

VR and AR hold significant potential in revolutionizing environmental education and promoting ecosystem health through technological advancements (Noss 1990). In the realm of VR, immersive experiences can transport students and educators to virtual ecosystems, allowing them to explore diverse habitats and witness environmental phenomena firsthand. This experiential learning can enhance understanding and appreciation for the interconnectedness of ecosystems, fostering a sense of responsibility for environmental stewardship. VR can simulate scenarios such as deforestation, pollution, or climate change impacts, enabling users to witness the consequences of human actions on ecosystems. On the other hand, AR can overlay digital information onto the real-world environment, offering interactive and engaging educational experiences. For example, AR applications can provide real-time data about local flora and fauna, geological features, or pollution levels, allowing users to interact with their surroundings and gain a deeper understanding of environmental issues. By integrating VR and AR into environmental education, we can leverage technology to inspire a new generation of environmentally conscious individuals who are equipped to contribute to the well-being of ecosystems.

These technologies also have the potential to address challenges in reaching diverse audiences and overcoming geographical constraints. VR and AR platforms can democratize access to environmental education by providing virtual field trips or simulations that are not limited by physical location or resource availability (Walpole et al. 2009). This inclusivity can empower individuals from various backgrounds to engage with environmental issues and contribute to the collective effort of promoting ecosystem health. Additionally, the interactive and immersive nature of VR and AR experiences can enhance learning retention and motivation, making environmental education more effective and enjoyable. As technology continues to advance, the integration of VR and AR in environmental education will play a pivotal role in fostering a deeper connection between individuals and their natural surroundings, ultimately contributing to a more sustainable and ecologically aware society.

# 6 Case Studies: Successful Implementation of Technologies

Several case studies exemplify the successful implementation of cutting-edge technologies to foster technological advancements and promote ecosystem health (Naidoo and Adamowicz 2005). For example, in precision agriculture, farmers have incorporated sensors, drones, and data analytics to maximize resource utilization, minimize ecological consequences, and improve crop productivity. Remote sensing technologies have also been pivotal in monitoring and managing ecosystems, enabling scientists to assess biodiversity, detect deforestation, and track the impact of climate change. Additionally, smart monitoring systems in water bodies leverage IoT devices to collect real-time data on water quality, aiding in early detection of pollution and facilitating prompt intervention. These examples showcase how the strategic adoption of advanced technologies contributes to the sustainable management and preservation of ecosystems, ultimately fostering a harmonious balance between technological progress and environmental well-being.

## *6.1 Smart Agriculture and Precision Farming*

The use of precision agriculture technologies, including sensors, drones, and data analytics, has significantly improved resource efficiency and reduced environmental impacts. Farmers can now monitor soil health, optimize water usage, and minimize the use of pesticides. Companies like John Deere and startups like FarmLogs have contributed to the adoption of precision farming practices.

## *6.2 Remote Sensing for Forest Management*

Governments and organizations are utilizing satellite imagery and remote sensing technologies to monitor and manage forests effectively. For example, the Global Forest Watch platform employs satellite data to detect deforestation in near real-time, enabling timely interventions to protect ecosystems.

## *6.3 Marine Conservation with AI and Robotics*

Saildrone, an ocean data collection company, deploys autonomous sailboats equipped with sensors to gather data on ocean health. These saildrones can monitor parameters like sea surface temperature, salinity, and carbon dioxide levels, aiding in better understanding and conservation of marine ecosystems.

### 6.4 Blockchain for Sustainable Supply Chains

The utilization of blockchain technology enables the establishment of supply chains that are both transparent and traceable, thereby guaranteeing the sustainability of items. The IBM Food Trust utilizes blockchain technology to track the entire process of food production, from the farm to the table. This allows consumers to access detailed information regarding the source and ecological impact of their food.

### 6.5 Urban Green Spaces Monitoring

Cities worldwide are implementing technologies to monitor and enhance urban green spaces. For instance, the city of Melbourne, Australia, uses a network of sensors to monitor air quality, soil moisture, and plant health in its urban forests. This data informs better management practices and contributes to overall urban ecosystem health.

### 6.6 Waste Management Through IoT

Smart waste management systems, utilizing the IoT, help optimize waste collection routes and reduce environmental impact. Barcelona, Spain, implemented a smart waste management system with sensors in bins that alert collection trucks when they need emptying, leading to more efficient waste collection and reduced fuel consumption.

### 6.7 Renewable Energy Integration

The incorporation of renewable sources of energy, such as wind and solar power, into the electricity network facilitates the utilisation of cleaner energy and mitigates the ecological repercussions of energy generation. Nations such as Denmark and Germany have effectively enacted regulations and adopted technologies to augment the proportion of green energy in their power systems.

# 7 Policy Implications and Regulatory Frameworks

The policy implications and regulatory frameworks for technological advancements in promoting ecosystem health must prioritize a balance between innovation and environmental sustainability (Vitousek et al. 1997). Governments should establish comprehensive policies that encourage the development and adoption of environmentally friendly technologies while mitigating potential negative impacts on ecosystems. Regulatory frameworks should be dynamic, capable of adapting to the rapidly evolving technological landscape, and emphasize transparency, accountability, and stakeholder engagement. Collaboration between governmental bodies, industry leaders, and environmental experts is essential to establish guidelines that ensure the responsible deployment of technologies. Additionally, incentives for businesses to invest in sustainable practices and the development of eco-friendly solutions should be incorporated into regulatory frameworks. Striking a harmonious equilibrium between technological progress and ecosystem well-being requires a forward-thinking approach that fosters innovation while safeguarding the delicate balance of the natural environment.

## *7.1 Integrating Technology into Environmental Policies*

By incorporating cutting-edge tools such as remote sensing, artificial intelligence, and data analytics, policymakers can enhance their ability to monitor, assess, and respond to environmental changes effectively (Estes et al. 2011). These technologies provide real-time data, enabling more informed decision-making and proactive measures to address ecological challenges. Additionally, by embracing digital solutions, governments can streamline communication, collaboration, and coordination among stakeholders, fostering a more integrated and efficient approach to environmental management. Furthermore, using technology can enable individuals and enterprises to actively engage in sustainable behaviors, fostering a collaborative endeavor to conserve biodiversity, mitigate the effects of climate change, and safeguard the long-term well-being of our ecosystems. In order to successfully address the intricate difficulties of the twenty-first century and attain a healthy equilibrium between human progress and ecological welfare, it is crucial to establish a strong connection between technology and regulations pertaining to the environment.

## *7.2 International Collaborations for Ecosystem Health*

As nations grapple with complex environmental challenges, collaborative efforts leverage the collective expertise and resources of diverse stakeholders to address issues such as climate change, biodiversity loss, and pollution on a global scale.

Through shared research initiatives, data exchange, and the development of cutting-edge technologies, countries can collectively monitor, analyze, and manage ecosystems more effectively (Tittensor et al. 2014). These collaborations facilitate the transfer of knowledge and best practices, fostering the rapid implementation of sustainable solutions. Through the utilization of technological progressions, such as remote sensing, data analytics, and artificial intelligence, international collaborations not only augment our comprehension of ecosystem dynamics but also enable nations to make well-informed decisions and policies that foster long-term ecological resilience and health.

### *7.3 Ethical Considerations in Technological Interventions*

As we harness technological advancements to address environmental challenges, it is imperative to ensure that these interventions align with ethical principles such as sustainability, biodiversity conservation, and equitable distribution of benefits. The potential risks, unintended consequences, and long-term impacts of technological solutions on ecosystems and their inhabitants must be thoroughly assessed and mitigated. Furthermore, it is crucial to prioritize inclusivity and stakeholder participation in order to prevent marginalization and guarantee that a wide range of viewpoints and information are taken into account throughout decision-making procedures. Ensuring a harmonious link between technology and the environment and protecting ecosystems' well-being necessitates maintaining a delicate equilibrium between technical advancement and ethical accountability.

## 8 Future Prospects and Emerging Trends

Advanced sensor technologies, coupled with the IoT, are enabling real-time monitoring of ecosystems, allowing for more accurate and timely data collection on biodiversity, climate patterns, and pollution levels. AI and machine learning algorithms are enhancing our ability to analyze vast datasets, identifying patterns and predicting ecological changes. Drones and satellite imaging contribute to comprehensive and efficient surveillance, aiding in the mapping and management of ecosystems. Additionally, blockchain technology is being explored to create transparent and secure systems for tracking and verifying sustainable practices, promoting accountability across supply chains. As these technologies continue to evolve, they hold the potential to provide innovative solutions for conservation, biodiversity preservation, and sustainable resource management, paving the way for a healthier and more resilient global ecosystem.

## 8.1 *Potential Impact on Achieving Sustainable Development Goals*

Technological advancements like as remote sensing, AI, and improved monitoring systems allow for the collecting and analysis of data in real-time, hence enabling more efficient and focused conservation initiatives. These technologies enhance our understanding of ecosystems, track biodiversity patterns, and identify potential threats, allowing for proactive interventions. Furthermore, advancements in clean energy technologies contribute to mitigating climate change, a crucial factor in achieving SDG 13 (Climate Action). Integrating these technological solutions into sustainable development strategies can enhance resource efficiency, promote responsible consumption and production (SDG 12), and ultimately foster a balance between human activities and the health of ecosystems, supporting the broader agenda of sustainable development.

## 8.2 *Challenges and Opportunities*

Technological advancements present both challenges and opportunities in promoting ecosystem health. On the one hand, the rapid development and deployment of technology can contribute to environmental degradation. For instance, the manufacturing and disposal of electronic devices often lead to significant pollution and resource depletion. In addition, some technical practices, such as the over use of energy-intensive data centres, contribute to the release of carbon emissions and hence contribute to climate change. Achieving a harmonious equilibrium between technological advancement and ecological preservation is a substantial obstacle, necessitating careful evaluation of the entire lifespan and ecological consequences of emerging technology.

On the other hand, there are considerable opportunities for technological advancements to positively impact ecosystem health. Remote sensing technologies, for example, enable more effective monitoring of ecosystems, helping to detect and address environmental issues such as deforestation, pollution, and habitat loss. Furthermore, AI and machine learning can be utilized to analyze vast amounts of environmental data, providing valuable insights into ecosystem dynamics and facilitating more informed conservation and restoration efforts. The development of eco-friendly technologies, renewable energy sources, and sustainable practices also offers the potential to reduce the overall ecological footprint of human activities. The key lies in leveraging technological advancements responsibly, incorporating eco-friendly design principles, and fostering innovation that prioritizes long-term environmental sustainability.

# 9 Conclusion

In conclusion, the integration of technology in ecosystem health initiatives stands as an imperative step towards fostering a sustainable and resilient environment. As we confront the escalating challenges posed by climate change, habitat loss, and biodiversity decline, harnessing the power of technology becomes not only a necessity but a strategic advantage. By promoting innovation in monitoring, data analysis, and communication, we can gain deeper insights into ecosystem dynamics, enabling more informed decision-making. The call to action is clear—stakeholders across government, academia, industry, and civil society must collaborate to invest in and implement cutting-edge technologies that enhance our capacity to understand, protect, and restore ecosystems. This concerted effort will not only preserve biodiversity but also ensure the well-being of human populations that are intricately linked to the health of our planet.

To propel this transformative agenda forward, it is essential that we prioritize education and awareness campaigns, fostering a global understanding of the interdependence between technology and ecosystem health. Governments should incentivize the adoption of sustainable practices and technologies, offering support and resources to initiatives that integrate technological solutions into conservation and restoration efforts. Additionally, fostering international collaboration and knowledge exchange will allow for the development and dissemination of best practices, ensuring a collective and effective approach to promoting ecosystem health. The time to act is now, and by embracing technology, we can embark on a journey towards a harmonious coexistence between humanity and the natural world, safeguarding the diversity and vitality of our ecosystems for generations to come.

## References


Akbar R, Seyedin S (2019) The role of information and communication technology in environmental sustainability: a systematic literature review. J Clean Prod 232:1318–1331

Bakshi P, Bhambri P, Thapar V (2021) A review paper on wireless sensor network techniques in Internet of Things (IoT). In: International conference on contemporary issues in engineering and technology

Barbier EB (2012) Progress and challenges in valuing coastal and marine ecosystem services. Rev Environ Econ Policy 6(1):1–19

Bhambri P, Khang A (2024) Biosensor applications and principles of agricultural and aquacultural sectors. In: Khang A (ed) Agriculture and aquaculture applications of biosensors and bioelectronics. IGI Global, pp 1–17. https://doi.org/10.4018/979-8-3693-2069-3.ch001

Bonney R, Shirk JL, Phillips TB, Wiggins A, Ballard HL, Miller-Rushing AJ, Parrish JK (2014) Next steps for citizen science. Science 343(6178):1436–1437

Bose MM, Yadav D, Bhambri P, Shankar R (2021) Electronic customer relationship management: benefits and pre-implementation considerations. J Maharaja Sayajirao Univ Baroda 55(01(VI)):1343–1350

Buytaert W, De Bievre B (2012) Water for cities: the impact of climate change and demographic growth in the tropical Andes. Water Resour Res 48(1):W01508

Carrington D (2019) How technology is transforming conservation. The Guardian. Retrieved from https://www.theguardian.com/environment/2019/aug/20/how-technology-transforming-conservation

Daily GC, Polasky S, Goldstein J, Kareiva PM, Mooney HA, Pejchar L, Shallenberger R (2009) Ecosystem services in decision making: time to deliver. Front Ecol Environ 7(1):21–28

Dandois JP, Ellis EC (2010) Remote sensing of vegetation structure using computer vision. Remote Sens 2(4):1157–1176

Estes JA, Terborgh J, Brashares JS, Power ME, Berger J, Bond WJ, Wardle DA (2011) Trophic downgrading of planet Earth. Science 333(6040):301–306

Ficetola GF, Padoa-Schioppa E (2009) Towards a tool to assess the vulnerability of ecosystems and species to climate change. Biodivers Conserv 18(12):3181–3190

Folke C, Carpenter S, Walker B, Scheffer M, Elmqvist T, Gunderson L, Holling CS (2004) Regime shifts, resilience, and biodiversity in ecosystem management. Annu Rev Ecol Evol Syst 35:557–581

Gärdenfors U, Hilton-Taylor C, Mace GM, Rodríguez JP (2001) The application of IUCN red list criteria at regional levels. Conserv Biol 15(5):1206–1212

Hilty JA, Merenlender AM (2003) Use of riparian corridors and vineyards by mammalian predators in northern California. Conserv Biol 17(1):265–273

IPCC (2019) IPCC special report on the ocean and cryosphere in a changing climate. Retrieved from https://www.ipcc.ch/srocc/download-report/

Kapos V, Balmford A, Aveling R, Bubb P, Carey P, Entwistle A, Scharlemann JP (2008) Calibrating conservation: new tools for measuring success. Conserv Lett 1(4):155–164

Kati V, Devillers P, Dufrene M, Legakis A, Vokou D (2004) Testing the value of six taxonomic groups as biodiversity indicators at a local scale. Conserv Biol 18(3):667–675

Kissling WD, Ahumada JA, Bowser A, Fernandez M, Fernández N, García EA, Espadaler X (2018) Building essential biodiversity variables (EBVs) of species distribution and abundance at a global scale. Biol Rev 93(1):600–625

Kremen C, Miles A (2012) Ecosystem services in biologically diversified versus conventional farming systems: benefits, externalities, and trade-offs. Ecol Soc 17(4):40

Lai HR, He D, Zheng HB (2019) The impact of information and communication technology on environmental efficiency: an empirical study of the Chinese manufacturing industry. Technol Forecast Soc Chang 141:103–113

Lemieux CJ, Clarke JL, Hamilton DP, Baker M (2008) Remote sensing of chlorophyll—a as a measure of cyanobacterial biomass in three large shallow lakes. Hydrobiologia 610(1):225–235

Liu J, Dietz T, Carpenter SR, Alberti M, Folke C, Moran E, Taylor WW (2007) Complexity of coupled human and natural systems. Science 317(5844):1513–1516

Millennium Ecosystem Assessment (2005) Ecosystems and human well-being: synthesis. Island Press, London

Mora C, Sale PF (2011) Ongoing global biodiversity loss and the need to move beyond protected areas: a review of the technical and practical shortcomings of protected areas on land and sea. Mar Ecol Prog Ser 434:251–266

Naidoo R, Adamowicz WL (2005) Biodiversity and nature-based tourism in the Sabie-Sand region of South Africa. Environ Dev Econ 10(6):759–782

Noss RF (1990) Indicators for monitoring biodiversity: a hierarchical approach. Conserv Biol 4(4):355–364

Pimm SL, Russell GJ, Gittleman JL, Brooks TM (1995) The future of biodiversity. Science 269(5222):347–350

Rachna BP, Chhabra Y (2022) Deployment of distributed clustering approach in WSNs and IoTs. In: Cloud and fog computing platforms for Internet of Things. Chapman and Hall/CRC, London, pp 85–98

Riggio J, Jacobson A, Dollar L, Bauer H, Becker M, Dickman A, Lindsey P (2013) The size of savannah Africa: a lion's (Panthera leo) view. Biodivers Conserv 22(1):17–35

Sala OE, Chapin FS, Armesto JJ, Berlow E, Bloomfield J, Dirzo R, Lubchenco J (2000) Global biodiversity scenarios for the year 2100. Science 287(5459):1770–1774
Sayer J, Sunderland T, Ghazoul J, Pfund JL, Sheil D, Meijaard E, Boedhihartono AK (2013) Ten principles for a landscape approach to reconciling agriculture, conservation, and other competing land uses. Proc Natl Acad Sci 110(21):8349–8356
Schmitt RJ, Holbrook SJ (2003) Local-scale patterns of larval settlement in a planktivorous damselfish (Pomacentridae: Chromis multilineata). Mar Ecol Prog Ser 256:233–242
Secretariat of the Convention on Biological Diversity (2010) Global biodiversity outlook 3. Retrieved from https://www.cbd.int/gbo/gbo3/publication/gbo3-final-en.pdf
Secretariat of the Convention on Biological Diversity (2020) Global Biodiversity Outlook 5. Retrieved from https://www.cbd.int/gbo/gbo5/publication/gbo-5-en.pdf
Singh G, Bhambri P (2023) Simulation analysis of AODV and DSDV routing protocols for secure and reliable service in mobile adhoc networks (MANETs). In: Integration of AI-based manufacturing and industrial engineering systems with the Internet of Things. CRC Press, London, pp 205–216
Sodhi NS, Koh LP, Clements R, Wanger TC, Hill JK, Hamer KC, Posa MR (2010) Conserving Southeast Asian forest biodiversity in human-modified landscapes. Biol Cons 143(10):2375–2384
Sousa WTZ, Baessa PAM, de Carvalho RB, Rego RDS (2013) Infrared thermography as a tool to assess animal health and welfare in dairy farms. Sci World J 2013:1
Su L, Ran J (2018) Ecological footprint accounting for supporting decision-making in China's Silk Road Economic Belt. Environ Sci Pollut Res 25(28):28069–28078
Tallis H, Kareiva P, Marvier M, Chang A (2008) An ecosystem services framework to support both practical conservation and economic development. Proc Natl Acad Sci 105(28):9457–9464
Tittensor DP, Walpole M, Hill SL, Boyce DG, Britten GL, Burgess ND, Mace GM (2014) A mid-term analysis of progress toward international biodiversity targets. Science 346(6206):241–244
Turner W, Rondinini C, Pettorelli N, Mora B, Leidner AK, Szantoi Z, Burgess ND (2015) Free and open-access satellite data are key to biodiversity conservation. Biol Cons 173:173–176
Vitousek PM, Mooney HA, Lubchenco J, Melillo JM (1997) Human domination of Earth's ecosystems. Science 277(5325):494–499
Walpole M, Almond REA, Besançon C, Butchart SHM, Campbell-Lendrum D, Carr GM, Deutz A (2009) Tracking progress toward the 2010 biodiversity target and beyond. Science 325(5947):1503–1504
Watson JE, Dudley N, Segan DB, Hockings M (2014) The performance and potential of protected areas. Nature 515(7525):67–73

# Universal Access to the Internet for Attaining Responsible Consumption-Based Sustainable Development Goal (SDG)

**Kalidas Kaman and Arun Kumar Deshmukh**

**Abstract** Nearly every country establishes a multitude of local and national policies and initiatives to address the social, economic, and environmental concerns mentioned in the SDGs. It has been noted that one of the main obstacles to reaching the SDGs outlined by the United Nations is the absence of public participation in the implementation of policies. Online shopping, or buying products and services over the Internet, is a huge industry that has changed the face of modern retailing and greatly impacted the world economy. Access to the universal internet is a key enabler in extending the scope of online shopping. It offers convenience, wider choices, affordable prices, flexibility, ubiquity, and accessibility to consumers. By engaging in mindful consumption, individuals can become more aware of and responsible for their choices as consumers. It also encourages individuals to take their time, think about their morals, and make decisions consistent with their ethical and environmental values, eventually encouraging more sustainable and fulfilling lifestyles. This exploratory study utilized a qualitative approach to conceptualize the role of universal access to the Internet in online shopping; the relationship between online shopping and responsible consumption; and the part that mindful consumption plays using the Stimulus–Organism–Response (S–O–R) framework. Propositions are generated to express the linkages as a result of the study.

**Keywords** Online shopping · Universal access to the internet · SDG · Mindful consumption · Responsible consumption

K. Kaman · A. K. Deshmukh (✉)
Institute of Management Studies, Banaras Hindu University, Varanasi 221 005, India
e-mail: akdeshmukh@fmsbhu.ac.in

K. Kaman
e-mail: kalidaskaman@fmsbhu.ac.in

 
W. Leal Filho et al. (eds.), *Digital Technologies to Implement the UN Sustainable Development Goals*, World Sustainability Series,
https://doi.org/10.1007/978-3-031-68427-2_22

# 1 Introduction

Sustainable Development Goals (SDGs) are a set of global objectives established by the United Nations (UN) in 2015 (United Nations, n.d.). The SDGs offer a thorough framework for tackling the most important social, economic, and environmental issues facing the globe to ensure a sustainable future for all. The objectives aim to address a wider variety of problems, including poverty, inequality, climate change, environmental degradation, and access to healthcare and education. They build on the achievement of the Millennium Development Goals (MDGs). There are 17 interrelated SDGs with individual goals that must be met by 2030 (United Nations, n.d.). These objectives acknowledge the need for a comprehensive strategy that takes into account the social, economic, and environmental facets of human well-being to achieve sustainable development (Scharlemann et al. 2020). The 17 SDGs cover a wide range of topics, such as eradicating poverty and hunger, advancing quality education, achieving gender equality, ensuring access to clean water and sanitation, fostering innovation and infrastructure development, promoting sustainable economic growth, combating climate change, safeguarding biodiversity and ecosystems, and fostering inclusive and peaceful societies (United Nations, n.d.). The SDGs strongly emphasize the value of alliances and cooperation between governments, civil society organizations, the private sector, and people. All stakeholders must work together to achieve these goals, and there must be transformative local, national, and international policies and initiatives (Banerjee et al. 2020).

Online shopping has revolutionized the retail sector by giving consumers throughout the world convenience, choice, and accessibility (Umit Kucuk and Krishnamurthy 2007). It has created new business opportunities for entrepreneurs and expanded market reach for retailers. As technology advances, online shopping is expected to further evolve and shape how we shop and engage in commerce. A key enabler of sustainable development across many SDGs is universal Internet access (Onyango and Ondiek 2021). Access to education is improved, gender equality is encouraged, economic growth is stimulated, infrastructure development is supported, sustainable cities are encouraged, peace and justice are supported, and international collaborations are made easier (Cernev and Fenner 2020). We can advance towards attaining the SDGs and creating a more inclusive and sustainable future for all by ensuring that everyone has access to the Internet. The expansion and accessibility of online shopping are fundamentally dependent on widespread access to the Internet. It broadens the market's reach, improves convenience and accessibility, makes it easier to compare products and make secure purchases, encourages the development of online marketplaces, makes it possible for efficient logistics, and gives users access to customer care and reviews (Martinsons 2008). These factors collectively contribute to the flourishing of the online shopping industry and its benefits for both businesses and consumers. Online shopping is anticipated to further contribute to economic growth and change the business environment as the digital economy develops (Liu 2013).

Online buying gives customers access to product information, encourages comparison shopping, offers environmentally friendly product options, lessens carbon emissions, helps consumers avoid impulse buys, offers product reviews and ratings, and supports recycling and donation initiatives. Utilizing these benefits enables people to make ethical consumption decisions while shopping online.

The research questions (RQs) below are designed to help in examining the connections between universal internet access, online shopping, mindful consumption, and responsible consumption.

**RQ1**: What part does universal internet access play in online shopping?
**RQ2**: How are responsible consumption and online shopping related?
**RQ3**: How can online shopping and responsible consumption go hand in hand? What part does mindful consumption play?

It has been noted that the SDGs cannot be realized if the public is not involved in government initiatives. Although it is thought that having access to the Internet everywhere encourages economic growth, but its sustainability is uncertain. Online purchasing and universal internet access encourage public participation in global initiatives, which can be accomplished through encouraging mindful consumption and engaging in responsible consumption. This chapter is divided into the following sections to respond to the framed RQs above. The sections cover universal internet access and sustainable economic development, online shopping and economic growth, online shopping and responsible consumption, the role of mindful consumption in online shopping and responsible consumption, discussions, future research directions, and conclusion.

## 2 Universal Internet Access and Sustainable Economic Development

Access to the Internet for everybody is essential for promoting sustainable economic growth (Hoosain et al. 2020). It gives people and communities access to knowledge, assets, and opportunities previously unavailable. Universal internet access makes it easier for disconnected and digitally connected communities to communicate (Kushlev et al. 2017). By ensuring equitable access to information and communication technologies, it encourages people from all socioeconomic backgrounds to participate in the digital economy. The knowledge and educational resources available on the internet are immense. Students and learners can access educational resources, online programs, and distance learning options because of widespread internet connection (Gillis and Krull 2020). It aids in raising literacy rates, developing skills, and enabling people to take part in a knowledge-based economy. A global consumer base can be reached by beginning an online business with access to the internet (Juith and Mark 2003).

It lowers entrance barriers, promotes creativity, and supports the expansion of startups. When people have access to online markets, financial services, and e-commerce platforms, they may build long-term businesses and support economic expansion (Chen et al. 2021). Small enterprises and individuals in rural locations have access to larger markets because of universal Internet connectivity (Aziz and Naima 2021). They may advertise their goods and services online, expand their clientele, and establish connections with suppliers and distributors all over the world. Internet connectivity also makes it easier for people to access financial services like online banking, digital payments, and crowdfunding platforms, enabling them to engage in economic activity (Aziz and Naima 2021). The Internet fosters remote work and opens up new job prospects (Anwar and Graham 2021). People can access freelance marketplaces, online recruitment platforms, and employment portals because of widespread internet access. They can participate in the global gig economy, work remotely, and communicate with clients and colleagues anywhere in the world.

Access to the Internet makes it easier for the government to supply services like e-governance, online tax filing, and digital documentation (Ullah et al. 2021). It improves administrative efficiency, eliminates bureaucracy, and increases openness, resulting in improved governance and sustainable development. The ability to share information, concepts, and best practices internationally is made possible by universal Internet access (Tanhua et al. 2019). Collaboration between researchers, scientists, and professionals is facilitated, which promotes creativity, problem-solving, and industry growth. Connectivity to telemedicine, online consultation services, and healthcare information is improved by internet connectivity (Jnr et al. 2021). It makes it possible to provide vital services to isolated or underserved locations, as well as remote monitoring and health education. Better health outcomes and general well-being are encouraged, both of which are essential for economic growth. It is crucial to remember that to achieve universal internet access, infrastructure issues must be resolved along with the promotion of digital literacy and affordability (Tarasova and Rohracher 2023). Governments, the corporate sector, and international organizations must work together to design policies that support equitable digital development, invest in broadband infrastructure, and offer training and support (Clark et al. 2022).

## 3 Universal Internet Access and Online Shopping

For people to engage in online purchasing, having universal access to the internet is essential. People would not be able to access e-commerce platforms and online marketplaces to browse and buy things without dependable and reasonably priced internet connectivity. The potential client base for online shops is increased by universal internet access (Reisdorf et al. 2022). A wider audience may shop online when more individuals have access to the internet, which increases business sales and revenue. Online merchants can reach customers outside of their local marketplaces because of widespread internet access (Chawla and Kumar 2022). With an internet connection, companies can reach new markets and client groups on a national and even international scale (Eferin et al. 2019). Online purchasing is more accessible and convenient because of a universal internet connection. When consumers have reliable internet connectivity, they can shop whenever and wherever they choose, whether they're at home or on the move (Belbağ 2022). Consumers are empowered by having access to the Internet at all times because it gives them knowledge and choices (Barth et al. 2019). While purchasing online, consumers may compare costs, read product reviews, and come to wise selections (Yang et al. 2021). This informational accessibility helps to create a more open and aggressive online market (Ahearne et al. 2022). Secure digital payment options, such as Internet banking, mobile wallets, and credit card transactions, are made easier by widespread Internet access (Zhu et al. 2010). For customers to conduct transactions quickly and securely, these payment alternatives are essential for online purchasing. Due to easy access to the Internet, online shopping has increased, which benefits the economy (Chen et al. 2021). Online shopping promotes entrepreneurship, opens up job opportunities, and stimulates technological and logistical innovation (Modgil et al. 2022). It's vital to remember that while having access to the Internet is a requirement for online shopping, the demand for a better Internet connection can also be fuelled by online purchasing (Gu et al. 2021). More efforts are to be made to boost internet infrastructure and accessibility as more people engage in online shopping, which increases the demand for dependable and fast internet connectivity (Billewar et al. 2022). Online purchasing is more accessible and convenient because of a universal internet connection. Together, they have altered the retail scene, bringing convenience, variety, and access to a worldwide market to consumers while also promoting innovation and economic progress.

Considering the justifications, it can be inferred that having access to the internet always serves as a stimulus for online buying, as shown in Fig. 1 of the S–O–R framework. As a result, the following Proposition 1 is made.

**Proposition 1** *Online Sopping and universal internet access are mutually reliant, with online shopping generating demand for more accessible internet access and universal internet access facilitating online purchasing.*

**Fig. 1** S–O–R framework of universal access to the Internet, online shopping, and economic growth. (*Source*: Authors' compilation)

## 4 Online Shopping and Economic Growth

Recent economic growth has been significantly impacted by online buying. Online retailers provide consumers with convenience, affordable access to a variety of goods, and competitive pricing (Chopra 2016). Due to the ease with which people can now compare prices, read reviews, and make purchases from the comfort of their homes, consumer spending has soared. The expansion in overall consumer expenditure due to online shopping has aided economic expansion. The e-commerce sector has grown as a result of the rise of online shopping (Sardjono et al. 2021). E-commerce sites like Amazon, Alibaba, eBay, etc. have produced a lot of jobs and economic activity.

These platforms need workers for warehouses, logistics, customer support, and other positions, which helps to create jobs and money (Hosseini et al. 2023). Online buying has made it easier for startups and small companies to get started (Susanto et al. 2021). E-commerce platforms and marketplaces make it very simple for people and companies to launch online stores and access a worldwide clientele. Small firms now have more opportunities, which has promoted entrepreneurship, job development, and innovation (Chege and Wang 2020). Market development beyond traditional geographic limits has been made possible by online shopping (Sheth 2020). Because of this, physical store locations are no longer restricted in how businesses may reach their customers. This has expanded commerce, given businesses access to new markets, and promoted global economic integration. Technology developments in fields including e-payment systems, logistics, data analytics, and supply chain management have been sparked by online shopping (Altarturi et al. 2023). These developments have boosted productivity and economic growth in various areas of the economy in addition to increasing the effectiveness of online purchasing. The popularity of online shopping has increased demand in several ancillary sectors (Faccia et al. 2023). Further, the demand for digital marketing, web development, and cybersecurity services, also refers to the requirement for packaging, shipping, and delivery services (Graesch et al. 2021). These supporting industries have expanded and given rise to more job possibilities. It's crucial to remember that there are many different factors at play in the intricate relationship between internet shopping and economic growth (Gu et al. 2021). To achieve sustainable and inclusive economic growth, concerns about data privacy, cybersecurity, and the concentration of market power in the hands of a few dominating businesses also need to be carefully addressed (D'Cruz et al. 2022).

Considering the foregoing arguments, it is concluded that online shopping functions as a beneficial organism between widespread internet access and economic growth, as shown in Fig. 1 of the S–O–R framework. Accordingly, Proposition 2 that follows is made.

**Proposition 2** *Online shopping has undeniably boosted the economy, but it has also presented problems for conventional brick-and-mortar stores and caused disruptions in several industries.*

## 5 Online Shopping and Responsible Consumption

The relationship between online buying and sensible consumption can be complicated. Customers have a broad variety of product options while buying online, and the product descriptions are frequently in-depth (Nguyen et al. 2018). As a result, consumers may be more equipped to make wiser choices when making purchases, such as choosing products that are ethically produced, ecologically friendly, or sustainable (Young et al. 2010). Online shopping sites frequently provide competitive prices and a variety of promotions, which can trigger impulsive spending (Dawson and Kim 2010; Zhang et al. 2022). Customers could feel pressured to buy products that are not sustainable or things they don't necessarily need (Gonzalez-Arcos et al. 2021). Making thoughtful decisions while taking into account a purchase's necessity, quality, and environmental impact is encouraged by responsible consumption (Vandenbroele et al. 2020).

Due to unique shipments and extra packaging materials, online purchasing can increase packaging waste (Kader et al. 2023). Consumers who practice responsible consumption are urged to think about the effects of packaging on the environment and to support businesses that prioritize sustainable packaging methods (Purcărea et al. 2022). Online store return or exchange policies may have an impact on sensible consumption (Radhi and Zhang 2019). Customers might be more likely to return items or exchange them, which could lead to increased packaging and shipping waste. Responsible shoppers should be aware of their return practices and work to minimize needless exchanges and returns (Hjort et al. 2019). Local businesses may be impacted by online shopping, particularly from huge e-commerce platforms (Beckers et al. 2021).

To promote local economies and preserve a diversified retail environment, responsible consumption may include making purchases from craftsmen and shopkeepers in the area. Online buyers have the chance to look up and support companies that value moral production processes, just labor laws, and ethical supplier networks (Alexander 2020). These brands can be actively sought out and supported by ethical consumers, who can stay away from businesses with dubious business practices. By offering information, certifications, and labels that identify sustainable and ethical items, online shopping platforms can contribute to the promotion of responsible consumption (Palakshappa and Dodds 2020). They can inform customers about sensible consumption habits and promote environmentally friendly options. It's crucial to remember that responsible buying involves both consumers and retailers.

To encourage responsible consumption, online shopping platforms can integrate sustainability efforts, provide eco-friendly options, and offer transparent information (Riva et al. 2022). Consumers should be aware of their spending patterns, think about the social and environmental effects of their decisions, and actively look for brands and goods that uphold ethical standards. Online purchasing presents a potential for educated decision-making, supporting sustainable practices, and promoting ethical choices, even while it can pose challenges to responsible consumption. People may make use of online shopping's advantages while reducing its detrimental effects on the environment and society by being responsible shoppers.

The following Proposition 3 is made in the light of the aforementioned arguments;

**Proposition 3** *Online purchasing has some benefits in terms of accessibility and convenience, but it also has difficulties encouraging responsible consumer behaviour.*

## 6 Role of Mindful Consumption in Online Shopping and Responsible Consumption

Online shopping and ethical consumption both heavily rely on mindful consumption (Torkkeli et al. 2018). It entails being aware of the consumer decisions we make and taking their social, environmental, and economic effects into account. Consumers who practice mindful consumerism are prompted to think about the effects of their purchases (Milne et al. 2020). This entails assessing a product's utility, robustness, and potential social and environmental effects over the course of its existence. Consumers can make more responsible decisions while shopping online by keeping these things in mind (Rosa et al. 2006). Online shopping makes it easy to find a variety of goods, but it can also encourage impulsive purchases and overconsumption (Fook and McNeill 2020). People who practice mindful consumerism become more conscious of their purchasing habits and steer clear of pointless purchases (Rosenberg 2016).

Consumers can aid in lowering waste and excessive consumption by exercising control and concentrating on true requirements (Papargyropoulou et al. 2014). Ethics-conscious customers actively look for businesses that appreciate sustainable business practices. When people shop online, they consider topics like a company's social responsibility programs, environmental responsibilities, and fair work practices (Hayat et al. 2020). By supporting these businesses, customers may encourage moral business practices in the Internet marketplace. Transparency is encouraged by mindful consumption in the online retail sector (Ikram 2022). Consumers want to know exactly where a product comes from, how it was made, and what effect it has on the environment. Businesses may become more open and accountable as a result of this demand, which will improve sustainability practices throughout the supply chain. When shopping online, conscientious customers actively look for environmentally beneficial options (Grebitus et al. 2020). They give higher priority to items that use fewer resources, less packaging, or are energy-efficient. By selecting these options, customers can have an impact on the market by driving up demand for environmentally friendly goods and incentivizing companies to create more environmentally friendly items. Knowledge of the social and environmental problems caused by consumerism is necessary for mindful consumption (Milne et al. 2020). Online purchasing offers a multitude of resources that customers may use to educate themselves on ethical issues, sustainable practices, and responsible consumption

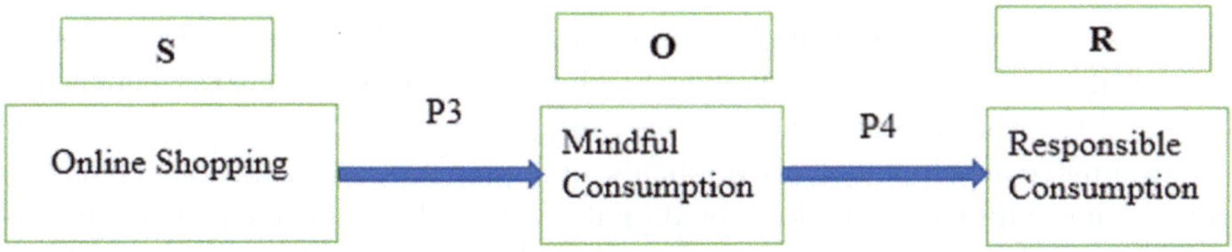


**Fig. 2** S–O–R framework of online shopping, mindful consumption, and responsible consumption. (*Source*: Authors' compilation)

(Lavicza et al. 2021). Consumers are better equipped to make educated decisions and actively participate in responsible online buying as a result of this information. To advance moral standards in the Internet retail industry, conscientious consumers can band together and take collective action (Gheihman 2021). This can be done by contributing to campaigns, signing petitions, or becoming a member of consumer groups that promote sustainability and sensible consumerism. Consumers can exercise collective power and promote change by banding together (Hatanaka 2020). The goal of mindful consumption and responsible consumption is to empower people to make decisions that are consistent with their values and work towards a more sustainable and ethical marketplace.

Considering the justifications, it is determined that mindful consumption serves as a positive intermediary between responsible and online buying, as depicted in Fig. 2 of the S–O–R framework. Proposition 4 that follows is therefore made.

**Proposition 4** *Consumers have the power to influence good change in online shopping and help create a more sustainable future by engaging in mindful consumption.*

## 7 Discussion

The Sustainable Development Goals (SDGs) and economic growth are intertwined and reinforce one another in their link to universal internet access. People and organizations may access a wealth of information, communication tools, and online services through universal Internet access (Reglitz 2020). This makes it possible for workers, small enterprises, and entrepreneurs to participate in the digital economy, access markets throughout the world, and conduct online business. The accessibility of the Internet encourages innovation, entrepreneurship, and knowledge exchange, which boosts economic growth, productivity, and job creation (Setini et al. 2020).

By encouraging entrepreneurship, boosting productivity, expanding financial inclusion, and facilitating education and skill development, universal internet access plays a crucial role in economic development (Kumar and Raj 2019; Sharma 2020; Thathsarani et al. 2021). By empowering people and communities, lowering inequities, enhancing healthcare, and promoting sustainable infrastructure and innovation, it also aids in the fulfilment of the SDGs (Hajian and Kashani 2021; Pathak and Deshkar 2023; van Niekerk 2020). To maximize the potential benefits for economic and sustainable development, governments, entrepreneurs, and stakeholders need to give priority to initiatives to close the digital divide and guarantee equal and affordable access to the Internet (Kerras et al. 2020; Shkabatur, Bar-El and Schwartz 2022; Tay, Tai and Tan 2022). Online purchasing has a complex and nuanced link with conscious and responsible consumption. Online shopping is becoming more and more common and practical, providing access to a variety of goods and services from the comfort of one's home. However, this ease of use can occasionally result in careless consumption and reckless spending patterns.

The practice of mindful consumption involves being aware of and intentional about the effects of our purchasing decisions on the self, economy, society, and the environment (Kang et al. 2021; Sheth et al. 2011). It entails being conscious of our needs versus wants, taking the durability and quality of things into account, and backing brands that are ethical and sustainable. Online purchasing can encourage conscious consumption while also posing challenges. Further, it gives consumers access to a greater selection of goods, enabling them to make educated decisions by comparing pricing, reading reviews, and learning about the sustainability and ethical standards of various businesses. It provides the ease of online buying, which lessens the environmental impact of traveling to physical stores. Online buying, though, can sometimes encourage mindless consumption. The convenience of Internet shopping, the effect of targeted advertising, and the "one-click" purchase option may encourage impulsive purchases and an excessive accumulation of items (Ho and Ito 2019; Schomburgk and Hoffmann 2023). Overconsumption, waste, and environmental damage may come from this.

Consideration of the social and environmental effects of our purchase decisions is referred to as "responsible consumption," which is closely tied to the concept of "mindful consumption" (Falcão and Roseira 2022; Teufer 2023). It entails backing moral and ethical behaviour, such as purchasing goods from businesses that emphasize worker rights, fair trade, and green manufacturing. Reducing waste, recycling, and considering a product's longevity and durability are other aspects of responsible consumption. Online shopping can encourage responsible consumption by giving consumers access to brands that are ethical and sustainable. To make it simpler for customers to make ethical decisions, several online shops now include filters or categories expressly for eco-friendly, fair trade, or ethically sourced products. Online platforms also give customers a chance to investigate and evaluate the sustainability initiatives of businesses, which aids in their decision-making. Online buying, though, also poses difficulties for ethical consumption. Online orders' packaging and shipping practices may increase carbon emissions and environmental pollution (Guo and Bloemhof-Ruwaard 2019; Rai 2021). The convenience of online shopping has made "fast fashion" and disposable consumer products more popular, which has increased waste and negatively impacted the environment (Degenstein et al. 2020; Stenton et al. 2021).

## 8 Directions for Future Research

The relationship between aspirations for sustainable economic development, online shopping, responsible consumption, and universal Internet access is examined in this chapter using qualitative research methods. Other approaches, like quantitative and mixed-method analysis, may be considered. Other factors, such as environmental and social concerns, may be considered in online shopping and mindful consumption. This chapter primarily focused on the economic aspects of sustainable development. The relationship between online shopping, mindful consumption, and responsible consumption is a dynamic and evolving field of research (Haider et al. 2022). To further understand this relationship, several potential research avenues can be explored. Look into the variables that affect online shopper behaviour and how that affects thoughtful and responsible consumption. Examine how a person's personality, sociocultural influences, and psychological motivations affect their decision-making and behaviour in the digital marketplace. Analyze the efficacy of various approaches, such as providing information on sustainability and moral behaviour, including nudges and prompts for moral decision-making, and creating user interfaces that promote conscious choice. Comparing the environmental effects of traditional brick-and-mortar retail versus those of online buying requires thorough analysis. Examine the carbon footprint resulting from the fulfilment of online orders, the use of packaging, and the logistics of shipment. Compare the environmental effects of various delivery alternatives (such as same-day delivery and consolidated shipments) and consider ways to reduce the negative environmental effects of online purchasing. Examine the working conditions and supply-chain policies used by online merchants, paying particular attention to matters of fair trade, employee rights, and the treatment of gig workers. Investigate how local companies, communities, and socioeconomic disparities are affected by online buying. Analyze how well consumer education programs are doing in encouraging responsible and thoughtful online purchases. Examine how information campaigns, certification labels, and internet tools may increase consumer knowledge and provide people the power to make wise decisions. Investigate the effects of various types of information on consumer choices, such as supply chain transparency and product sustainability ratings. Analyze how rules and policy changes might promote responsible consumption in the context of online purchasing. Examine the success of initiatives like eco-labeling, expanded producer accountability, and rewards for eco-friendly packaging. Examine the long-term impacts of online purchasing on consumer behaviour and environmentally friendly consumption habits. Analyze the lifetime and robustness of products bought online, the influence of peer and product reviews on long-term consumption decisions, and the possibility for habit formation and behavioural change in the setting of digital purchasing. These directions for research can help us comprehend the connections between online purchasing, mindful consumption, and responsible consumption better.

## 9 Conclusion

Several steps can be implemented to promote a favourable relationship between online purchasing, conscious consumption, and responsible consumption. To avoid mindless consumption, people should first put their requirements ahead of their wants and exercise discipline. Further, they ought to investigate and assist businesses that support moral and ethical principles and sustainable practises. Furthermore, online merchants can contribute by offering greater details about product origins, production techniques, and environmental effects. Governments and regulatory organizations can also put regulations into place to support ethical online retail practices, such as rewarding eco-friendly packaging and fostering supply chain transparency. There is a dynamic and interwoven relationship between online purchasing, aware consumption, and responsible consumption. Online buying presents issues that call for cooperation between people, corporations, and politicians to advance ethical and sustainable consumption in the digital age, even while it can help consumers make educated decisions and encourage responsible purchasing.

## References

Ahearne M, Atefi Y, Lam SK, Pourmasoudi M (2022) The future of buyer–seller interactions: a conceptual framework and research agenda. J Acad Mark Sci 50(1):22–45. https://doi.org/10.1007/s11747-021-00803-0

Alexander R (2020) Emerging roles of lead buyer governance for sustainability across global production networks. J Bus Ethics 162(2):269–290. https://doi.org/10.1007/s10551-019-04199-4

Altarturi HHM, Nor ARM, Jaafar NI, Anuar NB (2023) A bibliometric and content analysis of technological advancement applications in agricultural e-commerce. In: Electronic commerce research (Issue 0123456789). Springer, London. https://doi.org/10.1007/s10660-023-09670-z

Anwar MA, Graham M (2021) Between a rock and a hard place: freedom, flexibility, precarity and vulnerability in the gig economy in Africa. Compet Chang 25(2):237–258. https://doi.org/10.1177/1024529420914473

Aziz A, Naima U (2021) Rethinking digital financial inclusion: Evidence from Bangladesh. Technol Soc 64(January):101509. https://doi.org/10.1016/j.techsoc.2020.101509

Banerjee A, Murphy E, Walsh PP (2020) Perceptions of multistakeholder partnerships for the sustainable development goals: a case study of Irish non-state actors. Sustainability (switzerland) 12(21):1–15. https://doi.org/10.3390/su12218872

Barth S, de Jong MDT, Junger M, Hartel PH, Roppelt JC (2019) Putting the privacy paradox to the test: online privacy and security behaviors among users with technical knowledge, privacy awareness, and financial resources. Telemat Inform 41(November 2017):55–69. https://doi.org/10.1016/j.tele.2019.03.003

Beckers J, Weekx S, Beutels P, Verhetsel A (2021) Journal of retailing and consumer services COVID-19 and retail: the catalyst for e-commerce in Belgium? J Retail Consum Serv 62(June):102645. https://doi.org/10.1016/j.jretconser.2021.102645

Belbağ AG (2022) Impacts of Covid-19 pandemic on consumer behavior in Turkey: a qualitative study. J Consum Aff. https://doi.org/10.1111/joca.12423

Billewar SR, Jadhav K, Sriram VP, Arun DA, Mohd Abdul S, Gulati K, Bhasin DNKK (2022) The rise of 3D E-commerce: the online shopping gets real with virtual reality and augmented reality during COVID-19. World J Eng 19(2):244–253. https://doi.org/10.1108/WJE-06-2021-0338

Cernev T, Fenner R (2020) The importance of achieving foundational Sustainable Development Goals in reducing global risk. Futures 115:102492. https://doi.org/10.1016/j.futures.2019.102492

Chawla N, Kumar B (2022) E-commerce and consumer protection in India: the emerging trend. J Bus Ethics 180(2):581–604. https://doi.org/10.1007/s10551-021-04884-3

Chege SM, Wang D (2020) Information technology innovation and its impact on job creation by SMEs in developing countries: an analysis of the literature review. Technol Anal Strategic Manag 32(3):256–271. https://doi.org/10.1080/09537325.2019.1651263

Chen Y, Kumara EK, Sivakumar V (2021) Invesitigation of finance industry on risk awareness model and digital economic growth. Ann Oper Res. https://doi.org/10.1007/s10479-021-04287-7

Chopra S (2016) How omni-channel can be the future of retailing. Decision 43(2):135–144. https://doi.org/10.1007/s40622-015-0118-9

Clark S, Maclachlan M, Marshall K, Morahan N, Carroll C, Hand K, Boyle N, O'Sullivan K (2022) Including digital connection in the united nations sustainable development goals: a systems thinking approach for achieving the SDGs. Sustainability (switzerland). https://doi.org/10.3390/su14031883

D'Cruz P, Du S, Noronha E, Parboteeah KP, Trittin-Ulbrich H, Whelan G (2022) Technology, megatrends and work: thoughts on the future of business ethics. J Bus Ethics 180(3):879–902. https://doi.org/10.1007/s10551-022-05240-9

Dawson S, Kim M (2010) Cues on apparel web sites that trigger impulse purchases. J Fash Mark Manag 14(2):230–246. https://doi.org/10.1108/13612021011046084

Degenstein LM, McQueen RH, McNeill LS, Hamlin RP, Wakes SJ, Dunn LA (2020) Impact of physical condition on disposal and end-of-life extension of clothing. Int J Consum Stud 44(6):586–596. https://doi.org/10.1111/ijcs.12590

Eferin Y, Hohlov Y, Rossotto C (2019) Digital platforms in Russia: competition between national and foreign multi-sided platforms stimulates growth and innovation. Digital Policy Regul Gover 21(2):129–145. https://doi.org/10.1108/DPRG-11-2018-0065

Faccia A, Le Roux CL, Pandey V (2023) Innovation and e-commerce models, the technology catalysts for sustainable development: the emirate of Dubai case study. Sustainability (switzerland). https://doi.org/10.3390/su15043419

Falcão D, Roseira C (2022) Mapping the socially responsible consumption gap research: review and future research agenda, April. https://doi.org/10.1111/ijcs.12803

Fook LA, McNeill L (2020) Click to buy: the impact of retail credit on over-consumption in the online environment. Sustainability. https://www.mdpi.com/819920

Gheihman N (2021) Veganism as a lifestyle movement. Sociol Compass 15(5):1–14. https://doi.org/10.1111/soc4.12877

Gillis A, Krull LM (2020) COVID-19 remote learning transition in spring 2020: class structures, student perceptions, and inequality in college courses. Teach Sociol 48(4):283–299. https://doi.org/10.1177/0092055X20954263

Gonzalez-Arcos C, Joubert AM, Scaraboto D, Guesalaga R, Sandberg J (2021) "How do i carry all this now?" understanding consumer resistance to sustainability interventions. J Mark 85(3):44–61. https://doi.org/10.1177/0022242921992052

Graesch JP, Hensel-Börner S, Henseler J (2021) Information technology and marketing: an important partnership for decades. Ind Manag Data Syst 121(1):123–157. https://doi.org/10.1108/IMDS-08-2020-0510

Grebitus C, Roscoe RD, Van Loo EJ, Kula I (2020) Sustainable bottled water: how nudging and Internet Search affect consumers' choices. J Clean Prod 267:121930. https://doi.org/10.1016/j.jclepro.2020.121930

Gu S, Ślusarczyk B, Hajizada S, Kovalyova I (2021) Impact of the covid-19 pandemic on online consumer purchasing behavior. *J Theor…* https://www.mdpi.com/1262126
Guo X, Bloemhof-Ruwaard J (2019) On integrating crowdsourced delivery in last-mile logistics: a simulation study to quantify its feasibility
Haider M, Shannon R, Moschis GP (2022) Sustainable consumption research and the role of marketing: a review of the literature (1976–2021). Sustainability. https://www.mdpi.com/2071-1050/14/7/3999
Hajian M, Kashani SJ (2021) Evolution of the concept of sustainability. From Brundtland Report to sustainable development goals. In: Sustainable resource management: modern approaches and contexts. Elsevier Inc. https://doi.org/10.1016/B978-0-12-824342-8.00018-3
Hatanaka M (2020) Beyond consuming ethically? Food citizens, governance, and sustainability. J Rural Stud 77(May):55–62. https://doi.org/10.1016/j.jrurstud.2020.04.006
Hayat K, Jianjun Z, Zameer H, Iqbal S (2020) Understanding the influence of corporate social responsibility practices on impulse buying. Corp Soc Responsib Environ Manag 27(3):1454–1464. https://doi.org/10.1002/csr.1898
Hjort K, Hellström D, Karlsson S, Oghazi P (2019) Typology of practices for managing consumer returns in internet retailing. Int J Phys Distrib Logist Manag 49(7):767–790. https://doi.org/10.1108/IJPDLM-12-2017-0368
Ho H, Ito K (2019) Consumption-oriented engagement in social network sites. Eur J Mark 53(7):1355–1377. https://doi.org/10.1108/EJM-11-2017-0809
Hoosain MS, Paul BS, Ramakrishna S (2020) The impact of 4IR digital technologies and circular thinking on the united nations sustainable development goals. Sustainability (switzerland) 12(23):1–16. https://doi.org/10.3390/su122310143
Hosseini Z, Nyholm S, Le Blanc PM, Preenen PTY, Demerouti E (2023) Assessing the artificially intelligent workplace: an ethical framework for evaluating experimental technologies in workplace settings. AI Ethics. https://doi.org/10.1007/s43681-023-00265-w
Ikram M (2022) Transition toward green economy: technological Innovation's role in the fashion industry. Curr Opin Green Sustain Chem 37:100657. https://doi.org/10.1016/j.cogsc.2022.100657
Jnr BA, Nweke LO, Al-Sharafi MA (2021) Applying software-defined networking to support telemedicine health consultation during and post Covid-19 era. Heal Technol 11(2):395–403. https://doi.org/10.1007/s12553-020-00502-w
Judith G, Mark G (2003) The US wine industry and the internet: an analysis of success factors for online business models. Electron Mark 13(1):59–66. https://doi.org/10.1080/1019678032000039877
Kader MS, Rashaduzzaman M, Huang X, Kim S (2023) Influencing factors toward e-shoppers' adoption of green last-mile delivery. Int J Retail Distrib Manag 51(2):220–237. https://doi.org/10.1108/IJRDM-10-2021-0480
Kang J, Martinez CMJ, Johnson C (2021) Minimalism as a sustainable lifestyle: its behavioral representations and contributions to emotional well-being. Prod Consump. https://www.sciencedirect.com/science/article/pii/S2352550921000397
Kerras H, Sánchez-Navarro JL, López-Becerra EI, de-Miguel Gómez MD (2020) The impact of the gender digital divide on sustainable development: comparative analysis between the european union and the maghreb. Sustainability (switzerland) 12(8):1–30. https://doi.org/10.3390/SU12083347
Kumar R, Raj T (2019) Role of entrepreneurship in boosting economic growth and employment in India. SEDME (Small Enterpr Develop Manag Extens J) Worldwide Window MSME Stud 46(4):273–281. https://doi.org/10.1177/0970846419894750
Kushlev K, Proulx JDE, Dunn EW (2017) Digitally connected, socially disconnected: the effects of relying on technology rather than other people. Comput Hum Behav 76:68–74. https://doi.org/10.1016/j.chb.2017.07.001

Lavicza Z, Fenyvesi K, Lieban D, Park H, Hohenwarter M, Mantecon JD, Prodromou T (2021) This is a self-archived version of an original article. This version may differ from the original in pagination and typographic details. Bus Soc 60(2):1

Liu X (2013) Full-text citation analysis: a new method to enhance. J Am Soc Inform Sci Technol 64(July):1852–1863. https://doi.org/10.1002/asi

Martinsons MG (2008) Relationship-based e-commerce: theory and evidence from China. Inf Syst J 18(4):331–356. https://doi.org/10.1111/j.1365-2575.2008.00302.x

Milne GR, Villarroel Ordenes F, Kaplan B (2020) Mindful consumption: three consumer segment views. Aust Mark J 28(1):3–10. https://doi.org/10.1016/j.ausmj.2019.09.003

Modgil S, Dwivedi YK, Rana NP, Gupta S, (2022) Has Covid-19 accelerated opportunities for digital entrepreneurship? An Indian perspective. In: Forecasting and social. Elsevier, London. https://www.sciencedirect.com/science/article/pii/S0040162521008465

Nguyen DH, de Leeuw S, Dullaert WEH (2018) Consumer behaviour and order fulfilment in online retailing: a systematic review. Int J Manag Rev 20(2):255–276. https://doi.org/10.1111/ijmr.12129

Onyango G, Ondiek JO (2021) Digitalization and integration of sustainable development goals (SGDs) in public organizations in Kenya. Public Organ Rev 21(3):511–526. https://doi.org/10.1007/s11115-020-00504-2

Palakshappa N, Dodds S (2020) Mobilising SDG 12: co-creating sustainability through brands. Mark Intell Plan. https://doi.org/10.1108/MIP-08-2018-0360

Papargyropoulou E, Lozano RK, Steinberger J, Wright N, Ujang ZB (2014) The food waste hierarchy as a framework for the. J Clean Prod 76:106–115. https://doi.org/10.1016/j.jclepro.2014.04.020. This

Pathak V, Deshkar S (2023) Transitions towards sustainable and resilient rural areas in revitalising India: a framework for localising SDGs at gram panchayat level. Sustainability (switzerland). https://doi.org/10.3390/su15097536

Purcărea T, Ioan-Franc V, Ionescu ŞA, Purcărea IM, Purcărea VL, Purcărea I, Mateescu-Soare MC, Platon OE, Orzan AO (2022) Major shifts in sustainable consumer behavior in Romania and retailers' priorities in agilely adapting to It. Sustainability (switzerland) 14(3):1–51. https://doi.org/10.3390/su14031627

Radhi M, Zhang G (2019) Optimal cross-channel return policy in dual-channel retailing systems. Int J Prod Econ 210(January):184–198. https://doi.org/10.1016/j.ijpe.2019.01.014

Rai HB (2021) The net environmental impact of online shopping, beyond the substitution bias. J Transp Geogr 93(October 2020):103058. https://doi.org/10.1016/j.jtrangeo.2021.103058

Reglitz M (2020) The human right to free internet access. J Appl Philos 37(2):314–331. https://doi.org/10.1111/japp.12395

Reisdorf BC, Fernandez L, Hampton KN, Shin I, Dutton WH (2022) Mobile phones will not eliminate digital and social divides: how variation in internet activities mediates the relationship between type of internet access and local social capital in Detroit. Soc Sci Comput Rev 40(2):288–308. https://doi.org/10.1177/0894439320909446

Riva F, Magrizos S, Rubel MRB, Rizomyliotis I (2022) Green consumerism, green perceived value, and restaurant revisit intention: Millennials' sustainable consumption with moderating effect of green perceived quality. Bus Strateg Environ 31(7):2807–2819. https://doi.org/10.1002/bse.3048

Rosa JA, Garbarino EC, Maker AJ (2006) Keeping the body in mind: the influence of body esteem and body boundary aberration on consumer beliefs and purchase intentions. J Consum Psychol 16(1):79–91. https://doi.org/10.1207/s15327663jcp1601_10

Rosenberg EL (2016) Mindfulness and consumerism, September. https://doi.org/10.1037/10658-007

Sardjono W, Selviyanti E, Mukhlis M, Tohir M (2021) Global issues: utilization of e-commerce and increased use of mobile commerce application as a result of the covid-19 pandemic. J Phys Conf Ser. https://doi.org/10.1088/1742-6596/1832/1/012024

Scharlemann JPW, Brock RC, Balfour N, Brown C, Burgess ND, Guth MK, Ingram DJ, Lane R, Martin JGC, Wicander S, Kapos V (2020) Towards understanding interactions between Sustainable Development Goals: the role of environment–human linkages. Sustain Sci 15(6):1573–1584. https://doi.org/10.1007/s11625-020-00799-6

Schomburgk L, Hoffmann A (2023) How mindfulness reduces BNPL usage and how that relates to overall well-being. Eur J Mark 57(2):325–359. https://doi.org/10.1108/EJM-11-2021-0923

Setini M, Nyoman N, Yasa K, Supartha IWG, Ayu IG, Giantari K (2020) The-passway-of-women-entrepreneurship-Starting-from-social-capital-with-open-innovation-through-to-knowledge-sharing-and-innovative-performanceJournal-of-Open-Innovation-Technology-Market-and-Complexity.pdf.

Sharma RR (2020) Financial inclusion and economic growth: evidence-based research. Vision 24(2):139. https://doi.org/10.1177/0972262920933503

Sheth JN (2020) Borderless media: rethinking international marketing. J Int Mark. https://doi.org/10.1177/1069031X19897044

Sheth JN, Sethia NK, Srinivas S (2011) Mindful consumption: a customer-centric approach to sustainability mindful consumption: a customer-centric approach to sustainability, February. https://doi.org/10.1007/s11747-010-0216-3

Shkabatur J, Bar-El R, Schwartz D (2022) Innovation and entrepreneurship for sustainable development: lessons from Ethiopia. Prog Plan 160(December 2020):100599. https://doi.org/10.1016/j.progress.2021.100599

Stenton M, Kapsali V, Blackburn RS, Houghton JA (2021) From clothing rations to fast fashion: utilising regenerated protein fibres to alleviate pressures on mass production. Energies 14(18):1–18. https://doi.org/10.3390/en14185654

Susanto H, Yie LF, Mohiddin F, Setiawan AAR, Haghi PK, Setiana D (2021) Revealing social media phenomenon in time of COVID-19 pandemic for boosting start-up businesses through digital ecosystem. Appl Syst Innov 4(1):1–21. https://doi.org/10.3390/asi4010006

Tanhua T, Pouliquen S, Hausman J, O'Brien KM, Bricher P, de Bruin T, Buck JJ, Burger EF, Carval T, Casey KS, Diggs S, Giorgetti A, Glaves H, Harscoat V, Kinkade D, Muelbert JH, Novellino A, Pfeil BG, Pulsifer P, Zhao Z (2019) Ocean FAIR data services. Front Mar Sci. https://doi.org/10.3389/fmars.2019.00440

Tarasova E, Rohracher H (2023) Marginalising household users in smart grids. Technol Soc 72(August 2022):102185. https://doi.org/10.1016/j.techsoc.2022.102185

Tay LY, Tai HT, Tan GS (2022) Digital financial inclusion: a gateway to sustainable development. Heliyon 8(6):e09766. https://doi.org/10.1016/j.heliyon.2022.e09766

Teufer B (2023) How consumer networks contribute to sustainable mindful consumption and well-being, July 2022, 1–28. https://doi.org/10.1111/joca.12536

Thathsarani U, Wei J, Samaraweera G (2021) Financial inclusion's role in economic growth and human capital in South Asia: an econometric approach. Sustainability (switzerland) 13(8):1–18. https://doi.org/10.3390/su13084303

Torkkeli L, Kuivalainen O, Saarenketo S, Puumalainen K (2018) Int n er ar tin g R iew. November, 3–6

Ullah A, Pinglu C, Ullah S, Abbas HSM, Khan S (2021) The role of E-governance in combating COVID-19 and promoting sustainable development: a comparative study of China and Pakistan. Chin Polit Sci Rev. https://doi.org/10.1007/s41111-020-00167-w

Umit Kucuk S, Krishnamurthy S (2007) An analysis of consumer power on the Internet. Technovation 27(1–2):47–56. https://doi.org/10.1016/j.technovation.2006.05.002

United Nations (n.d.) Department of Economic and Social Affairs Sustainable Development. Retrieved May 28, 2023, from https://sdgs.un.org/

van Niekerk AJ (2020) Inclusive economic sustainability: SDGs and global inequality. Sustainability (switzerland). https://doi.org/10.3390/su12135427

Vandenbroele J, Vermeir I, Geuens M, Slabbinck H, Van Kerckhove A (2020) Nudging to get our food choices on a sustainable track. Proc Nutr Soc 79(1):133–146. https://doi.org/10.1017/S0029665119000971

Yang W, Zhang J, Yan H (2021) Impacts of online consumer reviews on a dual-channel supply chain. Omega (UK). https://doi.org/10.1016/j.omega.2020.102266
Young W, Hwang K, McDonald S, Oates CJ (2010) Sustainable consumption: green consumer behaviour when purchasing products. Sustain Dev 18(1):20–31. https://doi.org/10.1002/sd.394
Zhang L, Shao Z, Zhang J, Li X (2022) The situational nature of impulse buying on mobile platforms: a cross-temporal investigation. Electron Commer Res Appl 56(September):101204. https://doi.org/10.1016/j.elerap.2022.101204
Zhu F, Zhang XM (2010) Impact of online consumer reviews on sales: the moderating role of product and consumer characteristics. J Mark 74:133–148

# Digital Divide and Sustainable Development

**Pankaj Misra and Rajeev Srivastava**

**Abstract** At the dawn of twenty-first century, the world is witnessing an unprecedented surge in technological advancements, shaping every aspect of human life. Specifically, rapid advancements in digital technologies have brought about a fundamental shift in how we communicate, work, learn, entertain ourselves and navigate the world. Advancements in digital technology have had a significant impact on various aspects of society, including communication, education, healthcare, entertainment, business, and governance. They have facilitated greater connectivity, increased efficiency, expanded access to information and services and transformed the way we live, work, and interact with each other. However, the benefits of rapid advancement in digital technology has, unfortunately, been restricted to only those having access to it. This inequality among the people towards access to digital technologies is expressed by the term "digital divide" that refers to the gap between those who have access to and can effectively use digital technologies and those who do not. It encompasses disparities in internet access, technology skills, and the ability to utilize digital resources and services. The digital divide has significant implications for individuals and societies. Those who lack access to digital technologies may face difficulties in accessing educational resources, employment opportunities, healthcare services, and participating in civic and democratic processes. It can exacerbate existing social and economic inequalities and further widen the "knowledge gap" between those who are digitally connected and those who are not. This chapter attempts to drive our focus on the digital divide phenomenon and further draws a connection between digital divide and efforts to accomplish sustainable development goals for the humankind. Meanwhile, the discussion throws light upon the phenomenal impact of digital divide on human life, specifically, in the aftermath of the recent pandemic. Also, various factors responsible for digital divide formation,

P. Misra · R. Srivastava (✉)
School of Management, IMS Unison University, Dehradun, India
e-mail: rajeev.srivastava@iuu.ac

P. Misra
e-mail: pankaj.misra@iuu.ac

 
W. Leal Filho et al. (eds.), *Digital Technologies to Implement the UN Sustainable Development Goals*, World Sustainability Series,
https://doi.org/10.1007/978-3-031-68427-2_23

particularly, in the context of developed and developing countries are discussed and multiple government, non-government and private sector initiatives addressing to bridge it are brought to the notice.

**Keywords** Digital divide · Digitalization · COVID-19 · Sustainable development

# 1 Introduction

## *1.1 Origins of Digital Divide*

The term "Digital Divide" was introduced in the mid-1990s, which emerged in the studies of American Department of Commerce and referred to widespread inequalities among different demographic groups towards access to telephones, personal computers (PCs) and the Internet (NTIA 1997). The subsequent studies further showcased existence of widening digital divide in United States of America as well as other countries.

The advancement in digital technologies and prolific production and diffusion of digital devices/services further influenced its embracement in human life, however, simultaneously creating further gap in accessing these between different set of people (Heeks 2022) in different societies and countries. This gap, as envisaged through the term "Digital Divide", has taken varied forms of (in)accessibility with passage of time, as is indicated in the definition provided by OECD. The OECD (2001) defined "Digital Divide" in terms of gap between individuals, households, businesses and geographic areas towards having access to ICTs and use of the internet. Thus, in a broader sense, the Digital Divide refers to the inequality with reference to the access and use of the ICT (Aissaoui 2022). However, traditionally being conceptualized in form of "having or not having access to", the "Digital Divide" now concerns with multifaceted aspects of access, skills, use and performance observed at multiple levels including society, regions and countries (Aissaoui 2022; Park 2017; Thorsen 2006).

## *1.2 Digital Divide Today*

As the use and impact of digital technologies is being overwhelmingly witnessed today in human life, the digital divide presents a grave challenge in front of people, society, regions and nations. Across the world there are approximately four billion people unable to access the digital technology (Source: World Economic Forum 2021), thus accessibility becoming the greatest cause of digital divide. However, beyond the accessibility, there remain lack of knowledge and know how to use the technology, as critical factors further adding to the challenge of digital divide in present times. On one hand, the digital technology, around the world, is providing immense opportunities to the people to pursue their professional growth and to the nations to follow the path for economic growth and development and on the other hand the lack of access to it is rendering the people, regions and countries more vulnerable to become dependent on those having access and knowledge to use the information and communication technologies. During recent years, the world is bracing up with the consequences of one of the history's most fatal pandemic and the signs of digital divide are more visible at societal level, as serving vital human requirements has taken the digital route in order to combat the pandemic. As a result, people having access to the internet and ICT could easily avail critical services like online education and digital healthcare, while others could not.

## *1.3 Global Digital Divide*

The global digital divide reflects the inequality in access and/or capability to use the internet technology among people on an international scale. According to the International Telecommunication Union Report, the internet penetration rate is 87% in developed countries, whereas it is down to 47% and 19% in developing and least developed countries respectively. These worldwide internet penetration figures exhibit the extent of digital divide among developed, developing and least developed countries.

### 1.3.1 Digital Divide in Developing Countries

Though the emergence of information and communication technology has proven to be a boon for developing countries, for, it being used to address various public welfare concerns including financial inclusion, healthcare, education and many more, however, the issue of making the technology accessible by citizens from all sections of society and building the skills to use it effectively, has been a major hurdle in form of the digital divide. In recent years a lot of work has been done to develop internet infrastructure and boost affordability, the lack of internet literacy among various demographic and socio-economic groups still persists. Studies show that specific to the developing countries, there lies an overlapping pattern in literacy divide and digital divide. Furthermore, the usage of technology is conditional upon several individual and socio-economic factors as well like age, employment, education and healthcare, social support. Therefore, the conceptualization of digital divide is not limited to the (in)accessibility to digital technology (called first order digital divide), but also include various associated factors that lead to (in)accessibility. This wider approach to understand digital divide is termed as second order digital divide. For developing economies, the factors contributing to first order digital divide, like lack of reliable infrastructure are as significant as those socio-economic factors culminating into second order digital divide. Moreover, the digital divide renders detrimental effects on people living in developing countries. As we would see in subsequent sections, with major economic activities directly linked with serving common people are increasingly moving on to digital channels, people unable to use digital modes of service channels lag behind in utilizing the benefits of such activities. For example, major sectors like banking, healthcare, education, hospitality and tourism, legal services, retail and transportation etc. facilitate ease of transactions and experience of convenience for their customers, but inaccessibility to such facilities due to first order or second order digital divide, devoid people from living a quality life.

The disparity in ability to use digital technology has repercussions for people aspiring to avail good employment and job opportunities as well. The individuals without access to information and communication technology lack digital skills required to execute a job demanding such skills and thus fail to attain such job opportunities forcing themselves to compromise with less paying type of jobs. Such instances further fuel the economic disparity within the society.

The field of education too is not untouched by inaccessibility issues to digital technologies. During and post COVID-19 period, enhanced emphasis is placed by governments on providing online education to students, thus, access to internet and computing devices is playing a major role in dispensing education to different sections of population. However, lack of reliable IT infrastructure and connectivity issues put vast majority of students at disadvantage, even with certain instances of affordable mobiles/laptops being made available by government.

Different forms of digital divide are reflected across host of the developing economies across the globe. In South East Asian nations like Cambodia, Indonesia, Myanmar and Laos, majority of population lack internet connectivity and are forced to live offline. As around 40% of the population in those countries reside in rural areas (World Bank 2021), lack of digital infrastructure in rural areas gives birth to an imminent urban–rural digital divide. Despite the fact that recent years have witnesses increased internet penetration in those countries, the urban–rural digital divide still persists due to large population still struggling with digital infrastructure and digital literacy issues in rural areas. Moreover, in south east developing economies, though people have smart phones, however, the internet usage does not come by virtue of increased digital literacy, as most of the internet usage is for social media sites only (CNBC 2021). Following figure presents the internet penetration figures in South East Asian developing countries (Fig. 1).

The World bank report (2022) for South Asian developing economies comprising of India, Bangladesh, Bhutan, Sri Lanka, Maldives, Pakistan and Nepal puts forth an interesting fact about the internet utilization in the region. It observes that about 61% people in South Asia could not use the internet despite residing within the range of a telecom network due to lack of affordability. Further, various critical issues like low levels of digital literacy, especially among the women and data protection and cyber security concerns are proving to be counter productive in the process of enhancing internet penetration in the region. However, the problem of resource scarcity could be overcome by sharing the infrastructural resources like mobile towers. Proper design

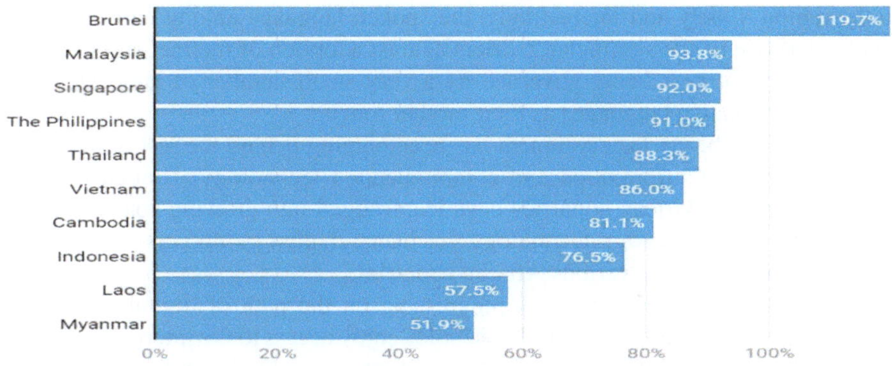


**Fig. 1** Internet Penetration Rate in ASEAN Countries (*Source*: Statistica, 2022)

and implementation of legislative frameworks towards expanding and sharing data networks as well as enabling skill building among people for effective usage of digital resources through bringing in changes in education policies would help tapping the unexplored potential in the region and provide immense growth opportunities through digitalization.

Embracement and usage of digital technology show not-so-different state of affairs in Central Asian and Middle East developing countries than the rest of the developing economies discussed above. Countries in Central Asia though are on the path of expanding their digital transformation set-ups, the adoption of internet services has not gone as far as expected. With around 40% population residing in rural areas, the connectivity problem hovering over 50 Million people has put them at risk of losing out the intended benefits. The urban–rural digital divide could conspicuously be seen in these countries too. People in the Middle east countries use the internet, however, mostly for social media activities. The use of digital technology by the people for larger benefit of economy is far more less, that could be seen in meagre financial transactions taking place through digital technology. Lot of work needs to be done in this area by bringing in transformational changes in policy regulations to boost up the technology usage by people in areas like digital payment.

The Latin American developing countries have witnessed enhanced internet coverage in the region, however the growth rate of internet adoption has been slowing down. The reducing subscription of internet services which leaves more than 200 Million people off the internet is owed to multiple reasons. The variation in income level of people effects internet adoption within and among the countries. The Higher income groups have more internet subscribers as against low income groups. Also, in the same income group there are more percentage of internet subscribers from high income countries like Brazil and Uruguay as against low income countries like Bolivia, Honduras and Paraguay. The digital gender divide exists in Latin American countries as well, manifested in significant proportions in countries like Boliiva, Peru and Ecuador. The Urban–rural divide representing inequality in access to internet too is present in larger magnitude in low income countries as against the richer countries in the region. Lastly and interestingly, the spoken language among native citizens emerged as one of the significant barriers against adoption of internet services. The fact of online content being governed by a very small number of languages bars people having native languages as their mother tongue.

Historically, the African continent has been lagging behind the rest of the world in development of information and communication technology. Though over the preceding decade African countries have leapfrogged into growing space of information and communication technology, lot many challenges do persist producing the digital divide among various African countries.

There are only a few countries progressing on the path of digital revolution. Overall, only around 30 percent of households enjoy access to the internet, as significant barriers in terms of absence of infrastructure, inequality in access and inability to afford internet prevail in the continent. The digital infrastructure deficit in African countries traces its root into lack of basic infrastructure like electricity and roads that hamper the prospect of investment in mobile and wireless infrastructure that needs support for availability of critical resources like energy. Also the nature of terrain comprising deep forest areas require relatively higher investment into setting up of mobile towers by already reluctant investors searching for ease in doing business. Apart from infrastructural issues, the non-availability of skilled users further fuel digital divide. Barring few countries like Nigeria, South Africa and Kenya which have modified their education system to incorporate courses and programs on digital technology, many of the rest of African countries have not been able to respond to this crisis effectively. Since majority of African population is yet to be brought in the ambit of ICT and digital technology usage, African continent provides huge potential and vast opportunities for development and progress of economies of African countries with equitable and effective usage of ICT and digital technology, provided certain essential reforms are worked upon. These reforms need to be focused upon regulatory frameworks, ease of financing in ICT projects, public–private partnerships and skill development.

### 1.3.2 Digital Divide Between Developed and Developing Countries

Nations with relatively developed economies, despite having one of the best internet coverage in world, have their own share of issues while dealing with digitalization. Australia embarked upon the digitalization journey way back in 2009 and has been able to provide 8 million internet connections to its population by 2021, with 2.5 million people still out of ambit of internet. Thus, the disparity in access to internet does exist there, attributed to various factors like age, income, employment status, geographic conditions and language ability. The digital divide between urban and remote areas also exist. On the usage part, the quality issues like slow speed of internet services further induce digital divide among Australian people over and above the demographic and economic variables. As a result, considerable size of population experience struggling with outdated technologies providing far slow speed of the internet. On the other side, a mixed result is observed towards internet coverage and adoption in European countries. Though, overall, 70% households have access to high peed internet connection in the European Union, the member countries have varied amount of access towards high speed internet. While Malta tops the rank with 100% of its household connected to the internet, Spain and Denmark lag behind with 94% and 95% coverage respectively. On the downside, Italy has been able to cover 44%, Cyprus 41% and Greece only 20% of their respective households with internet connections. The urban–rural digital divide is observed in Europe too, as in any other part of the world. Countries like Czech republic and Finland have 7% and 12% of rural coverage figures respectively. At aggregate level, the internet coverage of rural areas in European Union stands at 37%. Due to the geographical conditions and low population density, the rural areas face difficulties in spread of internet coverage in Europe. The United Kingdom, which left European Union in 2020, too wishes to bridge its urban–rural digital connectivity gap by aiming to provide internet coverage to 95% of its population by 2025.

The North American continent comprising some of the world's biggest developed economies, has internet coverage at 92%, wherein the United States of America stands at 92% and Canada at 93%. The Americans realized the necessity of High speed internet during the recent COVID-19 pandemic and since then they have been working hard to ensure availability of high speed internet connections at further reducing cost to everyone. However, as per the latest report available, there are still around 42 Million people lacking the broadband internet connections. The Canadian government is also looking forward to achieve its predetermined goal of providing high speed internet connections to all of its population by 2030, using the latest cutting edge communication technologies like Fixed Wireless Access (FWA) and Satellite connectivity.

Japan, the Asian advanced economy, has its own peculiarity towards the existence of digital divide. The Japanese are facing digital gap between its aging and young generations. The older generation in Japan are largely unfamiliar with the use of digital devices. To overcome this problem, the Japanese government is working on a 5 years aim of training upto 10 Million people to eradicate the unfamiliarity. Emerging Asian economies like China and India have made monumental progress in enhancing the internet coverage and availability of digital devices across its population. As per the data available from Chinese government, around 98% of Chinese population is provided internet coverage through fibre optics and 4G network. Thus going by the government claim, the first level digital divide accessibility issues towards internet might have been overcome in China, however, the urban–rural digital divide could not be ruled out in totality. Further, towards the usage of internet, several factors still barricade the progress, including digital literacy/skills and cost of internet services. The digital movement in India has taken off recently, however, in an impressive mode. For a country as large and as diversified, as India is, the digitalization process has never been easy. Though the Indian government launched digitalization movement in 2015 through its flagship program called "Digital India", the rapid pace of internet penetration came as a response-effect to the COVID-19 pandemic, which enhanced the internet coverage across the country. In 2022, more than half of the Indian population had access to internet with total number of internet users standing at 720 Million. The interesting fact about Indian internet growth story is that it has been led by rural users, more than their urban counterparts. By the end of the year 2022, the rural India came at the forefront of internet usage with 425 million rural users surpassing 295 million urban users in daily internet usage. Thus, the Indian digitalization process presents us the potential of rural users in enhancing internet usage, turning the urban–rural gap story upside-down as against many other developing and developed economies. Further, though India also claims to be the world's second largest internet market, it has many challenges way ahead in its kitty, including internet speed, lack of digitally skilled manpower and potential threat from cyber security issues and data privacy.

The digital divide, as discussed above, could be seen present within nations across the world, in some or the other form and in varied extent. When looked at global level, the world appears to be fragmented between two conspicuous categories of nations: Developed and developing countries. Though the developed countries have their own set of challenges for eradicating the domestic digital divide, they clearly are faring well ahead in coping up with digital divide challenge as against developing nations, on the back of abundantly available resources and advancement in technology. Towards attaining the sustainability goals, it is, however, essential to reduce this gap in the capability to access and use the internet and digital technology, between developed and developing countries. As inequality, in any form, would work against the basic premise of sustainability, the removal of inequality in accessing and using the digital resources between developed and developing nations would be the only way forward towards attaining universal sustainability. Thus, the digital divide "between" developed and developing nations is a great area of concern, as has been recognized by various governmental and non-governmental bodies across the world. Today, around 3 billion people across the world are unable to connect to the internet and majority of these belong to developing countries. This itself underlines the need to bridge the gap, as is asserted in an UNITED NATIONS E-GOVERNMENT SURVEY 2022, showcasing developed countries overwhelmingly ranked higher in the list of nations implementing the digital projects. Certainly, lot before than the income and resources availability, the developing countries would need political will and fundamental commitment to expedite their efforts to boost up infrastructure development and adoption of digital technology and internet among its citizens, resulting in closing the digital divide with developed nations.

## 2 Literature Review

The recent literature on digital divide phenomenon has largely been focused upon developing perspectives of digital divide concept and identifying the factors affecting it (Lythreatis et al. 2022). Further, different studies have examined the effect of multiple facets of digital divide on human life (Vassilakopoulou and Hustad 2023). Cheshmehzangi et al. (2022) attempted to find the impact of digital divide on mental health of people in specific context of Covid-19. Morris et al. (2022) reflected their concern about digital divide impact on rural small scale entrepreneurs and emphasized upon the availability of reliable internet connections for proper functioning of rural SMEs. The social exclusion engendered by digital divide was brought to the fore by Liao et al. (2022) who suggested an inclusive approach to deal with it. Mubarak and Suomi (2022) extended the social exclusion concern for elderly aged people, who, due to their inaccessibility and lack of skills to use digital devices and services, run the risk of getting socially disconnected. The digital divide affecting elderly people has been termed as 'grey digital divide'. Responding to the 'grey digital divide' challenge, Charness and Boot (2022) and Božić (2023) proposed to utilize artificial intelligence and extended reality to facilitate reducing age related digital divide.

Another impact of digital divide was made visible by Kolotouchkina et al. (2022) on urban city development measures, which are largely driven by digital technologies now a days,, showcasing the urgent need to address the driving factors of digital divide. However, amongst all, the most effected dimensions of human life has been learning and education that has taken a major hit due to the digital divide. Specifically, in the backdrop of the pandemic, the sudden and disorderly enforcement of online education has left marginalized and underprivileged people, out of the mainstream education due to the inequality in access to the IT infrastructure, i.e. digital divide (Krishnaswami et al. 2022;Van De Werfhorst et al. 2022). The setback has been more grave for female students, indicating the presence of gender based digital divide (Mathrani et al. 2022).

Additionally, several other studies in the extant literature have worked to examine the impact of digital divide on different modern day essential activities related primarily to finance (Odei-Appiah et al. 2022), healthcare (Frydman et al. 2022; Kim and Kwon 2022; Kozelka et al. 2023; Watson et al. 2022), work from home (Budnitz and Tranos 2022) and education (Al Husseiny 2023; Baidoo-Anu et al. 2023; Miras et al. 2023).

In furtherance to the above-mentioned scholarly attempts studying digital divide phenomenon in specific context, this chapter attempts to build a comprehensive understanding of readers on emergence of digital divide and its intensifying influence across the world posing great challenges amid pandemic situation engulfing both developing and developed countries. The chapter also underscores its future implications towards attainment of sustainable development goals.

## 3 Contemporary Issues and Challenges: COVID-19 and Digital Divide

When sudden spread of COVID-19 pandemic engulfed the whole world and human life seemed to be pushed down to halt due to imposed restrictions on socio-economic activities, the resumption of economic growth was made possible through transformation of major activities into online mode. However, such a transformation required major boost in setting up of digital infrastructure, that was missing or insufficient in most of the non-developed countries. Further, the need for digital revamp to support the increasing demand of online activities was felt across the countries along with the realization of a big gap in accessibility to digital resources between developed and developing countries.

## *3.1 Digital Divide Exposure Under the Pandemic*

The COVID-19 pandemic has certainly influenced acceleration in equipping the nations with digital technology (Pandey and Pal 2020), as needed to support increasing requirement of digital services and activities. However, it has also ended up in bringing an underlying fact of inaccessibility to the digital resources of majority of world population, to the fore.

As the pandemic has not only pushed the vast array of essential services and public utilities, like education, healthcare, transportation, legal and retail services etc. to take delivery path through the digital route, but also has created new employment opportunities in "work from home" mode, the criticality of access to internet and other digital resources has immensely intensified. Therefore, the global scenario of unavailability or inequality in access to digital resources among different sets of world populations would be unimaginable, while visualising the path to the sustainable economic growth in future.

The present time, however, presents certain hard facts about inequality in availability/usage of the digital resources, also called the digital divide within and among the different countries of the world. These hard facts have been encountered as a consequence of spread of pandemic in recent years.

In a survey conducted in New York city, named "2021 Unheard Third survey on the digital divide", it was found that the COVID-19 pandemic has aggravated the inequality in accessing.

reliable internet connections that has suddenly become a basic requirement to learn and work effectively. A graphical excerpt from the survey (Fig. 2) shows the disparity in affordability and access to the internet/ digital devices among people from low and moderately high-income segments.

The access to the internet has been hit the most for economically vulnerable people across the countries. The following Fig. 3 shows the percentage of such people from the world as well as from developing and developed countries having access to the internet during pandemic times.

Thus, we see that highly economically vulnerable people, whether from developed or developing countries have had least access to the internet.

Apart from income, the other demographic variables like age and gender also reflect the presence of digital divide in corresponding population. For example, in Japan, there exists significant difference in the internet usage among people belonging to low age and old age groups, as shown in the Fig. 4.

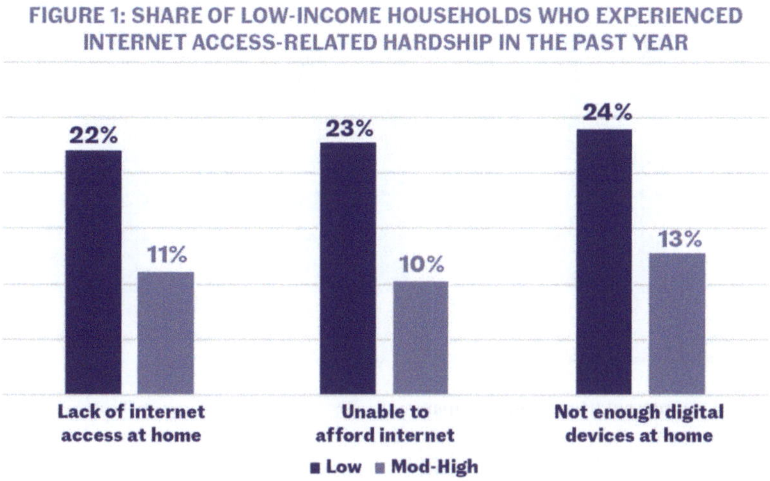


**Fig. 2** Disparity in internet access based on affordability. *Source*: Joseph and Lew (2022). Retrieved from: https://www.cssny.org/news/entry/disconnected-the-digital-divide-disrupted-schooling-in-nyc

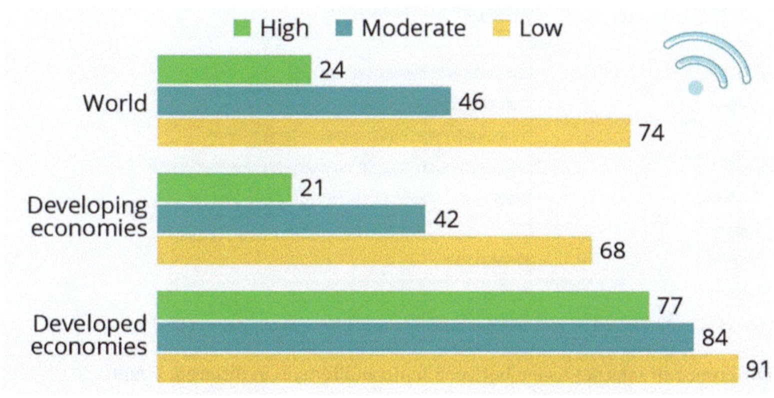


**Fig. 3** Internet access among different economic sections (*Source*: Statistica, 2021)

The digital divide is reflected in internet usage by people from different gender as well. Females are found to be less users of computer and internet as against their male counterparts in different countries (Acilar and Sæbø 2023), due to multiple reasons of trust, digital literacy, digital device ownership, data security and cyber crime instances, as shown below in Fig. 5.

The urban–rural digital divide is another form of digital divide that exists in almost all the countries across the world (Mammen et al. 2023). Though, the gap in internet coverage between urban and rural areas is increasingly getting narrowed down, the

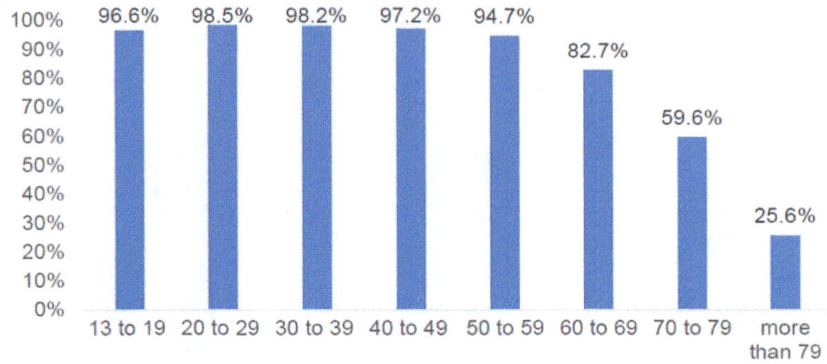


Source: Ministry of Internal Affairs and Communications, Government of Japan, July 2021.

**Fig. 4** Age-wise internet usage data in Japan

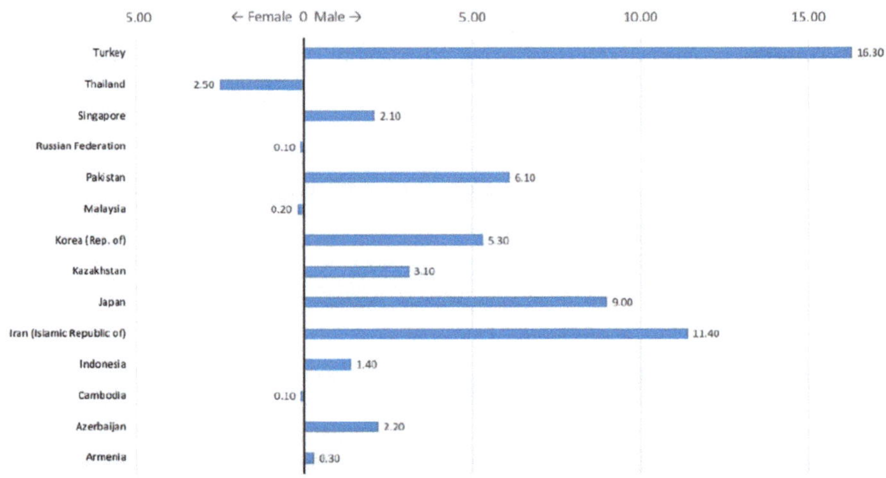


Source: Produced by ESCAP using ITU, "World Telecommunication/ICT Indicators Database 2021", July 2021.

**Fig. 5** Difference in internet usage between male and female in different countries

major challenge remains in terms of less usage of internet in rural areas in comparison with the urban ones.

Thus, we see that the inequality in internet access/ usage serves as a major constituent of digital divide that is reflected across different sets of people in the world. The digital divide, that the world got exposed to during the pandemic time, now needs to be bridged, as the world wishes to move on ahead on the path of sustainable economic progress with collective contribution and collaboration among developed and developing countries.

### 3.2 *Essentiality of Bridging the Gap Post Pandemic*

During the pandemic times, restrictions on physical movement forced the world to look for and invent the use of digitally enabled processes for continuance of hassle-free economic activities. The use of information and communication technology came to rescue the halted momentum of human life and extended the ray of hope to the entire world. It, however, also unearthed a big gap that existed between those who have had access and ability to use the technology and those who have not. This gap, also termed as "digital divide" made the governments over the world realize people's unmet need to have access to the internet and digital devices to navigate through the life that has never been the same, post the pandemic.

Today, as the world has moved on to accommodate a new way of life, that hugely relies on remote access (Smite et al. 2023) to almost any thing that people need, it is the time to ensure that the availability and access to digital technologies, especially the internet, is provided to all to facilitate this proposition.

Though the COVID-19 pandemic has subsided, it has brought an issue pertaining and critical to every human being: the health care and medical services, to the fore. The experience learnt through the pandemic has shaped our wisdom towards extending the coverage of effective healthcare services to the last mile. The digital technology has been able to overcome age old difficulties coming in the way of extending the healthcare reach to all intended recipients. On the back of unprecedented scale of advancement in digital technology and its application in healthcare, the world has now become capable to envisage "Health for all", as reflected on the World Health Day in April 2023. The intervention of digital technology in healthcare is not only able to generate the much needed reach through provisions like remote diagnosis and telemedicine (Ahmad et al. 2022), but also is helpful in bringing down the overcall healthcare costs. For example, as per the World Health Organization's latest estimates, 77% of the people effected from fatal non-communicable diseases worldwide, belong to low- and middle-income countries and the related healthcare cost is more than Two Trillian Dollars per year, that could be effectively reduced with the help of digital healthcare solutions.

However, as we have discussed, since the advancement and usage of digital technologies has been limited to certain parts of the world, the access to digital healthcare solutions is still barricaded by the digital divide phenomenon (Gallegos-Rejas et al. 2023), which includes lack of high speed internet access, low levels of digital literacy and variation in cultural norms pertaining to health treatment.

Apart from the healthcare, various other aspects that are essential to the sustainable economic growth are not untouched from the prospects of being positively affected with the use of digital technology. For example, providing financial services to the underprivileged people in order to bring them into mainstream economy has become possible with the help of information and communication technology based digital financial services, that are extended to the people living in far flung areas facing difficulty in reaching out to financial set ups located in urban locations. Thus, the use of digital technology in delivering financial services has enabled financial security for underprivileged and underserved people and communities, by serving them the means of savings, investment and insurance. However, the digital divide would deprive other people of such benefits due to lack of digital infrastructure and resources.

Another important dimension of human life that has become one of the largest beneficiaries of digitalization, is learning and education. During the COVID-19 pandemic, when most of the essential activities were either slowed down or came to halt, the learning process continued to progress because of seamless delivery of education through digital mediums (Guo and Wan 2022). Be it schools, colleges or universities, implementation of ICT had made it possible to run the classes over the internet that could be attended by the students without stepping out and of course, from the comfort of their homes. The trend of taking classes from anywhere has now become a new norm, however, only for the people who have access to the internet. Even during the pandemic times, the educational institutions across the African countries witnessed drastic decrease in enrolment due to difficulties in internet access and usage.

Such instances of digital divide across the regions and the world population still persist, that are needed to be dealt with to bring about the benefits of digitalization to everyone. The digitalization process, that has rescued the human life from ill effects of pandemic in the past, should also become the pathway that would help deliver a bright, promising and sustainable future to the whole world.

## 4 Initiatives: Bridging the Digital Divide

As humans are becoming more dependent on technology usage, the chances of widening of the gap in access and usage of technology among them also increase.

Even in present times, the digital divide persists not only in developing/least developed countries, but also in the developed economies, albeit in different forms. The United States of America, wherein internet penetration rate stands at 91.8%, there are 8.3 million Americans who don't have access to high-speed internet. Similarly, in Australia around 2.8 million people, residing in regional places, do not have access and know-how to use an affordable internet. For the developing/least developed countries, the scenario of digital divide is more worrisome. Therefore, it is highly imperative for the leading institutions across the world to envision and implement consistent efforts for reducing the digital divide by making available the required infrastructure in place and sharing it with those having lack of access and affordability by extending support and cooperation with them.

In the above context, a lot of effort is seen at the end of the government, non-government and private bodies across the world.

## 4.1 Government and Non-government Initiatives

Undoubtedly, overcoming the digital divide is in best of the interests of not only the common population, but also for the government and commercial entities. The public, at large, intend to get benefitted from government's public welfare policies delivered with help of digital channels. As per the latest EY Connected Citizens survey, people in emerging economies like India, Brazil, South Africa etc. carry lot of expectations and interest towards the use of digital technology to gain advantage of digital government utilities. To fulfil the public expectations, governments further need to upgrade the ICT infrastructure. Furthermore, keeping in view the fact that majority of population having lack of access to the internet are clustered in developing economies, the government in such countries carry a larger responsibility of developing and boosting up their digital infrastructure. In this respect, the Indian government has taken major leap forward by launching multiple significant initiatives one after the other targeting different fronts concerning the digital divide. For example, Digital India and DISHA (Digital Saksharta Abhiyan) projects towards enhancing digital literacy among citizens, BharatNet project for connecting rural areas through optical fibre and DIKSHA (Digital infrastructure for knowledge sharing) project for extending digital transformation in education sector are some of the milestone projects that are shaping the course of digital transformation in the country. The Non-government (NGOs) and private sector organizations have also put their efforts to bridge the digital divide effecting people from marginalized sections, by delivering digital literacy programs to the children which now spans across multiple states in the country.

The Brazilian government has also responded to its digital divide challenge that has largely affected the students and people residing in rural areas. The country, which captures 40% of Latin American IT market, has invested heavily in strengthening digital infrastructure to increase the coverage of internet across rural and far flung areas. With the focus on 5G network services and increased investment in data centres, the government of Brazil came up with memorandum of understanding with SpaceX for operating Starlink internet services. It has greatly helped the government in connecting educational institutions with hi-speed internet services and covering the remote areas in the country.

Beyond the infrastructure development, the government has further focused upon building up and enhancing digital skills among low-income girls and women to bridge the gender digital divide. The E-commerce regulation framework introduced by the government is also well applauded.

The government of South Africa has been so far successful in bringing social development in the country on the front of critical areas like health, education, housing and energy. However in order to achieve inclusive economic progress, the government has recognized the importance of digital channels through gradually setting up digital infrastructure in the country. Recently, it has planned to invest 6 billion ZAR to set up a nationwide broadband internet network for connecting various government departments and public utility institutions. Further, with the aim to make every south African digitally skilled, the government has been working on "Digital and Future Skills Strategy" in partnership with educational institutions and other private sector stakeholders. The government has also facilitated multiple initiatives with the help of international organizations like United Nations and World Economic Forum. To check in the gender digital divide, the "African Girls Can Code Initiative (AGCCI)" was undertaken in partnership with United Nations that helped girls from public schools to learn programming skills and enabled them to sharpen their problem solving skills. For making the business grow, the use of advanced digital technology is worked out with the help of deliberations with the World Economic Forum.

However, at continental level, in Africa, that contains most of the least developed countries, other governments have relatively slowly started investing into setting up digital infrastructure, primarily in the field of public financial management to monitor and control the public expenditure. For them, the next logical step would be to build up digital services focused on its citizens.

Worldwide, the governmental setups from different countries, non-government organizations and agencies from United Nations are collaborating to bring about significant initiatives to reduce different forms of digital divide in different parts of the world. For example, realizing the gender digital gap to be the major factor thwarting women's economic development, the U.S. Agency for International Development (USAID) in collaboration with the Bill and Melinda Gates Foundation has launched "Digital Economy Fund" to reduce digital gender divide, specifically in African countries. Later, the fund has received further support from countries like Republic of South Korea. Also the global leading institutions like World Bank and International Monetary Fund are working to facilitate the worldwide collaboration into bringing digital connectivity in developing economies so as to reduce the digital divide and help the world economy recover from the slowdown caused by recent pandemic situations.

### *4.2 Private Sector Initiatives*

Though the government led initiatives towards bringing in digital transformation are crucial for reducing the digital divide, the rapid pace needed for such efforts to achieve the said objective would be possible with participation of the private sector. Also, post the pandemic, since the internet access is realized more as a necessity than a luxury, ensuring everyone to have it would need more than the government led initiatives. Taking note of it, the private technology giants have stepped up their efforts in this direction. For example, Google has not only announced contribution of $10 billion for digitisation in emerging economy like India, it has also entered into agreement with Indian companies to manufacture affordable smart phones. Realizing the lack of computer skills as a major constituent of digital divide, another technology major has Amazon launched "Amazon Future Engineer Program" for indigenous African students to enable them learn computer skills needed to fulfil the future demand of computer science professionals. Amazon's effort to reduce digital divide are not limited to developing countries. Back home at the world's most advanced country, the United States, it has invested more than ten billion dollars dedicated to "Project Kuiper", aimed for providing affordable broadband network for people who lack internet access in United states and around. On similar lines, another tech giant Facebook claimed to have invested in subsea Fiber projects to bring internet access to African, Middle East and Asia pacific regions. The Microsoft has launched "Airband initiative" to provide internet access as well as raise digital skills among marginalized communities across the world.

The above mentioned initiatives led by major global tech companies are certainly exerting catalytic influence on worldwide efforts to reduce the digital divide at mass level, however, the local level small businesses also are coming forward to empower people by raising their digital skills and preparing them to be the part of future workforce. Such a contribution by local businesses will not only help minimize the digital divide further, but also benefit future economy which is increasingly becoming digitalized. For example, taking the support of local NGOs, many small scale businesses are imparting digital literacy among people from marginalized and low income communities. Initiatives like "Computing for all", "Internet essentials", "Rhode Island Latino Web Project" are some of the examples from the United States of America that replicate the efforts by small businesses in other parts of the world aimed at reducing digital divide and building capacity to embrace digitalization in future.

## 5 Future Implications: Digitalization and Sustainable Development

As the deadline to attain Sustainable Development Goals (SDGs) set by world's nations is reaching close and fast, the presence of digital divide appears to be a threatening reminder to intensify our efforts to bring the world population under the ambit of internet coverage, that would set the stage for attaining the SDGs in stipulated timeline.

Post the COVID-19 pandemic, sudden changes forced into human life altered the economic activities all across the world and made the digitalization the basic need of human life. Therefore, until the "newly formed basic need" is fulfilled and met by all people, the future development goals are hard to be met. In the light of this argument, the materialization of the envisaged SDGs needs significant progress in ensuring equal access to the internet by people from different sets of geographic, social and economic background. As the critical aspects of human life, including, health, education, movement and logistics and availability of essential goods and services have strongly been influenced by digital transformation, the common man is left with no choice except to embrace these changes by becoming digitally literate and adaptive to digital devices.

As we proceed towards the timeline set by United Nations member countries in 2015 for achieving the 17 identified sustainable development goals, a recollect of the underlying interconnections between specific targets relating to the different SDGs becomes highly essential. Each of these 17 SDGs relate to different dimensions of human life ranging from poverty, hunger, health and education to clean energy, economic growth and so on and collectively cover economic, social and environmental aspects of human life. When we look at the recent steps undertaken for bringing in development and progress in human life, it is observed that a range of initiatives, be it various schemes promoting social welfare and poverty reduction, delivery of health awareness programs and medical services at distant places, imparting distance education in far flung areas, promoting cashless transactions or bringing in seamless experience of transparent and corruption free essential public services, all of these relate to the economic, social and environmental concerns of human life and seem to be feasible and deliverable at large scale, only through the driving force of digitalization.

Thus, the digitalization process is a clearly visible driving factor that has the capability to address different targets set under the 17 SDGs.

For example, with the help of digital infrastructure, affordable digital financial services could help people secure and invest their savings and thus help their upliftment from poverty. The ICT enabled agricultural practices could assist farmers in increase of agricultural outputs resulting in eradication of hunger. With hi-speed internet connectivity, the digital health solutions like remote medical counselling and online pharmacy services would facilitate improvement in health and wellbeing of people. The digital connectivity further assists in imparting quality and affordable education beyond the time and space limitations and thus extend the reach of opportunities to women and marginalized people. The advanced digital technologies have the capability not only to address the new age challenges like clean water and energy, climate change, smart city infrastructure and e-waste reduction but also to take cognizance of changing life patterns on land and water for recommending protection measures of endangered species. Worldwide, e-governance services could transparently deliver the government services to respective citizens of different countries, and could be availed without any discrimination, thus helping in restoration of peace and justice. This, in turn, would help establish and promote an environment conducive for economic growth and development and would facilitate nations to come together with mutually beneficial partnerships and collaborations aiming for a sustainable future for people and humanity.

## 6 Conclusion

The progression of digital technologies in recent times has simultaneously created an undesirable gap between quality of life of people that have differential access to it. This gap is termed as "digital divide" and needs an immediate attention of policy makers towards bridging it at all possible levels. Worldwide, all the developed and developing countries need to realise that in absence of collectively dedicated efforts towards removal of factors responsible for forming digital divide, the benefits of digitalization could not be ascertained for the citizens. In modern times, the digital technology has become essential for ascertaining quality of life for all human beings irrespective of their demographics and hence has the capability to play a pivotal role in realizing the attainment of worldwide sustainable development goals. It becomes clear that the present time is the most opportune for intensifying the efforts to provide internet access to the worldwide uncovered population as the reach of digitalization to everyone would help to attain the sustainable development goals within the stipulated time before the year 2030. Therefore, bridging the digital divide, that has served as a hinderance in presenting the benefits of digitalization to the worldwide human population, through collective and effective efforts by world nations, is now the need of the hour.

## References

### *Published Resources*


Acilar A, Sæbø Ø (2023) Towards understanding the gender digital divide: a systematic literature review. Global Knowl Memory Commun 72(3):233–249

Ahmad KAB, Khujamatov H, Akhmedov N, Bajuri MY, Ahmad MN, Ahmadian A (2022) Emerging trends and evolutions for smart city healthcare systems. Sustain Cities Soc 80:103695

Aissaoui N (2022) The digital divide: a literature review and some directions for future research in light of COVID-19. Global Knowl Memory Commun 71(8/9):686–708

Al Husseiny F (2023) Pandemic aftermath on education in the UK: evolving digital divide. In: Perspectives on enhancing learning experience through digital strategy in higher education. IGI Global, pp 207–225

Baidoo-Anu D, Gyamerah K, Munezhi M (2023) Digital divide in higher education in Sub-Saharan Africa: evidence from online learning during the COVID-19 pandemic. SN Soc Sci 3(8):122

Božić V (2023) Artifical intelligence as the reason and the solution of digital divide. Lang Educ Technol 3(2):1

Budnitz H, Tranos E (2022) Working from home and digital divides: resilience during the pandemic. Ann Am Assoc Geogr 112(4):893–913

Charness N, Boot WR (2022) A grand challenge for psychology: reducing the age-related digital divide. Curr Dir Psychol Sci 31(2):187–193

Cheshmehzangi A, Zou T, Su Z (2022) The digital divide impacts on mental health during the COVID-19 pandemic. Brain Behav Immun 101:211–213

Frydman JL, Gelfman LP, Goldstein NE, Kelley AS, Ankuda CK (2022) The digital divide: Do older adults with serious illness access telemedicine? J General Internal Med 2022:1–3

Gallegos-Rejas VM, Thomas EE, Kelly JT, Smith AC (2023) A multi-stakeholder approach is needed to reduce the digital divide and encourage equitable access to telehealth. J Telemed Telecare 29(1):73–78

Guo C, Wan B (2022) The digital divide in online learning in China during the COVID-19 pandemic. Technol Soc 71:102122

Heeks R (2022) Digital inequality beyond the digital divide: conceptualizing adverse digital incorporation in the global South. Inf Technol Dev 28(4):688–704

Kim SH, Kwon IWG (2022) Digital divide in the rural healthcare market in the United States: health equity and inclusiveness. Int J Bus Manag 17(7):1

Kolotouchkina O, Barroso CL, Sánchez JLM (2022) Smart cities, the digital divide, and people with disabilities. Cities 123:103613

Kozelka EE, Acquilano SC, Al-Abdulmunem M, Guarino S, Elwyn G, Drake RE, Carpenter-Song E (2023) Documenting the digital divide: identifying barriers to digital mental health access among people with serious mental illness in community settings. SSM-Mental Health 4:100241

Krishnaswami M, Iyer LS, John C, Devanathan M (2022) Countering educational disruptions through an inclusive approach: bridging the digital divide in distance education. In: Socioeconomic inclusion during an era of online education. IGI Global, pp 204–224

Liao SC, Chou TC, Huang CH (2022) Revisiting the development trajectory of the digital divide: a main path analysis approach. Technol Forecast Soc Chang 179:121607

Lythreatis S, Singh SK, El-Kassar AN (2022) The digital divide: a review and future research agenda. Technol Forecast Soc Chang 175:121359

Mammen JT, Rugmini Devi M, Girish Kumar R (2023) North-South digital divide: a comparative study of personal and positional inequalities in USA and India. Afr J Sci Technol Innov Dev 15(4):482–495

Mathrani A, Sarvesh T, Umer R (2022) Digital divide framework: online learning in developing countries during the COVID-19 lockdown. Glob Soc Educ 20(5):625–640

Miras S, Ruiz Bañuls M, Gómez Trigueros IM, Mateo Guillén C (2023) Implications of the digital divide: a systematic review of its impact in the educational field

Morris J, Morris W, Bowen R (2022) Implications of the digital divide on rural SME resilience. J Rural Stud 89:369–377

Mubarak F, Suomi R (2022) Elderly forgotten? Digital exclusion in the information age and the rising grey digital divide. Inquiry J Health Care Organ Provis Financ 59:00469580221096272

NTIA (1997) Falling through the net: defining the digital divide. Department of Commerce, Washington, DC. Available online at: https://www.ntia.doc.gov/report/1999/falling-through-net-defining-digital-divide

Odei-Appiah S, Wiredu G, Adjei JK (2022) Fintech use, digital divide and financial inclusion. Digital Policy Regul Govern 24(5):435–448

OECD (2001) Rapport annuel. OECD Publication, Paris

Pandey N, Pal A (2020) Impact of digital surge during Covid-19 pandemic: a viewpoint on research and practice. Int J Inf Manage 55:102171

Park S (2017). Digital capital. Palgrave MacMillan, London. ISBN 978-1-137-59332-0. OCLC 1012343673

Smite D, Moe NB, Hildrum J, Gonzalez-Huerta J, Mendez D (2023) Work-from-home is here to stay: call for flexibility in post-pandemic work policies. J Syst Softw 195:111552

Thorsen C (2006) Tech tactics: technology for teachers, 2nd edn. Pearson Education Inc., Boston

Van De Werfhorst HG, Kessenich E, Geven S (2022) The digital divide in online education: Inequality in digital readiness of students and schools. Comput Educ Open 3:100100

Vassilakopoulou P, Hustad E (2023) Bridging digital divides: a literature review and research agenda for information systems research. Inf Syst Front 25(3):955–969

Watson A, Mellotte H, Hardy A, Peters E, Keen N, Kane F (2022) The digital divide: factors impacting on uptake of remote therapy in a South London psychological therapy service for people with psychosis. J Ment Health 31(6):825–832

## *Web Resources*

https://www.itu.int/en/mediacentre/backgrounders/Pages/icts-to-achieve-the-united-nations-sustainable-development-goals.aspx

https://jointsdgfund.org/event/digital-divide-actions-towards-2030-agenda

https://sdgactionzone.org/class/leave-no-one-behind-bridging-the-digital-divide/
https://europeansting.com/2023/03/27/why-we-cant-meet-the-sdgs-without-ending-the-digital-divide/
https://www.un.org/sustainabledevelopment/blog/2023/08/what-is-sustainable-development/?gclid=CjwKCAjwsKqoBhBPEiwALrrqiMU5SEkJIA-tNzsPKs1TAnke0y4QatnyVtrT_svY5KKTAaAax2waVhoCwssQAvD_BwE
https://www.streetbusinessschool.org/blog/economic-empowerment-sustainable-development-goals/?gclid=CjwKCAjwsKqoBhBPEiwALrrqiCuiZ4MBkCUkbNI7YRI8lB7gPtNTt7z7nwP7luR
https://ideas.repec.org/a/eee/tefoso/v150y2020ics0040162519310637.html
https://digileaders.com/bridging-the-digital-divide-how-digital-inclusion-breeds-sustainable-development/
https://www.wix.com/blog/bridging-the-digital-divide
https://tcglobal.com/bridging-the-digital-divide-an-urgent-need/
https://www.thenationalnews.com/business/economy/sustainable-technology-required-to-address-gender-inequality-at-workplace-say-experts-1.1072910
https://www.thenationalnews.com/business/technology/pandemic-highlights-the-need-to-bridge-the-digital-divide-experts-say-1.1
https://papers.ssrn.com/sol3/papers.cfm?abstract_id=4520785
https://www.imf.org/en/Blogs/Articles/2020/11/05/blog-bridging-digital-divide-to-scale-up-covid19-recovery
https://www.weforum.org/agenda/2016/05/4-billion-people-still-don-t-have-internet-access-here-s-how-to-connect-them
https://www.weforum.org/agenda/2021/01/covid-digital-divide-learning-education/
https://www.itic.org/news-events/techwonk-blog/latin-americas-digital-transformation-accelerating-momentum
https://blogs.iadb.org/salud/en/what-are-the-challenges-for-digital-health-in-latin-america-and-the-caribbean/
https://www.adlittle.com/en/insights/viewpoints/digital-transformation-developing-countries
https://oecd-development-matters.org/2022/03/02/a-faster-path-to-digital-transformation-in-latin-america/
https://www.ntu.edu.sg/cas/news-events/news/details/the-challenges-and-opportunities-of-bridging-the-africa-s-digital-divide
www.worldbank.org/en/programs/de4lac#:~:text=35%20percent%20of%20adults%20in,slow%20transition%20to%20digital%20economy
https://www.uschamber.com/on-demand/technology/how-government-and-business-leaders-can-grow-africas-digital-economy
https://www.weforum.org/agenda/2022/09/eu-high-speed-internet-digital-divide/?DAG=3&gclid=CjwKCAjwo9unBhBTEiwAipC116bnovAhfGHREYSJSW7M9b1-1ykn7iAVO7GnBOvZ0XdZt30Fbfv7YhoCIgcQAvD_BwE
http://blog.govnet.co.uk/technology/the-uks-digital-divide-how-is-it-intensifying-after-covid-19
http://global.chinadaily.com.cn/a/202006/01/WS5ed47cc5a310a8b241159e69.html
https://www.chinahighlights.com/travelguide/article-internet-access-in-china.htm
https://economictimes.indiatimes.com/tech/technology/rural-india-pips-urban-india-in-internet-usage-with-44-more-users-report/articleshow/98704031.cms
https://press.un.org/en/2021/dsgsm1579.doc.htm
https://www.ey.com/en_gl/international-development/how-governments-in-developing-countries-can-close-the-digital-gap
https://www.undp.org/blog/evolving-digital-divide
https://www.weforum.org/agenda/2023/04/world-health-day-how-ai-and-digital-health-healthcare/
https://www.intersystems.com/pulse-blog/bridging-the-digital-divide-in-healthcare/

https://parispeaceforum.org/en/2021-debate-insights-disconnected-from-the-future-bridging-the-digital-divide-for-a-universal-post-covid-recovery/
https://www.cnet.com/home/internet/fccs-new-coverage-maps-show-8-3-million-americans-lack-high-speed-internet/
https://www.drishtiias.com/blog/bridging-the-digital-divide-empowering-youth#:~:text=To%20overcome%20these%20challenges%2C%20governments,the%20advantages%20of%20digital%20empowerment.
https://networkreadinessindex.org/brazil-in-the-nri-2022-how-does-the-country-position-itself-to-the-world/
https://www.trade.gov/country-commercial-guides/brazil-ict-information-and-communications-technologies-and
https://www.csmonitor.com/World/Americas/2023/0316/Starlink-internet-spurs-schools-movies-and-illegal-mines-in-Brazil
https://ts2.space/en/starlink-and-education-in-brazil-bridging-the-digital-divide/
https://dig.watch/updates/bridging-the-digital-divide-south-africas-nationwide-broadband-connectivity-initiative
https://www.imf.org/en/Publications/fandd/issues/2021/03/africas-digital-future-after-COVID19-duarte
https://africa.unwomen.org/en/stories/news/2023/07/empowering-south-african-girls-for-the-digital-age-african-girls-can-code-initiative-unleashes-the-potential-of-africas-future-tech-leaders
https://www.itu.int/en/ITU-D/Innovation/Documents/Publications/Brochure%E2%80%93DIP%20South%20Africa.pdf
https://www.gov.za/sites/default/files/gcis_document/202009/43730gen513.pdf
https://www.worldbank.org/en/news/podcast/2022/05/17/the-digital-revolution-fostering-inclusion-and-resilient-growth-highlights-from-the-wbg-imf-spring-meetings-2022
https://reliefweb.int/report/world/fact-sheet-us-government-collaboration-g20-countries-and-partners-launches-global-initiative-women-digital-economy-totaling-over-515-million
https://mynbc15.com/features/interviews/how-amazon-is-bridging-the-growing-digital-divide
https://engineering.fb.com/2021/06/28/connectivity/digital-divide/
https://www.aboutamazon.com/news/policy-news-views/lets-work-together-broadband-connectivity-for-all-americans
https://www.gsma.com/mobileeconomy/eurasia/
https://www.cnbc.com/2023/06/01/aseans-digital-economy-has-great-potential-but-roadblocks-remain.html
https://www.worldbank.org/en/news/press-release/2022/03/16/digital-economy-could-reap-huge-benefits-for-middle-east-and-north-africa

# Retraction Note to: Responsible Digital Technologies: Shaping Sustainable Practices

Rachna Rana, Pankaj Bhambri, and Sandeep Kautish

## Retraction Note to: Chapter 20 in: W. Leal Filho et al. (eds.), *Digital Technologies to Implement the UN Sustainable Development Goals, World Sustainability Series*, https://doi.org/10.1007/978-3-031-68427-2_20

The Editors have retracted this book chapter due to a significant overlap with a previously published article [1] by different authors. The authors failed to provide a satisfactory explanation for the overlap.

Pankaj Bhambri, on behalf of all the authors, has stated that they disagree with the retraction.

Reference: [1] Pappas, I.O., Mikalef, P., Dwivedi, Y.K. et al. Responsible Digital Transformation for a Sustainable Society. Inf Syst Front 25, 945–953 (2023). https://doi.org/10.1007/s10796-023-10406-5

---

The retracted version of this chapter can be found at
https://doi.org/10.1007/978-3-031-68427-2_20



W. Leal Filho et al. (eds.), *Digital Technologies to Implement the UN Sustainable Development Goals*, World Sustainability Series,
https://doi.org/10.1007/978-3-031-68427-2_24

MIX
Papier aus verantwortungsvollen Quellen
Paper from responsible sources
FSC® C105338
FSC
www.fsc.org

If you have any concerns about our products,
you can contact us on
**ProductSafety@springernature.com**

In case Publisher is established outside the EU,
the EU authorized representative is:
**Springer Nature Customer Service Center GmbH**
**Europaplatz 3, 69115 Heidelberg, Germany**

Printed by Libri Plureos GmbH
in Hamburg, Germany